# Sustainability Engineering

*Sustainability Engineering: Challenges, Technologies, and Applications* focuses on emerging topics within sustainability science and engineering, including the circular economy, advanced recycling technologies, decarbonization, renewable energy, and waste valorization. Readers will learn the trends driving today's sustainability research and innovation as well as the latest in sustainable process technologies.

This book:

- Addresses emerging sustainability development challenges, progress, and disruptive technologies.
- Discusses biological sustainability, recycling technologies, and sustainable process design and manufacture.
- Features a comprehensive view from renowned experts who are leaders in their respective research areas.

This work is aimed at an interdisciplinary audience of engineers and scientists working on solutions to advance the development and application of sustainable technologies, including – but not limited to – chemical and environmental engineers.

# Sustainability Engineering

## Challenges, Technologies, and Applications

Edited by
Eric C.D. Tan

CRC Press
Taylor & Francis Group
Boca Raton London New York

CRC Press is an imprint of the
Taylor & Francis Group, an **informa** business

First edition published 2024
by CRC Press
6000 Broken Sound Parkway NW, Suite 300, Boca Raton, FL 33487-2742

and by CRC Press
4 Park Square, Milton Park, Abingdon, Oxon, OX14 4RN

*CRC Press is an imprint of Taylor & Francis Group, LLC*



ISBN: 978-0-367-76600-9 (hbk)
ISBN: 978-0-367-76601-6 (pbk)
ISBN: 978-1-003-16769-3 (ebk)

DOI: 10.1201/9781003167693

Typeset in Times
by MPS Limited, Dehradun

# *Dedication*

*"We cannot solve our problems with the same thinking we used when we created them."*

*– Albert Einstein, Physicist (1879–1955)*

*To my family, once more and always.*

# Contents

# Preface

Humanity today is facing tremendous sustainability challenges—natural resource depletion, out-of-control waste generation (e.g., plastic waste and food waste), global warming, water crisis, and many others—brought about primarily by human behavior and actions. Sustainability is an emerging field that has resulted from the realization that there has been an incongruity between humans and the natural environment, which caused us to face thresholds and tipping points within planetary boundaries, such as biodiversity loss, ocean acidification, and climate change.

After decades of debating among international, national, and subnational bodies, two widely accepted definitions for sustainability have emerged. They are centered around the themes of "intergenerational equity" and "triple bottom line." The intergenerational equity theme emphasizes that we (the current generation) must properly manage natural resources and ecosystems so that future generations can also or still have those resources and systems. As for the triple bottom line, it is a widely accepted definition of sustainability; this sustainability framework encompasses the three core aspects: environmental, economic, and social. Based on the core principles of intergenerational equity and the triple bottom line, it is clear that engineering plays a critical role in sustainability advancement, and one could interpret sustainability engineering as *a deliberate attempt to minimize the negative impact of human activities on areas of protection, namely, natural environment, human health, and natural resources, to ensure human well-being can exist continuously.*

To move forward, we must have a wide discussion within the scientific community. The world has become more integrated, more complex, and more uncertain. The solutions to the problems will have to be transdisciplinary approaches, paying particular attention to the interactions between the resource systems, their users, and the governance system.

*Sustainability Engineering: Challenges, Technologies, and Applications* provides a collection of a variety of sustainability topics, perspectives, and insights from the global scientific community with researchers from 11 countries. This book presents and discusses the advancement of sustainability science and engineering, focusing on emerging topics, including decarbonization, renewable energy, bioeconomy, circular economy, waste valorization, water sustainability, human behavior in sustainability, sustainable urban travel, electrification, sustainability assessment methods, and policies in sustainability. Readers will learn the trends driving today's sustainability research and innovation and the latest sustainable process technologies.

**Eric C.D. Tan**
*Littleton, Colorado, December 2022*

# Preface

# Editor Biographies

**Dr. Eric C.D. Tan** is a senior research engineer at the National Renewable Energy Laboratory (NREL) in Golden, Colorado. He provides leadership spanning many research areas; he is the principal investigator for marine biofuels, direct air capture, and other projects; and plays a key role in the conceptual process design and evaluation of various biomass conversion pathways to produce infrastructure-compatible, low-carbon, cost-competitive liquid hydrocarbon fuels and chemicals. Prior to NREL, he designed and developed engineered carbon nanostructural materials, industrial fuel processing technology, hydrogen production processes, and commercial fuel cell systems. With 25 years of experience in the industry, he has published extensively in areas encompassing biorefinery, process intensification, decarbonization, bioeconomy, hydrogen economy, circular economy, and sustainability. Additionally, he actively engages in various professional working groups and committees, such as serving as an associate editor of *Frontiers in Sustainability*'s *Sustainable Chemical Process Design*, a reviewer for many scientific journals, and a technical advisor and elected chair of the American Institute of Chemical Engineers (AIChE)'s Sustainable Engineering Forum Leadership Team. Tan is also an American Center for Life Cycle Assessment (ACLCA) certified LCA Professional. He received a PhD in Chemical Engineering from the University of Akron and has a master's degree in Sustainability from Harvard University.

# Acknowledgments

This book is a product of the COVID-19 pandemic and owes its existence to the valuable contributions from the authors of the chapters. Their exceptional efforts in providing a diverse range of sustainability topics, perspectives, and insights made the preparation of this book an immensely enriching and enjoyable experience. Collectively, we have taken a step towards making our planet more sustainable and are playing our part in creating a better future.

# Contributors

**Selorme Agbleze**
Department of Chemical and Biomedical Engineering
West Virginia University
Morgantown, WV, USA

**Halima Abu Ali**
Department of Petroleum Chemistry
American University of Nigeria
Yola, Nigeria

**Resy Norma Annisa**
Department of Chemistry
Faculty of Mathematics and Natural Sciences
Gadjah Mada University
Yogyakarta, Indonesia

**Robert M. Baldwin**
National Renewable Energy Laboratory
Golden, Colorado, USA

**Rick Becker**
V-Grid Energy Systems, Inc.
Camarillo, California, USA

**Daniel Carpenter**
National Renewable Energy Laboratory
Golden, Colorado, USA

**Karen Acurio Cerda**
Department of Chemical and Biomolecular Engineering
University of Nebraska-Lincoln
Lincoln, Nebraska, USA

**Li-Heng Chen**
Industrial Technology Research Institute
Chutung, Hsinchu, Taiwan

**Tse-Lun Chen**
ETH Zurich
Zurich, Switzerland

**Greg Campbell**
V-Grid Energy Systems
Camarillo, California, USA

**Singfoong "Cindy" Cheah**
Tourmaline Forest LLC
Lakewood, Colorado, USA

**Fei Ding**
National Renewable Energy Laboratory
Golden, Colorado, USA

**Shudipto K. Dishari**
Department of Chemical and Biomolecular Engineering
University of Nebraska-Lincoln
Lincoln, Nebraska, USA

**Abhijit Dutta**
National Renewable Energy Laboratory
Golden, Colorado, USA

**Seefat Farzin**
Department of Chemical and Biomolecular Engineering
University of Nebraska-Lincoln
Lincoln, Nebraska, USA

**Kristina Fürst**
Research Institute for Sustainability – Helmholtz Centre
Potsdam, Germany

**Jean Baptiste Habyarimana**
Department of Petroleum Chemistry
American University of Nigeria
Yola, Nigeria

**Neeraj Hanumante**
Department of Chemical Engineering
Indian Institute of Technology Bombay
Mumbai, India

**Kylee Harris**
National Renewable Energy Laboratory
Golden, Colorado, USA

**Carrie Hartford**
Jenike & Johanson, Inc.
San Luis Obispo, California, USA

**Hasanudin Hasanudin**
Department of Chemistry
Faculty of Mathematics and Natural Sciences
Sriwijaya University
Indralaya, South Sumatra, Indonesia

**Wan Jin Jahng**
Department of Petroleum Chemistry
American University of Nigeria
Yola, Nigeria

And

Department of Energy
Department of Ophthalmology
Julia Laboratory
Suwon, Korea

**David John**
Department of Petroleum Chemistry
American University of Nigeria
Yola, Nigeria

**Matthew L. Jordan**
Argonne National Laboratory
Lemont, Illinois, USA

**Jayant Khambekar**
Jenike & Johanson, Inc.
Houston, Texas, USA

**Shuyun Li**
Department of Chemical and Biomedical Engineering
West Virginia University
Morgantown, WV, USA

**Fernando V. Lima**
Department of Chemical and Biomedical Engineering
West Virginia University
Morgantown, WV, USA

**Yupo J. Lin**
Argonne National Laboratory
Lemont, Illinois, USA

**Thomas Lippert**
Argonne National Laboratory
Lemont, Illinois, USA

**Michael Narodoslawsky**
Technical University of Graz
Graz, Austria

**Peter Eyinnaya Nwankwor**
Department of Petroleum Chemistry
American University of Nigeria
Yola, Nigeria

**Michael S. O'Banion**
The Wonderful Company
Los Angeles, California, USA

**Oghenetega Allen Obewhere**
Department of Chemical and Biomolecular Engineering
University of Nebraska-Lincoln
Lincoln, Nebraska, USA

**Erendira T. Quintanar-Orozco**
Centro de Investigaciones Químicas
Universidad Autónoma del Estado de Hidalgo
Mineral de la Reforma, Mexico

**Gerardo J. Ruiz-Mercado**
Office of Research and Development
U.S. Environmental Protection Agency
26 W Martin L. King Dr., Cincinnati, Ohio, USA

And

Chemical Engineering Graduate Program
Universidad del Atlántico
Puerto Colombia, Colombia

**Raffaele Russo**
Joby Aviation
Santa Cruz, California, USA

**Reinhard Seiser**
National Renewable Energy Laboratory
Golden, Colorado, USA

**Ani Setyopratiwi**
Department of Chemistry
Faculty of Mathematics and Natural Sciences
Gadjah Mada University
Yogyakarta, Indonesia

**Yogendra Shastri**
Department of Chemical Engineering
Indian Institute of Technology Bombay
Mumbai, India

**Avantika Singh**
National Renewable Energy Laboratory
Golden, Colorado, USA

**Shubham Sonkusare**
Department of Chemical Engineering
Indian Institute of Technology Bombay
Mumbai, India

**Akhmad Syoufian**
Department of Chemistry
Faculty of Mathematics and Natural Sciences
Gadjah Mada University
Yogyakarta, Indonesia

**Hassan Tajudeen**
Department of Petroleum Chemistry
American University of Nigeria
Yola, Nigeria

**Eric C.D. Tan**
National Renewable Energy Laboratory
Golden, Colorado, USA

**Sonja Thielges**
Research Institute for Sustainability –
Helmholtz Centre
Potsdam, Germany

**Stephen M. Tifft**
National Renewable Energy Laboratory
Golden, Colorado, USA

**Anna Trendewicz**
Future Ventures Management
Berlin, Germany

**Maisari Utami**
Department of Chemistry
Faculty of Mathematics and Natural Sciences
Universitas Islam Indonesia
Yogyakarta, Indonesia

**Tebaldo Vinciguerra**
Libera Università Maria SS. Assunta (LUMSA)
Rome, Italy

**Karna Wijaya**
Department of Chemistry
Faculty of Mathematics and Natural Sciences
Gadjah Mada University
Yogyakarta, Indonesia

**Geetanjali Yadav**
National Renewable Energy Laboratory
Golden, Colorado, USA

**Neal Yancey**
Idaho National Laboratory
Idaho Falls, Idaho, USA

# 1 Lignin-based Materials for Energy Conversion and Storage Devices

*Seefat Farzin, Karen Acurio Cerda, Oghenetega Allen Obewhere, and Shudipto K. Dishari*
Department of Chemical and Biomolecular Engineering, University of Nebraska-Lincoln, Lincoln, Nebraska, USA

## CONTENTS



DOI: 10.1201/9781003167693-1

## 1.1 INTRODUCTION

Proton exchange membrane fuel cells (PEMFCs), rechargeable Li-/Na-ion batteries (LIBs/NIBs), redox flow batteries (RFBs), and supercapacitors (SCs) are the most prominent energy conversion and storage devices that use electrochemistry to generate electricity in an eco-friendly manner. They can meet our clean energy needs not only by powering stationary/portable devices, light-to-heavy duty electric vehicles, and large-scale applications but also by storing the excess electric power from grid/renewable energy sources (wind/solar) to discharge during peak time. The global market of fuel cells is projected to reach up to \$14.6 billion by 2027 [1]. There have been ongoing efforts and investments at the government level and by automakers across the globe (including the United States, European Union, Japan, and Germany) to sketch a hydrogen roadmap to implement [2,3] and mass-scale commercialization of fuel cell-based vehicles (FCVs). On the other hand, the global LIB market was valued at \$55.02 billion in 2020 and is projected to reach up to \$181.6 billion by 2027 [4]. To meet the global demand for clean energy, we, therefore, need to improve the performance of sustainable energy-driven technologies through the development of efficient materials. Some of the major needs to be addressed to make these electrochemical devices efficient and commercialize them at a large scale are: (i) achieving high energy and power density simultaneously; (ii) achieving high electronic- and ionic conductivity, (iii) minimizing ion-transport limitations at ionomer-catalyst interfaces of electrodes; (iv) enabling high-temperature operations; (v) increasing material durability and device life; and (vi) lowering the materials and device production costs. As a cheap and biodegradable polymer with unique chemical structure, lignin can help to achieve some of these material needs for energy conversion and storage devices.

Lignin is the most abundant aromatic polymer in nature [5] and constitutes ~15–40% dry weight of most terrestrial plants [6]. Moreover, the waste streams of pulp and paper industries and biorefineries contain lignin. Every year, ~100 million tonnes of lignin are produced worldwide [7]. Of this, ~50–70 million tonnes come from pulp and paper industries in the form of black liquor, [6,8] and ~100,000–200,000 tonnes from cellulosic ethanol plants [8,9]. However, ~98–99% of this lignin is either combusted to generate steam and process energy or simply disposed of [6,7]. Utilizing these untapped, lignin-rich, conversion process wastes and plant/crop wastes, and converting the lignin to value-added materials can make these industries and refineries more economically viable. Over the last two decades, there has been a rise in research interest in lignin valorization and the development of products, such as concrete additives, [10] plasticizers, [11,12] stabilizing [13]/dispersing agents, [14,15] corrosion inhibitors, [16] liquid fuels [17,18] and chemicals (benzene, toluene, xylene, phenolic acids, diols), [7,17,19] sequestering agents, [7] and building blocks of thermoplastics [19–21], thermosets [22,23] and block copolymers [24,25]. If the waste lignin is also valorized by making functional materials for energy conversion and storage devices, the cost of the devices can be significantly cut down without compromising the device efficiency. This way, both energy- and bioeconomy can be boosted simultaneously [26]. Moreover, using lignin will help achieve eco-friendliness of the device materials alongside a low-carbon-footprint operation of these devices. This book chapter attempts to review the current research progress on the applications of lignin and its derivatives in making some of the key components of sustainable energy conversion and storage devices.

### 1.1.1 Chemical Structure and Extraction Processes of Lignin

In a plant cell wall, lignin (along with hemicellulose) wraps around and covalently bonds with cellulose microfibrils (Figure 1.1a) and imparts mechanical stability to the cell wall. Lignin is a three-dimensional, hyperbranched, complex polymer that is rich in aromatic benzene rings functionalized with groups like ether (-O-), hydroxyl (-OH), methoxy ($-OCH_3$), and more. It mainly

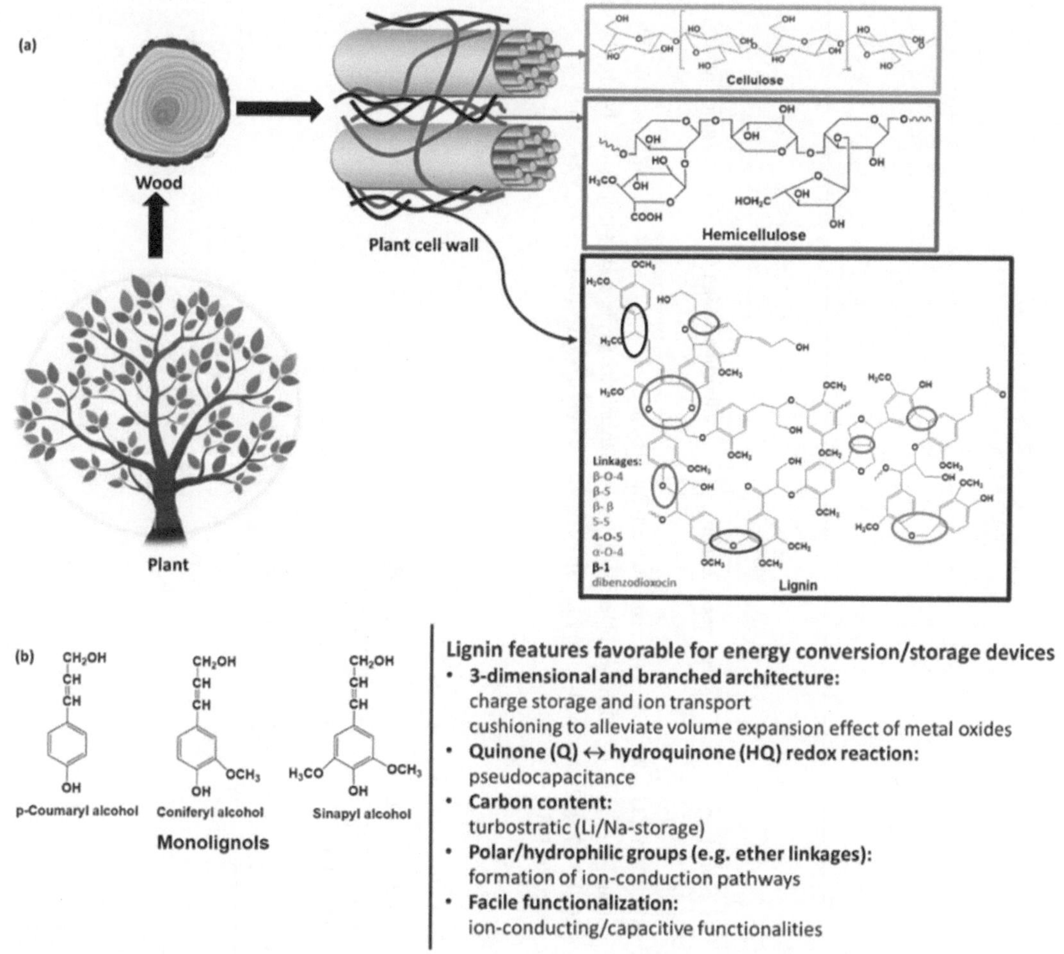


**FIGURE 1.1** (a) Chemical structure of plant cell wall components: cellulose, hemicellulose, and lignin. (b) Chemical structure of monolignols, the building blocks of lignin (p-coumaryl, coniferyl, and sinapyl alcohol) (left), and unique features of lignin that makes it suitable for energy conversion and storage devices (right).

constitutes three monolignol units: p-coumaryl alcohol, coniferyl alcohol, and sinapyl alcohol (Figure 1.1b). The monolignols are connected to each other via a range of interunit linkages (such as β-O-4, α-O-4, β-β, β-5, β-1, 5–5, and 4-O-5 (Figure 1.1a)) to give the primary building blocks of lignin, such as p-hydroxyphenyl [H], guaiacyl [G] and syringyl [S] units. In every plant, these primary building blocks vary depending on the plant species; for example, coniferyl alcohol is the more dominant monolignol in softwood (SW) lignin, and both coniferyl and sinapyl alcohol are present in hardwood (HW) lignin in a ratio of 4:1 to 1:2; while grass comprises all three types of monolignols (5–33%, 33–80%, and 20–54% of p-coumaryl, coniferyl, and sinapyl alcohols, respectively) [27].

The structure, reactivity, molecular weight (MW), and chemical composition of lignin are diverse and vary from source to source. Based on the extraction process, lignin has been classified into the following major classes: [28] kraft lignin, soda lignin, lignosulfonate, organosolv lignin, acid-hydrolyzed lignin, and ionic lignin (Table 1.1). ***Kraft lignin***, also known as alkali lignin, is obtained from the sulfate cooking process where sodium hydroxide and sodium sulfide are used as treating agents [28]. This process dissolves ~90–95% of the lignin present in the wood, [29] but during this process, the α-aryl ether and β-aryl ether bonds are also cleaved, which breaks the

**TABLE 1.1**
**Properties of Lignin**

| Lignin-Type Properties | Kraft Lignin | Lignosulfonate | Soda Lignin | Organosolv Lignin | Hydrolysis Lignin | Ionic Lignin |
|---|---|---|---|---|---|---|
| **Source** | SW, [42] HW [42] | SW, [42] HW [42] | HW, [32,42] G [42] | SW, [42] HW, [42] G [42] | HW [43] | SW, [32] HW [32] |
| **Solubility** | Alkali, Organic solvents [42] | Water [42] | Alkali [42] | Organic solvents [42] | Organic/Inorganic acid [44] | Ionic solvents [41,45] |
| **$M_w$ (g.mol$^{-1}$)** | 1,500–2,500 [32]<br>2,000–3,000 [6,46]<br>2,950–5,900 [47]<br><25,000 [48]<br>800–53,360 [49]<br>429,000 [50] | 1,000–150,000 [32,48,51]<br>20,000–50,000 [6]<br>64,000 [36] | 1,000–3,000 [32]<br><15,000 [48] | 500–5,000 [32,48] | 5,000–10,000 [32,48] | 900–4,983 [52] |
| **PDI** | 2.0–4.0 [32,42,46,48] | 4.2–9.0 [32,36,42,48,53] | 2.5–3.5 [32,42,48] | 1.5–2.5 [32,42,48] | 1.5–11 [32,48] | 1.7–2.1 [53] |
| **$T_g$ (°C)** | 108–165 [42,46,48,50] | 127–154 [48] | 140–155 [42,48] | 89–110 [42,48] | 75–90 [48] | −13–115 [54] |
| **Sulfur (%)** | 1.0–3.0 [32,48] | 3.5–8.0 [32,48] | 0 [32,48] | 0 [32,48] | 0–1.0 [32,48] | 1.5 [32] |

** HW = hardwood, SW = softwood, G = grass, $M_w$ = weight average molecular weight, $T_g$ = glass transition temperature.

lignin into smaller fragments with different molecular weights and increases its solubility due to the increase in hydroxyl (-OH) and carboxylic acid (-COOH) groups [28–31]. The kraft process is predominant in the pulp and paper industries and generates ~85% of the lignin produced worldwide [28]. ***Soda lignin*** is obtained from the soda or soda-anthraquinone pulping process [32]. This process uses an alkaline solution (sodium hydroxide) to precipitate lignin from the pulping liquor [33]. While there are structural similarities between kraft and soda lignin, soda lignin (unlike kraft lignin) is free of sulfur [30] and undergoes lower degree modification upon treatment [33,34]. The soda cooking process is predominantly used for non-wood plants (grass, bamboo, etc.) and is only rarely applied to hardwood species [28]. ***Lignosulfonate*** is obtained *via* the sulfite cooking process [28,35]. By treating the wood chips with the solutions of sulfite ($SO_3^{-2}$) and bisulfite ($HSO_3^{-}$) ions, [28] this process breaks down the bonds between lignin and cellulose components of lignocellulose. As a result, almost pure cellulose is obtained. At the same time, this process breaks the lignin down [36] and solubilizes it in water [37,38] via functionalizing it with sulfonate (-$SO_3Na$), phenolic -OH, and -COOH groups [32]. The ***organosolv lignin*** is produced in a pulping process in which water and a combination of organic solvents (e.g., ethanol, formic acid, acetic acid, peroxiorganic acids) are used as the cooking liquor [28,39]. The dissolved lignin has a low molecular weight, [28] high purity, [28] and is sulfur-free [28]. ***Hydrolysis or hydrolyzed lignin*** is the product of the enzymatic activity on lignocellulosic biomass [40]. ***Ionic lignin*** is the lignin extracted using ionic liquids. This is a less harsh method compared to what is used for obtaining kraft lignin/lignosulfonate [41].

## 1.2 LIGNIN FOR ENERGY CONVERSION AND STORAGE DEVICES

Lignin has some unique structural features which make it suitable for energy conversion and storage devices (Figure 1.1b), e.g., three-dimensional and branched architecture, polar and hydrophilic functionalities (like ether linkages (-O-), hydroxyl (-OH), sulfonic acid/sulfonate (-$SO_3H$/$SO_3^-$) groups), sites for facile functionalization, [5,55] redox-active quinone groups which favor faradaic/non-faradaic charge storage/ion transport [56–58]. Also, the carbonization of lignin provides a non-graphitic and turbostratic, disordered structure which creates facile $Li^+$/$Na^+$-ion intercalation sites [59–65]. That is why lignin derivatives are reported as promising candidates for making core components of energy conversion and storage devices, such as carbon precursors for electrodes, membrane separators, and ion-conducting/redox-active materials. These will be discussed in the following sections in a relevant manner. Also, the potential applications of lignin-based compounds in LIB-, NIB- and SC-electrodes are summarized in Tables 1.2a, b.

### 1.2.1 Lithium-ion Batteries (LIBs)

Rechargeable LIBs are currently dominating the market for powering electric vehicles, portable electronics, and smart grids. A typical LIB setup (Figure 1.2b) uses a layer-structured, Li-transition metal oxide-based cathode (e.g., lithium cobalt oxide ($LiCoO_2$), lithium manganese oxide ($LiMn_2O_4$)), graphite-based anode, porous separator, and Li-salt dissolved in the non-aqueous solvent electrolyte. In LIB, $Li^+$-ions de-intercalate from the anode, shuttle towards the cathode through a separator, and intercalate within the porous cathode during discharging. $Li^+$-ions take the opposite route during battery charging (Figure 1.2b). LIBs offer high power density (Figure 1.2a), high charge-discharge efficiency, and long life. However, the energy density of LIBs is still lower than PEMFCs and internal combustion engines (ICEs) (Figure 1.2a). Energy density is the product of voltage and specific capacity of reversible Li-insertion/extraction into/from electrodes. We, therefore, need high-capacity electrode materials to improve the energy density of LIBs [144,145]. Lignin cannot only be an ideal carbon precursor [5,83,144,146–150] for LIB electrodes but also can be used to make gel-[5,30,55,151–159]/solid [5,30,160] polymer electrolytes (GPEs/SPEs), separators, [5,159–161] and binders [162–166] for LIBs.

**TABLE 1.2A**
**Overview of Lignin-derived LIB and NIB Electrodes**

| Materials | Source and Type of Lignin | Synthesis Conditions | Specific Surface Area, Pore Volume | Electrochemical Performance | | | | Additional Remarks | Ref. |
|---|---|---|---|---|---|---|---|---|---|
| | | | | Capacity (mAh/g) | Current Density | Capacity Retention | CE (%) | | |
| **LIB anodes** | | | | | | | | | |
| Hierarchical porous carbon | Alkali lignin | KOH-based activation in the presence of HCl followed by carbonization at 700°C, 2 h, $N_2$ | 907 $m^2/g$, 0.51 $cm^3/g$ | 795 | 200 mA/g | 470 mAh/g at 200 mA/g after 400 cycles | – | – | [59] |
| Hierarchical porous carbon | Enzymatic hydrolysis lignin | Carbonization with the help of $ZnCO_3$ at 600°C, $N_2$ | – | 1,218 | 200 mA/g | 550 mAh/g at 200 mA/g after 200 cycles | 99.5 | – | [66] |
| Hierarchical porous carbon | Alkali lignin (SW) | $K_2CO_3$-based activation in the presence of HCl followed by carbonization at 1,000°C, Ar | 2,300 $m^2/g$ | 1,693 | 200 mA/g | 520 mAh/g at 200 mA/g after 200 cycles | 99.9 | – | [60] |
| Carbon microsphere | Sodium lignosulfonate | Carbonization at 900°C, 3 h, $Ar/H_2$ | 328 $m^2/g$, 0.17 $cm^3/g$ | 180.6 | 20 mA/g | 169 mAh/g at 20 mA/g after 100 cycles | – | – | [67] |
| Hard carbon | Sodium lignosulfonate | Pre-oxidization at 200°C, 24 h, air followed by carbonization at 600°C, 2 h, air | 676 $m^2/g$, 0.79 $cm^3/g$ | 214 | 500 mA/g | 150 mAh/g at 500 mA/g after 400 cycles | – | – | [61] |
| Hard carbon | – | Carbonization at 900°C, 2 h, Ar followed by KOH-based activation | 1,534 $m^2/g$ | 250 | 25 mA/g | 235 mAh/g at 25 mA/g after 75 cycles | – | – | [68] |
| N-doped carbon nanosphere | Alkali lignin | Carbonization of lignin-based azo polymer at 750°C, 2 h, Ar | 419 $m^2/g$, 0.19 $cm^3/g$ | 301.3 | 60 mA/g | 225 mAh/g at 60 mA/g after 50 cycles | 100 | – | [69] |
| N-doped hard carbon | – | Carbonization of lignin-melamine resin at 1,000°C with $Ni(NO_3)_2.6H_2O$, 4 h, Ar | 79.2 $m^2/g$, 0.216 $cm^3/g$ | 345 | 100 mA/g | 248 mAh/g at 100 mA/g after 300 cycles | – | – | [62] |
| Hard carbon | By-product of alcohol production from corn stalks | Carbonization at 800°C, 2 h, $N_2$ | 188 $m^2/g$ | 882.2 | 37.2 mA/g | 228.8 mAh/g at 744 mA/g after 200 cycles | – | – | [63] |
| $MnO_2$ and lignin composite | Kraft lignin | Grinding the $MnO_2$-lignin mixture followed by ionic liquid-based activation | 45 $m^2/g$, 0.03 $cm^3/g$ | 760 | 50 mA/g | 665 mAh/g at 50 mA/g after 20 cycles | 89 | Superior performance than pure $MnO_2$ anode | [70] |

| | | | | | | | | | |
|---|---|---|---|---|---|---|---|---|---|
| $SiO_x$-carbon composite | Alkali lignin | Carbonization at 600°C, 2 h, Ar | – | 901 | 100 mA/g | 900 mAh/g at 200 mA/g after 250 cycles | >99 | No requirement of binder/ conductive agent | [71] |
| CNFs with $Fe_2O_3$ | Alkali lignin | Electrospinning lignin-PVA mixture and carbonization at 1,200°C, 1 h, Ar followed by surface functionalization with $Fe_2O_3$ nanoparticles (nps) at 90°C, 6 h, urea | 583 $m^2/g$ | 951 | 50 mA/g | 715 mAh/g at 50 mA/g after 80 cycles | – | – | [72] |
| Mesoporous carbon-NiO np | Sodium lignosulfonate | Carbonization of $Ni(OH)_2$-lignin at 450°C, 2 h, $N_2$ | 852 $m^2/g$ | 1,022 | 100 mA/g | 863 mAh/g at 100 mA/g after 100 cycles | – | – | [73] |
| Si/C composite | Kraft lignin (HW) | Carbonization of Si-lignin-PEO at 800°C, 2 h, Ar | – | 1,251 | 540 mA/g | 1,391 mAh/g at 540 mA/g after 100 cycles | 99.8 | No requirement of binder/ conductive carbon | [74] |
| Carbon-$SnO_2$ composite | Enzymatic hydrolysis lignin | Ultrasonication of porous carbon-$SnO_2$ dispersion in ethanol | 508 $m^2/g$ | 1,549 | 100 mA/g | 620 mAh/g at 100 mA/g after 100 cycles | – | Superior performance than pure $SnO_2$ anode | [75] |
| Carbon-ZnO composite | Lignin from rice husk | Carbonization at 500°C, 3 h, Ar | – | 2,137.3 | 0.2 C[a] | 898.1 mAh/g at 0.2 C after 110 cycles | 99.6 | Discharge capacity comparable to theoretical specific capacity of ZnO | [76] |
| Carbon-coated Si np | Lignin/lignocellulose from pulp industry | Carbonization of lignin np at 600°C, 5 h, Ar | – | 2,287 (lignin/Si) 2,242 (lignocellulose/ Si) | 300 mA/g | 881 mAh/g (lignin/Si) 955 mAh/g at 300 cycles mA/g after 51 cycles | – | Capacity retention higher than recycled Si np (200 mAh/g) | [77] |
| N-doped carbon-coated Si np | Alkali lignin | Carbonization at 600°C, 3 h, Ar | – | 2,290 | 200 mA/g | 882 mAh/g at 200 mA/g after 150 cycles | 99 | Capacity higher than undoped carbon-coated Si np (700 mAh/g) | [64] |
| Lignin-coated Si np | – | Self-assembly in phytic acid | – | 3,000 | 300 mA/g | 2,670 mAh/g at 300 mA/g after100 cycles | 100 | Capacity retention higher than Si np (200 mAh/g) | [78] |

(*Continued*)

**TABLE 1.2A** *(Continued)*
**Overview of Lignin-derived LIB and NIB Electrodes**

| Materials | Source and Type of Lignin | Synthesis Conditions | Specific Surface Area, Pore Volume | Electrochemical Performance | | | | Additional Remarks | Ref. |
|---|---|---|---|---|---|---|---|---|---|
| | | | | Capacity (mAh/g) | Current Density | Capacity Retention | CE (%) | | |
| $Fe_3O_4$-graphene nanosheet | Lignin from soda pulping | thermal reduction followed by carbonization at 700°C, 8 h, $N_2$ | 258.7 $m^2/g$ | 3,829 | 50 mA/g | 750 mAh/g at 1000 mA/g after 1,400 cycles | 100 (after 3rd cycle) | Ultrahigh 1st discharge capacity (3,829 mAh/g), charge capacity (2,250 mAh/g) 4 times higher than the theoretical value of $Fe_3O_4$-anode at 50 mA/g | [79] |
| CNF | – | Electrospinning lignin-based carbon precursor, carbonization at 1,000°C, 1 h, $N_2$ | – | 310 | 158 mA/g | 150 mAh/g at 3162 mA/h after 50 cycles | 100 | – | [80] |
| Mesoporous CNF | – | Electrospinning, thermally annealing at 250°C, 3 h, Air followed by carbonization at 1,200°C, 1 h, $N_2$ | 68.8 $m^2/g$, 0.081 $cm^3/g$ | 384.4 | 1,000 mA/g | 177.7 mAh/g at 1,000 mA/g after 400 cycles | 100 | No requirement of binder and current collector | [81] |
| Graphite | Kraft lignin, cellulose, glucose, organosolv lignin, hydrolysis lignin, raw SW, raw HW | Graphitization at 1,200°C, $N_2$ | 26 $m^2/g$ (similar to commercial graphite) | 399 | 0.1 C[a] | 300 mAh/g at 0.5 C after 100 cycles | >99 | – | [82] |
| CNF mat | Organosolv lignin | Electrospinning lignin-PEO, carbonization at 900°C, 1h, Ar (without doping) or 500°C, 2 h (with N-doping), dipping into urea solution, re-carbonization at 900°C, 2 h | 473 $m^2/g$ (undoped); 381 $m^2/g$ (N-doped) | 650 (undoped & doped) | 30 mA/g (undoped & doped) | 430 mAh/g (undoped) 550 mAh/g (doped) at 30 mA/g after 50 cycles cycles | >99 | – | [83] |

| | | | | | | | | | |
|---|---|---|---|---|---|---|---|---|---|
| CNF | Alkali lignin | KOH-based activation of lignin-PVA, electrospinning, stabilizing at 100°C, 12 h, $I_2$ vapor followed by carbonization at 600°C, 1 h, Ar | 1,049 $m^2/g$, 0.046 $cm^3/g$ | 925 | 50 mA/g | 290 mAh/g at 50 mA/g after 100 cycles | 100 | No requirement of binder | [84] |
| CNF | Alkali lignin | KOH-based activation carbonization of lignin/PVA at 600°C, 1 h, Ar | 93.1 $m^2/g$, 0.376 $cm^3/g$ | 1,470 | 50 mA/g | 272 mAh/g at 50 mA/g after 100 cycles | 125 | – | [85] |
| CNF | Organosolv lignin (HW) | Electrospinning lignin-PAA, carbonization at 900°C, 0.5 h, $N_2$ | 670 $m^2/g$ | 1,750 | 186 mA/g | 572 mAh/g at 186 mA/g after 500 cycles | – | – | [86] |
| Graphite | Lignin from HW | Carbonization at 1,000 and 2,000°C, Ar | – | 350 | 15 mA/g | 126.1 mAh/g at 15 mA/g after 70 cycles | 99.89 | – | [87] |
| Layered and holey carbon scaffold | Lignosulfonate | Lignin-based exfoliation of boron nitride, carbonization at 240°C, 2h, then at 900°C, 2 h, $Ar/H_2$ | 261.9 $m^2/g$ | 152 | 0.2 C[a] | 90 mAh/g at 10 C (4.5 $mA/cm^2$) after 1,800 cycles | > 99 | – | [88] |
| Lignin-exfoliated $MoS_2$ nanosheet | Alkali lignin | Ultrasonication to make lignin-$MoS_2$ dispersion, vacuum filtration followed by sonication, 2 h | – | 196 | 100.2 mA/g | 164 mAh/g at 100.2 mA/g after 40 cycles | >99 | | [89] |
| Lithiated aromatic biopolymer | – | Reaction of lignin and lithium hydroxide at room temperature, 24 h | 30.6 $m^2/g$ | 135 | 170 mA/g | 126.5 mAh/g at 170 mA/g after 600 cycles | 100 | – | [90] |
| **NIB anodes** | | | | | | | | | |
| Amorphous carbon | – | Carbonization lignin-pitch mixture at 1,400°C, 2 h, Ar | 1.34 $m^2/g$ | 254 | 30 mA/g | 226 mAh/g at 30 mA/g after 150 cycles | 100 | – | [91] |
| Hard carbon | Lignin from rice husk | Carbonization at 1,600°C, $N_2$ | 224.6 $m^2/g$, 0.41 $cm^3/g$ | 280 | 30 mA/g | 276 mAh/g at 30 mA/g after 100 cycles | – | – | [92] |
| Lignin-derived hard carbon microsphere | Byproduct of ethanol production (corn stalks) | Emulsion-solvent evaporation followed by pre-carbonization at 400°C, 2 h, $N_2$ and carbonization at 1,300°C, 3 h, $H_2/Ar$ | 5.3 $cm^2/g$ | 296.5 | 50 mA/g | 276.8 mAh.g at 50 mA/g after 100 cycles | 100 | – | [93] |
| Hard carbon microsphere | Sodium lignin sulfonate | Carbonization of spray-dried lignin at 1,300°C | 11.9 $m^2/g$ | 300 | 0.1 C[a] | 312 mAh/g at 0.1 C after 100 cycles | 100 | – | [94] |
| Hard carbon | Lignin from scrap wood | Carbonization at 1,000°C, 6 h, Ar | 30 $m^2/g$ | 308 | 20 mA/g | 270 mAh/g at 20 mA/h after 80 cycles | 99.5 | – | [95] |

*(Continued)*

**TABLE 1.2A *(Continued)***

**Overview of Lignin-derived LIB and NIB Electrodes**

| Materials | Source and Type of Lignin | Synthesis Conditions | Specific Surface Area, Pore Volume | Electrochemical Performance | | | | Additional Remarks | Ref. |
|---|---|---|---|---|---|---|---|---|---|
| | | | | Capacity (mAh/g) | Current Density | Capacity Retention | CE (%) | | |
| Hard carbon | – | Carbonization of lignin-epoxy resin mixture at 1,400°C, 1 h, Ar | 65 $m^2/g$ | 316 | 30 mA/g | 285 mAh/g at 30 mA/g after 150 cycles | 100 | – | [96] |
| Hard carbon | Lignin from platyphylla and sophora japonica leaves | Carbonization at 800°C, 2 h, Ar | – | 318 | 20 mA/g | 318 mAh/g at 20 mA/g after 10 cycles | – | – | [97] |
| Hard carbon nanosphere | – | Resinification and carbonization at 1,300°C, 1 h, $N_2$ | 16.3 $m^2/g$ | 347 | 0.1 C[a] | 210 mAh/g at 0.5C after 300 cycles | >99 | – | [98] |
| Hard carbon | – | Carbonization at 1,300°C, 6 h, $N_2$ | 55 $m^2/g$, 0.113 $cm^3/g$ | 378 | 50 mA/g | 141 mAh/g at 300 mA/g after 700 cycles | 100 | – | [99] |
| Hard carbon | Acid hydrolysis lignin (oak wood) | Carbonization at 1,300°C, 6 h, $N_2$ | 249.8 $m^2/g$, 0.207 $cm^3/g$ | 450 | 50 mA/g | 297 mAh/g at 50 mA/g after 50 cycles | 100 | – | [100] |
| Hard carbon microsphere | By-product of alcohol production from corn stalks | Carbonization at 1,300°C, $H_2$ | 11.1 $m^2/g$, 0.01 $cm^3/g$ | 494.4 | 250 mA/g | 446.44 mAh/g at 100 mA/g after 250 cycles | >99 | – | [101] |
| Hard carbon | Peat moss lignin | Carbonization at 1,100°C, Ar and KOH-based activation at 300°C, air | 196.6 $m^2/g$, 0.18 $cm^3/g$ | 532 | 50 mA/g | 298 mAh/g at 50 mA/g after 10 cycles | 100 | Larger inter-graphene spacing (0.388 nm) than graphite (0.335 nm) | [102] |
| CNF | Kraft lignin | Electrospinning of lignin-cellulose acetate mixture and carbonization at 100°C, 1 h, $N_2$ | 540.9 $m^2/g$, 0.27 $cm^3/g$ | 290 | 50 mA/g | 340 mAh/g at 50 mA/g after 200 cycles | >95 | – | [103] |
| CNF | HW lignosulfonate | Electrospinning lignin-PAN based nanofibers, carbonization at 1,300°C, 0.5 h, $N_2$ | 26.6 $m^2/g$ | 292.6 | 20 mA/g | 247 mAh/g at 100 mA/g after 300 cycles | – | Free-standing and binder-free anode | [104] |
| CNF | Kraft lignin (SW) | Electrospinning and carbonization at 1,200°C, 0.33 h, $N_2$ | 94 $m^2/g$ | 310 | 30 mA/g | 272 mAh/g at 100 mA/g after 100 cycles | >99.9 | No requirements of current collectors and additives | [105] |

| | | | | | | | | | |
|---|---|---|---|---|---|---|---|---|---|
| PEDOT/lignin cathode | Alkali lignin | Chemical oxidative polymerization of EDOT and lignin with $FeCl_3$ as a catalyst | – | 70 | 25 mA/g | 46 mAh/g at 25 mA/g after 100 cycles | – | – | [106] |
| N-doped hierarchical porous carbon | Alkaline lignin | Carbonization at 400°C, 2 h and 800°C, 4 h, $N_2$ | – | 320.5 | 30 mA/g | 130.8 mAh/g at 150 mA/g after 4000 cycles | 99 | – | [107] |
| Hierarchical porous carbon | Lignin from oak leaves | Carbonization at 1,000°C, 1 h, Ar | 161 $m^2/g$ | 360 | 10 mA/g | 288 mAh/g at 20 mA/g after 200 cycles | 75 | No requirement of binders, carbon additives, and current collector | [108] |
| N-doped carbon microsphere | – | Hydrothermal reaction at 250°C, 12 h and carbonization at 1,100°C, 2 h, $N_2$ | 151 $m^2/g$ | 374 | 25 mA/g | 223 mAh/g at 100 mA/g after 300 cycles | >90 | – | [109] |

**TABLE 1.2B**
**Overview of Lignin-derived SC Electrodes**

| Materials | Source and Type of Lignin | Synthesis Conditions | Specific Surface Area ($m^2/g$) | Electrochemical Performance | | | | | Additional Remarks | Ref. |
|---|---|---|---|---|---|---|---|---|---|---|
| | | | | Specific Capacitance | Current Density | Capacitance Retention | Energy Density | Power density | | |
| **Activated, templated, and composite carbon materials** | | | | | | | | | | |
| Porous carbon | Black liquor-derived lignin | KOH-chemical activation, carbonization at 900 °C, 2 h | 3,089.2 | 41.4 F/g | 0.01 A/g | – | – | – | Specific surface area 3 times higher than a commercial activated carbon from charcoal (943.5 $m^2/g$) | [110] |
| Porous carbon | – | Mixing carbonaceous mudstone and lignin, activation with 16 M $HNO_3$ at 60 °C, 24 h | 4.5 | 155.6 F/g | – | 25.70% | – | – | – | [111] |
| Hierarchical porous carbon | Lignin from the butanol fermentation of corn straw | Hydrothermal carbonization at 800 °C for 3 h | 1,660 | 420 F/g | 0.1 A/g | 99% after 10,000 cycles at 5 A/g | 22.9 Wh/kg | 25,400 W/kg | – | [112] |
| Porous carbon material | Alkali-lignin from corn straw | Carbonization at 900 °C at 5 °C/min, 2 h, $N_2$, green bacterial activation | 1,831 | 428 F/g | 1 A/g | 96.7% after 10,000 cycles at 5 A/g | 66.18 Wh/kg | 312 W/kg | – | [113] |
| O-enriched, hierarchical porous carbon | Lignin residue from bioethanol production | Microwave heating (800 W, 30 min), humidified $N_2$ | – | 173 F/g | – | – | 23 Wh/kg | 1,100 W/kg | – | [114] |

| | | | | | | | | | | |
|---|---|---|---|---|---|---|---|---|---|---|
| N, P-doped Nanoporous carbon | Enzymatic hydrolysis lignin from corncob | Soft templating and N, P-doping using deep eutectic solvents (choline chloride/$ZnCl_2$), carbonization at 700°C, 2 h | 220 | 177.5 F/g | 0.5 A/g | Stable over 3,000 cycles at 20 A/g | – | – | – | [115] |
| N-S co-doped carbon | Lignin amine | KOH-activation, carbonization at 700°C | 1,199 | 241 F/g | 1 A/g | 95% after 3,000 cycles | 27.2 Wh/kg | 10,000 W/kg | Specific capacity 260% higher than lignin amine carbon without $Fe_3O_4$ | [116] |
| N-doped hierarchical porous carbon | Lignin-derived byproducts from the solid residues of poplar bioethanol fermentation | N-doping, carbonization (200°C, 24 h), chemical activation (KOH at 800°C, 1 h, Ar) | 2,218 | 312 F/g | 1 A/g | 98% after 20,000 cycles | 8.8 Wh/kg | 21,300 W/kg | – | [117] |
| N-O co-doped carbon | Lignin from coconut shell | Pyrolyzation at 500°C for 1 h, chemical activation with KOH at 725°C, 30 min | 3,000 | 368 F/g | 0.2 A/g | – | – | – | – | [118] |
| Heteroatom-doped porous carbons | Organosolv lignin | Hydrothermal carbonization at 300°C, 100 bar, 30 min, KOH and adenine mixing at 90°C, 10 h | 2,957 | 372 F/g | – | ~ 100 % after 30,000 cycles | – | – | Higher specific capacitance and surface area than pristine porous carbon | [119] |

(*Continued*)

**TABLE 1.2B** ***(Continued)***

**Overview of Lignin-derived SC Electrodes**

| Materials | Source and Type of Lignin | Synthesis Conditions | Specific Surface Area ($m^2/g$) | Electrochemical Performance | | | | | Additional Remarks | Ref. |
|---|---|---|---|---|---|---|---|---|---|---|
| | | | | Specific Capacitance | Current Density | Capacitance Retention | Energy Density | Power density | | |
| N-doped porous carbon | Sodium lignosulfonate | Hydrothermal crosslinking reaction (1,6-hexanediamine as a crosslinking agent and N-source) and KOH activation | 1,867.4 | 440 F/g | 0.5 A/g | 94.8% after 3,000 cycles | 18.5 Wh/kg | 200 W/kg | Increase charge storage owing to a higher fraction of disordered carbon | [120] |
| CNF | Alkali lignin | Electrospinning with PVA, carbonization at 1,200°C, 1 h | 583 | 64 F/g | 0.4 A/g | 90% over 6,000 cycles | 5.67 Wh/kg | 94.19 W/kg | Free-standing, binder-free electrode | [121] |
| CNFs | Corn stover lignin | Electrospinning, carbonization 800°C, 2 h | 538 | 151 F/g | 0.1 A/g | 95% after 20,000 cycles at 4 A/g | 6.68 Wh/kg | 100.2 W/kg | – | [122] |
| N-O co-doped CNF | Harwood kraft lignin | Electrospinning, impregnation of Pluronic F127, carbonization at 1,000°C, 30 min and activation with $CO_2$ | 1,148 | 102.3 F/g | 1 A/g | 94.5% after 5,000 cycles | 17.92 Wh/kg | 800 W/kg | – | [123] |
| Lignin-cellulose acetate CNFs | Poplar lignin | Electrospinning, carbonization at 800°C for 1 h | 837.4 | 346.6 F/g | 0.1 A/g | 82.30% | 31.5 Wh/kg | 400 W/kg | | [124] |
| CNFs | Poplar lignin | Electrospinning with PAN, carbonization at 950°C for 2 h | 1,062.5 | 349.2 F/g | 0.5 A/g | 90.5% after 5,000 cycles | 39.6 Wh/kg | 5,000 W/kg | – | [125] |

| | | | | | | | | | | |
|---|---|---|---|---|---|---|---|---|---|---|
| CNFs | Lignin from corn stalk refining | Electrospinning with PAN, carbonization at 100°C, 4 h | 2,042.9 | 442.2 F/g | 1 A/g | 97.1% after 10,000 | 37.1 Wh/kg | 400 W/kg | Specific capacitance 4 times higher than unmodified lignin | [126] |
| 2D-carbon nanosheet | – | Freeze-casting aq. lignin dispersion, carbonization without activation | 854.7 | 281 F/g | 0.5 A/g | 91% after 5,000 | 14.3 Wh/kg | 28,611 W/kg | – | [127] |
| rGO composite | Alkaline lignin | Carbonization without activation at 800°C | 2,630 | 103 F/g | – | – | – | – | – | [128] |
| **Lignin-derived carbon and metal oxide-based composites** | | | | | | | | | | |
| Activated carbon coated with $MnO_2$ | Alkali lignin | Hydrothermal treatment of activated carbon-lignin-$KMnO_4$ mixture at 160°C, 1 h, slurry coating on foil, heating at 100°C, 4 h, under vacuum | – | 5.52 $mF/cm^2$ | 6.01 × $10^{-3}$ A/g | 97.5% up to 2,000 cycles | 14.11 Wh/kg | 1,000 W/kg | – | [129] |
| NiO-embedded mesoporous carbon | Sodium lignosulfonate | Dissolution of Ni-salt in Pluronic F127 and incorporation of lignosulfonate, glutaraldehyde crosslinking, carbonization in flowing $N_2$ at 600°C, 2 h | 802 | 882 F/g | 1 A/g | 93.7% after 1,000 cycles | – | – | – | [130] |
| CNF decorated with $MnO_2$ | Alkali lignin | Electrospinning with PVA, carbonization at 1,200°C, 1 h | 583 | 83.3 F/g | 0.25 A/g | 70% at 2 A/g after 10,000 cycles | 84.3 Wh/kg | 5,720 W/kg | Freestanding and flexible electrode. No need for additional binder or current collector | [131] |

*(Continued)*

**TABLE 1.2B** ***(Continued)***
**Overview of Lignin-derived SC Electrodes**

| Materials | Source and Type of Lignin | Synthesis Conditions | Specific Surface Area ($m^2/g$) | Electrochemical Performance | | | | | Additional Remarks | Ref. |
|---|---|---|---|---|---|---|---|---|---|---|
| | | | | Specific Capacitance | Current Density | Capacitance Retention | Energy Density | Power density | | |
| CNF with ZnO | Hardwood lignin | One-step lignin and pitch electrospinning air at 280°C, 1 h and carbonized in an inert atmosphere at 800°C, 1 h | 1,194 | 165 F/g | – | 94% after 3,000 cycles | 22–18 Wh/kg | 400–10,000 W/kg | – | [132] |
| CNF decorated with $NiCo_2O_4$ | Alkali lignin | Electrospinning with PAN, CNF nanoweb coated with $NiCo_2O_4$ at 120°C, 10 h. | – | 1,757 F/g | 2 mA/$cm^2$ | 138% after 5,000 cycles | 47.75 Wh/kg | 799.53 W/kg | – | [133] |
| **Lignin (non-carbonized)-electron-conducting polymer-based composites** | | | | | | | | | | |
| PEDOT-based composite | Lignosulfonate | Polymerization of EDOT/lignin in the presence of $Na_2S_2O_8$ for 8 h. | – | 170.4 F/g | 1 A/g | 83% after 1,000 cycles | – | – | Specific capacitance higher than PEDOT (80 F/g) | [134] |
| PEDOT-based electrode and SC | lignosulfonate | Deposition of a layer of poly(aminoanthraquinone) (PAAQ) on PEDOT/lignin-based electrode | | 418 F/g (in electrode format) 74 F/g (in SC format) | 1 A/g | 80% over 10,000 cycles | 8.2 Wh/kg | 700 W/kg | Specific capacitance higher than lignin-free, PEDOT/PAAQ electrode (383 F/g). [135] | [136] |
| PANI-based composite | Lignosulfonate | Polymerization of aniline and lignosulfonate at 0 °C for 24 h. | – | 377.2 F/g | 10 A/g | 74.3% after 10,000 cycles | – | – | Specific capacitance higher than | [137] |

| | | | | | | | | | | |
|---|---|---|---|---|---|---|---|---|---|---|
| | | | | | | | | | pure PANI (183.7 F/g) | |
| PPy-based composite coated Au electrode | Lignosulfonate | In-situ oxidative polymerization of pyrrole | – | 1,200 (2,500 with AQS) F/g | 0.83 A/g | | – | – | Specific capacitance higher than pure PPy (250 F/g at 0.1 A/g) | [138] |
| **Lignin (non-carbonized)-graphite/carbon nanotube (CNT)-based composites** | | | | | | | | | | |
| rGO composite | Technical soda bagasse lignin | Two-zone vapor transport technique, heating at 360°C, 24 h, saturation in lignosulfonate solution, 24 h | – | 211 F/g | 1 A/g | 88% after 15,000 cycles | – | – | – | [139] |
| Polymer-incorporated into rGO | Lignin-derived aromatic oligomers from eucalyptus | Polymerization of lignin-derived oligomers, then incorporated with rGO | – | 250 F/g | 10 A/g | – | – | – | Specific capacitance 3 times higher than rGO (87 F/g) | [140] |
| Nanocrystal-entrapped into rGO | sodium lignosulfonate (SW) | Modified Hummers method for graphene production, mixing graphene and lignin in a hydrazine solution at 60°C, 12 h | – | 432 F/g | 1 A/g | 96% after 3,000 cycles | – | – | Capacitance 6 times higher than rGO (93 F/g) | [141] |
| Multi-walled CNT | Kraft lignin | Catalytic decomposition of acetylene to form multi-walled CNTs, redispersion in methyl sulfoxide | – | 181 F/g | 2.5 A/g | 93% after 500 cycles at 1 A/g | – | – | Capacitance higher than lignin-free CNT electrodes (~85 F/g) | [142] |
| Lignosulfonate/single-walled-CNT/holey rGO | Sodium lignosulfonate | Hydrothermal treatment at 100°C, 10 h | – | 287 F/g | 1 A/g | – | 0.077Wh/$cm^2$ | 2.5 $W/cm^2$ | – | [143] |

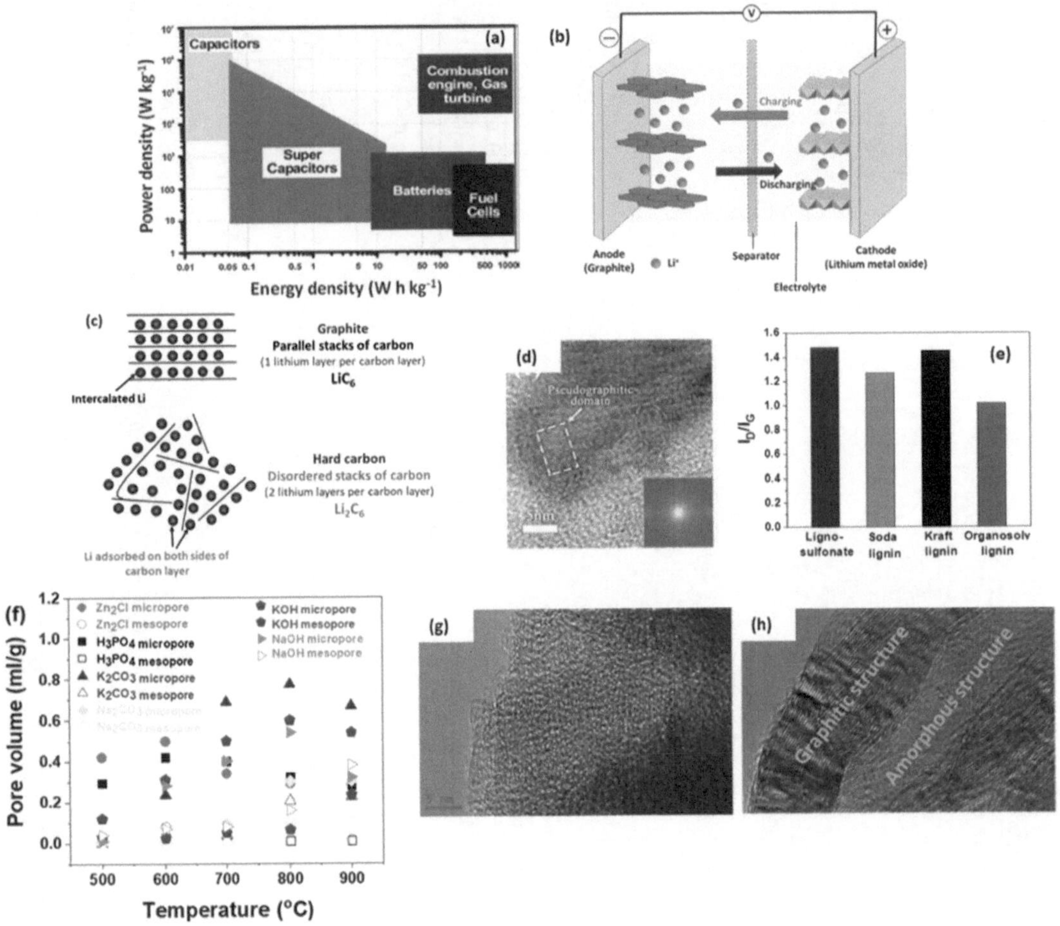

**FIGURE 1.2** (a) Ragone plot for different types of electrochemical energy conversion and storage devices (Reproduced with permission from ref. [167]). (b) Schematic diagram of a lithium-ion battery (LIB). (c) Schematic showing the basic differences in lithium intercalation mechanisms between graphite (top) and single-layer type, disordered hard carbon (house of card model, bottom) (redrawn with permission from ref. [168]). The higher lithium accommodation sites explain the higher charge storage capacities of hard carbon as compared to graphite. (d) TEM image of pristine lignin-derived carbon revealing the amorphous nature with significant turbostratic carbon and few ordered structures (taken with permission from ref. [61]). (e) The ratio of Raman peak intensities for disordered carbon ($I_D$ at ~1,350 $cm^{-1}$) to graphitic, ordered carbon ($I_G$, at ~1,600 $cm^{-1}$); i.e., $I_D/I_G$ of carbons obtained from different sources of lignin-based carbon precursors (the data taken and redrawn from ref. [28]). (f) Effect of carbonization temperature and the nature of activation agent on the pore volume of the derived activated carbon (redrawn with permission from ref. [65]). (g, h) HRTEM images of lignin-melamine resin-derived hard carbon without (g) and with (h) N-doping. The hard carbon obtained without N-doping was amorphous; while the N-doped carbon sample showed graphitic structure embedded in amorphous carbon; scale bar 5 nm (reprinted with permission from ref. [62]).

#### 1.2.1.1 Carbonaceous Materials for LIB Electrodes

Some of the prominent anode materials for LIBs are graphite and carbon-based materials, silicon, transition metal oxides, and alloys [73,79,169]. However, each of them has some specific challenges, such as low specific capacity (graphite (372 mAh/g)), [144,170,171] large volume expansion (Si), [172,173] irreversible incorporation of Li into a layer between electrode and electrolyte (known as solid-electrolyte interphase (SEI)), [174] and irregular deposition of Li (known as dendrites) during charging (graphite) [170,172,175,176], which compromise the capacity, battery lifetime, and safety. Especially, the protruding dendrites can lead to short-circuiting, failure of batteries, and fire [145,175,176]. We, therefore, need high-capacity anode materials, which can also help to alleviate some of the issues mentioned above.

Hard carbon materials with appropriate porosity are beneficial for Li-/Na-storage [68]. Hard carbons offer a turbostratic structure [61,87,176–178] which is somewhere between amorphous carbon and crystalline graphite. When Li is intercalated within crystalline graphite, there is one layer of Li per carbon layer and one Li-atom coordinated with six carbon atoms ($LiC_6$) (Figure 1.2c). But in hard carbon, the Li can be adsorbed on each side of a single graphene layer, resulting in up to 2 Li-atoms adsorbed per 6 carbon atoms ($Li_2C_6$) (Figure 1.2c) [168,171]. These hard carbons (unlike soft carbons) also comprise a large number of nanosized internal pores [168,178] in between single layers of graphene sheets. $Li^+$-ions can, therefore, intercalate not only into the graphene sheet-based domains but also fill the nanopores [87,168,176,177]. As a result, a capacity >700 mAh/g [59,60,66,144,179–181] can be attained using hard carbon–based anodes. Additionally, the turbostratic structure [87,168,176,177] can inhibit the extensive intercalation of aggressive electrolytes within the anode [168,176,182] and maintain the chemical stability of the electrode [87,176,177]. Since lignin-derived hard carbon can offer such turbostratic disorder [61] (Figure 1.2d) and other favorable characteristics; it has high potential as anode material for LIBs [61] and NIBs [183].

The carbonization of lignin precursor is done at a high temperature, leading to an increase in %C-content typically from ~48–62% to ~99.2–99.6% while keeping the percent of other elements (N, S) very low [28]. The nature of this carbon is studied using Raman spectra. The ratio of disordered (turbostratic)-to-graphitic carbon ($I_D/I_G$) obtained from lignin-based carbon precursors (>1, Figure 1.2e) confirms the high potential of lignin-based carbon for Li-/Na-storage [28]. It has to be noted that the structure and properties of lignin precursors, carbonization temperature, and additional conditions of pre-/post-carbonization determine the carbonization yield, composition, micro/nanostructure, porosity, specific surface area, density, and electronic conductivity of the carbon materials. These characteristics, in turn, control the electrochemical performance (such as capacity, rate capability, cyclability, and coulombic efficiency) of the carbon anodes.

Simple carbonization of lignin may not produce suitable pore structure [60] and/or generate buried pores, [68] which are not beneficial for Li-storage in electrodes. Therefore, additional steps are taken to gain control over pore structure, specific surface area, and degree of graphitization of lignin-derived carbons, such as carbonization followed by activation, [59,60,65] reduction after carbonization [63]/oxidation before carbonization, [61] and heteroatom (N, S, P) doping [62,64].

The chemical activation can induce different extents of open micro/mesopores depending on the activation agent (KOH, $K_2CO_3$, $ZnCl_2$, $H_3PO_4$) [60,65,184] and temperature [65] of subsequent carbonization step. Huang et al. [59] mixed KOH with alkali lignin in water, then dried and carbonized it. As the calcination temperatures exceeded the melting point of KOH (380°C), the spaces occupied by KOH were freed. This resulted in a 3D, macroporous carbon network with mesopores and micropores decorated on the carbon walls [59]. The KOH-activated carbon offered a specific surface area (907 $m^2/g$) and capacity (470 mAh/g at the 400th cycle at a current density of 200 mA/g) higher than the carbon without activation (77.1 $m^2/g$, 180 mAh/g) [59]. $K_2CO_3$-activation was even more effective in increasing the specific surface area (2,300 $m^2/g$ at 900°C) [60] and pore volume (0.8 mL/g at 800°C, Figure 1.2f) [65] of carbon. Activation of enzymatic hydrolysis lignin gave carbon which offered a reversible $Li^+$-ion storage capacity of 520 mAh/g at a current density of 200 mA/g over 200 cycles, and the capacity retained to 260 mAh/g at 1 A/g for over 1,000 cycles [60].

$H_2$ reduction after carbonization (in $N_2$ at 800°C) [63] of acetone-extracted lignin (corn stalks) yielded hard carbon giving a specific capacity of 882.2 mAh/g at a current density of 372 mA/g (Coulombic efficiency (CE) ~ 62.4%) [63]. Moreover, the charge capacity value retained as 228.8 mAh/g at 744 mA/g after 200 cycles [63]. Oxidation of lignin precursor (sodium lignosulfonate) before carbonization, on the other hand, introduced oxygen-containing functional groups (such as carbonyl, alcohol, and peroxide) into hard carbon [61]. These functional groups ensured cross-linking during the low-temperature pre-oxidation step and inhibited the melting and rearrangement of carbon structure during the high-temperature carbonization process [61]. The pre-oxidation thus helped to inhibit graphitization and gain a highly disordered carbon structure with a specific capacity of 584 mAh/g at 50 mA/g [61].

The presence of heteroatoms (N, S, P) in carbon can enhance the electrochemical reactivity as well as ion/electron diffusion through the carbon materials. For example, when carbonized at 1,000°C, melamine-grafted lignin offered a carbon containing both graphitic and turbostratic structure (Figure 1.2g, h), which stored Li synergistically [62]. This N-doped carbon-based anode offered a reversible capacity of 345 mAh/g at a current density of 100 mA/g. About 69% of this capacity retained when the current density increased to 1 A/g. This value was higher than the corresponding undoped carbon [62]. N-doping also boosted the capacity of organosolv lignin-based carbon fibrous mats from 445 mAh/g to 576 mAh/g [83].

A combination of N-doped carbon and Si led to a very high specific capacity, likely due to the synergistic activities of Si with high theoretical capacity (4,200 mAh/g) [185] and N-doped carbon [77]. For example, when an alkali lignin or alkali-lignin-derived azo polymer and Si-nanoparticle (dia. ~50 nm) dispersion was carbonized, carbon-coated Si-nanoparticle was obtained, which gave a specific capacity of 882 mAh/g at a current density of 200 mA/g over 150 cycles and CE of 99% [64]. On the other hand, an initial specific capacity of 2,286 mAh/g at a current density of 300 mA/g was achieved from a Si-carbon composite made using recycled silicon-slicing slurry and lignin/lignocellulose [77]. This carbon-Si composite retained a specific capacity (881 mAh/g) higher than the recycled Si (~200 mAh/g) used in this work after 51 cycles [77]. When lignin was used as a crosslinking agent for Si-nanoparticles, a conformal network of Si-nanoparticles was achieved, which offered one of the highest reported capacities (3,000 mAh/g) [78] for sustainable biomass-based electrodes for LIBs. The functional conformal network accommodated a large volume change of Si-nanoparticles during charge/discharge processes. Also, the lignin prevented the crack of the Si-nanoparticles and promoted the formation of a stable SEI layer on the Si surface [78].

Like Si-based anodes, a challenge for transition-metal oxide-based anodes is the volume expansion during cycling. If the metal oxide nanoparticles are embedded within a mesoporous carbon structure, the carbon skeleton can act as a cushion and effectively alleviate the volume expansion effect of the metal nanoparticles. The examples include (i) nickel oxide (NiO)-embedded carbon nanospheres from sodium lignosulfonate giving a discharge capacity of 863 mAh/g at a current density of 100 mA/g; [73] and (ii) ferric oxide ($Fe_3O_4$)-graphene nanosheet (from lignin from soda pulping) composites giving a discharge capacity of 3,829 mAh/g at a current density of 50 mA/g [79].

#### 1.2.1.2 Carbon Nanofibers (CNFs) for LIB Electrodes

Carbon fiber–reinforced composites find substantial applications in lightweight, automotive applications [186,187]. Nanometer-sized carbon fibers (CNFs) with structural stability and adjustable porosity can also be promising anode materials for LIBs [80,187]. Polyacrylonitrile (PAN), with excellent mechanical strength and relatively low density, is the state-of-the-art precursor material for CNFs [187–189]. However, the high cost of PAN and the final product ($21.7/kg), [190] as well as the environmental concerns (petroleum-based polymer, high $CO_2$ emission, solvent use), [86] hinder the wide use of PAN as a carbon precursor [80]. Lignin can act as an inexpensive, single-component-based carbon precursor, [191] or a low-cost companion of structure-reinforcing polymers or carbon materials (such as PAN, [80,192,193] polyethylene, [192] polypropylene, [192] polylactide, [194] or carbon nanotube (CNT) [192]) for making CNFs. As per a report from 2018, lignin-based CNF production can cost ~$6.2/kg which is considerably lower than the production cost of PAN-based CNFs [190]. Lignin-based CNFs can also offer mechanical properties comparable to traditional CNFs [192].

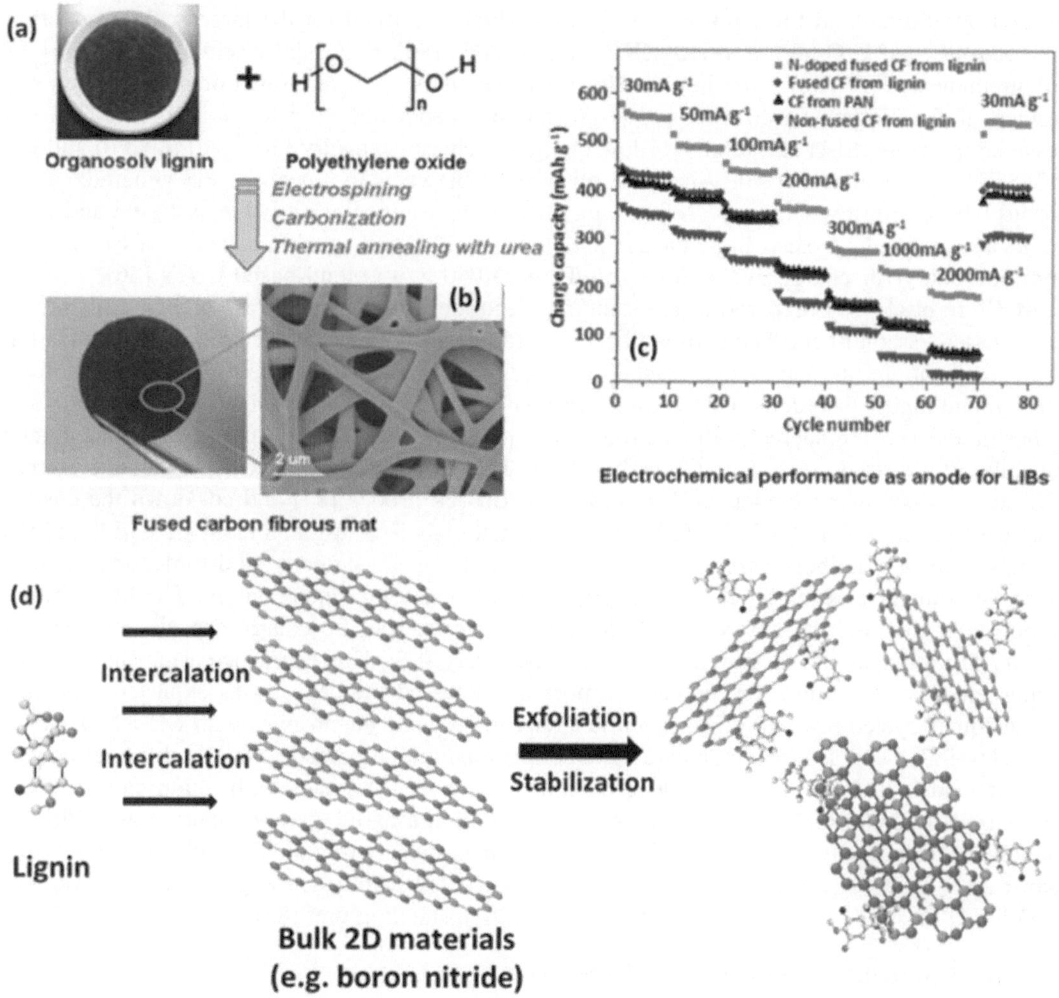


**FIGURE 1.3** (a) Procedure to fabricate lignin-PEO-based fused, free-standing electrospun CNF mats. (b) FESEM image of electrospun CNFs derived from lignin-PEO (9:1) blends. (c) Charge capacities of a range of carbon fiber-based samples at different current rates. (Reprinted with permission from ref. [83]). (d) Schematic illustration of the exfoliation of bulk 2D materials using lignin/lignin-based compounds.

Lignin-based CNFs can be fabricated via melt spinning, [186,187] gel spinning, [192] solution spinning, [187,193] or electrospinning, [187,195] followed by thermostabilization and carbonization. Electrospun carbon fibers can form highly porous free-standing mats that could be directly used as electrode materials accessible to electrolyte ions without the addition of any conducting agent or binder, largely simplifying the fabrication process [83]. He et al. [83] fabricated free-standing, fused, CNF mats by electrospinning a blend of organosolv lignin and polyethylene oxide (PEO) (lignin: PEO = 9: 1) followed by carbonization (Figure 1.3a, b). This lignin/PEO-based CNF exhibited a specific capacity (445 mAh/g at a current density of 50 mA/g), [83] similar to pure PAN-based CNF (450 mAh/g) and higher than graphite (372 mAh/g). Thermal annealing in the presence of urea gave an N-doped variety of this PEO-lignin-based CNF, which improved the capacity to 576 mAh/g (Figure 1.3c) [83]. Large quantities of micropores in these lignin-based CNF mats gave more accessibility to electrolytes and ions within the carbon structure. The added advantage is that this is an economical and straightforward method for electrode fabrication which is critical for the large-scale production of electrodes [83]. Other reported CNFs used PAN, [80] poly(lactic acid) (PLA), [86] or polyurethane (TPU) [86] with lignin. The presence of oxygenated functional groups on the surface of PAN/lignin-based CNFs was considered responsible for the formation of Li-ion passivation layers (SEI) and relatively low charge-discharge capacity (300–310 mAh/g) and CE (52–55%) of the anodes [80]. Also, the miscibility between lignin and TPU generated non-porous CNFs with low capacity (280 mAh/g); while the immiscibility between lignin and PLA led to well-defined porosity, high surface area (~670 $m^2$/g), and specific capacity of 611 mAh/g after 500 charge/discharge cycles for lignin/PLA (50/50 w/w) blend-based CNFs [86].

McGuire et al. [87] prepared a free-standing electrode by engineering the carbon structure of lignin-based carbon fibers. The carbon fibers (~8–15 μm dia.), prepared via oxidative stabilization and carbonization, had nanoscale, crystalline graphitic domains surrounded by continuous, low-density, and highly disordered (turbostratic) carbon matrices. Such a combination of graphitic and turbostratic carbons rendered both electron-conducting (graphitic region) and Li-insertion/storage capability (turbostratic region) to the carbon fiber. This eliminated the need for a metallic current collector and polymeric binder used in traditional LIB-electrodes [87] and cut down the cost of electrodes (price of Cu ~$3.30/lb, full cost of raw materials for Cu-graphite anode ~16/lb [175]). Lignin-based carbon fibers, carbonized at 1,000°C, had the greatest degree of disorder and smallest graphitic domains and inhibited solvent (propylene carbonate) co-intercalation. The lignin-based carbon fiber mat and the slurry-coated electrode offered a specific charge capacity of 193 and 350 mAh/g (comparable to graphite anode), respectively [87]. However, increasing the carbonization temperature graphitized the carbon more and lowered the capacity as expected. The carbonized mats cycled reversibly in conventional aprotic organic electrolytes with CE >99.9% over 70 cycles, [87] a characteristic comparable to commercial graphite anodes [87]. Stable cycling was also attained in 1M $LiPF_6$ in propylene carbonate for over 40 cycles with lignin carbon fibers carbonized at 2,000°C [87]. The projected cost of lignin-based carbon fibers was $3/lb for monolithic, electrically connected, self-supporting mats, which was significantly less expensive than battery-grade graphite powers ($12/lb for unprocessed battery-grade graphite) [87]. Lignin-PEO-based electrode (discussed earlier) was also a free-standing one [83].

#### 1.2.1.3 Exfoliating Agents for LIB Electrodes

Amphiphilic lignin [56,89,196,197] and lignosulfonate [88] can act as surfactant and exfoliate bulk 2D materials (like, graphite, [56,196] molybdenum sulfide ($MoS_2$), [89] tungsten disulfide, [89] or boron nitride [88,89]) into nanosheets. Carbonization of these exfoliated materials creates layered and holey carbon structures where the interlayer gaps are larger than typical layered structures. These holey structures favor Li insertion (Figure 1.3d), [88] electrolyte accessibility and ion transport, volume expansion, and thermal expansion during cycling [88]. Such materials can be beneficial for hosting Li in anodes of both lithium-ion (LIBs) and lithium metal batteries (LMBs). For example, lignin-exfoliated $MoS_2$ nanosheets [89] retained a specific capacity of 110 mAh/g at 2 C and CE (82%) higher than that of pristine $MoS_2$ due to enhanced ion/electron transfer kinetics [89]. Aqueous-phase exfoliation of graphite using alkali lignin gave 20% graphene as a monolayer that was stabilized *via* π-π interactions between graphene and alkali lignin. This aqueous process was not only economic and eco-friendly but also offered electrical conductivity (26,000 mS/cm (unannealed) and 93,000 mS/cm (annealed)) [197] similar to or higher than graphene exfoliated in organic solvents [89,197].

#### 1.2.1.4 Gel-/Solid Polymer Electrolytes (GPEs/SPEs) and Separators for LIBs

In LIBs, $Li^+$-ions travel through an electrolyte system between two Li-intercalating electrodes. The use of organic liquid electrolytes alone poses severe safety concerns in LIBs, such as leakage, flame, and blast [146,198]. To reduce electrolyte leakage and produce safe LIBs, gel polymer electrolytes (GPEs) and solid polymer electrolytes (SPEs) are introduced. In a traditional GPE, the organic liquid electrolyte is immobilized within a polymer matrix and requires a plasticizer (or solvent) [199]. The materials conventionally used as polymer matrix include, poly(ethylene oxide) (PEO), polyacrylonitrile (PAN), poly(vinylidene fluoride) (PVDF), poly(vinyl alcohol) (PVA), and poly (methyl methacrylate) (PMMA) [146,151,200]. The interconnected pores in the polymer framework are filled with electrolytes, creating a swollen gel phase through which $Li^+$ ions are transported. This way, GPE can act as an electrolyte as well as a separator [198]. On the other hand, SPEs in LIBs are solvent-free (dry), $Li^+$ ion-conducting materials [199,201]. They are nonvolatile, less flammable, and less likely to suffer from severe degradation (unlike liquid electrolytes). The most common examples of SPEs are PEO and poly(propylene oxide) (PPO) with Li-salts dissolved in the polymer. In PEO, the $Li^+$ cations coordinate with flexible ethylene oxide segments, [160,199,202] and the intra- and inter-chain ion hopping and segmental motion of the ethylene oxide assist in ionic conductivity in SPEs [160,199]. These SPEs can also act both as an ion-conductor and separator.

Some of the typical requirements of GPEs and SPEs are common, like sufficient mechanical rigidity to slow down the growth kinetics of metal deposits (dendrites) while exhibiting high bulk and interfacial $Li^+$-ion transport [203,204]. It was demonstrated that SPEs with modulus ~0.1 MPa can suppress the dendrite growth, a major safety concern of LIBs [204]. It was also shown that the majority of the conventional GPEs can exhibit ionic conductivity on the order of 1 mS/cm [198]. On the other hand, a range of SPEs has demonstrated $Li^+$-ion conductivity in the range of 0.1–1 mS/cm [201]. Additionally, GPEs/SPEs with $Li^+$-ion transference number ($t_{Li+}$) approaching 1 are highly desired as it implies that the ion-conducting performance of the GPE/SPE is majorly accomplished by $Li^+$-ion.

With the porous, branched structure, and aliphatic and polar functional groups compatible with polar organic solvents, [5] lignin can serve as a polymer matrix for GPEs. Lignin can also fulfill the criteria of SPEs as the hydrophilic groups in lignin can interact with salt anions, enhance Li-salt solubility in polymer, and facilitate cation ($Li^+$) conduction [5]. Moreover, lignin can be easily functionalized with sulfonate ($-SO_3^-$) or other anionic functionalities to make them cation-conductive by itself [5,55]. Finally, lignin, unlike many conventional GPEs and SPEs, is biodegradable. Considering all these, lignin can be attractive for SPEs, polymer matrix for GPEs as well as separators.

Li et al. [152] were the first to demonstrate a lignin-based GPE by immersing a cast, dried lignin membrane into the liquid electrolyte (1M lithium hexafluorophosphate ($LiPF_6$) in ethylene carbonate (EC)/dimethyl carbonate (DMC)/ethyl methyl carbonate (EMC) (1:1:1, w/w/w)). The liquid electrolyte uptake by this GPE reached up to 230 wt%. The mechanical property of the GPE was 0.85 MPa and 1.16 MPa before and after absorbing the electrolyte, respectively. The room-temperature, ionic conductivity (3.73 mS/cm), and $Li^+$-ion transference number ($t_{Li+}$ ~0.85) of this membrane [152] were higher than those of the commercial Celgard 2730 separator with liquid electrolyte (0.21 mS/cm, $t_{Li+}$ ~0.27) [153] and traditional polymer matrices ($t_{Li+}$ ~0.20–0.70). Moreover, the lignin-based GPE was electrochemically stable up to 7.5 V [152] with no electrolyte loss up to a temperature of 100°C, likely due to a strong interaction between -OH groups of lignin and carbonyl and ester groups of the electrolyte [152].

Blending lignin with poly(N-vinylimidazole)-co-poly(ethylene glycol)methyl ether methacrylate led to a physically cross-linked network structure which enabled casting a free-standing film [154]. Upon activation with organic electrolyte, this GPE-film exhibited a $t_{Li+}$ of 0.63, and a tensile modulus (4 MPa) 10 times higher than that of pure lignin film (0.35 MPa) [154]. The ionic conductivity was 0.63 and 1.70 mS/cm at 30 and 80°C, respectively. A specific capacitance of 150 mAh/g at 1 C over 450 cycles was achieved when this GPE was used in $LiFePO_4$/Li-based LIB cells [154]. In another work, rather than physical cross-linking, alkali lignin was covalently linked with polyvinylpyrrolidone (PVP) [153]. Lignin/PVP was then cast into a membrane and immersed into a liquid electrolyte. Adding PVP (22 wt%) to lignin significantly improved the mechanical properties of the membrane (670%). The ionic conductivity of this GPE membrane

(2.52 mS/cm) was higher than that of pure PVP-polymer electrolyte (0.0142 mS/cm) and was attributed to high liquid electrolyte uptake (237 wt%) by lignin/PVP-based GPE. In a LIB setup ($LiFePO_4$/GPE/Li), the reversible capacity (145 mAh/g at 0.1 C) was higher than that of Celgard 2730 (50 mAh/g) [153]. Also, 95% of its capacity was retained after 100 cycles with a high CE (>99%), demonstrating high charge reversibility at the electrode-electrolyte interface [153].

Several GPEs were made by converting a mix of lignin and a high- MW polymer to nanofiber-based [161] or composite, [159] non-woven membranes for subsequent liquid electrolyte uptake. For example, alkali lignin ($M_w$ ~10,000)/PVA ($M_w$ ~146,000–186,000) nanofiber-based GPE had a very high electrolyte uptake (488–533 wt%), lower shrinkage (15%) than Celgard separator (45 %), and charge transfer resistance ($R_{ct}$ ~98 Ω) slightly higher than Celgard ($R_{ct}$ ~84 Ω). A LIB based on lignin/PVA-based GPE gave an average specific discharge capacity of 133.3, 117.9, 91.2, and 33.4 mAh/g at 0.5C, 1C, 2C, and 5C, respectively, whereas it was 122.6, 88.8, 47.8, and 7.6 mAh/g at the same C-rates for Celgard separator. This indicated improved ionic transport in lignin/PVA-based GPEs [161]. On the other hand, with high porosity (74%) and a very high electrolyte uptake (790 wt%), a lignin/PAN-based separator with 1M $LiPF_6$ in EC-DMC offered a relatively higher specific capacity (148.9 mAh/g after 50 cycles at 0.2C, 95% capacity retention) [159].

In some cases, lignocellulose was used instead of lignin to make GPEs. In these cases, PAN, [157] PVA [155] or polyethylene glycol (PEG) [205] was mixed with lignocellulose to make a composite membrane for electrolyte sorption. Thermal and mechanical stabilities improved with the use of lignocellulose due to its denser network structure [157]. Importantly, both lignin and cellulose in the lignocellulose network underwent H-bonding interactions with electrolyte anions ($PF_6^-$ ions of $LiPF_6$) and solvents (EC, DMC, DEC) synergistically. Such interactions likely limited the movement of $PF_6^-$ anions while facilitating the movement of $Li^+$ ions [205]. The ionic conductivity of these GPEs varied between 2.94 (lignocellulose/PAN) [157] and 3.92 mS/cm (lignocellulose/PVA), [155] and were higher than that of Celgard (0.2 mS/cm). The $t_{Li+}$ ranged between 0.81 (lignocellulose/ PEG) [205] and 0.84 (lignocellulose/PAN). The specific capacities of these lignocellulose-based GPEs in $LiFePO_4$/GPE/Li-based cells were also not very different. A PAN-based one gave 143.90 mAh/g at 74.4 mA/g after the 90th cycle [157]. The PVA [155]- and PEG [205]-based GPEs offered specific discharge capacities of 153 mAh/g [155] and 171 mAh/g, [205] respectively, at 0.2 C. An unconventional GPE-formation approach used potato starch with lignocellulose, [151] rather than a well-defined polymer (like PAN, [157] PEG [205]), but still was able to achieve an ionic conductivity (2.52 mS/cm) and specific capacity (160 mAh/g at 0.2 C) [151] comparable to lignocellulose/PEG-based GPEs [205].

Low-MW lignin fractions ($M_w$~2,923–3,484) [30] or small molecules derived from lignin depolymerization [158] were also used to make GPEs [30,158] and SPEs [30]. In one work, softwood pine kraft lignin was used as the precursor of GPE. To make the SPE, alkylated wheat straw/ Sarkanda grass soda lignin was mixed with a thiol-based curing agent (for thiol-ene polymerization) and lithium bis(trifluoromethanesulfonyl)imide (LiTFSI) salt (as needed) [30]. Both of these GPEs (0.0072 mS/cm) and SPEs (0.05–0.07 mS/cm) offered low ionic conductivity. Nevertheless, the lignin-based SPE exhibited a high $t_{Li+}$ (0.9) in the presence of 1 wt% LiTFSi within the SPE [30]. In another work, [158] rather than starting from the lignin, GPEs were made using lignin-derived small molecules, like vanillyl alcohol and gastrodigenin. First, these lignin-derived small units were allylated and then UV-polymerized with multifunctional thiol monomer via thiol-ene reaction. The resulting polymer films displayed ionic conductivity up to 0.01 mS/cm and high storage moduli of up to 10.08 MPa [158]. Thiol-ene polymerization was also employed to react thiol-functionalized PEG with alkene-functionalized PEG [160]. To this resulting lignin-graft PEG, LiTFSI salt was added to obtain an SPE. The SPE offered ionic conductivity (0.14 mS/cm at 35°C) higher than the homo-polymeric PEG [160]. Also, the presence of lignin moderately impacted conductivity at elevated temperature compared to homopolymer PEG [160].

Another work attempted to use lignin as an electrolyte rather than just a polymer matrix for a GPE [55]. Sulfonation and subsequent chlorination made lignin anionic, and it thus worked as a cation conductor. The fixed acid groups on lignin were more likely to contribute to the stability of LIBs. The acid groups of lignin were then converted to Li-salt (by reacting with LiOH), mixed with PVA (to impart mechanical properties) by dissolving in dimethyl sulfoxide (DMSO), poured onto Teflon dishes and dried to make GPEs. This lignosulfonate/PVA composite-based GPE offered ionic conductivity of 0.248 mS/cm and $t_{Li+}$ of 0.89 [55].

#### 1.2.1.5 Binders for LIBs

The binder material in a LIB electrode acts like an electrode-electrolyte interface. The binder material bridges the active component, conductive additive, and current collector as well as ensures the mechanical integrity of the electrode without significantly affecting its ionic/electronic conductivity [206]. PVDF, the most commonly used binder for LIBs, is expensive, not easy to recycle, and requires the use of toxic and not-environment-friendly solvents (like N-methyl pyrrolidone (NMP)) to produce slurry for the coating process [162,207]. Most importantly, PVDF can react violently with metallic Li or lithiated carbon ($Li_xC_6$) at a high temperature, which is detrimental to the cycling performance of LIBs [146]. In a plant cell, lignin plays a role similar to a binder (like imparting mechanical rigidity and structural integrity). Therefore, cheap and environmentally benign lignin is a desirable binder material for LIBs.

Lignin extracted from black liquor (a by-product of Kraft paper mills) was used as a binder material for electrodes [163]. The lignin was pretreated to remove the low-MW fractions ($M_w$~700) to maintain good battery performance. Also, lignin was dissolved in acetone, a better solvent than NMP from an environmental perspective. The electrodes were pressed to reduce the porosity of the electrodes and increase the interparticle contacts. With this pretreated lignin ($M_w$ ~16,000, PDI ~6.8)-based binder, the $LiFePO_4$ and graphite electrodes exhibited specific capacities of 140 mAh/g and 305 mAh/g, respectively, at a current density of 37.2 mA/g [163]. Rodríguez et al. [162] used three lignin-rich fractions ($M_n$~1,490–1,970, $M_w$~2,520–4,170, PDI~1.7–2.3) obtained from three different pulping processes (kraft, soda, and organosolv pulping) from wheat straw to make binder materials for electrodes in LIBs. Electrodes made of the different lignin as binders showed specific capacities (260–280 mAh/g) similar to the electrode made with PVDF as a binder [162].

Lignin was used as a binder for $LiNi_{0.5}Mn_{1.5}O_4$ cathodes [164]. Carbonate-based electrolytes are known to form free radicals (at the cathode-electrolyte interface) and cathode/electrolyte interphase (CEI) in the presence of Li-salt. This CEI formation happens at the cost of massive decomposition of the electrolytes and leads to poor cycling performance of 5V-LIBs. While PVDF-based binders could not, with the phenolic groups, lignin acted like a free-radical scavenger and terminated the free-radical chain reaction which helped to suppress the oxidative decomposition of the electrolyte [164]. With a lignin-based binder, this battery showed high capacity retention of 94.1% (110.8 mAh/g) after 1,000 cycles [164].

An alkali lignin-based, water-soluble binder (lignin-graft-sodium polyacrylate) was utilized in Si-microparticle-based anode [165] that offered a specific capacity (1,914 mAh/g at a current density of 840 mA/g until the 100th cycle) higher than that of the electrode using carboxymethyl cellulose-based binder (1,544 mAh/g). The lignin-based binder also accommodated the volume expansion of the anode by forming octopus-tentacle-like chains to grasp the pulverized units of Si-microparticles [165]. Unlike this approach, where lignin was used as a binder at a non-carbonized state, [165] Cheng et al. [166] varied the lignin carbonization temperature (400–600°C) so that sufficient graphitization could be achieved (for good electronic conductivity) while retaining the flexibility of the precursor lignin to some extent within the carbon materials. This flexibility was crucial for the carbon (surrounding the Si-nanoparticles) to cope with the volume expansion of Si-based anodes. This method allowed lignin-derived carbon to serve as both binder and conductive additive. The resulting electrode offered an excellent initial discharge capacity of 3,086 mAh/g and a retaining capacity of 2,378 mAh/g after 100 cycles at 1 A/g [166].

## 1.2.2 Other Lithium-based Batteries

### 1.2.2.1 Lithium Metal Batteries (LMBs)

Lithium metal, used as anode material in LMBs, has a very high theoretical capacity (3,680 mAh/g) and low redox potential (−3.04 V vs. standard hydrogen electrode) [208,209]. While this could be theoretically one of the best ways to design high-energy-density batteries, LMBs have safety concerns due to the formation of unstable SEI and Li-dendrites during cycling. An ideal SPE should inhibit the dendrite growth by serving as a mechanical barrier and providing liquid-free interphase to lessen the formation of unstable SEI. Lee et al. [208] synthesized a lignin-based star-shaped polymer that acted as both SPE and cathode binder for LMBs. Lignin was first grafted with $Li^+$-ion-conducting side chains (i.e., having ether linkages) via atom-transfer radical polymerization (ATRP). One block of these grafted lignins had epoxy-terminated side chains, which inter-crosslinked lignin chains further with the help of di-epoxy-functionalized polymer (poly(ethylene glycol)diglycidyl ether (PEGDE)) and UV-radiation [208]. The low glass transition temperature ($T_g \sim -18°C$) of this material (LPGP) ensured the room-temperature chain mobility needed for $Li^+$-ion conduction [208]. This cross-linked SPE network offered ionic conductivity (0.065 mS/cm) higher than lithium phosphorus oxynitride (LiPON) (0.002 mS/cm) [210,211], the most frequently used solid electrolyte in LMBs. Having said that, control over the degree of cross-linking of LPGP might help to achieve higher ionic conductivity as uncross-linked lignin showed twoshowed two times highertimes higher $Li^+$ ion conductivity as compared to grafted and cross-linked LPGP [208]. The storage modulus ~5–7 MPa of this lignin-based SPE was also larger than that needed to impede dendrite penetration (0.1 MPa minimum) [204;208]. Moreover, the lignin-based binder, used in the cathode, was able to create cross-linked networks with ion-conducting pathways due to the ion-conducting and cross-linkable epoxy moieties present in the lignin-based polymer. The cells containing lignin-based binders showed higher capacity and more stable cycling performance as compared to the ones with PVDF-based binders [208]. Lignin was also considered as an active component of the cathodes of LMBs [212]. A hydrolysis lignin-based LMB, investigated at room temperature using 1M $LiBF_4$ in γ-butyrolactone electrolyte, exhibited a specific capacity of 450 mAh/g at a discharge current density of 25 $\mu A/cm^2$ [212].

To cope with the cost and safety concerns, thin Li-metal anodes and/or anodes with high surface area/porosity were also proposed to homogenize the Li-nucleation site and guide uniform Li-metal deposition [209,213]. An oxygen-rich lignin-derived, lightweight carbon membrane was considered as a low-cost, lithiophilic skeleton to host Li-metal in anodes of LMBs. The resulting anode offered >98% coulombic efficiency over 230 cycles and long cycle life (>1,000 h) [213]. In another work, thermally conductive boron nitride nanosheets, exfoliated via lignosulfonate, [88] homogeneously distributed heat on the surface of the carbon layer, enabled uniform nucleation and growth of Li, and suppressed Li-dendrite growth over electrodes. In a full cell with this exfoliated carbon-based anode (hosting Li) and $LiFePO_4$ cathode, a capacity of 90 mAh/g was achieved at 4.5 $mA/cm^2$ with 92% capacity retention after 1,800 cycles. In contrast, the capacity decayed to 15 mAh/g after 1,800 cycles when Li-foil was used as an anode [88].

### 1.2.2.2 Lithium-sulfur Batteries (LSBs)

The micropores in the hierarchical porous carbon, derived from lignin, were utilized to confine sulfur (S) in the cathodes of LSBs [214]. Another approach [215] used a composite cathode composed of non-carbonized lignin, CNT, and graphene. The -OH groups of lignin helped to adsorb strongly electronegative polysulfides (by H-bonding) and trap those on the electrode surface, while the porous carbon acted as an S-host [215]. The composite electrode offered a specific capacity of 1,632 mAh/g [215] which was close to the theoretical capacity of elemental S (1,675 mAh/g) [216] in LSBs. The shuttling of polysulfides across the separator was also prevented by placing an interlayer between the cathode and separator in the battery. Chen et al. [214] prepared such a protective layer by mixing low-cost lignin with multi-walled CNT. This composite protective layer limited the transport of polysulfide but promoted $Li^+$-ion transport *via* the porous structure of lignin facilitating electrolyte uptake. With this layer, the LSB exhibited an energy density of ~637 mAh/g (at a current density of ~3,350 mA/g) [214]. A coating of lignin nanoparticles over a commercial Celgard separator chemically trapped the polysulfide to -OH groups of lignin and improved the cycling stability of the separator [217]. Also, a lignocellulose fiber-based membrane, soaked in LiTFSI, was reported as GPE for LSB [218]. The GPE suppressed the polysulfide shuttling and offered an ionic conductivity of 4.52 mS/cm and a Li-ion transference number of 0.79. A capacity of 653 mAh/g (at 20 mA/g after 100 cycles) was attained when this GPE was used in a LSB [5,218].

### 1.2.3 Na-ion Batteries (NIBs)

Since sodium (Na) exists just below Li in the periodic table as an alkali metal, Na in NIBs shows electrochemical properties similar to Li in LIBs. Moreover, the natural abundance, global distribution, and low cost of Na make NIBs a potential alternate to LIBs for large-scale grid energy storage [100,144]. The working mechanism of LIB and NIB are also similar, only in NIBs, $Na^+$ ions (instead of $Li^+$-ions) shuttle from cathode to anode during charging and the opposite during discharging. However, the capacity of graphite as NIB anode (30 mAh/g) [144] is significantly lower than that of LIB anode (372 mAh/g). This is because the interlayer distance within the crystallographic structure of graphite is too small [93,219] to intercalate and store $Na^+$-ions (116 pm) with an ionic radius larger than $Li^+$-ions (90 pm) [220]. Therefore, there is a need for low-cost, but energy-dense, and safe, carbon-based anode materials for NIBs.

#### 1.2.3.1 Carbonaceous Materials for NIB Anodes

Hard carbon (non-graphitizable carbon), with suitable interlayer spacing, porosity, and binding with $Na^+$-cations, has been attractive for high-capacity NIB anodes [93,144]. As discussed in the LIB anodes section, biomass can be a low-cost, eco-friendly, and efficient precursor of hard carbon and a potential alternative to graphite for NIB anodes [91,93,95,104,105,221,222]. Moreover, hard carbons derived from lignin-rich precursors can offer capacity higher than their hemicellulose- and pectin-rich counterparts [222,223]. The experimental evidence [105,223,224] indicated nanopore filling, intercalation of Na between graphene sheets, and adsorption of $Na^+$-ion on the pore surfaces as the most common mechanisms of Na-storage on hard carbon. Reactive molecular dynamics simulations suggested that Na is preferentially localized on the surface of the curved graphene fragments, while Li prefers to bind to the hydrogen-dense interfaces of crystalline and amorphous carbon domains [225]. Lv et al. [226] suggested that micropores (< 2 nm) hinder $Na^+$ ion diffusion and hardly accommodate $Na^+$ ions; but mesopores (2–50 nm) facilitate $Na^+$-ion intercalation and shorten the ion diffusion pathways (capacity ~283.7 mAh/g). These suggested the importance of pore size and structural regulation to attain control over the Na-storage capacity of hard carbon [226].

A Na-storage capacity ranging between 250–300 mAh/g was achieved in general by mixing lignin with different types of carbon precursors [93,100,104,223,224]. For example, hard carbons derived from pitch and lignin (1:1) upon calcination at 1,400°C exhibited a reversible capacity of 254 mAh/g at a current density of 30 mA/g after 150 cycles, an initial coulombic efficiency of 82%, and high cycling stability [100]. Carbonizing strong acid hydrolysis lignin (oak wood) at 1,300°C [100] offered small graphitic domains with well-developed graphene layers, large interlayer spacing (0.403 nm), and high micropore surface area (207.5 $m^2/g$). With these features, this anode exhibited a reversible capacity of 297 mAh/g at a current density of 50 mA/g after 100 cycles [100]. Importantly, a stable capacity of 114 mAh/g was obtained at a continuous, high charge-discharge rate of 2.5 A/g, even after ~500 cycles, with almost 100% CE [100].

Achieving a balance between the graphitic and turbostratic fraction of carbon is critical to achieving efficient metal storage (turbostratic) and electrical conductivity (graphitic) simultaneously [91,93,224]. This was often achieved by appropriately choosing the carbon precursors in a mixed system. For example, a solution of pitch-derived amphiphilic carbonaceous materials and lignin was used to form microspheres, then carbonized further at 1,300°C to get hard carbon [93]. Here, the disordered amorphous carbon regions, formed from 3D, low-density, branched lignin, served as active sites for $Na^+$-ion insertion, while the pitch-derived carbon component was liable for graphite-like crystalline domains contributing to electrical conductivity [93]. A reversible capacity of 296 mAh/g and an initial coulombic efficiency of 82% were achieved from this carbon-based electrode. Moreover, the spherical shape and amphiphilic nature of lignin and its carbon partner facilitated interactions to give high packing density, thermal stability, and mechanical strength, all beneficial for the NIB-electrodes [93].

Experimental evidence suggested that when lignin was oxidized before carbonization, the graphite domains tended to bend under high temperature instead of growing straight and thus increased the interlayer distance facilitating Na-ion insertion. Also, the voids created this way contributed to a large number of micropores. In the pre-oxidation approach, when used for NIB anodes, the capacity and coulombic efficiency improved [223] as compared to graphite anode (30 mAh/g) and carbon samples obtained via direct carbonization, but the improvement was not substantial as compared to other strategies reported to improve Na-storage capacity (around 250–300 mAh/g range [93,100,104,223]).

#### 1.2.3.2 Carbon Nanofibers (CNFs) for NIB Anodes

As seen for LIBs, lignin-based electrospun CNFs for NIBs are mostly prepared by mixing lignin ($M_w$ ~1,000–4,000 [103,227] in reported ones) with a high-MW polymer (e.g., PAN ($M_w$ ~80,000 [104]–150,000 [227]). This is because the MW of the polymers influences polymer chain entanglements which is critical for electrospinning and uniform fiber formation. Thus in a PAN/lignin-based CNF, PAN improved the viscosity and spinnability, while lignin/lignosulfonate acted as a bio-based, eco-friendly carbon precursor to reduce the cost of CNFs [104]. Electrospinning of a mixture of PAN and refined lignin (RL, extracted from hardwood lignosulfonate), thermal stabilization, and subsequent carbonization yielded CNF webs with sufficient turbostratic carbon domains (Figure 1.4a) [100,104]. The CNFs were interconnected to create an interpenetrating 3D network with porous channels (Figure 1.4b). Such long, continuous, and robust morphology prevented any significant fracture or pulverization, as evident in the electrode morphology even after 200 charge-discharge cycles (Figure 1.4c). These free-standing, binder-free electrodes offered a reversible capacity of ~293 mAh/g at 20 mA/g, initial CE of 71%, and good cycling stability (reversible capacity of 247 mAh/g at a current density of 100 mA/g, 90.2% capacity retention over 300 cycles). The PAN-to-lignin ratio did not impact the cycling performance significantly (Figure 1.4d, e) [104]. A slightly improved capacity was achieved when cellulose acetate ($M_n$ ~30,000)/kraft lignin ($M_n$ ~1,000–3,900)-based CNFs (~340 mAh/g at a current density of 50 mA/g after 200 cycles) were made [103]. Lindbergh et al. [105] found a way to work around the low electrospinnability of lignin. They fractionated kraft lignin to obtain a fraction with a relatively larger MW ($M_n$ ~3,000, $M_w$ ~8,600) which made a CNF out of lignin without any need for another high MW-component. The capacity of these pure lignin-based CNFs (310 mAh/g) [105] was similar to that of other mixed-matrix CNFs.

A lignin-based composite cathode was made by electropolymerizing electron-conducting additive 3,4-ethylenedioxythiophene (EDOT) in lignin, followed by mixing with carbon black [106]. However, the approach yielded a capacity (70 mAh/g) [106] lower than other lignin-derived carbon-based electrodes and needs a further understanding of the origin of the low performance.

### 1.2.4 Lead-acid Batteries

Lead-acid batteries are the most widely used rechargeable power sources in the world and find applications in automobiles (starting, lighting, ignition), large backup power supplies, grid energy

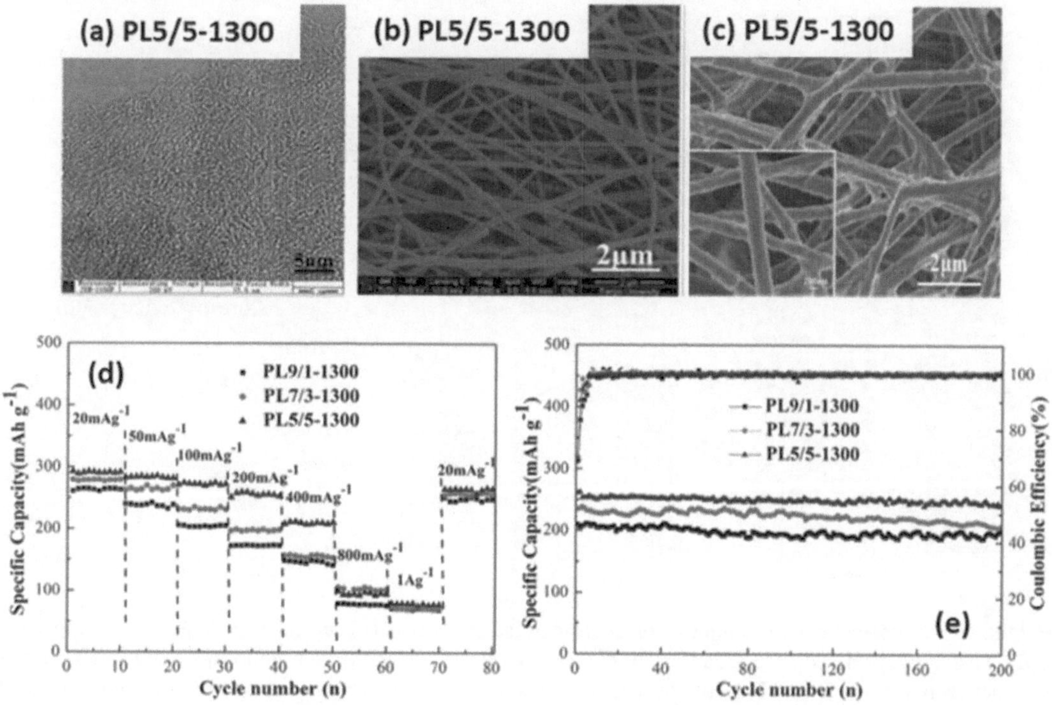


**FIGURE 1.4** (a) HRTEM images of polyacrylonitrile (PAN)/rectified lignin (RL)-based CNFs (PAN: RL = 1:1, carbonized at 1,300°C) showing turbostratic structure. FESEM image of PAN/RL-based CNFs (1:1) (b) before and (c) after 200 charge-discharge cycles. (d) Rate and (e) cycling performances of the CNFs made using different PAN-to-RL ratios (9:1, 7:3, 5:5) and carbonized at 1,300°C. Reproduced with permission from ref. [104].

storage, off-grid household electric supply, and more [6]. Lead-acid batteries use metallic lead (Pb) as an anode, lead (IV) oxide ($PbO_2$), or $PbO_2$-coated Pb as a cathode, and $H_2SO_4$ as electrolyte. During discharging, the anode and cathode produce lead sulfate ($PbSO_4$) and deposit it on the electrode surface. The battery can then be charged by applying a voltage sufficient to cause $PbSO_4$ to convert back to $PbO_2$ at the cathode and Pb at the anode. In these batteries, a mixture of lignosulfonate, barium sulfate, and carbon black has been used as battery expanders since the 1950s, where lignosulfonate suppresses the growth of large $PbSO_4$ crystals during charge/discharge cycles [228] and enables the formation of long needle-like dendrites [229,230]. Unlike the bulk crystals, these long, loosely packed crystals offer more surface area for electrochemical reaction and help $PbSO_4$ to easily convert back to the active materials during charging [230]. It is claimed that the application of lignosulfonate in batteries can improve the life span of batteries to years, while the battery works for days when no lignosulfonate was used due to corrosion of the lead plates [6]. Lignosulfonate ensures low compaction of the negative active materials, [58] stabilizes their physical structure, [231] and impedes their degradation during battery operation [231]. Also, the low sulfur content (0.1–0.2 wt% of the lead compound) in lignosulfonate can reduce the acidic foam formation inside the battery, limit the dissolution of lignosulfonate in $H_2SO_4$ and water and lower the degradation of active materials [6]. A combination of lignosulfonate and π-conjugated polymer PANI was employed where the 3D structure of lignosulfonate helped to create ion-transporting channels, while PANI maintained the electron conductivity [58]. The PANI/lignosulfonate composite electrode made the conversion of $PbSO_4$ to Pb facile and increased the active material utilization and cycle life (seven times higher than pure lignosulfonate) [58].

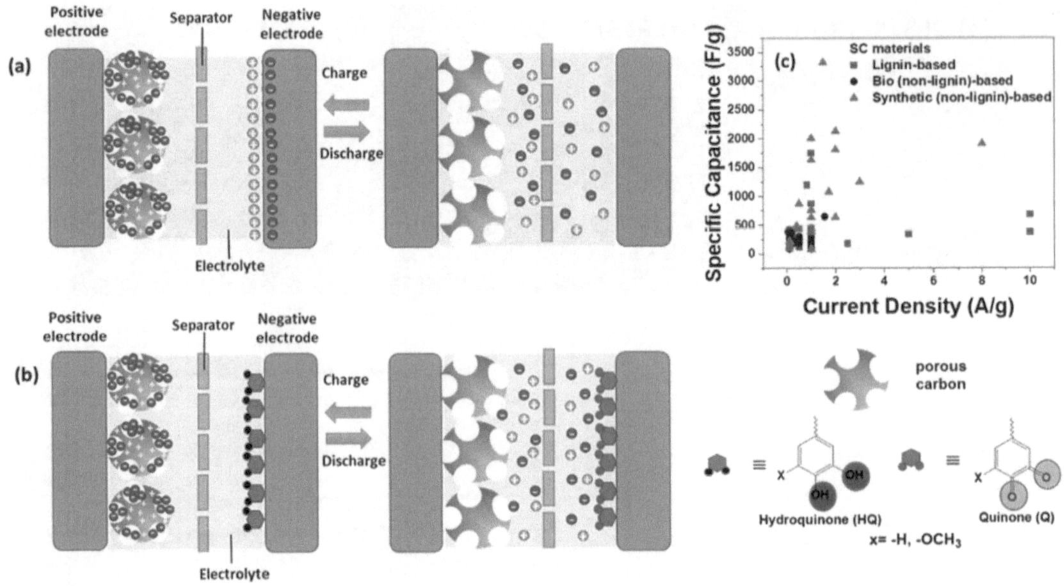


**FIGURE 1.5** Schematic representation of charging/discharging of (a) an electric double layer capacitor (EDLC) using lignin-based porous carbon for the positive electrode, and (b) a pseudocapacitor using lignin as redox-active material (Q ↔ HQ) for the negative electrode. (c) A comparison of specific capacitance of lignin-based, other bio-based (excluding lignin), and synthetic, non-lignin-based electrode materials for supercapacitors reported in literature. [111–113,115–120,124,126,129,130,132–134,136–138,141,235, 237–242].

### 1.2.5 Supercapacitors (SCs)

Capacitance is the ability to store charge. At low voltage conditions, supercapacitors (SCs) offer capacitance (C = q/V) higher than other capacitors. They can provide power density as high as $10^6$ W/kg, [167,232] charge and discharge faster than batteries, [233] and have a much longer lifetime (> 100,000 cycles, or ~15 years) than batteries (~ 200 cycles) [233]. Also, SCs can maintain a power density-energy density balance (power density of ~1,000 W/kg at an energy density of ~10 Wh/kg [167]) and bridge the gap between electrolyte capacitors (showing high power density) and fuel cells/batteries (showing high energy density) [167,234,235] as can be seen in the Ragone plot (Figure 1.2a). SCs are thus attractive for applications requiring a burst of power momentarily or many rapid charge/discharge cycles (rather than long-term compact energy storage); e.g., high-power backup supply in emergency electronic systems, auxiliary power units, instantaneous electricity compensators, short-distance transportation, regenerating braking systems for automobiles, trains, cranes, elevators, and more [233,236].

Supercapacitors can be of three types: electric double-layer capacitors (EDLCs), pseudocapacitors, and hybrid capacitors. EDLCs (Figure 1.5a) store charge electrostatically, i.e., via electrostatic adsorption of ions at the electrode-electrolyte interface in a Helmholtz double layer, while pseudocapacitors (Figure 1.5b) store charges electrochemically via Faradaic reaction using redox-active materials. The design improvement goal for high power-density supercapacitors is to achieve high energy density simultaneously to compete with batteries (Figure 1.2a). Also, to speed up their market penetration, materials costs must be reduced (< $10/kg). Lignocellulosic biomass or lignin can be beneficial as low-cost but efficient carbonaceous, [112,146,237] and redox-active, [56–58] materials for supercapacitors. This is apparent in the comparison of specific capacitance of lignin- and non-lignin-based electrodes (Figure 1.5c).

#### 1.2.5.1 Activated, Templated, and Composite Carbon Materials for SC Electrodes

Carbon materials offering a high specific surface area are critical for the formation of electric double layer (EDL) in EDLCs [117,235,237]. The specific surface area is dependent on the structure and shape of pores, pore-size distribution, and functional groups attached to the surface of the carbon materials [235]. The theoretical capacitance of graphene is ~550 F/g, [128] but it is difficult to achieve such a high capacitance from graphene as single-layer graphene sheets tend to restack with each other via van der Waals interactions and limit the surface area for charge storage [128]. This is where lignin-derived carbons with less-stacking tendency are beneficial, alongside its 3-D, lessdense, loose structure [146] that can be tuned to obtain desired pore size upon pyrolysis [146]. Three major classes of carbon materials are explored for lignin-based electrodes for SCs: activated carbon, [112,146,235,237] templated carbon, [115,116,243,244] and composite carbon [146].

The synergistic actions of micropores (< 2 nm), mesopores (2–50 nm), and macropores (> 50 nm) make activated carbons suitable for charge storage in the electrodes of EDLCs [235]. While pores should store charges (capacitive action), the pores should also be accessible to electrolyte ions for carrying the charges to the storage sites. The micropores can offer an ion-accessible surface area for the construction of EDL and electrostatic charge storage thereby, [68,146,237,245] mesopores can facilitate the electrolyte ion diffusion, and macropores can act as the storage area for the electrolytes to help ions enter into micro- and mesopores [244]. It is, therefore, considered important to have pores with different dimensions simultaneously in a carbon material. To design hierarchical porous carbon, lignin is an ideal precursor material [112,146]. The pre-/post-carbonization approaches for making carbons for SCs are similar to what is used for LIBs and NIBs, like physical [146]/chemical [112,146,237] activation, heteroatom-doping, [117] and templating [115,116,243,244].

For example, hierarchical 3D porous carbon was obtained from enzymatic hydrolysis lignin via hydrothermal carbonization (at 800°C) followed by activation using KOH [112,146]. With abundant micropores, small mesopores, and macropores, this hierarchical porous carbon in electrode exhibited a high specific surface area of up to 1,660 $m^2$/g, an electrical conductivity of 5,400 mS/cm, a high specific capacitance of 420 F/g at 0.1 A/g, and excellent cycle stability (99% capacitance retention after 10,000 cycles at 5 A/g) in 6M KOH electrolyte [112,146]. When this porous carbon electrode was assembled into a symmetric capacitor using an ionic liquid-based electrolyte, a high energy density of 46.8 Wh/kg was achieved. In fact, an energy density of 22.9 Wh/kg was maintained at a high power density of 25,400 W/kg [112,146]. A similar carbonization-KOH activation of steam explosion lignin offered carbon with a relatively higher specific surface area (3,775 $m^2$/g) [237]. The specific capacitance of the carbon was 286.7 F/g at a current density of 0.2 A/g in a 6M KOH electrolyte. The maximum power density achieved by these materials was 1,897.6 W/kg, at which the energy density was 5.90 Wh/kg [237]. These and many other reports [112,146,237] made it evident that the final structure of activated carbon and the resulting electrochemical properties are strongly dependent on the structure and composition of the precursor plant or lignin sources [146]. Moreover, the high specific surface area does not necessarily warrant high specific capacitance; e.g., a black-liquor-derived, KOH-activated carbon offered a higher specific surface area (3,089.2 $m^2$/g), but low specific capacitance (41.4 F/g) [110]. A correlation between pore size and specific capacitance suggested that this carbon mainly had mesopores (~1.9 nm), [110] while for EDLC, we need both meso- (ion transport) and micropores (charge/ion storage) [245].

Several approaches bypassed the traditional chemical activation methods [113,128,243,246] as they are complex, expensive, and not environmentally friendly [113]. One approach carbonized a composite of lignin and graphene oxide without any activation step [128]. Here, lignin carbonization produced biochar which prevented the re-stacking of graphene oxide, increased the specific surface area (2,630 $m^2$/g), and offered a specific capacitance of 103 F/g [128]. The 2nd approach avoided activation but thermostabilized the precursor before carbonization (surface area ~1,195 $m^2$/g, specific capacitance ~114 F/g, power density ~1,006 W/kg at an energy density of 12.8 Wh/kg) [243,246].

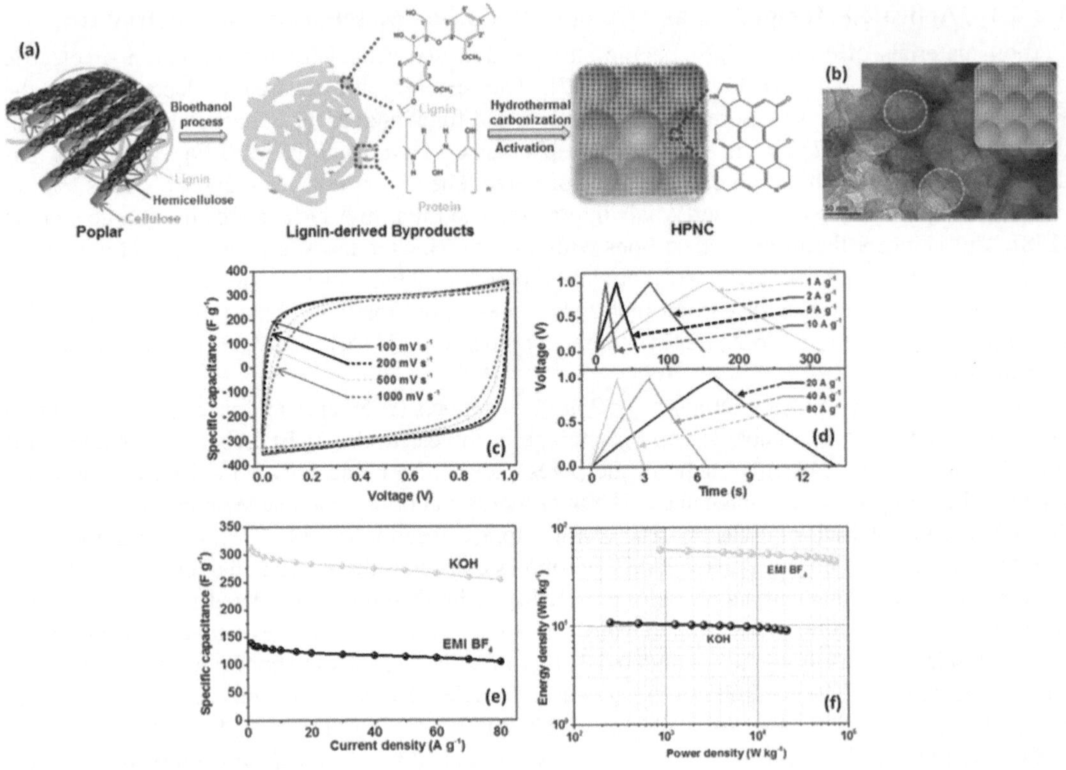


**FIGURE 1.6** (a) Schematic representation of the fabrication process of hierarchical porous N-doped carbon (HPNC) from lignin-derived by-products. (b) FE-TEM image of HPNC. The orange rings showed the bowl-like pore structures. (c) CV curves and (d) charge/discharge voltage profile in 6M KOH electrolytes. (e) Specific capacitance at different current densities and (f) Ragone plot of HPNC supercapacitors operated in different electrolytes. Reproduced with permission from ref. [117].

The 3rd approach used bacterial activation before carbonization, [113] which yielded carbon with a specific surface area of up to 1,831 $m^2/g$ and specific capacitance (428 F/g at 1 A/g) [113] higher than many chemical activation-based [110,237]/activation-free [128] approaches. The same is true for its energy density (66.18 Wh/kg at a power density of 312 W/kg) [113] in an SC setup.

Heteroatom (N, S, O)-doped carbons can take part in the reversible faradaic reaction and act as pseudocapacitive materials [117,118,146,247]. Lignin-derived byproducts from bioethanol production contain proteins (with amino acids) and polysaccharides, which are suitable precursors of N-doped carbon [117]. Hydrothermal treatment and activation of this by-product yielded N (3.4%)-doped carbon with hierarchical, bowl-like, porous structures (HPNC, Figure 1.6a,b), large specific surface area (2,218 $m^2/g$), and electronic conductivity (4,800 mS/cm) [117]. The symmetrical, rectangular CV curves (Figure 1.6c) and symmetric, triangular galvanostatic charge-discharge voltage curves with small voltage drop (Figure 1.6d) indicated symmetric EDLC behavior of these porous, N-doped carbon electrode-based SCs [117]. In 6M KOH electrolyte, the SC exhibited a specific capacitance of 312 F/g at a current density of 1 A/g (Figure 1.6e) as well as an excellent cycle life (98% of initial capacitance retained after 20,000 cycles at a current density of 10 A/g). The energy and power densities of the SC improved significantly when an aqueous electrolyte (8.8 Wh/kg, 21,300 W/kg) was replaced by an ionic liquid electrolyte (44.7 Wh/kg, 73,100 W/kg) (Figure 1.6f) [117].

An N-rich compound is sometimes used to crosslink the carbon precursor and obtain N-doped, porous carbon; e.g., 1,6-hexanediamine was used as a cross-linking agent and N-source for sodium

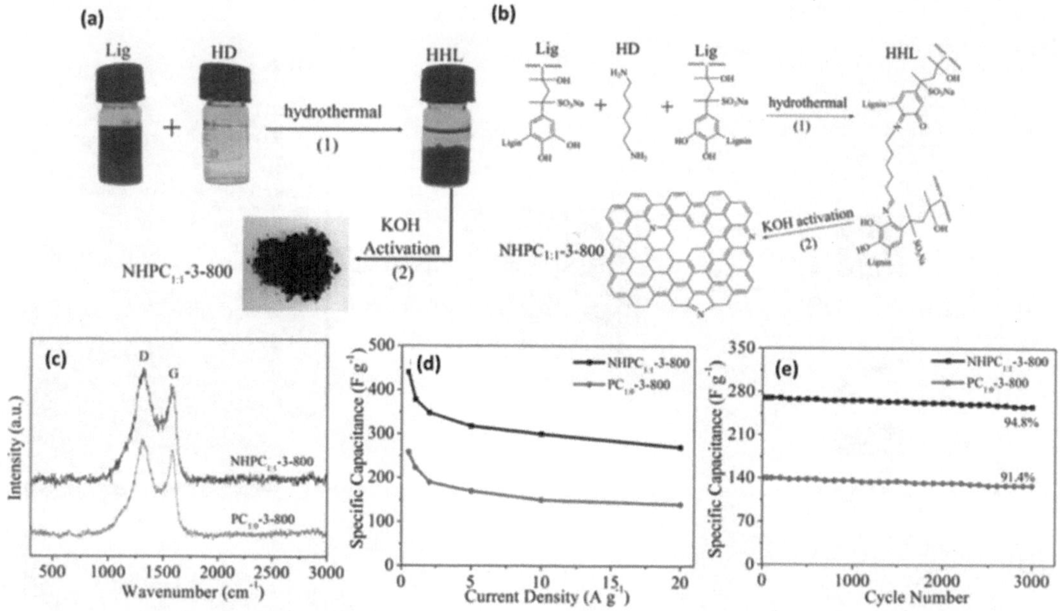


**FIGURE 1.7** (a), (b) Schematic representation of the synthesis of N-doped hierarchical porous carbon using sodium lignosulfonate as carbon precursor, and 1,6 hexanediamine as a nitrogen source and crosslinking agent. (c) Raman spectra show the relative fraction of disordered (D) and graphitized (G) carbon in the sample. (d) Specific capacitance and (e) cycle stability of the carbon when used as electrode materials for supercapacitor. Reproduced with permission from ref. [120].

lignosulfonate (Figure 1.7a, b) [120]. In the obtained carbon, the fraction of disordered carbon (D) was higher than that of graphitized (G) carbon (Figure 1.7c) as desired for charge storage [120]. The material exhibited a specific capacitance of 440 F/g at a current density of 0.5 A/g (Figure 1.7d), high cycle stability (up to 94.8% of initial capacitance after 3,000 cycles, Figure 1.7e), and energy density of 18.5 Wh/kg at a power density of 200 W/kg [120]. Another approach [115] used deep eutectic solvents (choline chloride/$ZnCl_2$) as a soft template and dual heteroatom (N, P)-source for derived carbon (Figure 1.8a). In this material, urea was also added as an additional N-source and porogen [115]. The capacitance of this material was relatively low (177.5 F/g at a current density of 0.5 A/g) but stable over ~3,000 cycles [115]. Other examples of dual doping include (i) N-S co-doped carbon from the $Fe_3O_4$-based templated, carbonized, and KOH-activated lignin amine (Figure 1.8b) [116] (specific surface area ~1,199 $m_2$/g, specific capacitance 241 F/g at 1A/g of current density, 95% capacitance retention after 3,000 cycles, the energy density of 27.2 Wh/kg at a power density of 10,000 W/kg); (ii) N-O co-doped carbon from lignin-containing coconut shell and melamine (N-source) [118] (specific surface area~ 3,000 $m^2$/g, specific capacitance 368 F/g at 0.2 A/g of current density, low self-discharge, low leakage current); and (iii) N-O co-doped CNF from PAN (N, O-source) and lignin [241] (specific capacitance 320 F/g at 1A/g of current density, 94.5% capacitance retention after 5,000 cycles, the energy density of 17.92 Wh/kg at a power density of 800 W/kg).

The carbon derived from $Fe_3O_4$-based template [116] was also an example of carbon prepared to leverage the dual action of templating and activation. There are other examples along this line [123,244]. In a work, templating and pore-forming agent Pluronic F127 was impregnated into solvent-swollen gel of pre-cross-linked hardwood kraft lignin gel [123]. The impregnated gel was then carbonized and activated physically (using $CO_2$)/chemically (using KOH) to make mesoporous carbon (specific surface area ~1,148 $m^2$/g, specific capacitance ~102.3 F/g ($CO_2$-activated carbon), 91.7 F/g (KOH-activated carbon)) [123]. In another work, $K_2CO_3$ played the role of

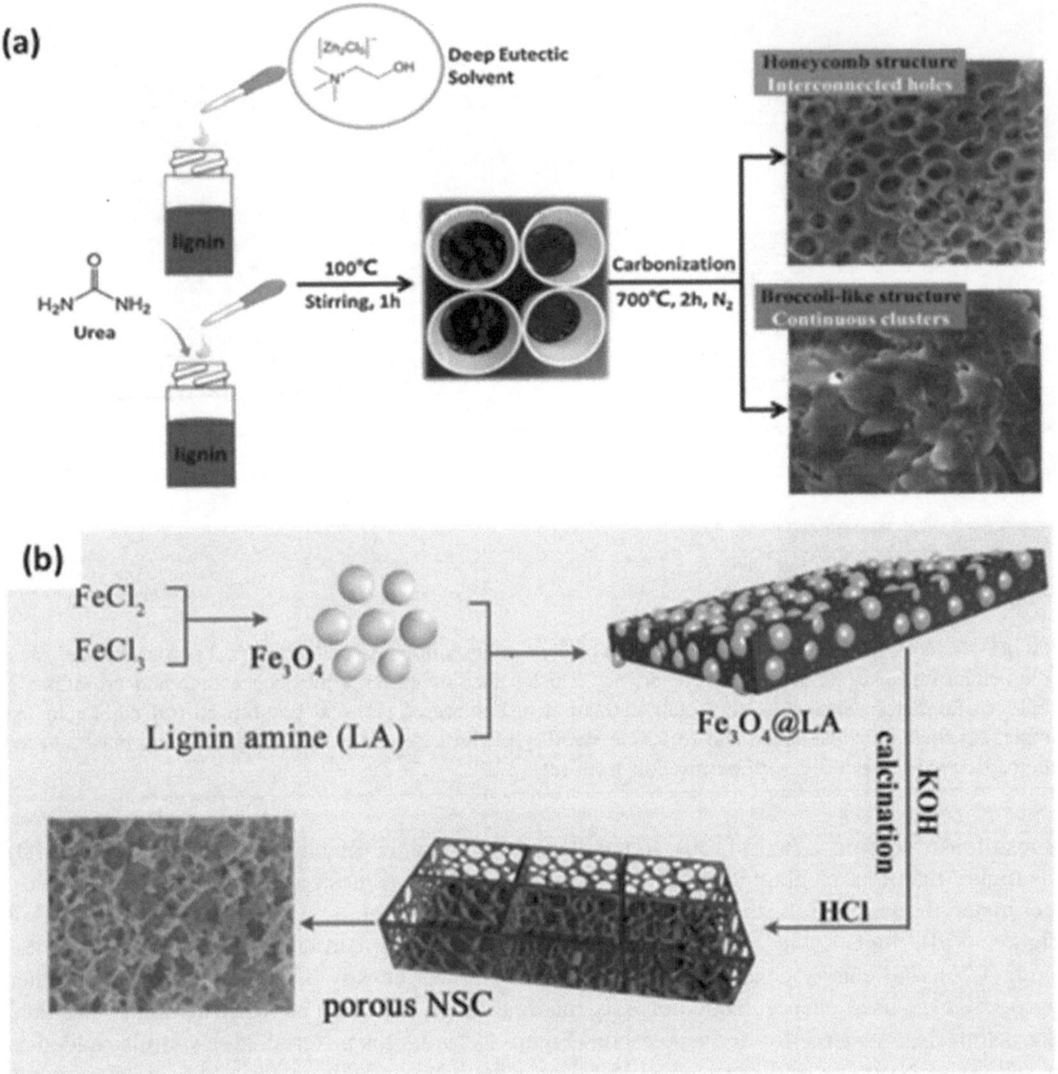


**FIGURE 1.8** (a) Porous carbon synthesis using deep eutectic solvent as a soft template and the resulting carbon nanostructure. Reproduced with permission from ref. [115]. (b) Procedure of making dual-doped (N-S) porous carbon via $Fe_3O_4$-based templating and KOH-based activation. Reproduced with permission from ref. [116].

mesopore template, activation agent (to produce micropores with size ~4 nm), and pH-regulator (to create an alkaline aqueous environment and dissolve lignin) in the making of porous carbon microspheres (specific surface area ~1,529 $m^2/g$, specific capacitance 140 F/g) [244]. On the other hand, a chemical activation-free but template-based synthesis of porous carbons from lignin gave specific surface areas of 418 $m^2/g$ (Pluronic F127 as template) [243,248] and 650–1,085 $m^2/g$ (zeolite as template) [243,249]. While templating allows a tailorable and hierarchical porous structure in carbon, templates are often expensive and must be completely removed at the end of the process. To address this challenge, inspired by ice-templating, aqueous dispersion of lignin was freeze-cast to 2D lignin nanosheets, which were then carbonized without any further activation step [127]. In an SC setup, this carbon (specific surface area ~854.7 $m^2/g$) offered a specific capacitance of 281 F/g at a current density of 0.5 A/g, 91% capacitance retention after

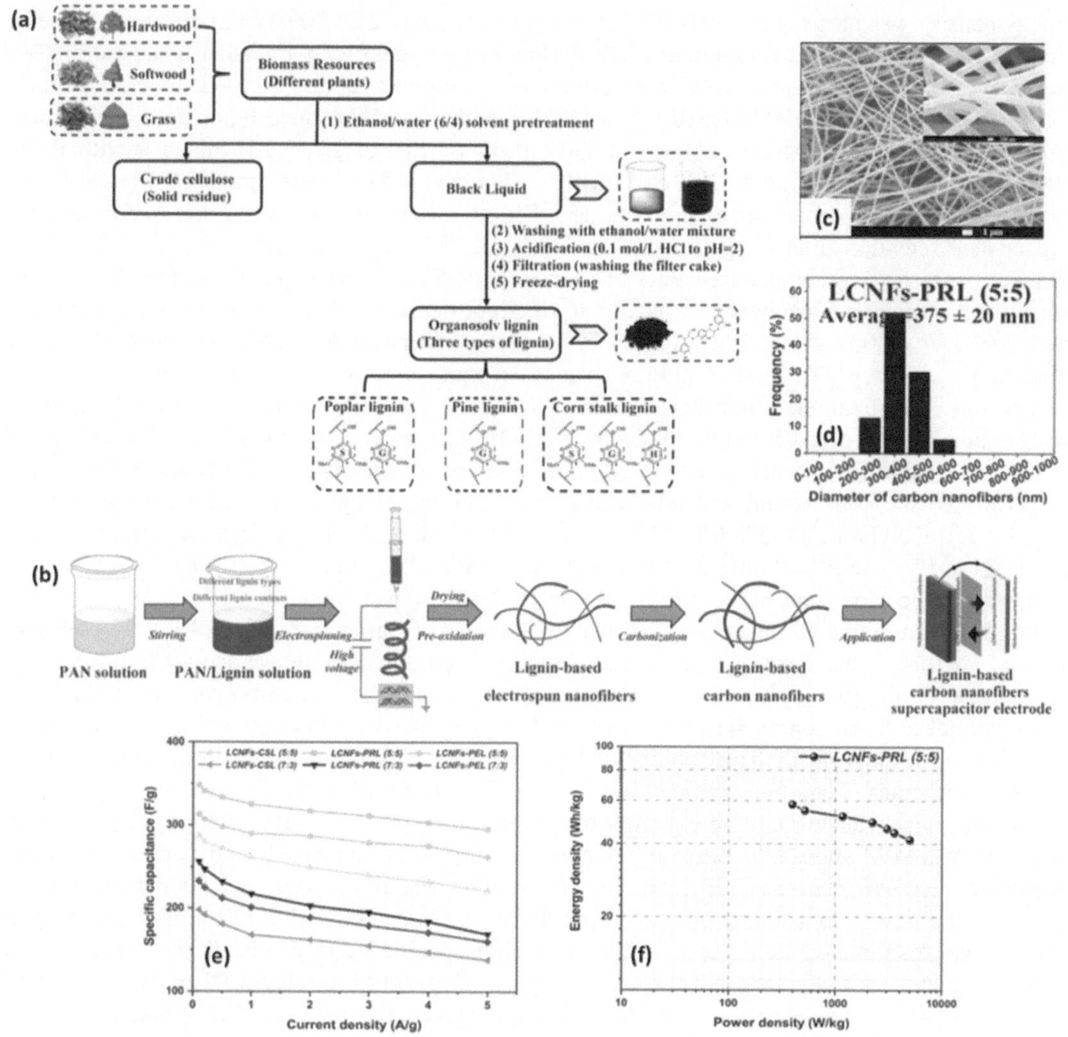


**FIGURE 1.9** Schematic representation of (a) organosolv lignin extraction from different plants (poplar (PR); pine (PE), corn stalk (CS)) and (b) procedure to make PAN/lignin-based CNFs. (c) SEM image, (d) fiber diameter, (e) specific capacitance, and (f) energy and power densities of poplar lignin/PAN-based CNFs (LCNFs-PRL (5:5)). Reproduced with permission from ref. [125].

5,000 cycles, and energy density dependent on power density (energy density ~14.3 Wh/kg at a power density of 28,611 W/kg) [127].

Low-cost CNFs were electrospun for SCs using lignin and fiber-strengthening, high-MW polymers (e.g., PAN), [122,125] polyvinyl alcohol (PVA), [121] polyethylene oxide (PEO), [233] cellulose acetate [124]) as precursor typically. For example, Chen et al. [125] extracted 3 types of lignin from the organosolv process (poplar ($M_w$ ~3,910, from hardwood), pine ($M_w$ ~2,870, from softwood), and cornstalk ($M_w$ ~2,810, from grass)) (Figure 1.9a)) and blended those lignins individually with PAN ($M_w$ ~150,000) to make CNFs (Figure 1.9b, c, d). Irrespective of the lignin source, there was a predominant presence of turbostratic carbon, which is needed for metal/charge storage [125]. But of the three, the poplar lignin/PAN-based CNFs (lignin: PAN=50:50 (w/w)) exhibited the highest tensile strength (35.32 MPa), specific surface area (1,062.5 $m^2/g$), and specific capacitance (349.2 F/g) with 90.52% capacitance retention after 5,000 cycles (Figure 1.9e). The energy density was also high (~39.6 Wh/kg at a power density of 5,000 W/kg) (Figure 1.9f). The study identified that more syringyl units (58.8%) (Figure 1.9a) and β-*O*-4 aryl ether linkages (33.7%), larger $M_w$, smaller PDI, and less -COOH groups in the lignin structure enabled the high SC performance of poplar lignin/PAN-based CNFs [125]. Another work obtained lignin from corn stover *via* methanol/ionic liquid-based extraction and enzymatic hydrolysis, [122] mixed it with PAN, electrospun it to polymer nanofibers, and finally carbonized [122]. These lignin/PAN-based CNFs exhibited a specific capacitance of up to 151 F/g at a current density of 0.1 A/g with 95% capacitance retention after 20,000 cycles at 4 A/g [122]. Alkali lignin/PVA-based CNFs [121] formed a free-standing, binder-free electrode for SC. The SC offered a specific surface area of 583 $cm^2/g$, and a specific capacitance of 64 F/g at a current density of 0.4 A/g with 90% capacitance retention over 6,000 cycles [121]. Titirici et al. [233] fabricated lignin/PEO-based CNFs using PEO as a plasticizer. The as-spun polymer nanofiber mats were uniaxially compressed (up to 40 bar) before carbonization, which densified the CNF mats and reduced the inter-fiber, large (μm-scale) pores not involved in EDL capacitance [233]. The process resulted in lignin/PEO-based CNFs with a volumetric capacitance of 130 $F/cm^3$ and energy density of 6 Wh/L at 0.1 A/g, which outperformed the commercial and lab-scale porous carbons obtained via carbonization of bioresources (50–100 $F/cm^3$, 1–3 Wh/L) [233]. Zhou et al., [124] instead of using traditional polymers like PAN or PVA, used cellulose acetate as the companion of lignin to make CNFs. They obtained a higher specific capacitance (346.6 F/g) and energy density (31.5 Wh/kg at a power density of 400 W/kg). The resulting CNF leveraged the thermal stability of lignin and the flexibility of cellulose acetate. The phosphate-form of lignin reacted with -OH groups of cellulose acetate to form stable bonds between the two moieties. Also, the phosphating enabled P-doped polymer fibers [124].

In parallel to these approaches, where a high MW-polymer was blended with lignin, emphasis was given to improving the spinnability of lignin by tweaking the lignin modification and fractionation strategies [126]. For example, by reacting -OH groups of lignin and isocyanate groups in isophorone diisocyanate and subsequently dialyzing the product (molecular weight cut-off ~3.5 and 7 KDa), MW and spinnability of precursor lignin were improved, and carbon loss from precursor fiber was reduced [126]. The $T_g$ of this high $M_w$-lignin was high, which potentially protected the fibers from fusing and softening during thermostabilization [126]. The electrospun CNFs from fractionated lignin also exhibited a high specific surface area (2,042 $m^2/g$) and a specific capacitance (442.2 F/g at a current density of 1 A/g) four times higher than the unmodified lignin. The EDLC prepared from this CNF offered a high energy density (37.1 Wh/kg at a power density of 400 W/kg) [126].

Decorating or embedding the lignin-based CNFs further with other functional molecules (like metal oxides, [73,130,250,251] electron-conducting polymers, or other carbon materials) can also boost the performances of SCs (discussed in sections below).

#### 1.2.5.2 Lignin-based Electroactive Materials for SC Electrodes

While for electric double-layer capacitors (EDLCs), a large specific surface area of carbon is critical, [252] for pseudocapacitors, redox-active materials capable of efficient faradaic charge transfer are needed [252,253]. Milczarek was the first [254] to report the redox-active behavior of lignin [254] and lignosulfonic acid [255]. He showed that upon appropriate anodic treatment of lignosulfonic acid monolayer, deposited on a glassy carbon electrode, the guaiacyl [G] and syringyl [S] units of lignin structure could give quinone [255]. This quinone can then be reversibly cycled between its redox states; i.e., quinone (Q) ↔ hydroquinone (HQ) [255] (Figure 1.10a), by virtue of which lignin/lignosulfonic acid can exhibit pseudocapacitance (faradaic charge storage) [146,253,254,256,257]. Through this Q/HQ-type redox process, two electrons and two protons are stored in a structure of six carbons and two oxygen atoms, equivalent to an electronic charge density of 496 mAh/g [253].

Although lignin contains a high fraction of redox-active groups, the material itself is typically electronically insulating due to the absence of extended π-conjugation [56–58,256,258]. As a

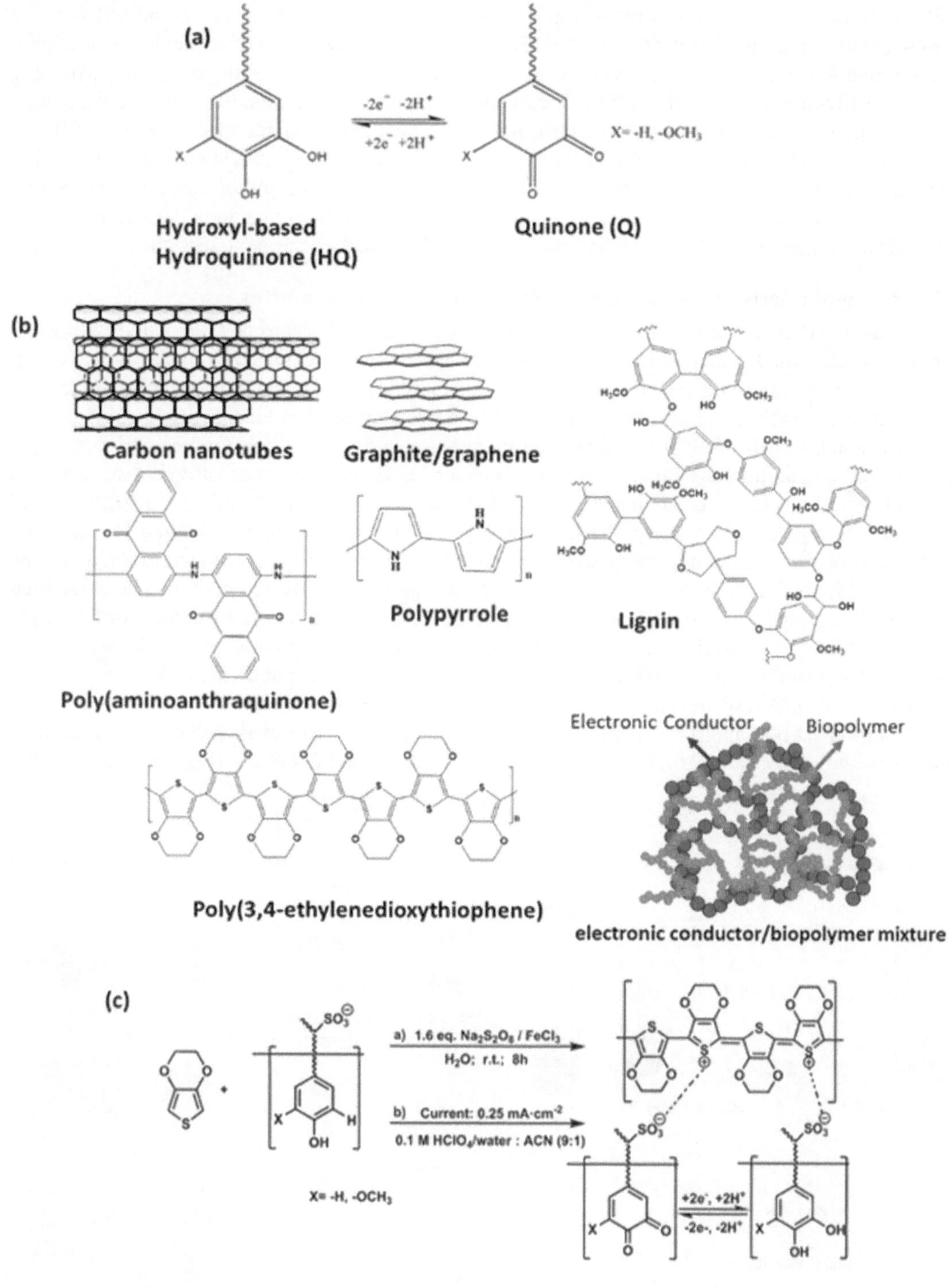


**FIGURE 1.10** (a) Redox reaction in lignin in which switching between quinone (Q) and phenol-based hydroquinone (HQ) structures involves 2 protons and 2 electrons. Reused with permission from ref. [146]. (b) The pseudocapacitance of lignin-based materials can be improved by incorporating carbon nanotubes (CNTs), graphite/graphene, and π-conjugated polymers within the three-dimensional lignin matrix. Reproduced with the permission of ref. [256]. (c) Chemical oxidative polymerization and electrochemical polymerization of PEDOT/lignin composite. Reproduced with permission from ref. [134].

result, only the redox-active groups of lignin close to the electrode interface can be utilized in a lignin-coated, thick electrode [56,258]. Inclusion of small quinone-based molecules as a dopant can improve redox behavior. However, small molecules may leach out of the materials easily and reduce the lifetime of the device [256]. Two common approaches adopted to improve pseudocapacitive charge storage and design low-cost, thick, practical, and scalable electrodes [56–58] are: making composites out of (i) lignin-derived carbon and redox-active materials (such as transition metal oxides [73,130,131,250,251]); and (ii) lignin and redox-active and/or electron-conducting, π-conjugated polymers [58,256,258] or other carbon materials (graphene, carbon nanotube) [56,57,256] (Figure 1.10b, c). Relevant examples are discussed in the next sections.

### 1.2.5.3 Lignin-derived Carbon and Metal Oxide-based Composites

Fong et al. [131] reported electrospun alkali lignin-derived CNF electrodes decorated with manganese dioxide ($MnO_2$) nanowhiskers. The theoretical gravimetric capacity of $MnO_2$ is 1,370 F/g, [131] but like any other metal oxide, their electrical conductivity is low. Combining lignin-derived carbon (having graphitic carbon fraction) with $MnO_2$ (rendering pseudocapacitive behavior) thus mutually benefits each other [131]. Since the prepared electrode was freestanding and flexible, it did not need any additional binder or current collector [131]. The electrode offered a surface area of ~583 $m^2/g$, a specific capacitance of 83.3 F/g (at a current density of 250 mA/g), and an exceptionally high energy density of 84.3 Wh/kg at a power density of 5,720 W/kg [131]. Liang et al., [129] on the other hand, attempted to increase the electroactive surface area of the anode by decorating $MnO_2$ over a mixed matrix made of alkali lignin and activated carbon. The assembled SC offered an initial aerial-specific capacitance of 5.52 $mF/cm^2$ (at a current density of 6.01 mA/g), of which 97.5% retained up to 2,000 cycles. The maximum energy density of 14.11 Wh/kg was obtained at a power density of 1,000 W/kg. A high CE (98%) was retained after 2,000 cycles [129].

Zhong et al. [73,130] developed two different approaches to prepare NiO-embedded lignin-based porous carbon matrix. In one approach,[73] lignosulfonate was carbonized first, then embedded with NiO nanoparticles (Figure 1.11a). In the second approach (Figure 1.11b), nickel

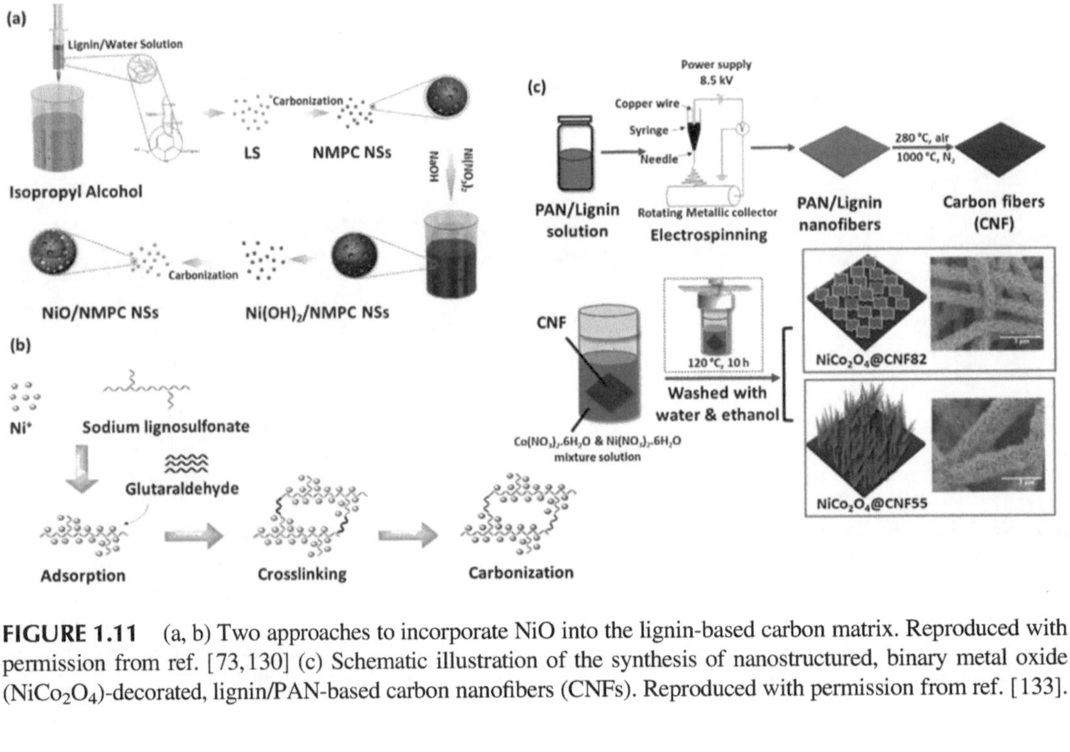


**FIGURE 1.11** (a, b) Two approaches to incorporate NiO into the lignin-based carbon matrix. Reproduced with permission from ref. [73,130] (c) Schematic illustration of the synthesis of nanostructured, binary metal oxide ($NiCo_2O_4$)-decorated, lignin/PAN-based carbon nanofibers (CNFs). Reproduced with permission from ref. [133].

salt (dissolved in Pluronic F127) was incorporated first into lignosulfonate, which was then crosslinked using glutaraldehyde and carbonized [130]. The specific surface area of these mesoporous materials obtained from these two approaches were not significantly different (851 $m^2/g$ (first), [73] 802 $m^2/g$ (second) [130]), but the second approach yielded higher metal oxide content (49–79 wt%) [130] as compared to the first one (11 wt%) [73]. The pore size distribution centered around the mesopore range with a considerable amount of micro- and macropores [73]. The carbon coating over NiO nanoparticles facilitated the construction of electron-conducting pathways and overcame the poor conductivity of metal oxides. Moreover, the mesoporous carbon shell prevented the inner metal oxide nanoparticles from aggregation and pulverization as well as adjusted with the volume changes during the charging/discharging cycles [130]. The second approach offered a specific capacitance of 882 F/g at a current density of 1 A/g, [130] while the first approach offered 508 F/g at a scan rate of 20 mV/s [73]. The first approach was more effective in retaining specific capacitance (~92–94%) over a larger number of charge/discharge cycles (2,000 cycles (first), [73] 1,000 cycles (second) [130]).

An exceptionally high specific capacitance (1757 F/g at 2 $mA/cm^2$ with 50% lignin) was achieved when lignin/PAN-based CNFs were decorated with $NiCo_2O_4$, a binary metal oxide (Figure 1.11c) [133]. SEM images confirmed the uniform growth of $NiCO_2O_4$ nanosheets and nanoneedles on the surface of the CNFs [133]. The materials also showed outstanding cycling stability (~138% capacitance retention after 5000 cycles at 7 $mA/cm^2$) and a high energy density (47.75 Wh/kg at a power density of 799.53 W/kg) [133].

#### 1.2.5.4 Lignin (Non-carbonized)-electron-conducting Polymer-based Composites

π-conjugated polymers with their π-electron-delocalized backbone can facilitate electron conduction if incorporated within an electrode. Moreover, π-conjugated oligo/polymers having redox-active quinone groups along their conjugation path [253,259–262] or anionic redox species as counter ions [253,263,264] can amplify the pseudocapacitive charge storage capability of the electrodes [253,259,261]. Also, in some cases, π-conjugated polymers exhibit good electron-conducting properties but suffer from ion diffusion resistance and large volume changes during the doping/de-doping process [137]. Therefore, combining redox-active and/or electron-conducting π-conjugated polymers with redox-active and ion-conducting (when sulfonated) lignin can produce low-cost, but efficient pseudocapacitive materials. The chemical/electrochemical polymerization of monomeric units of π-conjugated polymers, such as aniline (ANI), [137] pyrrole (Py), [138,253,257] or ethylene dioxythiophene (EDOT, Figure 1.10c) [134] in the presence of lignin allows lignin embedding within conductive polymer framework.

In polyaniline (PANI)/lignosulfonate-based composite, [137] lignosulfonate conducted cations, offered Q↔HQ-type charge storage capability, and helped to accommodate the volume expansion. The composite material displayed a specific capacitance (377.2 F/g at a current density of 10 A/g) higher than that of pure PANI (183.7 F/g). The capacity retention rate of PANI/lignosulfonate was 74.3% after 10,000 cycles, while it was 52.8% for pure PANI [137]. Inganäs et al. [138,253,257] synthesized polypyrrole (PPy)/lignosulfonate-coated Au electrodes via similar in-situ oxidation polymerization of pyrrole. The electrode offered a specific capacity of ~52–73 mAh/g (at a current density of 0.83 A/g), [138] and specific capacitance as high as ~1,200 F/g, [138] while pure PPy offered ~250 F/g at a current density of ~0.1 A/g [138]. The capacitance varied with the thickness of the electrode and charge/discharge rate [253]. The charge storage capacity and capacitance significantly improved when another electroactive dopant, anthraquinone sulfonate (AQS) was added (186 mAh/g, 2,500 F/g) [138]. Adding inorganic metal acid (phosphomolybdic acid) to the PPy/lignin composite electrode made the capacitance even higher (477–682 F/g at a discharge current density of 1 A/g) [242]. Inganäs et al. [134] also prepared poly(ethylene dioxythiophene (PEDOT)/lignin composite-based (Figure 1.10c) composite electrode, which exhibited specific capacitance (170.4 F/g) higher than that of a pure PEDOT-based electrode (80 F/g) [134]. Depositing a layer of poly(aminoanthraquinone) (PAAQ) on a PEDOT/lignin-based electrode increased the specific capacitance further (418 F/g) through synergistic activities of lignin, PEDOT, and PAAQ [136]. This capacitance value was also higher than that obtained from the lignin-free, PEDOT/PAAQ electrode (383 F/g) [135]. The SC assembled using PEDOT/lignin/PAAQ as a positive electrode and PEDOT/PAAQ as a negative electrode offered a specific capacitance of 74 F/g with 80% capacitance retention after 10,000 cycles [136].

#### 1.2.5.5 Lignin (Non-carbonized)-graphite/CNT-based Composites

Graphite and its derivatives are low-cost electron-conducting materials. Graphite, [57] partially reduced graphite oxide [265]/graphene oxide (rGO), [141] graphene, [139] and carbon nanotubes (CNTs) [142] have been used in conjunction with lignin/lignin-containing biomass in energy storage devices [56–58,266]. Inganäs et al. [57] designed biomass/graphite hybrid electrode materials without any attempt to extract/separate lignin. The electrodes exhibited a discharge capacity of 20 mAh/g with 68% capacity retention after 1,000 cycles, a low capacitance (80 F/g), [57] and the fraction of lignin in the biomass was shown to control the overall electrochemical performance of the electrode [57]. But the capacitance significantly improved when graphene or graphene oxide adopted different configurations; [139,141] e.g., redox-active lignosulfonate was confined within an electron-conducting, graphene cages that mimic venus flytraps [139] (Figure 1.12a). These $FeCl_3$-exfoliated graphene cages (Figure 1.12b-e) prevented the dissolution of water-soluble lignosulfonate; at the same time, graphene acted as a 3D current collector. This bioinspired design offered a specific capacitance of 211 F/g at a current density of 1 A/g, and exceptionally high cycling stability (88% capacitance retention after 15,000 cycles) [139].

Park et al., [141] instead of confining lignosulfonate, confined lignin nanocrystals on an electron-conducting rGO surface. The strong interaction between lignin and rGO allowed fast redox charge transfer and resulted in a capacitance (432 F/g) about six times higher than that of rGO (93 F/g) [141]. Zhong et al. [143] fabricated a lignosulfonate/single-walled CNT/holey reduced graphite oxide-based electrode and cellulose hydrogel-based separator. Here, the interaction between cellulose and lignin

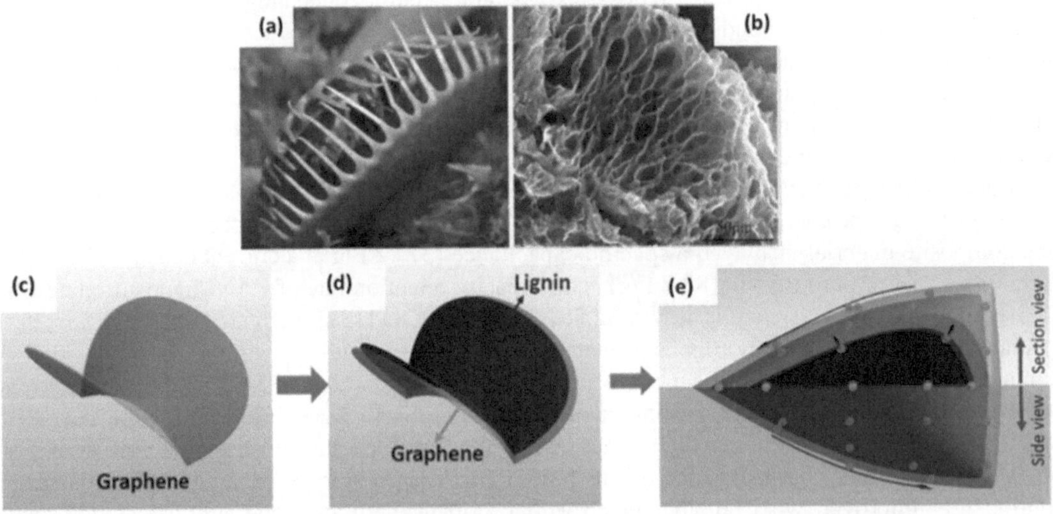


**FIGURE 1.12** (a) A photograph of venus flytrap plant, mimicking which redox-active lignosulfonate was confined within electron-conducting graphene cage. (b) SEM image of $FeCl_3$-exfoliated graphene cage. Schematic illustration of (c) open-mouth exfoliated graphene, (d) graphene (light grey) with open mouth trapping lignin (dark grey), and (e) electron transport pathways in the hybrid graphene-lignin materials. Here, the electrons are generated during the electrochemical reaction and transferred from lignin to graphene. Reproduced with permission from ref. [139].

(seen naturally in the plant cell wall) was leveraged to reduce the contact resistance between electrode and separator and prevent the delamination of these two components in SC. The SC exhibited a high tensile strength (112.3 MPa) and a specific capacitance of ~287 F/g [143].

Multi-walled CNTs were decorated with kraft lignin and deposited over the gold electrode [142]. The composite electrode showed a specific capacitance of 181 F/g (at 2.5 A/g) with a weak dependence on applied charge-discharge current, and 93% capacitance retention after 500 cycles (at 1A/g) [142]. This capacitance value was higher than lignin-free CNT electrodes (~85 F/g) and demonstrated the role of quinones in lignin in improving the redox behavior of the lignin/CNT composite electrode [142].

In another interesting approach, lignin-derived aromatic oligomers (from lignin depolymerization) were reacted with formaldehyde to repolymerize and then combine with rGO [140]. This lignin oligomer-based polymer had higher quinone content (3.76 mmol/g) than commercially available lignin (like organosolv/alkali lignin). The hybrid electrode exhibited a specific capacitance of 250 F/g, which was three times higher than that of rGO [140].

#### 1.2.5.6 Lignin-based Electrolytes for SCs

Park et al. [267] designed an all-lignin-based flexible SC where chemically cross-linked lignin hydrogel electrolyte and lignin/PAN-based CNF electrodes were used. Here, poly(ethylene glycol) diglycidyl ether (PEGDGE) acted as a chemical cross-linking agent for alkali lignin ($M_w$=10,000) to synthesize the electrolyte which gave an ionic conductivity of 10.35 mS/cm. The lignin-based electrode and electrolyte gave this SC a specific capacitance of 129.23 F/g with 95% capacitance retention over 10,000 cycles, an energy density of 4.49 Wh/kg, and a power density of 2,630 W/kg [267]. Liu et al. [268] chemically cross-linked lignin first using PEGDGE following a similar method as the previous one, [267] then physically cross-linked again using $H_2SO_4$ [268]. This doubly cross-linked, highly compressible lignin-based hydrogel electrolyte was used to make a flexible SC giving a relatively higher ionic conductivity (~80 mS/cm) and specific capacitance (190 F/g). The capacitance retained up to 500 cycles at 180° bending. The energy- and power densities were 15.24 Wh/kg and 2157.3 W/kg, respectively [268].

### 1.2.6 Proton Exchange Membrane Fuel Cells (PEMFCs)

PEMFCs generate electricity using $H_2$ as fuel (Figure 1.13a). During this process, they produce no harmful gas (like $CO_2$, or catalyst for PEMFC cathode, whileCO). Therefore, PEMFC-driven vehicles are considered eco-friendly alternates to traditional vehicles run by internal combustion engines. Despite having these attractive features, PEMFCs are highly expensive since expensive materials are used to make proton-conducting ionomer-based separators and ionomer-catalyst layers on electrodes. Moreover, the current benchmark ionomer Nafion, used in PEMFCs, is fluorocarbon-based and is thus not eco-friendly. Finally, the oxygen reduction reaction (ORR) at the cathode of PEMFC is sluggish owing to a high ion-transport limitation within the sub-μm thick ionomer binder layer at the ionomer-catalyst interface [269–276]. This negatively impacts the power density and efficiency of PEMFCs [277,278]. Therefore, some of the critical needs of PEMFCs are: (i) high proton conductivity across the membrane separator as well as ionomer-catalyst thin interfaces, [279,280] (ii) low cost, and (iii) environment-friendly materials. Low-cost lignin derivatives hold promise as membrane separators, [281] ionomers for catalyst binders, [26] precursors of carbon-based catalysts, and electron-conducting materials [282–284] for PEMFCs.

A lignin-derived S-doped carbon was demonstrated, where the S provided extra catalytic sites for ORR at PEMFC-cathode by transferring 3.4 electrons/molecule at 0.8 V and approaching the optimum four-electron pathway [282]. Ozaki et al. [283] prepared core-shell nanomaterials with N-rich, lignin-derived carbon as the core and cobalt phthalocyanine (CoPc) as the shell. CoPc acted here as an ORR catalyst for the PEMFC cathode, while the N in lignin-derived carbon pulled the electrons from the adjacent carbon atoms and made them more cationic by nature. This change in

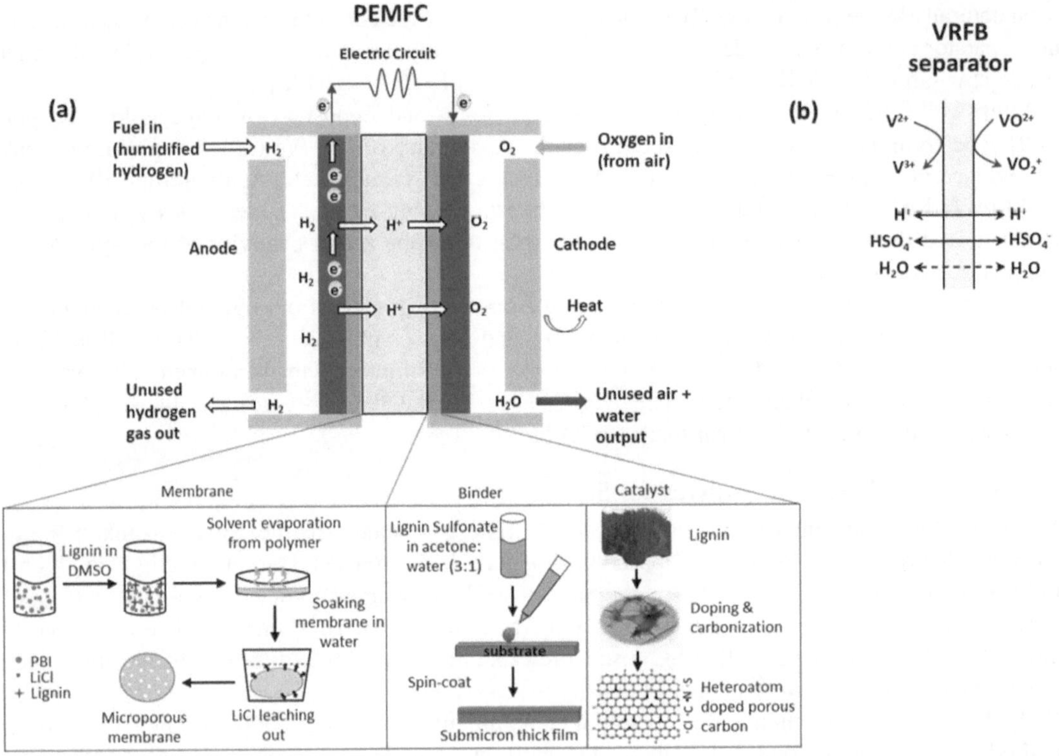


**FIGURE 1.13** (a) A schematic of PEMFC and how lignin can be used to make the proton-conducting membrane, catalyst binder, and catalyst for PEMFC. (b) The basic design criteria for a proton-exchange membrane separator for vanadium redox flow battery (VRFB). The membrane should selectively transport protons from anode to cathode but prevent the permeation of vanadium ions ($V^{+n}$) across the membrane.

the electronic distribution of carbons [284] made them attract electrons to the cathode. This way, lignin-derived carbon also acted as active sites for ORR [283]. The ORR activity of this core-shell material-based cathode was higher than that of phenol-formaldehyde resin-based core-shell materials with the same amount of CoPc added [283]. In another report, multi-heteroatom (N, S, Cl)-doped lignin-based carbon offered ORR activity superior to the state-of-the-art Pt/C catalysts and most of the non-noble metal catalysts [285] for similar reasons. This material with a high specific surface area (1,289 $m^2/g$) offered one of the highest power densities (779 $mW/cm^2$) among the non-metallic ORR catalysts reported so far (Figure 1.13a) [285].

3D porous structures of lignin and porogenic LiCl were utilized to make composite polybenzimidazole (PBI)-based proton-conducting membrane separator [281] (Figure 1.13a-bottom left). The pure PBI membrane showed a proton conductivity of $2.23 \times 10^{-9}$ mS/cm at 25°C, which increased to $2 \times 10^{-2}$ mS/cm when lignin (20 wt%) was added to the PBI matrix. The proton conductivity increased further when the porous structure of the membrane was utilized to dope phosphoric acid. The acid-doped membrane offered proton conductivities of 69 and 152 mS/cm at 25 and 160°C, respectively [281].

Dishari et al. [26] designed kraft lignin-based proton-conducting ionomers and demonstrated them as efficient catalyst binders for PEMFC-cathodes (Figure 1.13a). The 3D, less dense structure of lignin with abundant polar ether linkages and proton-conducting -$SO_3H$ groups favored the formation of larger and more well-connected proton-conducting domains than those in Nafion-based sub-μm thick films. The controlled sulfonation inhibited water solubility of these lignin-based ionomers (unlike commercial lignosulfonate), which made them suited for practical proton conduction in a humid environment. A ~180 nm thick lignin sulfonic acid film offered an ionic conductivity (~15 mS/cm) more than an order magnitude higher than that of a Nafion film with similar thickness [26]. This demonstrated the high prospect of lignin in minimizing ion-transport limitation at ionomer-catalyst interfaces and improving the ORR efficiency of PEMFCs.

### 1.2.7 Redox Flow Batteries (RFBs)

Redox flow batteries are promising energy storage systems for large-scale applications for their long cycle life, high reliability, deep discharge capability, and relatively low cost [5]. Lignin has been demonstrated to be able to play the role of a separator, electrode, and electrolyte in RFBs. In a vanadium redox flow battery (VRFB), a proton-exchange membrane (PEM, typically Nafion, sulfonated poly(sulfone), or sulfonated poly(ether ether ketone) (SPEEK)) is placed in between the anode and cathode compartments consisting of $V^{+2}/V^{+3}$ and $VO^{2+}/VO_2^{+}$ ions dissolved in sulfuric acid solutions, respectively. The crossover of vanadium ions ($V^{+n}$, in general) can be detrimental to the capacity and open-circuit voltage of VRFBs. To suppress $V^{+n}$ ion crossover while maintaining proton conduction across PEMs (Figure 1.13b), ion-conduction pathways have been attempted to narrow down by incorporating inorganic nanoparticles/organic materials [286] within the PEM matrix, layer-by-layer self-assembly of polyelectrolytes at PEM surface, [287] interfacial polymerization, [288] and more [5]. Lignin was used as an additive in SPEEK membrane with low ion exchange capacity (IEC) [289]. The abundant -OH groups of lignin facilitated the dispersion of lignin within the SPEEK matrix. The proton conductivity of SPEEK/lignin composite membrane (29.56 mS/cm) and Nafion membrane (32.26 mS/cm) was similar, [289] but the $V^{+n}$ ion selectivity of SPEEK/lignin composite membrane (173.86 × $10^4$ S.min/$cm^3$) was higher than those of pure SPEEK (130.65 × $10^4$ S min/$cm^3$) and Nafion (16.26 × $10^4$ S min/$cm^3$) membranes [289]. The high ion selectivity likely originated from the reduced size of water channels and prolonged ion transport pathways in the thicker SPEEK/lignin membranes [289]. The VRFB based on this composite membrane exhibited 71.47% capacity retention after 300 cycles under 120 mA/$cm^2$. The high $V^{+n}$ ion selectivity also offered high CE (99.5%) and energy efficiency (83.5%) [289]. Another work used a lignin/Nafion-based composite membrane, the $V^{+n}$ ion permeation of which was 25% of the pure Nafion membrane [290]. The proton conductivity was shown to depend on the lignin-to-Nafion ratio in the membrane [290].

Multiple groups explored the potential of CNFs obtained from redox-active lignin as electrode materials for VRFBs [147,291,292]. Lignin-derived carbon electrospun mats exhibited a higher activity towards $VO_2^{+}/VO^{2+}$ reaction as compared to commercial carbon papers (GDL29AA), which were attributed to the higher surface area (742 $m^2$/g (lignin-based CNF); 41 $m^2$/g (GDL29AA)) and higher amount of oxygen-containing functional groups at the CNF surface [147,292]. A comparison of kraft, organosolv, and phosphoric acid lignin-based CNFs (where PEO and dimethylformamide (DMF) were used as plasticizer and solvent, respectively) also showed that the oxidation of V(II)/V(III) is sensitive to the nature of the raw material [291].

Zhu et al. [293] utilized lignosulfonate in perchloric acid as an anolyte and hydrobromic acid in bromine ($Br_2$/HBr) as catholyte for aqueous RFB where the redox capability of lignin (Q↔HQ) was leveraged. Lignosulfonate underwent multiple reduction/oxidation cycles without any significant decrease in activity. The flow cell achieved current densities of up to 20 mA/$cm^2$ and charge polarization resistance of 15 ohms $cm^2$ [293,294].

### 1.2.8 Lignin in Other Energy Conversion and Storage Devices

Lignin has also been demonstrated for other varieties of batteries. For example, a Nafion/lignin (20 wt%)-based separator was reported as a potential alternative to brittle, glass fiber-based separators in zinc-ion batteries (ZIBs) [5,295]. ZIBs follow a similar working principle as LIBs but have the advantage of the stability of metallic zinc in an aqueous environment [5]. Soaking the membrane into an aq. 2M $ZnSO_4$ electrolyte for 12h at 90°C followed by $H_2O$-soaking for 2 h at 90°C gave the separator an ionic conductivity of 9.1 mS/cm [295]. As the $SO_3^-$ groups of Nafion electrostatically interacted with $Zn^{+2}$ ions, stable solid electrolyte interphase (SEI) layer on the Zn metal and a stable stripping/plating cycle life (~410 h) longer than glass fiber-based separators (~175 h) was achieved. At the same time, the -OH groups of lignin facilitated electrolyte uptake and formation of proton/cation-conducting channels [295]. Another demonstration of ZIB used a PPy/lignin-based cathode and an electrolytc composed of zinc acetate, choline acetate (ionic liquid), and lignosulfonate [296]. The PPy/lignin composite electrode led to a discharge capacity higher than the charge capacity over 100 charge/discharge cycling processes [296].

N-doped porous carbon, derived from raw wood, was used to make metal-free cathode materials for Zn-air batteries [297]. This cathode, when used in the battery, offered a specific capacity of 801 mAh/g, an energy density of 955 Wh/kg, and long-term stability (~110 h). Multi-heteroatom (Fe, N, P, S)-doped lignosulfonate [298] and lignocellulosic biomass [299] were also demonstrated as catalysts for oxygen reduction reaction (ORR) in Zn-air batteries [298,299] and oxygen evolution (OER) and hydrogen evolution (HER) reactions in water-splitting devices [298].

In direct methanol fuel cells (DMFCs), lignin was used as carbon dot- [300]/fiber [301]-based electrode materials to support Pt-catalysts and sodium lignosulfonate in a chitosan-based composite membrane separator [302]. As an electrode material, lignin worked to transport $O_2$ and catalyze methanol electro-oxidation; [300,301] while as membrane material, [302] lignosulfonate helped to improve the proton conductivity of the separator (to 64.7 mS/cm at 60°C).

A direct biomass fuel cell was reported [303] which worked like a PEMFC. The only difference was that the anode chamber contained redox-active lignin and polyoxometalate, which released protons during inter-redox processes. In a complementary approach, a biophotofuel cell [304]/ microbial fuel cell (MFC) set-up was used to simultaneously generate electricity and depolymerize lignin. Such photon- [304]/microbe [305,306]-assisted depolymerization was beneficial for biomass degradation [304] or production of low-MW, value-added products, like vanillin [305,306]. Another approach used electrochemical water-splitting setup to electrocatalytically hydrogenate/ hydrogenolyze guaiacol and related lignin-model monomers [307–309]. Such lignin-conversion processes were low cost and less complex compared to the classical $H_2$-hydrotreating method to prepare the same materials.

## 1.3 CHALLENGES, OPPORTUNITIES, PROSPECTS

The progress achieved over the past two decades on lignin-based materials research for energy conversion and storage devices is remarkable. This can potentially pave the way towards a sustainable energy economy and bioeconomy if we can (i) achieve consistency and scalability in manufacturing the products and (ii) establish lignin-derived materials as cost-effective co-products of biorefineries and pulp and paper industries.

The biggest challenge for lignin is its complex chemistry that varies from source to source. This, in addition to the variation in lignin extraction methods, can drastically alter the molecular weight, polydispersity, nanostructure, and ultimate performance of the materials and devices. These challenges often make it difficult to understand the origin of the superior performance of a lignin-based material as compared to other reported ones. Thus, the key to revolutionizing lignin-based materials research is to make rigorous analytical characterization efforts to explore ill-defined and complicated lignin molecules. We must develop a sound understanding of how the chemical structure of lignin alters the nanostructure and electrochemical properties of the derived materials. Not only that, but we also must put more effort to educate the research community through in-depth analysis and insights on where exactly a specific material went different from others. Such critical analysis and reports can inform and guide the design of next-generation materials.

A daunting but necessary task will be to identify bulk lignin sources with consistent chemical structure and extraction methods that can offer lignin fractions with narrow PDI and suitable MW repeatably for any particular application. For example, low MW fractions should be removed from the lignin stream to achieve spinnability. Such improvements in lignin quality can minimize the use of high-cost synthetic polymers currently mixed with lignin to spin CNFs. The same is true for lignin/$\pi$-conjugated polymer-based electrodes since up to 50 wt% of $\pi$-conjugated polymers are needed with lignin to make current pseudocapacitive materials [258]. Additionally, the $\pi$-conjugated polymers often degrade and produce maleimide-type molecules [258] which compromises the electroactivity of the electrodes upon repeated cycling (high self-discharge). Lowering the percentage of $\pi$-conjugated polymers by increasing the percentage of lignin in these composites is thus desired, but that effort also needs attention to the quality of lignin for desired electrochemical and mechanical properties.

Making free-standing materials out of lignin is still not very feasible as it forms brittle materials. Still, if a workaround is found, we can make free-standing SPEs and membrane separators from lignin without any need for additional matrix materials [258]. A lignin variety with a sufficiently high MW or a different stress-strain behavior can be the potential pathway to achieve that. On the other side, the carbonization of lignin-based precursors sometimes leads to the loss of oxygen atoms and open pores, which are needed for redox reactions and ion adsorption. It is thus important to optimize the carbonization and activation conditions to attain the right size and distribution of meso- and micropores for charge storage and ion conduction.

Although lignin as a raw material appears to be cheap, the processes of lignin fractionation and purification may cause the production cost to rise. More research and development efforts are thus needed for scalable and economic fractionation. The use of harsh chemicals for extraction should be limited to retain the environmental friendliness of lignin-based materials. Additionally, the reports on lignin-based materials still have a major focus on performance at a single component level (e.g., electrode or separator). Thus, the performance check upon integration of these components into devices is a critical need. At the end of the day, achieving high energy- and power density simultaneously using lignin-based materials can be the key to competing with the currently used materials for sustainable energy technologies. At the same time, more techno-economic analysis and customer discovery efforts can help to take these materials from the lab bench to the marketplace.

## ACKNOWLEDGMENTS

The work was primarily supported by the Nebraska Center for Energy Science Research (NCESR). SF, KAC, and OAO also acknowledge the partial support from NSF CAREER Award (NSF-DMR # 1750040) and DOE Office of Science Early CAREER Award (DE-SC0020336).

## REFERENCES

[1] *Global Fuel Cells Market Report 2021: Market to Reach $14.6 Billion by 2027- Polymer Electrolyte Membrane Fuel Cell Dominates the Fuel Cells Market.* Research and Markets.

[2] *Road map to a US Hydrogen Economy.*

[3] Cullen, D. A.; Neyerlin, K. C.; Ahluwalia, R. K.; Mukundan, R.; More, K. L.; Borup, R. L.; Weber, A. Z.; Myers, D. J.; Kusoglu, A. New Roads and Challenges for Fuel Cells in Heavy-Duty Transportation. *Nat. Energy* 2021, *6*, 462–474.

[4] *Global Li-ion battery industru research report, growth trends and competitive analysis 2021–2027.*

[5] Lizundia, E.; Kundu, D. Advances in Natural Biopolymer-Based Electrolytes and Separators for Battery Applications. *Adv. Funct. Mater.* 2021, *31*, 1–29. 10.1002/adfm.202005646

[6] Aro, T.; Fatehi, P. Production and Application of Lignosulfonates and Sulfonated Lignin. *ChemSusChem* 2017, *10*, 1861–1877. 10.1002/cssc.201700082

[7] Bajwa, D. S.; Pourhashem, G.; Ullah, A. H.; Bajwa, S. G. A Concise Review of Current Lignin Production, Applications, Products and Their Environment Impact. *Ind. Crop. Prod.* 2019, *139*, 1–11. 10.1016/j.indcrop.2019.111526

[8] Xu, R.; Zhang, K.; Liu, P.; Han, H.; Zhao, S.; Kakade, A.; Khan, A.; Du, D.; Li, X. Lignin Depolymerization and Utilization by Bacteria. *Bioresour. Technol.* 2018, *269*, 557–566. 10.1016/j.biortech.2018.08.118

[9] Ragauskas, A. J.; Beckham, G. T.; Biddy, M. J.; Chandra, R.; Chen, F.; Davis, M. F.; Davison, B. H.; Dixon, R. A.; Gilna, P.; Keller, M.; Langan, P.; Naskar, A. K.; Saddler, J. N.; Tschaplinski, T. J.; Tuskan, G. A.; Wyman, C. E. Lignin Valorization: Improving Lignin Processing in the Biorefinery. *Science (80-.)* 2014, *344*, 1246843-1–10. 10.1126/science.1246843

[10] Danner, T.; Justnes, H.; Geiker, M.; Lauten, R. A. Phase Changes during the Early Hydration of Portland Cement with Ca-Lignosulfonates. *Cem. Concr. Res.* 2015, *69*, 50–60. 10.1016/j.cemconres.2014.12.004

[11] Kalliola, A.; Vehmas, T.; Liitiä, T.; Tamminen, T. Alkali-O2 Oxidized Lignin-A Bio-Based Concrete Plasticizer. *Ind. Crop. Prod.* 2015, *74*, 150–157. 10.1016/j.indcrop.2015.04.056

[12] Naseem, A.; Tabasum, S.; Zia, K. M.; Zuber, M.; Ali, M.; Noreen, A. Lignin-Derivative Based Polymer, Blends and Composites: A Review. *Int. J. Biol. Macromol.* 2016, *93*, 296–313.

[13] Cerrutti, B. M.; Souza, C. S. De; Castellan, A.; Ruggiero, R.; Frollini, E. Carboxymethyl Lignin as Stabilizing Agent in Aqueous Ceramic Suspensions. *Ind. Crop. Prod.* 2012, *36*, 108–115. 10.1016/j.indcrop.2011.08.015

[14] Kai, D.; Tan, J.; Chee, L.; Chua, K.; Yap, L. Towards Lignin-Based Functional Materials in a Sustainable World. *Green Chem* 2016, *18*, 1175–1200. 10.1039/c5gc02616d

[15] Konduri, M. K.; Kong, F.; Fatehi, P. Production of Carboxymethylated Lignin and Its Application as a Dispersant. *Eur. Polym. J.* 2015, *70*, 371–383. 10.1016/j.eurpolymj.2015.07.028

[16] Abu-dalo, M. A.; Al-rawashdeh, N. A. F.; Ababneh, A. Evaluating the Performance of Sulfonated Kraft Lignin Agent as Corrosion Inhibitor for Iron-Based Materials in Water Distribution Systems. *Desalination* 2013, *313*, 105–114.

[17] Che, C.; Vagin, M.; Wijeratne, K.; Zhao, D.; Warczak, M.; Jonsson, M. P.; Crispin, X. Conducting Polymer Electrocatalysts for Proton-Coupled Electron Transfer Reactions: Toward Organic Fuel Cells with Forest Fuels. *Adv. Sustain. Syst.* 2018, *2* (7), 1–7. 10.1002/adsu.201800021

[18] Azadi, P.; Inderwildi, O. R.; Farnood, R.; King, D. A. Liquid Fuels, Hydrogen and Chemicals from Lignin: A Critical Review. *Renew. Sustain. Energy Rev.* 2013, *21*, 506–523. 10.1016/j.rser.2012.12.022

[19] Lange, H.; Decina, S.; Crestini, C. Oxidative Upgrade of Lignin – Recent Routes Reviewed. *Eur. Polym. J.* 2013, *49*, 1151–1173. 10.1016/j.eurpolymj.2013.03.002

[20] Xia, Q.; Chen, C.; Yao, Y.; Li, J.; He, S.; Zhou, Y.; Li, T.; Pan, X.; Yao, Y.; Hu, L. A Strong, Biodegradable and Recyclable Lignocellulosic Bioplastic. *Nat. Sustain.* 2021, *4*, 627–635. 10.1038/s41893-021-00702-w

[21] Reshmy, R.; Thomas, D.; Philip, E.; Paul, S. A.; Madhavan, A.; Sindhu, R.; Sirohi, R.; Varjani, S.; Pugazhendhi, A.; Pandey, A.; Binod, P. Bioplastic Production from Renewable Lignocellulosic Feedstocks: A Review. *Rev. Environ. Sci. Biotechnol.* 2021, *20*, 167–187. 10.1007/s11157-021-09565-1

[22] Xu, Y.; Odelius, K.; Hakkarainen, M. Recyclable and Flexible Polyester Thermosets Derived from Microwave-Processed Lignin. *ACS Appl. Polym. Mater.* 2020, *2*, 1917–1924. 10.1021/acsapm.0c00130

[23] Jawerth, M. E.; Brett, C. J.; Terrier, C.; Larsson, P. T.; Lawoko, M.; Roth, S. V.; Lundmark, S.; Johansson, M. Mechanical and Morphological Properties of Lignin-Based Thermosets. *ACS Appl. Polym. Mater.* 2020, *2*, 668–676. 10.1021/acsapm.9b01007

[24] Holmberg, A. L.; Stanzione, J. F.; Wool, R. P.; Epps, T. H. A Facile Method for Generating Designer Block Copolymers from Functionalized Lignin Model Compounds. *ACS Sustain. Chem. Eng.* 2014, *2*, 569–573. 10.1021/sc400497a

[25] Holmberg, A. L.; Reno, K. H.; Wool, R. P.; Epps, T. H. Biobased Building Blocks for the Rational Design of Renewable Block Polymers. *Soft Matter* 2014, *10*, 7405–7424. 10.1039/c4sm01220h

[26] Farzin, S.; Johnson, T. J.; Chatterjee, S.; Zamani, E.; Dishari, S. K. Ionomers From Kraft Lignin for Renewable Energy Applications. *Front. Chem.* 2020, *8*, 690. 10.3389/fchem.2020.00690

[27] Hasanov, I.; Raud, M.; Kikas, T. The Role of Ionic Liquids in the Lignin Separation from Lignocellulosic Biomass. *Energies* 2020, *13* (18), 1–24. 10.3390/en13184864

[28] Köhnke, J.; Gierlinger, N.; Mateu, B. P.; Unterweger, C.; Solt, P.; Mahler, A. K.; Schwaiger, E.; Liebner, F.; Gindl-Altmutter, W. Comparison of Four Technical Lignins as a Resource for Electrically Conductive Carbon Particles. *BioResources* 2019, *14*, 1091–1109. 10.15376/biores.14.1.1091-1109

[29] Chakar, F. S.; Ragauskas, A. J. Review of Current and Future Softwood Kraft Lignin Process Chemistry. *Ind. Crops Prod.* 2004, *20* (2), 131–141. 10.1016/j.indcrop.2004.04.016
[30] Baroncini, E. A.; Rousseau, D. M.; Strekis, C. A.; Stanzione, J. F. Viability of Low Molecular Weight Lignin in Developing Thiol-Ene Polymer Electrolytes with Balanced Thermomechanical and Conductive Properties. *Macromol. Rapid. Commun.* 2021, *42*, 1–9. 10.1002/marc.202000477
[31] Lora, J. H. Industrial Commercial Lignins: Sources, Properties and Applications. *Monomers, Polym. Compos. from Renew. Resour.* 2008, 225–241. 10.1016/B978-0-08-045316-3.00010-7
[32] Vishtal, A.; Kraslawski, A. Challenges in Industrial Applications of Technical Lignins. *BioResources* 2011, *6*, 3547–3568. 10.15376/biores.6.3.vishtal
[33] Wörmeyer, K.; Ingram, T.; Saake, B.; Brunner, G.; Smirnova, I. Comparison of Different Pretreatment Methods for Lignocellulosic Materials. Part II: Influence of Pretreatment on the Properties of Rye Straw Lignin. *Bioresour. Technol.* 2011, *102* (5), 4157–4164. 10.1016/j.biortech.2010.11.063
[34] Nadif, A.; Hunkeler, D.; Käuper, P. Sulfur-Free Lignins from Alkaline Pulping Tested in Mortar for Use as Mortar Additives. *Bioresour. Technol.* 2002, *84* (1), 49–55. 10.1016/S0960-8524(02)00020-2
[35] Fan, J.; Zhan, H. Optimization of Synthesis of Spherical Lignosulphonate Resin and Its Structure Characterization. *Chinese J. Chem. Eng.* 2008, *16* (3), 407–410. 10.1016/s1004-9541(08)60097-x
[36] Fredheim, G. E.; Braaten, S. M.; Christensen, B. E. Molecular Weight Determination of Lignosulfonates by Size-Exclusion Chromatography and Multi-Angle Laser Light Scattering. *J. Chromatogr. A.* 2002, *942*, 191–199. 10.1016/S0021-9673(01)01377-2
[37] Deng, Y.; Feng, X.; Yang, D.; Yi, C.; Qiu, X. π-π Stacking of the Aromatic Groups in Lignosulfonates. *BioResources* 2012, *7* (1), 1145–1156. 10.15376/biores.7.1.1145-1156
[38] Moacanin, J.; Felicetta, V. F.; Haller, W.; Mccarthy, J. L. Lignin. VI. Molecular Weights of Lignin Sulfonates by Light Scattering. *J. Am. Chem. Soc.* 1955, *77* (13), 3470–3475. 10.1021/ja01618a011
[39] Xu, F.; Sun, J. X.; Sun, R.; Fowler, P.; Baird, M. S. Comparative Study of Organosolv Lignins from Wheat Straw. *Ind. Crops Prod.* 2006, *23* (2), 180–193. 10.1016/j.indcrop.2005.05.008
[40] Carrott, P. J. M.; Suhas; Ribeiro Carrott, M. M. L.; Guerrero, C. I.; Delgado, L. A. Reactivity and Porosity Development during Pyrolysis and Physical Activation in CO2 or Steam of Kraft and Hydrolytic Lignins. *J. Anal. Appl. Pyrolysis* 2008, *82* (2), 264–271. 10.1016/j.jaap.2008.04.004
[41] Pu, Y.; Jiang, N.; Ragauskas, A. J. Ionic Liquid as a Green Solvent for Lignin. *J. Wood. Chem. Tech.* 2007, *27*, 23–33. 10.1080/02773810701282330
[42] Laurichesse, S.; Avérous, L. Chemical Modification of Lignins: Towards Biobased Polymers. *Prog. Polym. Sci.* 2014, *39*, 1266–1290. 10.1016/j.progpolymsci.2013.11.004
[43] Nimz, H. Beech Lignin-Proposal of a Constitutional Scheme. *Angew. Chem. Int. Ed.* 1974, *13*, 313–321.
[44] de Carvalho, D. M.; Colodette, J. L. Comparative Study of Acid Hydrolysis of Lignin and Polysaccharides in Biomasses. *BioResources* 2017, *12*, 6907–6923. 10.15376/biores.12.4.6907-6923
[45] Binder, J. B.; Gray, M. J.; White, J. F.; Zhang, Z. C.; Holladay, J. E. Reactions of Lignin Model Compounds in Ionic Liquids. *Biomass and Bioenergy* 2009, *33* (9), 1122–1130. 10.1016/j.biombioe.2009.03.006
[46] Kubo, S.; Kadla, J. F. Kraft Lignin/Poly(Ethylene Oxide) Blends: Effect of Lignin Structure on Miscibility and Hydrogen Bonding. *J. Appl. Polym. Sci.* 2005, *98*, 1437–1444. 10.1002/app.22245
[47] Pakkanen, H.; Alén, R. Molecular Mass Distribution of Lignin from the Alkaline Pulping of Hardwood, Softwood, and Wheat Straw. *J. Wood Chem. Technol.* 2012, *32*, 279–293. 10.1080/02773813.2012.659321
[48] Eraghi Kazzaz, A.; Fatehi, P. Technical Lignin and Its Potential Modification Routes: A Mini-Review. *Ind. Crop. Prod.* 2020, *154*, 1–13. 10.1016/j.indcrop.2020.112732
[49] Dessbesell, L.; Paleologou, M.; Leitch, M.; Pulkki, R.; Xu, C. (Charles) Global Lignin Supply Overview and Kraft Lignin Potential as an Alternative for Petroleum-Based Polymers. *Renew. Sustain. Energy Rev.* 2020, *123*, 1–11. 10.1016/j.rser.2020.109768
[50] Gellerstedt, G. Softwood Kraft Lignin: Raw Material for the Future. *Ind. Crop. Prod.* 2015, *77*, 845–854. 10.1016/j.indcrop.2015.09.040
[51] Yang, D.; Qiu, X.; Zhou, M.; Lou, H. Properties of Sodium Lignosulfonate as Dispersant of Coal Water Slurry. *Energy Convers. Manag.* 2007, *48*, 2433–2438. 10.1016/j.enconman.2007.04.007
[52] Achinivu, E. C. Protic Ionic Liquids for Lignin Extraction-A Lignin Characterization Study. *Int. J. Mol. Sci.* 2018, *19*, 1–14. 10.3390/ijms19020428
[53] Belesov, A. V.; Ladesov, A. V.; Pikovskoi, I. I.; Faleva, A. V.; Kosyakov, D. S. Characterization of Ionic Liquid Lignins Isolated from Spruce Wood with 1-Butyl-3-Methylimidazolium Acetate and

Methyl Sulfate and Their Binary Mixtures with DMSO. *Molecules* 2020, *25*, 1–12. 10.3390/molecules25112479
[54] Guterman, R.; Molinari, V.; Josef, E. Ionic Liquid Lignosulfonate: Dispersant and Binder for Preparation of Biocomposite Materials. *Angew. Chem. Int. Ed.* 2019, *58*, 13044–13050. 10.1002/anie.201907385
[55] Shabanov, N. S.; Rabadanov, K. S.; Gafurov, M. M.; Isaev, A. B.; Sobola, D. S.; Suleimanov, S. I.; Amirov, A. M.; Asvarov, A. S. Lignin-Based Gel Polymer Electrolyte for Cationic Conductivity. *Polymers (Basel)* 2021, *13*, 2306. 10.3390/polym13142306
[56] Liu, L.; Solin, N.; Inganäs, O. Scalable Lignin/Graphite Electrodes Formed by Mechanochemistry. *RSC Adv* 2019, *9*, 39758–39767. 10.1039/c9ra07507k
[57] Liu, L.; Wang, L.; Solin, N.; Inganäs, O. Quinones from Biopolymers and Small Molecules Milled into Graphite Electrodes. *Adv. Mater. Technol.* 2021, *2001042*, 1–15. 10.1002/admt.202001042
[58] Chen, C.; Liu, Y.; Chen, Y.; Li, X.; Cheng, J.; Chen, S.; Lin, J.; Zhang, X.; Zhang, Y. Effect of Polyaniline-Modified Lignosulfonate Added to the Negative Active Material on the Performance of Lead-Acid Battery. *Electrochim. Acta* 2020, *338*, 135859. 10.1016/j.electacta.2020.135859
[59] Zhang, W.; Yin, J.; Lin, Z.; Lin, H.; Lu, H.; Wang, Y.; Huang, W. Facile Preparation of 3D Hierarchical Porous Carbon from Lignin for the Anode Material in Lithium Ion Battery with High Rate Performance. *Electrochim. Act.* 2015, *176*, 1136–1142. 10.1016/j.electacta.2015.08.001
[60] Xi, Y.; Wang, Y.; Yang, D.; Zhang, Z.; Liu, W.; Li, Q.; Qiu, X. K2CO3 Activation Enhancing the Graphitization of Porous Lignin Carbon Derived from Enzymatic Hydrolysis Lignin for High Performance Lithium-Ion Storage. *J. Alloy. Compd.* 2019, *785*, 706–714. 10.1016/j.jallcom.2019.01.039
[61] Du, Y. F.; Sun, G. H.; Li, Y.; Cheng, J. Y.; Chen, J. P.; Song, G.; Kong, Q. Q.; Xie, L. J.; Chen, C. M. Pre-Oxidation of Lignin Precursors for Hard Carbon Anode with Boosted Lithium-Ion Storage Capacity. *Carbon N. Y.* 2021, *178*, 243–255. 10.1016/j.carbon.2021.03.016
[62] Yang, Z.; Guo, H.; Li, F.; Li, X.; Wang, Z.; Cui, L.; Wang, J. Cooperation of Nitrogen-Doping and Catalysis to Improve the Li-Ion Storage Performance of Lignin-Based Hard Carbon. *J. Energy Chem.* 2018, *27*, 1390–1396. 10.1016/j.jechem.2018.01.013
[63] Chang, Z. Z.; Yu, B. J.; Wang, C. Y. Influence of H2 Reduction on Lignin-Based Hard Carbon Performance in Lithium Ion Batteries. *Electrochim. Act.* 2015, *176*, 1352–1357. 10.1016/j.electacta.2015.07.076
[64] Du, L.; Wu, W.; Luo, C.; Zhao, H.; Xu, D.; Wang, R.; Deng, Y. Lignin Derived Si@C Composite as a High Performance Anode Material for Lithium Ion Batteries. *Solid State Ionics* 2018, *319*, 77–82. 10.1016/j.ssi.2018.01.039
[65] Hayashi, J.; Kazehaya, A.; Muroyama, K.; Watkinson, A. P. Preparation of Activated Carbon from Lignin by Chemical Activation. *Carbon N. Y.* 2000, *38*, 1873–1878. 10.1016/S0008-6223(00)00027-0
[66] Xi, Y.; Huang, S.; Yang, D.; Qiu, X.; Su, H.; Yi, C.; Li, Q. Hierarchical Porous Carbon Derived from the Gas-Exfoliation Activation of Lignin for High-Energy Lithium-Ion Batteries. *Green Chem* 2020, *22*, 4321–4330. 10.1039/d0gc00945h
[67] Fan, L.; Fan, L.; Yu, T.; Tan, X.; Shi, Z. Hydrothermal Synthesis of Lignin-Based Carbon Microspheres as Anode Material for Lithium-Ion Batteries. *Int. J. Electrchem. Sci.* 2020, *15*, 1035–1043. 10.20964/2020.02.16
[68] Navarro-Suárez, A. M.; Saurel, D.; Sánchez-Fontecoba, P.; Castillo-Martínez, E.; Carretero-González, J.; Rojo, T. Temperature Effect on the Synthesis of Lignin-Derived Carbons for Electrochemical Energy Storage Applications. *J. Power Sources* 2018, *397*, 296–306. 10.1016/j.jpowsour.2018.07.023
[69] Zhao, H.; Wang, Q.; Deng, Y.; Shi, Q.; Qian, Y.; Wang, B.; Lü, L.; Qiu, X. Preparation of Renewable Lignin-Derived Nitrogen-Doped Carbon Nanospheres as Anodes for Lithium-Ion Batteries. *RSC Adv.* 2016, *6*, 77143–77150. 10.1039/c6ra17793j
[70] Klapiszewski, Ł.; Szalaty, T. J.; Kurc, B.; Stanisz, M.; Skrzypczak, A.; Jesionowski, T. Functional Hybrid Materials Based on Manganese Dioxide and Lignin Activated by Ionic Liquids and Their Application in the Production of Lithium Ion Batteries. *Int. J. Mol. Sci.* 2017, *18*, 1–29. 10.3390/ijms18071509
[71] Chen, T.; Hu, J.; Zhang, L.; Pan, J.; Liu, Y.; Cheng, Y. T. High Performance Binder-Free SiOx/C Composite LIB Electrode Made of SiOx and Lignin. *J. Power Sources* 2017, *362*, 236–242. 10.1016/j.jpowsour.2017.07.049
[72] Ma, X.; Smirnova, A. L.; Fong, H. Flexible Lignin-Derived Carbon Nanofiber Substrates Functionalized with Iron (III) Oxide Nanoparticles as Lithium-Ion Battery Anodes. *Mat. Sci. Eng. B* 2019, *241*, 100–104. 10.1016/j.mseb.2019.02.013

[73] Zhou, Z.; Chen, F.; Kuang, T.; Chang, L.; Yang, J.; Fan, P.; Zhao, Z.; Zhong, M. Lignin-Derived Hierarchical Mesoporous Carbon and NiO Hybrid Nanospheres with Exceptional Li-Ion Battery and Pseudocapacitive Properties. *Electrochim. Acta* 2018, *274*, 288–297. 10.1016/j.electacta.2018.04.111

[74] Chen, T.; Zhang, Q.; Xu, J.; Pan, J.; Cheng, Y. T. Binder-Free Lithium Ion Battery Electrodes Made of Silicon and Pyrolized Lignin. *RSC Adv.* 2016, *6*, 29308–29313. 10.1039/c6ra03001g

[75] Xi, Y.; Yang, D.; Lou, H.; Gong, Y.; Yi, C.; Lyu, G.; Han, W.; Kong, F.; Qiu, X. Designing the Effective Microstructure of Lignin-Based Porous Carbon Substrate to Inhibit the Capacity Decline for SnO2 Anode. *Ind. Crop. Prod.* 2021, *161*, 1–9. 10.1016/j.indcrop.2020.113179

[76] Yu, K.; Liu, T.; Zheng, Q.; Wang, X.; Liu, W.; Liang, J.; Liang, C. Rice Husk Lignin-Based Porous Carbon and ZnO Composite as an Anode for High-Performance Lithium-Ion Batteries. *J. Porous Mater.* 2020, *27*, 875–882. 10.1007/s10934-019-00824-9

[77] Chou, C. Y.; Kuo, J. R.; Yen, S. C. Silicon-Based Composite Negative Electrode Prepared from Recycled Silicon-Slicing Slurries and Lignin/Lignocellulose for Li-Ion Cells. *ACS Sustain. Chem. Eng.* 2018, *6*, 4759–4766. 10.1021/acssuschemeng.7b03887

[78] Niu, X.; Zhou, J.; Qian, T.; Wang, M.; Yan, C. Confined Silicon Nanospheres by Biomass Lignin for Stable Lithium Ion Battery. *Nanotechnology* 2017, *28*, 1–9. 10.1088/1361-6528/aa84cd

[79] Yi, X.; He, W.; Zhang, X.; Yue, Y.; Yang, G.; Wang, Z.; Zhou, M.; Wang, L. Graphene-like Carbon Sheet/Fe3O4 Nanocomposites Derived from Soda Papermaking Black Liquor for High Performance Lithium Ion Batteries. *Electrochim. Act.* 2017, *232*, 550–560. 10.1016/j.electacta.2017.02.130

[80] Choi, D. I.; Lee, J. N.; Song, J.; Kang, P. H.; Park, J. K.; Lee, Y. M. Fabrication of Polyacrylonitrile/Lignin-Based Carbon Nanofibers for High-Power Lithium Ion Battery Anodes. *J. Solid State Electrochem.* 2013, *17*, 2471–2475. 10.1007/s10008-013-2112-5

[81] Shi, Z.; Jin, G.; Wang, J.; Zhang, J. Free-Standing, Welded Mesoporous Carbon Nanofibers as Anode for High-Rate Performance Li-Ion Batteries. *J. Electroanal. Chem.* 2017, *795*, 26–31. 10.1016/j.jelechem.2017.03.047

[82] Sagues, W. J.; Yang, J.; Monroe, N.; Han, S. D.; Vinzant, T.; Yung, M.; Jameel, H.; Nimlos, M.; Park, S. A Simple Method for Producing Bio-Based Anode Materials for Lithium-Ion Batteries. *Green Chem* 2020, *22*, 7093–7108. 10.1039/d0gc02286a

[83] Wang, S. X.; Yang, L.; Stubbs, L. P.; Li, X.; He, C. Lignin-Derived Fused Electrospun Carbon Fibrous Mats as High Performance Anode Materials for Lithium Ion Batteries. *ACS Appl. Mater. Interfaces* 2013, *5*, 12275–12282. 10.1021/am4043867

[84] Stojanovska, E.; Serife Pampal, E.; Kilic, A. Activated Lignin Based Carbon Nanofibers As Binder-Free Anodes for Lithium Ion Batteries. *ECS Trans* 2016, *73*, 121–127. 10.1149/07301.0121ecst

[85] Stojanovska, E.; Pampal, E. S.; Kilic, A.; Quddus, M.; Candan, Z. Developing and Characterization of Lignin-Based Fibrous Nanocarbon Electrodes for Energy Storage Devices. *Compos. B Eng.* 2019, *158*, 239–248. 10.1016/j.compositesb.2018.09.072

[86] Culebras, M.; Geaney, H.; Beaucamp, A.; Upadhyaya, P.; Dalton, E.; Ryan, K. M.; Collins, M. N. Bio-Derived Carbon Nanofibres from Lignin as High-Performance Li-Ion Anode Materials. *ChemSusChem* 2019, *12*, 4516–4521. 10.1002/cssc.201901562

[87] Tenhaeff, W. E.; Rios, O.; More, K.; McGuire, M. A. Highly Robust Lithium Ion Battery Anodes from Lignin: An Abundant, Renewable, and Low-Cost Material. *Adv. Funct. Mater.* 2014, *24*, 86–94. 10.1002/adfm.201301420

[88] Cao, D.; Zhang, Q.; Hafez, A. M.; Jiao, Y.; Ma, Y.; Li, H.; Cheng, Z.; Niu, C.; Zhu, H. Lignin-Derived Holey, Layered, and Thermally Conductive 3D Scaffold for Lithium Dendrite Suppression. *Small Methods* 2019, *3*, 1800539-1–10. 10.1002/smtd.201800539

[89] Liu, W.; Zhao, C.; Zhou, R.; Zhou, D.; Liu, Z.; Lu, X. Lignin-Assisted Exfoliation of Molybdenum Disulfide in Aqueous Media and Its Application in Lithium Ion Batteries. *Nanoscale* 2015, *7*, 9919–9926. 10.1039/c5nr01891a

[90] Lu, G.; Zheng, J.; Jin, C.; Yan, T.; Zhang, L.; Nai, J.; Wang, Y.; Liu, Y.; Liu, T.; Tao, X. Lithiated Aromatic Biopolymer as High-Performance Organic Anodes for Lithium-Ion Storage. *Chem. Eng. J.* 2021, *409*, 1–7. 10.1016/j.cej.2020.127454

[91] Li, Y.; Hu, Y. S.; Li, H.; Chen, L.; Huang, X. A Superior Low-Cost Amorphous Carbon Anode Made from Pitch and Lignin for Sodium-Ion Batteries. *J. Mater. Chem. A* 2016, *4*, 96–104. 10.1039/c5ta08601a

[92] Rybarczyk, M. K.; Li, Y.; Qiao, M.; Hu, Y. S.; Titirici, M. M.; Lieder, M. Hard Carbon Derived from Rice Husk as Low Cost Negative Electrodes in Na-Ion Batteries. *J. Energy Chem.* 2019, *29*, 17–22. 10.1016/j.jechem.2018.01.025

[93] Zhang, J.; Yu, B.; Zhang, Y.; Wang, C. Lignin-Derived Hard Carbon Microspheres Synthesized via Emulsion-Solvent Evaporation as Anode for Sodium Storage. *Energy Technol* 2020, *8*, 1901423 (1-9). 10.1002/ente.201901423

[94] Li, C.; Sun, Y.; Wu, Q.; Liang, X.; Chen, C.; Xiang, H. A Novel Design Strategy of a Practical Carbon Anode Material from a Single Lignin-Based Surfactant Source for Sodium-Ion Batteries. *Chem. Commun.* 2020, *56*, 6078–6081. 10.1039/d0cc01431a

[95] Marino, C.; Cabanero, J.; Povia, M.; Villevieille, C. Biowaste Lignin-Based Carbonaceous Materials as Anodes for Na Ion Batteries Biowaste Lignin Based Carbonaceous Materials as Anodes for Na-Ion Batteries. *J. Electrochem. Soc.* 2018, *165*, A1400–A1408. 10.1149/2.0681807jes

[96] Zhang, H.; Zhang, W.; Ming, H.; Pang, J.; Zhang, H.; Cao, G.; Yang, Y. Design Advanced Carbon Materials from Lignin-Based Interpenetrating Polymer Networks for High Performance Sodium-Ion Batteries. *Chem. Eng. J.* 2018, *341*, 280–288. 10.1016/j.cej.2018.02.016

[97] Zheng, P.; Liu, T.; Yuan, X.; Zhang, L.; Liu, Y.; Huang, J.; Guo, S. Enhanced Performance by Enlarged Nano-Pores of Holly Leaf-Derived Lamellar Carbon for Sodium-Ion Battery Anode. *Sci. Rep.* 2016, *6*, 1–9. 10.1038/srep26246

[98] Zhang, Y.; Zhu, Y.; Zhang, J.; Sun, S.; Wang, C.; Chen, M.; Zeng, J. Optimizing the Crystallite Structure of Lignin-Based Nanospheres by Resinification for High-Performance Sodium-Ion Battery Anodes. *Energy Technol.* 2020, *8*, 1–8. 10.1002/ente.201900694

[99] Alvin, S.; Yoon, D.; Chandra, C.; Cahyadi, H. S.; Park, J. H.; Chang, W.; Chung, K. Y.; Kim, J. Revealing Sodium Ion Storage Mechanism in Hard Carbon. *Carbon N. Y.* 2019, *145*, 67–81. 10.1016/j.carbon.2018.12.112

[100] Yoon, D.; Hwang, J.; Chang, W.; Kim, J. Carbon with Expanded and Well-Developed Graphene Planes Derived Directly from Condensed Lignin as a High-Performance Anode for Sodium-Ion Batteries. *ACS Appl. Mater. Interfaces* 2018, *10*, 569–581. 10.1021/acsami.7b14776

[101] Yu, X.; Yu, B.; Zhang, J.; Zhang, Y.; Zeng, J.; Chen, M.; Wang, C. An Attempt to Improve Electrochemical Performances of Lignin-Based Hard Carbon Microspheres Anodes in Sodium-Ion Batteries by Using Hexamethylenetetramine. *ChemistrySelect* 2018, *3*, 9518–9525. 10.1002/slct.201801980

[102] Ding, J.; Wang, H.; Li, Z.; Kohandehghan, A.; Cui, K.; Xu, Z.; Zahiri, B.; Tan, X.; Lotfabad, E. M.; Olsen, B. C.; Mitlin, D. Carbon Nanosheet Frameworks Derived from Peat Moss as High Performance Sodium Ion Battery Anodes. *ACS Nano* 2013, *7*, 11004–11015. 10.1021/nn404640c

[103] Jia, H.; Sun, N.; Dirican, M.; Li, Y.; Chen, C.; Zhu, P.; Yan, C.; Zang, J.; Guo, J.; Tao, J.; Wang, J.; Tang, F.; Zhang, X. Electrospun Kraft Lignin/Cellulose Acetate-Derived Nanocarbon Network as an Anode for High-Performance Sodium-Ion Batteries. *ACS Appl. Mater. Interfaces* 2018, *10*, 44368–44375. 10.1021/acsami.8b13033

[104] Jin, J.; Yu, B. J.; Shi, Z. Q.; Wang, C. Y.; Chong, C. Lignin-Based Electrospun Carbon Nanofibrous Webs as Free-Standing and Binder-Free Electrodes for Sodium Ion Batteries. *J. Power Sources* 2014, *272*, 800–807. 10.1016/j.jpowsour.2014.08.119

[105] Peuvot, K.; Hosseinaei, O.; Tomani, P.; Zenkert, D.; Lindbergh, G. Lignin Based Electrospun Carbon Fiber Anode for Sodium Ion Batteries Lignin Based Electrospun Carbon Fiber Anode for Sodium Ion Batteries. *J. Electrochem. Soc.* 2019, *166*, A1984–A1990. 10.1149/2.0711910jes

[106] Casado, N.; Hilder, M.; Pozo-Gonzalo, C.; Forsyth, M.; Mecerreyes, D. Electrochemical Behavior of PEDOT/Lignin in Ionic Liquid Electrolytes: Suitable Cathode/Electrolyte System for Sodium Batteries. *ChemSusChem* 2017, *10* (8), 1783–1791. 10.1002/cssc.201700012

[107] Chen, S.; Feng, F.; Ma, Z. F. Lignin-Derived Nitrogen-Doped Porous Ultrathin Layered Carbon as a High-Rate Anode Material for Sodium-Ion Batteries. *Compos. Commun.* 2020, *22*, 1–7. 10.1016/j.coco.2020.100447

[108] Li, H.; Shen, F.; Luo, W.; Dai, J.; Han, X.; Chen, Y.; Yao, Y.; Zhu, H.; Fu, K.; Hitz, E.; Hu, L. Carbonized-Leaf Membrane with Anisotropic Surfaces for Sodium-Ion Battery. *ACS Appl. Mater. Interfaces* 2016, *8*, 2204–2210. 10.1021/acsami.5b10875

[109] Fan, L.; Shi, Z.; Ren, Q.; Yan, L.; Zhang, F.; Fan, L. Nitrogen-Doped Lignin Based Carbon Microspheres as Anode Material for High Performance Sodium Ion Batteries. *Green Energy Environ.* 2021, *6*, 220–228. 10.1016/j.gee.2020.06.005

[110] Zhao, X. Y.; Cao, J. P.; Morishita, K.; Ozaki, J. I.; Takarada, T. Electric Double-Layer Capacitors from Activated Carbon Derived from Black Liquor. *Energy Fuels* 2010, *24*, 1889–1893. 10.1021/ef901299c

[111] Zeng, L.; Lou, X.; Zhang, J.; Wu, C.; Liu, J.; Jia, C. Carbonaceous Mudstone and Lignin-Derived Activated Carbon and Its Application for Supercapacitor Electrode. *Surf. Coat. Technol.* 2019, *357*, 580–586. 10.1016/j.surfcoat.2018.10.041

[112] Guo, N.; Li, M.; Sun, X.; Wang, F.; Yang, R. Enzymatic Hydrolysis Lignin Derived Hierarchical Porous Carbon for Supercapacitors in Ionic Liquids with High Power and Energy Densities. *Green Chem.* 2017, *19*, 2595–2602. 10.1039/c7gc00506g

[113] Zhang, K.; Liu, M.; Zhang, T.; Min, X.; Wang, X.; Chai, L.; Shi, Y. High-Performance Supercapacitor Energy Storage Using a Carbon Material Derived from Lignin by Bacterial Activation before Carbonization. *J. Mater. Chem. A* 2019, *7*, 26838–26848.

[114] Chen, W.; Wang, X.; Feizbakhshan, M.; Liu, C.; Hong, S.; Yang, P.; Zhou, X. Preparation of Lignin-Based Porous Carbon with Hierarchical Oxygen-Enriched Structure for High-Performance Supercapacitors. *J. Coll. Interface Sci.* 2019, *540*, 524–534. 10.1016/j.jcis.2019.01.058

[115] Chen, L.; Deng, J.; Song, Y.; Hong, S.; Lian, H. Deep Eutectic Solvent Promoted Tunable Synthesis of Nitrogen-Doped Nanoporous Carbons from Enzymatic Hydrolysis Lignin for Supercapacitors. *Mater. Res. Bull.* 2020, *123*, 110708. 10.1016/j.materresbull.2019.110708

[116] Yin, W. M.; Tian, L. F.; Pang, B.; Guo, Y. R.; Li, S. J.; Pan, Q. J. Fabrication of Dually N/S-Doped Carbon from Biomass Lignin: Porous Architecture and High-Rate Performance as Supercapacitor. *Intl. J. Biol. Macromol.* 2020, *156*, 988–996. 10.1016/j.ijbiomac.2020.04.102

[117] Zhang, L.; You, T.; Zhou, T.; Zhou, X.; Xu, F. Interconnected Hierarchical Porous Carbon from Lignin-Derived Byproducts of Bioethanol Production for Ultra-High Performance Supercapacitors. *ACS Appl. Mater. Interfaces* 2016, *8* (22), 13918–13925. 10.1021/acsami.6b02774

[118] Jurewicz, K.; Babeł, K. Efficient Capacitor Materials from Active Carbons Based on Coconut Shell/Melamine Precursors. *Energy Fuels* 2010, *24*, 3429–3435. 10.1021/ef901554j

[119] Demir, M.; Tessema, T. D.; Farghaly, A. A.; Nyankson, E.; Saraswat, S. K.; Aksoy, B.; Islamoglu, T.; Collinson, M. M.; El-Kaderi, H. M.; Gupta, R. B. Lignin-Derived Heteroatom-Doped Porous Carbons for Supercapacitor and CO2 Capture Applications. *Int. J. Energy Res.* 2018, *42*, 2686–2700. 10.1002/er.4058

[120] Zhang, W.; Yu, C.; Chang, L.; Zhong, W.; Yang, W. Three-Dimensional Nitrogen-Doped Hierarchical Porous Carbon Derived from Cross-Linked Lignin Derivatives for High Performance Supercapacitors. *Electrochim. Acta* 2018, *282*, 642–652. 10.1016/j.electacta.2018.06.100

[121] Lai, C.; Zhou, Z.; Zhang, L.; Wang, X.; Zhou, Q.; Zhao, Y.; Wang, Y.; Wu, X.; Zhu, Z.; Fong, H. Free-Standing and Mechanically Flexible Mats Consisting of Electrospun Carbon Nano Fi Bers Made from a Natural Product of Alkali Lignin as Binder-Free Electrodes for High-Performance Supercapacitors. *J. Power Sources* 2014, *247*, 134–141. 10.1016/j.jpowsour.2013.08.082

[122] Wang, X.; Xu, Q.; Cheng, J.; Hu, G.; Xie, X.; Peng, C.; Yu, X.; Shen, H.; Kent, Z.; Xie, H. Bio-Refining Corn Stover into Microbial Lipid and Advanced Energy Material Using Ionic Liquid-Based Organic Electrolyte. *Ind. Crop. Prod.* 2020, *145*, 112137. 10.1016/j.indcrop.2020.112137

[123] Saha, D.; Li, Y.; Bi, Z.; Chen, J.; Keum, J. K.; Hensley, D. K.; Grappe, H. A.; Meyer, H. M.; Dai, S.; Paranthaman, M. P.; Naskar, A. K. Studies on Supercapacitor Electrode Material from Activated Lignin-Derived Mesoporous Carbon. *Langmuir* 2014, *30*, 900–910. 10.1021/la404112m

[124] Cao, Q.; Zhu, M.; Chen, J.; Song, Y.; Li, Y.; Zhou, J. Novel Lignin-Cellulose-Based Carbon Nanofibers as High-Performance Supercapacitors. *ACS Appl. Mater. Interfaces* 2020, *12* (1), 1210–1221. 10.1021/acsami.9b14727

[125] Du, B.; Zhu, H.; Chai, L.; Cheng, J.; Wang, X.; Chen, X.; Zhou, J.; Sun, R. C. Effect of Lignin Structure in Different Biomass Resources on the Performance of Lignin-Based Carbon Nanofibers as Supercapacitor Electrode. *Ind. Crop. Prod.* 2021, *170*, 113745. 10.1016/j.indcrop.2021.113745

[126] Zhu, M.; Liu, H.; Cao, Q.; Zheng, H.; Xu, D.; Guo, H.; Wang, S.; Li, Y.; Zhou, J. Electrospun Lignin-Based Carbon Nanofibers as Supercapacitor Electrodes. *ACS Sustain. Chem. Eng.* 2020, *8* (34), 12831–12841. 10.1021/acssuschemeng.0c03062

[127] Liu, W.; Yao, Y.; Fu, O.; Jiang, S.; Fang, Y.; Wei, Y.; Lu, X. Lignin-Derived Carbon Nanosheets for High-Capacitance Supercapacitors. *RSC Adv.* 2017, *7*, 48537–48543. 10.1039/c7ra08531a

[128] Cai, Z.; Jiang, C.; Xiao, X. F.; Zhang, Y. S.; Liang, L. Lignin-Based Biochar/Graphene Oxide Composites as Supercapacitor Electrode Materials. *IOP Conf. Ser. Mater. Sci. Eng.* 2018, *359*, 012046. 10.1088/1757-899X/359/1/012046

[129] Jha, S.; Mehta, S.; Chen, Y.; Ma, L.; Renner, P.; Parkinson, D. Y.; Liang, H. Design and Synthesis of Lignin-Based Flexible Supercapacitors. *ACS Sustain. Chem. Eng.* 2020, *8*, 498–511. 10.1021/acssuschemeng.9b05880

[130] Chen, F.; Zhou, W.; Yao, H.; Fan, P.; Yang, J.; Fei, Z.; Zhong, M. Self-Assembly of NiO Nanoparticles in Lignin-Derived Mesoporous Carbons for Supercapacitor Applications. *Green Chem.* 2013, *15* (11), 3057–3063. 10.1039/c3gc41080c

[131] Ma, X.; Kolla, P.; Zhao, Y.; Smirnova, A. L.; Fong, H. Electrospun Lignin-Derived Carbon Nanofiber Mats Surface-Decorated with $MnO_2$ Nanowhiskers as Binder-Free Supercapacitor Electrodes with High Performance. *J. Power Sources* 2016, *325*, 541–548. 10.1016/j.jpowsour.2016.06.073

[132] Yun, S. I.; Kim, S. H.; Kim, D. W.; Kim, Y. A.; Kim, B. H. Facile Preparation and Capacitive Properties of Low-Cost Carbon Nanofibers with ZnO Derived from Lignin and Pitch as Supercapacitor Electrodes. *Carbon N. Y.* 2019, *149*, 637–645. 10.1016/j.carbon.2019.04.105

[133] Lei, D.; Li, X. D.; Seo, M. K.; Khil, M. S.; Kim, H. Y.; Kim, B. S. $NiCo_2O_4$ Nanostructure-Decorated PAN/Lignin Based Carbon Nanofiber Electrodes with Excellent Cyclability for Flexible Hybrid Supercapacitors. *Polymer (Guildf).* 2017, *132*, 31–40. 10.1016/j.polymer.2017.10.051

[134] Ajjan, F. N.; Casado, N.; Rebis, T.; Elfwing, A.; Solin, N.; Mercerreyes, D.; Inganäs, O. High Performance PEDOT/Lignin Biopolymer Composites for Electrochemical Supercapacitors. *J. Mater. Chem. A* 2016, *4*, 1838–1847. 10.1039/c5ta10096h

[135] Admassie, S.; Elfwing, A.; Inganäs, O. Electrochemical Synthesis and Characterization of Interpenetrating Networks of Conducting Polymers for Enhanced Charge Storage. *Adv. Mater. Interfaces* 2016, *3*, 1–6. 10.1002/admi.201500533

[136] Ajjan, F. N.; Vagin, M.; Rębiś, T.; Aguirre, L. E.; Ouyang, L.; Inganäs, O. Scalable Asymmetric Supercapacitors Based on Hybrid Organic/Biopolymer Electrodes. *Adv. Sustain. Syst.* 2017, *1*, 1–8. 10.1002/adsu.201700054

[137] Xu, H.; Jiang, H.; Li, X.; Wang, G. Synthesis and Electrochemical Capacitance Performance of Polyaniline Doped with Lignosulfonate. *RSC Adv.* 2015, *5*, 76116–76121. 10.1039/C5RA12292A

[138] Nagaraju, D. H.; Rebis, T.; Gabrielsson, R.; Elfwing, A.; Milczarek, G.; Inganäs, O. Charge Storage Capacity of Renewable Biopolymer/Conjugated Polymer Interpenetrating Networks Enhanced by Electroactive Dopants. *Adv. Energy Mater.* 2014, *4*, 1–7. 10.1002/aenm.201300443

[139] Geng, X.; Zhang, Y.; Jiao, L.; Yang, L.; Hamel, J.; Giummarella, N.; Henriksson, G.; Zhang, L.; Zhu, H. Bioinspired Ultrastable Lignin Cathode via Graphene Reconfiguration for Energy Storage. *ACS Sustain. Chem. Eng.* 2017, *5*, 3553–3561. 10.1021/acssuschemeng.7b00322

[140] Yang, W.; Qu, Y.; Zhou, B.; Li, C.; Jiao, L.; Dai, H. Value-Added Utilization of Lignin-Derived Aromatic Oligomers as Renewable Charge-Storage Materials. *Ind. Crop. Prod.* 2021, *171*, 1–9. 10.1016/j.indcrop.2021.113848

[141] Kim, S. K.; Kim, Y. K.; Lee, H.; Lee, S. B.; Park, H. S. Superior Pseudocapacitive Behavior of Confined Lignin Nanocrystals for Renewable Energy-Storage Materials. *ChemSusChem* 2014, *7* (4), 1094–1101. 10.1002/cssc.201301061

[142] Milczarek, G.; Nowicki, M. Carbon Nanotubes / Kraft Lignin Composite: Characterization and Charge Storage Properties. *Mat. Res. Bull.* 2013, *48*, 4032–4038. 10.1016/j.materresbull.2013.06.022

[143] Peng, Z.; Zhong, W. Facile Preparation of an Excellent Mechanical Property Electroactive Biopolymer-Based Conductive Composite Film and Self-Enhancing Cellulose Hydrogel to Construct a High-Performance Wearable Supercapacitor. *ACS Sustain. Chem. Eng.* 2020, *8*, 7879–7891. 10.1021/acssuschemeng.0c01118

[144] Zhu, H.; Luo, W.; Ciesielski, P. N.; Fang, Z.; Zhu, J. Y.; Henriksson, G.; Himmel, M. E.; Hu, L. Wood-Derived Materials for Green Electronics, Biological Devices, and Energy Applications. *Chem. Rev.* 2016, *116*, 9305–9374. 10.1021/acs.chemrev.6b00225

[145] Goodenough, J. B.; Kim, Y. Challenges for Rechargeable Li Batteries. *Chem. Mater.* 2010, *22*, 587–603. 10.1021/cm901452z

[146] Zhu, J.; Yan, C.; Zhang, X.; Yang, C.; Jiang, M.; Zhang, X. A Sustainable Platform of Lignin: From Bioresources to Materials and Their Applications in Rechargeable Batteries and Supercapacitors. *Prog. Energy Combust. Sci.* 2020, *76*, 100788. 10.1016/j.pecs.2019.100788

[147] Liu, H.; Xu, T.; Liu, K.; Zhang, M.; Liu, W.; Li, H.; Du, H.; Si, C. Lignin-Based Electrodes for Energy Storage Application. *Ind. Crop. Prod.* 2021, *165*, 1–15. 10.1016/j.indcrop.2021.113425

[148] Liu, L.; Solin, N.; Inganäs, O. Bio Based Batteries. *Adv. Energy Mater.* 2021, *2003713*, 1–12. 10.1002/aenm.202003713

[149] Espinoza-Acosta, J. L.; Torres-Chávez, P. I.; Olmedo-Martínez, J. L.; Vega-Rios, A.; Flores-Gallardo, S.; Zaragoza-Contreras, E. A. Lignin in Storage and Renewable Energy Applications: A Review. *J. Energy Chem.* 2018, *27*, 1422–1438. 10.1016/j.jechem.2018.02.015

[150] Baloch, M.; Labidi, J. Lignin Biopolymer: The Material of Choice for Advanced Lithium-Based Batteries. *RSC Adv.* 2021, *11*, 23644–23653. 10.1039/d1ra02611a

[151] Song, A.; Huang, Y.; Zhong, X.; Cao, H.; Liu, B.; Lin, Y.; Wang, M.; Li, X. Gel Polymer Electrolyte with High Performances Based on Pure Natural Polymer Matrix of Potato Starch Composite Lignocellulose. *Electrochim. Acta* 2017, *245*, 981–992. 10.1016/j.electacta.2017.05.176

[152] Gong, S. D.; Huang, Y.; Cao, H. J.; Lin, Y. H.; Li, Y.; Tang, S. H.; Wang, M. S.; Li, X. A Green and Environment-Friendly Gel Polymer Electrolyte with Higher Performances Based on the Natural Matrix of Lignin. *J. Power Sources* 2016, *307*, 624–633. 10.1016/j.jpowsour.2016.01.030

[153] Liu, B.; Huang, Y.; Cao, H.; Song, A.; Lin, Y.; Wang, M.; Li, X. A High-Performance and Environment-Friendly Gel Polymer Electrolyte for Lithium Ion Battery Based on Composited Lignin Membrane. *J. Solid State Electrochem.* 2018, *22*, 807–816. 10.1007/s10008-017-3814-x

[154] Wang, S.; Zhang, L.; Wang, A.; Liu, X.; Chen, J.; Wang, Z.; Zeng, Q.; Zhou, H. H.; Jiang, X.; Zhang, L. Polymer-Laden Composite Lignin-Based Electrolyte Membrane for High-Performance Lithium Batteries. *ACS Sustain. Chem. Eng.* 2018, *6*, 14460–14469. 10.1021/acssuschemeng.8b03117

[155] Deng, X.; Huang, Y.; Song, A.; Liu, B.; Yin, Z.; Wu, Y.; Lin, Y.; Wang, M.; Li, X.; Cao, H. Gel Polymer Electrolyte with High Performances Based on Biodegradable Polymer Polyvinyl Alcohol Composite Lignocellulose. *Mat. Chem. Phys.* 2019, *229*, 232–241. 10.1016/j.matchemphys.2019.03.014

[156] Xiong, W.; Yang, D.; Hoang, T. K. A.; Ahmed, M.; Zhi, J.; Qiu, X.; Chen, P. Controlling the Sustainability and Shape Change of the Zinc Anode in Rechargeable Aqueous Zn/LiMn2O4 Battery. *Energy Storage Mater.* 2018, *15*, 131–138. 10.1016/j.ensm.2018.03.023

[157] Ren, W.; Huang, Y.; Xu, X.; Liu, B.; Li, S.; Luo, C.; Li, X.; Wang, M.; Cao, H. Gel Polymer Electrolyte with High Performances Based on Polyacrylonitrile Composite Natural Polymer of Lignocellulose in Lithium Ion Battery. *J. Mater. Sci.* 2020, *55*, 12249–12263. 10.1007/s10853-020-04888-w

[158] Baroncini, E. A.; Stanzione, J. F. Incorporating Allylated Lignin-Derivatives in Thiol-Ene Gel-Polymer Electrolytes. *Intl. J. Biol. Macromol.* 2018, *113*, 1041–1051. 10.1016/j.ijbiomac.2018.02.160

[159] Zhao, M.; Wang, J.; Chong, C.; Yu, X.; Wang, L.; Shi, Z. An Electrospun Lignin/Polyacrylonitrile Nonwoven Composite Separator with High Porosity and Thermal Stability for Lithium-Ion Batteries. *RSC Adv.* 2015, *5*, 101115–101120. 10.1039/x0xx00000x

[160] Liu, H.; Mulderrig, L.; Hallinan, D.; Chung, H. Lignin-Based Solid Polymer Electrolytes: Lignin-Graft-Poly(Ethylene Glycol). *Macromol. Rapid. Commun.* 2021, *42*, 1–6. 10.1002/marc.202000428

[161] Uddin, M. J.; Alaboina, P. K.; Zhang, L.; Cho, S. J. A Low-Cost, Environment-Friendly Lignin-Polyvinyl Alcohol Nanofiber Separator Using a Water-Based Method for Safer and Faster Lithium-Ion Batteries. *Mat. Sci. Eng.* 2017, *223*, 84–90. 10.1016/j.mseb.2017.05.004

[162] Domínguez-Robles, J.; Sánchez, R.; Díaz-Carrasco, P.; Espinosa, E.; García-Domínguez, M. T.; Rodríguez, A. Isolation and Characterization of Lignins from Wheat Straw: Application as Binder in Lithium Batteries. *Int. J. Biol. Macromol.* 2017, *104*, 909–918. 10.1016/j.ijbiomac.2017.07.015

[163] Lu, H.; Cornell, A.; Alvarado, F.; Behm, M.; Leijonmarck, S.; Li, J.; Tomani, P.; Lindbergh, G. Lignin as a Binder Material for Eco-Friendly Li-Ion Batteries. *Materials (Basel)* 2016, *9*, 1–17. 10.3390/ma9030127

[164] Ma, Y.; Chen, K.; Ma, J.; Xu, G.; Dong, S.; Chen, B.; Li, J.; Chen, Z.; Zhou, X.; Cui, G. A Biomass Based Free Radical Scavenger Binder Endowing a Compatible Cathode Interface for 5 V Lithium-Ion Batteries. *Energy Environ. Sci.viron. Sci.* 2019, *12*, 273–280. 10.1039/c8ee02555j

[165] Luo, C.; Du, L.; Wu, W.; Xu, H.; Zhang, G.; Li, S.; Wang, C.; Lu, Z.; Deng, Y. Novel Lignin-Derived Water-Soluble Binder for Micro Silicon Anode in Lithium-Ion Batteries. *ACS Sustain. Chem. Eng.* 2018, *6*, 12621–12629. 10.1021/acssuschemeng.8b01161

[166] Chen, T.; Zhang, Q.; Pan, J.; Xu, J.; Liu, Y.; Al-Shroofy, M.; Cheng, Y. T. Low-Temperature Treated Lignin as Both Binder and Conductive Additive for Silicon Nanoparticle Composite Electrodes in Lithium-Ion Batteries. *ACS Appl. Mater. Interfaces* 2016, *8*, 32341–32348. 10.1021/acsami.6b11500

[167] Winter, M.; Brodd, R. J. What Are Batteries, Fuel Cells, and Supercapacitors? *Chem. Rev.* 2004, *104*, 4245–4269. 10.1021/cr020730k

[168] Liu, Y.; Xue, J. S.; Zheng, T.; Dahn, J. R. Mechanism of Lithium Insertion in Hard Carbons Prepared by Pyrolysis of Epoxy Resins. *Carbon N. Y.* 1996, *34*, 193–200. 10.1016/0008-6223(96)00177-7

[169] Qi, W.; Shapter, J. G.; Wu, Q.; Yin, T.; Gao, G.; Cui, D. Nanostructured Anode Materials for Lithium-Ion Batteries: Principle, Recent Progress and Future Perspectives. *J. Mater. Chem. A* 2017, *5*, 19521–19540. 10.1039/c7ta05283a

[170] Luo, J.; Wu, C. E.; Su, L. Y.; Huang, S. S.; Fang, C. C.; Wu, Y. S.; Chou, J.; Wu, N. L. A Proof-of-Concept Graphite Anode with a Lithium Dendrite Suppressing Polymer Coating. *J. Power Sources* 2018, *406*, 63–69. 10.1016/j.jpowsour.2018.10.002

[171] Winter, M.; Besenhard, J. O.; Spahr, M. E.; Novak, P. Insertion Electrode Materials for Rechargeable Lithium Batteries. *Adv. Mater.* 1998, *10*, 725–763.

[172] Wang, A.; Kadam, S.; Li, H.; Shi, S.; Qi, Y. Review on Modeling of the Anode Solid Electrolyte Interphase (SEI) for Lithium-Ion Batteries. *npj Comput. Mater.* 2018, *4*, 1–26. 10.1038/s41524-018-0064-0

[173] Zhang, Y.; Du, N.; Yang, D. Designing Superior Solid Electrolyte Interfaces on Silicon Anodes for High-Performance Lithium-Ion Batteries. *Nanoscale* 2019, *11*, 19086–19104. 10.1039/c9nr05748j

[174] Pender, J. P.; Jha, G.; Youn, D. H.; Ziegler, J. M.; Andoni, I.; Choi, E. J.; Heller, A.; Dunn, B. S.; Weiss, P. S.; Penner, R. M.; Mullins, C. B. Electrode Degradation in Lithium-Ion Batteries. *ACS Nano* 2020, *14*, 1243–1295. 10.1021/acsnano.9b04365

[175] Chatterjee, S.; Jones, E. B.; Clingenpeel, A. C.; McKenna, A. M.; Rios, O.; McNutt, N. W.; Keffer, D. J.; Johs, A. Conversion of Lignin Precursors to Carbon Fibers with Nanoscale Graphitic Domains. *ACS Sustain. Chem. Eng.* 2014, *2*, 2002–2010. 10.1021/sc500189p

[176] Chatterjee, S.; Clingenpeel, A.; McKenna, A.; Rios, O.; Johs, A. Synthesis and Characterization of Lignin-Based Carbon Materials with Tunable Microstructure. *RSC Adv.* 2014, *4*, 4743–4753. 10.1039/c3ra46928j

[177] Rios, O.; Tenhaeff, W. E.; McGuire, M. A.; Menchhofer, P. A.; Johs, A.; More, K. L. Carbon Nanotube Enhanced Functional Carbon Fibers from Renewable Resources. In *Honolulu PRiME 2012, The Electrochemical Society*; 2012.

[178] Kamiyama, A.; Kubota, K.; Nakano, T.; Fujimura, S.; Shiraishi, S.; Tsukada, H.; Komaba, S. High-Capacity Hard Carbon Synthesized from Macroporous Phenolic Resin for Sodium-Ion and Potassium-Ion Battery. *ACS Appl. Energy Mater.* 2020, *3*, 135–140. 10.1021/acsaem.9b01972

[179] Ni, J.; Huang, Y.; Gao, L. A High-Performance Hard Carbon for Li-Ion Batteries and Supercapacitors Application. *J. Power Sources* 2013, *223*, 306–311. 10.1016/j.jpowsour.2012.09.047

[180] Abe, T. Secondary Batteries-Lithium Rechargeable Systems-Lithium Ion-Negative Electrodes: Carbon. *Encyclopedia of Electrochemical Power Sources* 2021, 192–197.

[181] Abe, T. Secondary Batteries-Lithium Rechargeable Systems-Lithium-Ion-Negative Electrode: Graphite. *Encyclopedia of Electrochemical Power Sources* 2009, 198–208.

[182] Massé, R. C.; Liu, C.; Li, Y.; Mai, L.; Cao, G. Energy Storage through Intercalation Reactions: Electrodes for Rechargeable Batteries. *Natl. Sci. Rev.* 2017, *4*, 26–53. 10.1093/nsr/nww093

[183] Li, R.; Huang, J.; Xu, Z.; Qi, H.; Cao, L.; Liu, Y.; Li, W.; Li, J. Controlling the Thickness of Disordered Turbostratic Nanodomains in Hard Carbon with Enhanced Sodium Storage Performance. *Energy Technol* 2018, *6*, 1080–1087. 10.1002/ente.201700674

[184] Wu, X.; Jiang, J.; Wang, C.; Liu, J.; Pu, Y.; Ragauskas, A.; Li, S.; Yang, B. Lignin-Derived Electrochemical Energy Materials and Systems. *Biofpr* 2020, *14*, 650–672. 10.1002/bbb.2083

[185] Chen, Y.; Li, X.; Zhou, L.; Mai, Y. W.; Huang, H. High-Performance Electrospun Nanostructured Composite Fiber Anodes for Lithium-Ion Batteries. In *Multifunctionality of Polymer Composites: Challenges and New Solutions*; Elsevier Inc., 2015; pp. 662–689. 10.1016/B978-0-323-26434-1.00021-0

[186] Baker, D. A.; Gellego, N. C.; Baker, F. S. On the Characterization and Spinning of an Organic-Purified Llignin toward the Manufacture of Low-Cost Carbon Fiber. *J. Appl. Polym. Sci.* 2012, *124*, 227–234. 10.1002/app

[187] Wang, S.; Bai, J.; Innocent, M. T.; Wang, Q.; Xiang, H.; Tang, J.; Zhu, M. Lignin-Based Carbon Fibers: Formation, Modification and Potential Applications. *Green Energy Environ.* 2021, In press. 10.1016/j.gee.2021.04.006

[188] Fitzer, E. Pan-Based Carbon Fibers-Present State and Trend of the Technology from the Viewpoint of Possibilities and Limits to Influence and to Control the Fiber Properties by the Process Parameters. *Carbon N. Y.* 1989, *27*, 621–645. 10.1016/0008-6223(89)90197-8

[189] Chen, J. C.; Harrison, I. R. Modification of Polyacrylonitrile (PAN) Carbon Fiber Precursor via Post-Spinning Plasticization and Stretching in Dimethyl Formamide (DMF). *Carbon N. Y.* 2002, *40*, 25–45. 10.1016/S0008-6223(01)00050-1

[190] Nowak, A. P.; Hagberg, J.; Leijonmarck, S.; Schweinebarth, H.; Baker, D.; Uhlin, A.; Tomani, P.; Lindbergh, G. Lignin-Based Carbon Fibers for Renewable and Multifunctional Lithium-Ion Battery Electrodes. *Holzforschung* 2018, *72*, 81–90. 10.1515/hf-2017-0044

[191] Uraki, Y.; Nakatani, A.; Kubo, S.; Sano, Y. Preparation of Activated Carbon Fibers with Large Specific Surface Area from Softwood Acetic Acid Lignin. *J. Wood Sci.* 2001, *47*, 465–469. 10.1007/BF00767899

[192] Liu, H. C.; Chien, A. T.; Newcomb, B. A.; Liu, Y.; Kumar, S. Processing, Structure, and Properties of Lignin- and CNT-Incorporated Polyacrylonitrile-Based Carbon Fibers. *ACS Sustain. Chem. Eng.* 2015, *3*, 1943–1954. 10.1021/acssuschemeng.5b00562

[193] Husman, G. *Development and Commercialization of a Novel Low-Cost Carbon Fiber, Presentation at 2012 DOE Hydrogen and Fuel Cells Program and Vehicle Technologies Program Annual Merit Review and Peer Evaluation Meeting*; 2012.

[194] Thunga, M.; Chen, K.; Grewell, D.; Kessler, M. R. Bio-Renewable Precursor Fibers from Lignin/Polylactide Blends for Conversion to Carbon Fibers. *Carbon N. Y.* 2014, *68*, 159–166. 10.1016/j.carbon.2013.10.075

[195] Worarutariyachai, T.; Chuangchote, S. Carbon Fibers Derived from Pure Alkali Lignin Fibers through Electrospinning with Carbonization. *BioResources* 2020, *15*, 2412–2427. 10.15376/biores.15.2.2412-2427

[196] Ciesielski, A.; Samorì, P. Graphene via Sonication Assisted Liquid-Phase Exfoliation. *Chem. Soc. Rev.* 2014, *43*, 381–398. 10.1039/c3cs60217f

[197] Liu, W.; Zhou, R.; Zhou, D.; Ding, G.; Soah, J. M.; Yue, C. Y.; Lu, X. Lignin-Assisted Direct Exfoliation of Graphite to Graphene in Aqueous Media and Its Application in Polymer Composites. *Carbon N. Y.* 2015, *83*, 188–197. 10.1016/j.carbon.2014.11.036

[198] Zhao, X.; Tao, C. an; Li, Y.; Chen, X.; Wang, J.; Gong, H. Preparation of Gel Polymer Electrolyte with High Lithium Ion Transference Number Using GO as Filler and Application in Lithium Battery. *Ionics (Kiel)* 2020, *26*, 4299–4309. 10.1007/s11581-020-03628-z

[199] Yang, M.; Hou, J. Membranes in Lithium Ion Batteries. *Membranes (Basel)* 2012, *2*, 367–383. 10.3390/membranes2030367

[200] Zhu, M.; Wu, J.; Wang, Y.; Song, M.; Long, L.; Siyal, S. H.; Yang, X.; Sui, G. Recent Advances in Gel Polymer Electrolyte for High-Performance Lithium Batteries. *J. Energy Chem.* 2019, *37*, 126–142. 10.1016/j.jechem.2018.12.013

[201] Young, W. S.; Kuan, W. F.; Epps, T. H. Block Copolymer Electrolytes for Rechargeable Lithium Batteries. *J. Polym. Sci. B Polym. Phys.* 2014, *52*, 1–16. 10.1002/polb.23404

[202] Yao, P.; Yu, H.; Ding, Z.; Liu, Y.; Lu, J.; Lavorgna, M.; Wu, J.; Liu, X. Review on Polymer-Based Composite Electrolytes for Lithium Batteries. *Front. Chem.* 2019, *7*, 1–17. 10.3389/fchem.2019.00522

[203] Choudhury, S.; Stalin, S.; Vu, D.; Warren, A.; Deng, Y.; Biswal, P.; Archer, L. A. Solid-State Polymer Electrolytes for High-Performance Lithium Metal Batteries. *Nat. Commun.* 2019, *10*, 1–8. 10.1038/s41467-019-12423-y

[204] Khurana, R.; Schaefer, J. L.; Archer, L. A.; Coates, G. W. Suppression of Lithium Dendrite Growth Using Cross-Linked Polyethylene/Poly(Ethylene Oxide) Electrolytes: A New Approach for Practical Lithium-Metal Polymer Batteries. *J. Am. Chem. Soc.* 2014, *136*, 7395–7402. 10.1021/ja502133j

[205] Song, A.; Huang, Y.; Liu, B.; Cao, H.; Zhong, X.; Lin, Y.; Wang, M.; Li, X.; Zhong, W. Gel Polymer Electrolyte Based on Polyethylene Glycol Composite Lignocellulose Matrix with Higher Comprehensive Performances. *Electrochim. Act.* 2017, *247*, 505–515. 10.1016/j.electacta.2017.07.048

[206] Cholewinski, A.; Si, P.; Uceda, M.; Pope, M.; Zhao, B. Polymer Binders: Characterization and Development toward Aqueous Electrode Fabrication for Sustainability. *Polymers (Basel)* 2021, *No. 4*, 1–20. 10.3390/polym13040631

[207] Versaci, D.; Nasi, R.; Zubair, U.; Amici, J.; Sgroi, M.; Dumitrescu, M. A.; Francia, C.; Bodoardo, S.; Penazzi, N. New Eco-Friendly Low-Cost Binders for Li-Ion Anodes. *J. Solid. State Electrochem.* 2017, *21*, 3429–3435. 10.1007/s10008-017-3665-5

[208] Jeong, D.; Shim, J.; Shin, H.; Lee, J. C. Sustainable Lignin-Derived Cross-Linked Graft Polymers as Electrolyte and Binder Materials for Lithium Metal Batteries. *ChemSusChem* 2020, *13*, 2642–2649. 10.1002/cssc.201903466

[209] Jeong, J.; Chun, J.; Lim, W. G.; Kim, W. B.; Jo, C.; Lee, J. Mesoporous Carbon Host Material for Stable Lithium Metal Anode. *Nanoscale* 2020, *12*, 11818–11824. 10.1039/d0nr02258f

[210] Dudney, N. J. Solid-State Thin-Film Rechargeable Batteries. *Mat. Sci. Eng.* 2005, *116*, 245–249. 10.1016/j.mseb.2004.05.045

[211] Tron, A.; Nosenko, A.; Park, Y. D.; Mun, J. Enhanced Ionic Conductivity of the Solid Electrolyte for Lithium-Ion Batteries. *J. Solid State Chem.* 2018, *258* (November 2017), 467–470. 10.1016/j.jssc.2017.11.020

[212] Gnedenkov, S. V.; Opra, D. P.; Sinebryukhov, S. L.; Tsvetnikov, A. K.; Ustinov, A. Y.; Sergienko, V. I. Hydrolysis Lignin: Electrochemical Properties of the Organic Cathode Material for Primary Lithium Battery. *J. Ind. Eng. Chem.* 2014, *20*, 903–910. 10.1016/j.jiec.2013.06.021

[213] Tao, L.; Xu, Z.; Kuai, C.; Zheng, X.; Wall, C. E.; Jiang, C.; Esker, A. R.; Zheng, Z.; Lin, F. Flexible Lignin Carbon Membranes with Surface Ozonolysis to Host Lean Lithium Metal Anodes for Nickel-Rich Layered Oxide Batteries. *Energy Storage Mater* 2020, *24*, 129–137. 10.1016/j.ensm.2019.08.027

[214] Lai, Y. H.; Kuo, Y. T.; Lai, B. Y.; Lee, Y. C.; Chen, H. Y. Improving Lithium-Sulfur Battery Performance with Lignin Reinforced MWCNT Protection Layer. *Int. J. Energy Res.* 2019, *43*, 5803–5811. 10.1002/er.4680

[215] Liu, T.; Sun, S.; Song, W.; Sun, X.; Niu, Q.; Liu, H.; Ohsaka, T.; Wu, J. A Lightweight and Binder-Free Electrode Enabled by Lignin Fibers@carbon-Nanotubes and Graphene for Ultrastable Lithium-Sulfur Batteries. *J. Mater. Chem. A* 2018, *6*, 23486–23494. 10.1039/C8TA08521H

[216] Zhu, K.; Wang, C.; Chi, Z.; Ke, F.; Yang, Y.; Wang, A.; Wang, W.; Miao, L. How Far Away Are Lithium-Sulfur Batteries From Commercialization? *Front. Energy Res.* 2019, *7*, 1–12. 10.3389/fenrg.2019.00123

[217] Zhang, Z.; Yi, S.; Wei, Y.; Bian, H.; Wang, R.; Min, Y. Lignin Nanoparticle-Coated Celgard Separator for High-Performance Lithium-Sulfur Batteries. *Polymers (Basel)* 2019, *11*, 1–10. 10.3390/polym11121946

[218] Song, A.; Huang, Y.; Zhong, X.; Cao, H.; Liu, B.; Lin, Y.; Wang, M.; Li, X. Novel Lignocellulose Based Gel Polymer Electrolyte with Higher Comprehensive Performances for Rechargeable Lithium–Sulfur Battery. *J. Mem. Sci.* 2018, *556*, 203–213. 10.1016/j.memsci.2018.04.003

[219] Cao, Y.; Xiao, L.; Sushko, M. L.; Wang, W.; Schwenzer, B.; Xiao, J.; Nie, Z.; Saraf, L. V.; Yang, Z.; Liu, J. Sodium Ion Insertion in Hollow Carbon Nanowires for Battery Applications. *Nano Lett* 2012, *12*, 3783–3787. 10.1021/nl3016957

[220] Shannon, R. D. Revised Effective Ionic Radii and Systematic Studies of Interatomic Distances in Halides and Chalcogenides. *Acta Cryst* 1976, *A32*, 751–767. 10.1107/S0567739476001551

[221] Shen, F.; Luo, W.; Dai, J.; Yao, Y.; Zhu, M.; Hitz, E.; Tang, Y.; Chen, Y.; Sprenkle, V. L.; Li, X.; Hu, L. Ultra-Thick, Low-Tortuosity, and Mesoporous Wood Carbon Anode for High-Performance Sodium-Ion Batteries. *Adv. Energy Mater.* 2016, *6*, 1–7. 10.1002/aenm.201600377

[222] Dou, X.; Hasa, I.; Hekmatfar, M.; Diemant, T.; Behm, R. J.; Buchholz, D.; Passerini, S. Pectin, Hemicellulose, or Lignin? Impact of the Biowaste Source on the Performance of Hard Carbons for Sodium-Ion Batteries. *ChemSusChem* 2017, *10*, 2668–2676. 10.1002/cssc.201700628

[223] Lin, X.; Liu, Y.; Tan, H.; Zhang, B. Advanced Lignin-Derived Hard Carbon for Na-Ion Batteries and a Comparison with Li and K Ion Storage. *Carbon N. Y.* 2020, *157*, 316–323. 10.1016/j.carbon.2019.10.045

[224] Saurel, D.; Orayech, B.; Xiao, B.; Carriazo, D.; Li, X. From Charge Storage Mechanism to Performance: A Roadmap toward High Specific Energy Sodium-Ion Batteries through Carbon Anode Optimization. *Adv. Energy Mater.* 2018, *8*, 1–33. 10.1002/aenm.201703268

[225] Kizzire, D. G.; Richter, A. M.; Harper, D. P.; Keffer, D. J. Lithium and Sodium Ion Binding Mechanisms and Diffusion Rates in Lignin-Based Hard Carbon Models. *ACS Omega* 2021, *6*, 19883–19892. 10.1021/acsomega.1c02787

[226] Yang, L.; Hu, M.; Zhang, H.; Yang, W.; Lv, R. Pore Structure Regulation of Hard Carbon: Towards Fast and High-Capacity Sodium-Ion Storage. *J. Colloid Interface Sci.* 2020, *566*, 257–264. 10.1016/j.jcis.2020.01.085

[227] Ding, R.; Wu, H.; Thunga, M.; Bowler, N.; Kessler, M. R. Processing and Characterization of Low-Cost Electrospun Carbon Fibers from Organosolv Lignin/ Polyacrylonitrile Blends. *Carbon N. Y.* 2016, *100*, 126–136. 10.1016/j.carbon.2015.12.078

[228] Mahato, B. K. Lead-Acid Battery Expander: I. Electrochemical Evaluation Techniques. *J. Electrochem. Soc.* 1980, *127*, 1679–1687. 10.1149/1.2129980

[229] Mahato, B. K.; Laird, E. C. Battery Paste for Lead Acid Storage Batteries.Pdf, 1982.

[230] *Lead acid battery*. https://en.wikipedia.org/wiki/Lead–acid_battery.

[231] Boden, D. P. Lead Acid Battery Expanders with Improved Life at High Temperatures, 2008.

[232] Marichi, R. B.; Sahu, V.; Sharma, R. K.; Singh, G. Efficient, Sustainable, and Clean Energy Storage in Supercapacitors Using Biomass-Derived Carbon Materials. In *Handbook of Ecomaterials* 2018, 1–26. 10.1007/978-3-319-48281-1_155-1

[233] Hérou, S.; Bailey, J. J.; Kok, M.; Schlee, P.; Jervis, R.; Brett, D. J. L.; Shearing, P. R.; Ribadeneyra, M. C.; Titirici, M. High-Density Lignin-Derived Carbon Nanofiber Supercapacitors with Enhanced Volumetric Energy Density. *Adv. Sci.* 2021, *2*, 1–11. 10.1002/advs.202100016

[234] Da Silva, L. M.; Cesar, R.; Moreira, C. M. R.; Santos, J. H. M.; De Souza, L. G.; Pires, B. M.; Vicentini, R.; Nunes, W.; Zanin, H. Reviewing the Fundamentals of Supercapacitors and the Difficulties Involving the Analysis of the Electrochemical Findings Obtained for Porous Electrode Materials. *Energy Storage Mater.* 2020, *27*, 555–590. 10.1016/j.ensm.2019.12.015

[235] Poonam; Sharma, K.; Arora, A.; Tripathi, S. K. Review of Supercapacitors: Materials and Devices. *J. Energy Storage* 2019, *21*, 801–825. 10.1016/j.est.2019.01.010

[236] De, B.; Banerjee, S.; Pal, T.; Verma, K. D.; Tyagi, A.; Manna, P. K.; Kar, K. K. *Handbook of Nanocomposite Supercapacitor Materials II*; Kar, K. K. Ed.; Springer, 2020; Vol. 302. 10.1007/978-3-030-52359-6_15

[237] Zhang, W.; Zhao, M.; Liu, R.; Wang, X.; Lin, H. Hierarchical Porous Carbon Derived from Lignin for High Performance Supercapacitor. *Colloids Surf. A Physicochem. Eng. Asp.* 2015, *484*, 518–527. 10.1016/j.colsurfa.2015.08.030

[238] Najib, S.; Erdem, E. Current Progress Achieved in Novel Materials for Supercapacitor Electrodes: Mini Review. *Nanoscale Adv.* 2019, *1*, 2817–2827. 10.1039/c9na00345b

[239] Wang, F.; Ouyang, D.; Zhou, Z.; Page, S. J.; Liu, D.; Zhao, X. Lignocellulosic Biomass as Sustainable Feedstock and Materials for Power Generation and Energy Storage. *J. Energy Chem.* 2021, *57*, 247–280. 10.1016/j.jechem.2020.08.060

[240] Dessie, Y.; Admassie, S. Electrochemical Study of Conducting Polymer/Lignin Composites. *Orient. J. Chem.* 2013, *29*, 1359–1369. 10.13005/ojc/290411

[241] Dai, Z.; Ren, P. G.; He, W.; Hou, X.; Ren, F.; Zhang, Q.; Jin, Y. L. Boosting the Electrochemical Performance of Nitrogen-Oxygen Co-Doped Carbon Nanofibers Based Supercapacitors through Esterification of Lignin Precursor. *Renew. Energy* 2020, *162*, 613–623. 10.1016/j.renene.2020.07.152

[242] Adamassie, S.; Elfwing, A.; Jager, E. W. H.; Bao, Q.; Inganäs, O. A Renewable Biopolymer Cathode with Multivalent Metal Ions for Enhanced Charge Storage. *J.Mater.Chem.A* 2014, *2*, 1974–1979.

[243] Jeon, J. W.; Zhang, L.; Lutkenhaus, J. L.; Laskar, D. D.; Lemmon, J. P.; Choi, D.; Nandasiri, M. I.; Hashmi, A.; Xu, J.; Motkuri, R. K.; Fernandez, C. A.; Liu, J.; Tucker, M. P.; McGrail, P. B.; Yang, B.; Nune, S. K. Controlling Porosity in Lignin-Derived Nanoporous Carbon for Supercapacitor Applications. *ChemSusChem* 2015, *8* (3), 428–432. 10.1002/cssc.201402621

[244] Zhang, Y.; Yu, B.; Zhang, J.; Ding, X.; Zeng, J.; Chen, M.; Wang, C. Design and Preparation of Lignin-Based Hierarchical Porous Carbon Microspheres by High Efficient Activation for Electric Double Layer Capacitors. *ChemElectroChem* 2018, *5*, 2142–2149. 10.1002/celc.201800488

[245] Zhou, D. D.; Du, Y. J.; Wang, Y. G.; Wang, C. X.; Xia, Y. Y. Ordered Hierarchical Mesoporous /Microporous Carbon with Optimized Pore Structure for Supercapacitors. *J. Mater. Chem. A* 2013, *1*, 1192–1200.

[246] Ruiz-Rosas, R.; Bedia, J.; Lallave, M.; Loscertales, I. G.; Barrero, A.; Rodríguez-Mirasol, J.; Cordero, T. The Production of Submicron Diameter Carbon Fibers by the Electrospinning of Lignin. *Carbon N. Y.* 2010, *48*, 696–705. 10.1016/j.carbon.2009.10.014

[247] Hou, J.; Cao, C.; Idrees, F.; Ma, X. Hierarchical Porous Nitrogen-Doped Carbon Nanosheets Derived from Silk for Ultrahigh-Capacity Battery Anodes and Supercapacitors. *ACS Nano* 2015, *9*, 2556–2564. 10.1021/nn506394r

[248] Saha, D.; Payzant, E. A.; Kumbhar, A. S.; Naskar, A. K. Sustainable Mesoporous Carbons as Storage and Controlled-Delivery Media for Functional Molecules. *ACS Appl. Mater. Interfaces* 2013, *5*, 5868–5874. 10.1021/am401661f

[249] Ruiz-Rosas, R.; Valero-Romero, M. J.; Salinas-Torres, D.; Rodríguez-Mirasol, J.; Cordero, T.; Morallón, E.; Cazorla-Amorós, D. Electrochemical Performance of Hierarchical Porous Carbon Materials Obtained from the Infiltration of Lignin into Zeolite Templates. *ChemSusChem* 2014, *7*, 1458–1467. 10.1002/cssc.201301408

[250] Youe, W. J.; Kim, S. J.; Lee, S. M.; Chun, S. J.; Kang, J.; Kim, Y. S. MnO2-Deposited Lignin-Based Carbon Nanofiber Mats for Application as Electrodes in Symmetric Pseudocapacitors. *Intl. J. Biol. Macromol.* 2018, *112*, 943–950. 10.1016/j.ijbiomac.2018.02.048

[251] Yu, B.; Gele, A.; Wang, L. Iron Oxide/Lignin-Based Hollow Carbon Nanofibers Nanocomposite as an Application Electrode Materials for Supercapacitors. *Intl. J. Biol. Macromol.* 2018, *118*, 478–484. 10.1016/j.ijbiomac.2018.06.088

[252] Wang, J.; Dong, S.; Ding, B.; Wang, Y.; Hao, X.; Dou, H.; Xia, Y.; Zhang, X. Pseudocapacitive Materials for Electrochemical Capacitors: From Rational Synthesis to Capacitance Optimization. *Natl. Sci. Rev.* 2017, *4*, 71–90. 10.1093/nsr/nww072

[253] Milczarek, G.; Inganäs, O. Renewable Cathode Materials from Biopolymer/Conjugated Polymer Interpenetrating Networks. *Science (80-.)* 2012, *335*, 1468–1472.

[254] Milczarek, G. Preparation and Characterization of a Lignin Modified Electrode. *Electroanal.* 2007, *19*, 1411–1414. 10.1002/elan.200703870

[255] Milczarek, G. Lignosulfonate-Modified Electrodes: Electrochemical Properties and Electrocatalysis of NADH Oxidation. *Langmuir* 2009, *25*, 10345–10353. 10.1021/la9008575

[256] Liu, L.; Solin, N.; Inganäs, O. Bio Based Batteries. *Adv. Energy Mat.* 2021, *2003713*, 1–12. 10.1002/aenm.202003713

[257] Ajjan, F. N.; Jafari, M. J.; Rebis, T.; Ederth, T.; Inganäs, O. Spectroelectrochemical Investigation of Redox States in a Polypyrrole/Lignin Composite Electrode. *J. Mater. Chem. A* 2015, *3*, 12927–12937. 10.1039/C5TA00788G

[258] Admassie, S.; Ajjan, F. N.; Elfwing, A.; Inganäs, O. Biopolymer Hybrid Electrodes for Scalable Electricity Storage. *Mater. Horiz.* 2016, *3*, 174–185. 10.1039/c5mh00261c

[259] Häringer, D.; Novák, P.; Haas, O.; Piro, B.; Pham, M. Poly(5-amino-1,4-naphthoquinone), a Novel Lithium-Inserting Electroactive Polymer with High Specific Charge. *J. Electrochem. Soc.* 1999, *146*, 2393–2396. 10.1149/1.1391947

[260] Song, H. K.; Palmore, G. T. R. Redox-Active Polypyrrole: Toward Polymer-Based Batteries. *Adv. Mater.* 2006, *18*, 1764–1768. 10.1002/adma.200600375

[261] Naoi, K.; Suematsu, S.; Manago, A. Electrochemistry of Poly(1,5-Diaminoanthraquinone) and Its Application in Electrochemical Capacitor Materials. *J. Electrochem. Soc.* 2000, *147*, 420–426. 10.1149/1.1393212

[262] Hashmi, S. A.; Suematsu, S.; Naoi, K. All Solid-State Redox Supercapacitors Based on Supramolecular 1,5-Diaminoanthraquinone Oligomeric Electrode and Polymeric Electrolytes. *J. Power Sources* 2004, *137*, 145–151. 10.1016/j.jpowsour.2004.05.007

[263] Yoneyama, H.; Li, Y.; Kuwabata, S. Charge-discharge Characteristics of Polypyrrole Films Containing Charge-Discharge Characteristics of Polypyrrole Films Containing Incorporated Anthraquinone-1-Sulfonate. *J. Electrochem. Soc.* 1992, *139*, 28–32.

[264] Schiavon, G.; Sitran, S.; Zotti, G. A Simple Two-Band Electrode for in-Situ Conductivity Measurements of Polyconjugated Conducting Polymers. *Synth. Met.* 1989, *32*, 209–217.

[265] Navarro-Suarez, A. M.; Casado, N.; Carretero-Gonzalez, J.; Mecerreyes, D.; Rojo, T. Full-Cell Quinone/Hydroquinone Supercapacitors Based on Partially Reduced Graphite Oxide and Lignin/PEDOT Electrodes. *J. Mater. Chem. A* 2017, *5*, 7137–7143. 10.1039/c7ta00527j

[266] Chua, C. K.; Pumera, M. Chemical Reduction of Graphene Oxide: A Synthetic Chemistry Viewpoint. *Chem. Soc. Rev.* 2014, *43*, 291–312. 10.1039/c3cs60303b

[267] Park, J. H.; Rana, H. H.; Lee, J. Y.; Park, H. S. Renewable Flexible Supercapacitors Based on All-Lignin-Based Hydrogel Electrolytes and Nanofiber Electrodes. *J. Mater. Chem. A* 2019, *7*, 16962–16968. 10.1039/c9ta03519b

[268] Liu, T.; Ren, X.; Zhang, J.; Liu, J.; Ou, R.; Guo, C.; Yu, X.; Wang, Q.; Liu, Z. Highly Compressible Lignin Hydrogel Electrolytes via Double-Crosslinked Strategy for Superior Foldable Supercapacitors. *J. Power Sources* 2020, *449*, 227532. 10.1016/j.jpowsour.2019.227532

[269] Modestino, M. A.; Paul, D. K.; Dishari, S.; Petrina, S. A.; Allen, F. I.; Hickner, M. A.; Karan, K.; Segalman, R. A.; Weber, A. Z. Self-Assembly and Transport Limitations in Confined Nafion Films. *Macromolecules* 2013, *46*, 867–873. 10.1021/ma301999a

[270] Dishari, S. K.; Hickner, M. A. Confinement and Proton Transfer in Nafion Thin Films. *Macromolecules* 2013, *46*, 413–421. 10.1021/ma3011137

[271] Dishari, S. K.; Hickner, M. A. Antiplasticization and Water Uptake of Nafion® Thin Films. *ACS Macro Lett.* 2012, *1*, 291–295.

[272] Dishari, S. K. Current Understanding of Proton Conduction in Confined Ionomeric Systems. *J. Postdoc. Res.* 2014, 2, 30–39. 10.14304/SURYA.JPR.V2N4.3

[273] Dishari, S. K.; Rumble, C. A.; Maroncelli, M.; Dura, J. A.; Hickner, M. A. Unraveling the Complex Hydration Behavior of Ionomers under Thin Film Confinement. *J. Phys. Chem. C* 2018, *122*, 3471–3481. 10.1021/acsmacrolett.1c00110

[274] Farzin, S.; Zamani, E.; Dishari, S. K. Unraveling Depth-Specific Ionic Conduction and Stiffness Behavior across Ionomer Thin Films and Bulk Membranes. *ACS Macro Lett.* 2021, *10*, 791–798. 10.1021/acsmacrolett.1c00110

[275] Farzin, S.; Sarella, A.; Yandrasits, M. A.; Dishari, S. K. Fluorocarbon-Based Ionomers with Single Acid and Multiacid Side Chains at Nanothin Interfaces. *J. Phys. Chem. C* 2019, *123*, 30871–30884. 10.1021/acs.jpcc.9b10015

[276] Karan, K. Interesting Facets of Surface, Interfacial, and Bulk Characteristics of Perfluorinated Ionomer Films. *Langmuir* 2019, *35*, 13489–13520. 10.1021/acs.langmuir.8b03721

[277] Kusoglu, A.; Weber, A. Z. New Insights into Perfluorinated Sulfonic Acid Ionomers. *Chem. Rev.* 2017, *117*, 987–1104.

[278] Holdcroft, S. Fuel Cell Catalyst Layers: A Polymer Science Perspective. *Chem. Mater.* 2014, *26*, 381–393. 10.1021/cm401445h

[279] Chatterjee, S.; Zamani, E.; Farzin, S.; Obewhere, O. A.; Johnson, T.; Dishari, S. K. Molecular-Level Control Over Ionic Conduction and Ionic Current Direction by Designing Macrocycle-Based Ionomers. *JACS Au* 2022, *2*, 1144–1159. 10.1021/jacsau.2c00143

[280] Venugopalan, G.; Bhattacharya, D.; Kole, S.; Ysidron, C.; Angelopoulou, P. P.; Sakellariou, G.; Arges, C. G. Correlating High Temperature Thin Film Ionomer Electrode Binder Properties to Hydrogen Pump Polarization. *Mater. Adv.* 2021, *2*, 4228–4234. 10.1039/d1ma00208b

[281] Barati, S.; Abdollahi, M.; Khoshandam, B.; Mehdipourghazi, M. Highly Proton Conductive Porous Membranes Based on Polybenzimidazole/ Lignin Blends for High Temperatures Proton Exchange Membranes: Preparation, Characterization and Morphology- Proton Conductivity Relationship. *Intl. J. Hydrog. Energy* 2018, *43*, 19681–19690. 10.1016/j.ijhydene.2018.08.191

[282] Demir, M.; Farghaly, A. A.; Decuir, M. J.; Collinson, M. M.; Gupta, R. B. Supercapacitance and Oxygen Reduction Characteristics of Sulfur Self-Doped Micro/Mesoporous Bio-Carbon Derived from Lignin. *Mat. Chem. Phys.* 2018, *216*, 508–516. 10.1016/j.matchemphys.2018.06.008

[283] Kannari, N.; Takigami, M.; Maie, T.; Honda, H.; Kusadokoro, S.; Ozaki, J. Nanoshell-Containing Carbon Cathode Catalyst for Proton Exchange Membrane Fuel Cell from Herbaceous Plants Lignin. *Smart Grid Renew. Energy* 2013, *4*, 10–15. 10.4236/sgre.2013.47a002

[284] Lv, Q.; Si, W.; He, J.; Sun, L.; Zhang, C.; Wang, N.; Yang, Z.; Li, X.; Wang, X.; Deng, W.; Long, Y.; Huang, C.; Li, Y. Selectively Nitrogen-Doped Carbon Materials as Superior Metal-Free Catalysts for Oxygen Reduction. *Nat. Commun.* 2018, *9*, 1–11. 10.1038/s41467-018-05878-y

[285] Shen, Y.; Li, Y.; Yang, G.; Zhang, Q.; Liang, H.; Peng, F. Lignin Derived Multi-Doped (N, S, Cl) Carbon Materials as Excellent Electrocatalyst for Oxygen Reduction Reaction in Proton Exchange Membrane Fuel Cells. *J. Energy Chem.* 2020, *44*, 106–114. 10.1016/j.jechem.2019.09.019

[286] Mai, Z.; Zhang, H.; Li, X.; Xiao, S.; Zhang, H. Nafion/Polyvinylidene Fluoride Blend Membranes with Improved Ion Selectivity for Vanadium Redox Flow Battery Application. *J. Power Sources* 2011, *196*, 5737–5741. 10.1016/j.jpowsour.2011.02.048

[287] Xi, J.; Wu, Z.; Teng, X.; Zhao, Y.; Chen, L.; Qiu, X. Self-Assembled Polyelectrolyte Multilayer Modified Nafion Membrane with Suppressed Vanadium Ion Crossover for Vanadium Redox Flow Batteries. *J. Mater. Chem.* 2008, *18*, 1232–1238. 10.1039/b718526j

[288] Luo, Q.; Zhang, H.; Chen, J.; Qian, P.; Zhai, Y. Modification of Nafion Membrane Using Interfacial Polymerization for Vanadium Redox Flow Battery Applications. *J. Mem. Sci.* 2008, *311*, 98–103. 10.1016/j.memsci.2007.11.055

[289] Ye, J.; Cheng, Y.; Sun, L.; Ding, M.; Wu, C.; Yuan, D.; Zhao, X.; Xiang, C.; Jia, C. A Green SPEEK/Lignin Composite Membrane with High Ion Selectivity for Vanadium Redox Flow Battery. *J. Mem. Sci.* 2019, *572*, 110–118. 10.1016/j.memsci.2018.11.009

[290] Ye, J.; Yuan, D.; Ding, M.; Long, Y.; Long, T.; Sun, L.; Jia, C. A Cost-Effective Nafion/Lignin Composite Membrane with Low Vanadium Ion Permeation for High Performance Vanadium Redox Flow Battery. *J. Power Sources* 2021, *482*, 229023. 10.1016/j.jpowsour.2020.229023

[291] Vivo-Vilches, J. F.; Celzard, A.; Fierro, V.; Devin-Ziegler, I.; Brosse, N.; Dufour, A.; Etienne, M. Lignin-Based Carbon Nanofibers as Electrodes for Vanadium Redox Couple Electrochemistry. *Nanomater.* 2019, *9*, 1–13. 10.3390/nano9010106

[292] Ribadeneyra, M. C.; Grogan, L.; Au, H.; Schlee, P.; Herou, S.; Neville, T.; Cullen, P. L.; Kok, M. D. R.; Hosseinaei, O.; Danielsson, S.; Tomani, P.; Titirici, M. M.; Brett, D. J. L.; Shearing, P. R.; Jervis, R.; Jorge, A. B. Lignin-Derived Electrospun Freestanding Carbons as Alternative Electrodes for Redox Flow Batteries. *Carbon N. Y.* 2020, *157*, 847–856. 10.1016/j.carbon.2019.11.015

[293] Mukhopadhyay, A.; Hamel, J.; Katahira, R.; Zhu, H. Metal-Free Aqueous Flow Battery with Novel Ultrafiltered Lignin as Electrolyte. *ACS Sustain. Chem. Eng.* 2018, *6* (4), 5394–5400. 10.1021/acssuschemeng.8b00221

[294] Budnyak, T. M.; Slabon, A.; Sipponen, M. H. Lignin–Inorganic Interfaces: Chemistry and Applications from Adsorbents to Catalysts and Energy Storage Materials. *ChemSusChem* 2020, *13* (17), 4344–4355. 10.1002/cssc.202000216

[295] Yuan, D.; Manalastas, W.; Zhang, L.; Chan, J. J.; Meng, S.; Chen, Y.; Srinivasan, M. Lignin@ Nafion Membranes Forming Zn Solid–Electrolyte Interfaces Enhance the Cycle Life for Rechargeable Zinc-Ion Batteries. *ChemSusChem* 2019, *12*, 4889–4900. 10.1002/cssc.201901409

[296] Lahiri, A.; Yang, L.; Höfft, O.; Endres, F. Biodegradable Zn-Ion Battery with a Lignin Composite Electrode and Bio-Ionic Liquid Based Electrolyte: Possible: In Situ Energy Generation by Lignin Electrocatalysis. *Mater. Adv.* 2021, *2*, 2676–2683. 10.1039/d0ma00954g

[297] Peng, X.; Zhang, L.; Chen, Z.; Zhong, L.; Zhao, D.; Chi, X.; Zhao, X.; Li, L.; Lu, X.; Leng, K.; Liu, C.; Liu, W.; Tang, W.; Loh, K. P. Hierarchically Porous Carbon Plates Derived from Wood as Bifunctional ORR/OER Electrodes. *Adv. Mater.* 2019, *31*, 1–7. 10.1002/adma.201900341

[298] Li, P.; Wang, H.; Fan, W.; Huang, M.; Shi, J.; Shi, Z.; Liu, S. Salt Assisted Fabrication of Lignin-Derived Fe, N, P, S Codoped Porous Carbon as Trifunctional Catalyst for Zn-Air Batteries and Water-Splitting Devices. *Chem. Eng. J.* 2021, *421*, 1–10. 10.1016/j.cej.2021.129704

[299] Qiao, Y.; Zhang, C.; Kong, F.; Zhao, Q.; Kong, A.; Shan, Y. Activated Biochar Derived from Peanut Shells as the Electrode Materials with Excellent Performance in Zinc-Air Battery and Supercapacitance. *Waste Manag.* 2021, *125*, 257–267. 10.1016/j.wasman.2021.02.057

[300] Li, X.; Lv, Y.; Pan, D. Pt Catalysts Supported on Lignin-Based Carbon Dots for Methanol Electro-Oxidation. *Colloids Surf. A* 2019, *569*, 110–118. 10.1016/j.colsurfa.2019.02.051

[301] García-Mateos, F. J.; Cordero-Lanzac, T.; Berenguer, R.; Morallón, E.; Cazorla-Amorós, D.; Rodríguez-Mirasol, J.; Cordero, T. Lignin-Derived Pt Supported Carbon (Submicron)Fiber Electrocatalysts for Alcohol Electro-Oxidation. *Appl. Cat. B Environ.* 2017, *211*, 18–30. 10.1016/j.apcatb.2017.04.008

[302] Wang, W.; Shan, B.; Zhu, L.; Xie, C.; Liu, C.; Cui, F. Anatase Titania Coated CNTs and Sodium Lignin Sulfonate Doped Chitosan Proton Exchange Membrane for DMFC Application. *Carbohydr. Polym.* 2018, *187*, 35–42. 10.1016/j.carbpol.2018.01.078

[303] Zhao, X.; Zhu, J. Y. Efficient Conversion of Lignin to Electricity Using a Novel Direct Biomass Fuel Cell Mediated by Polyoxometalates at Low Temperatures. *ChemSusChem* 2016, *9*, 197–207. 10.1002/cssc.201501446

[304] Nemoto, J.; Horikawa, M.; Ohnuki, K.; Shibata, T.; Ueno, H.; Hoshino, M.; Kaneko, M. Biophotofuel Cell (BPFC) Generating Electrical Power Directly from Aqueous Solutions of Biomass and Its Related Compounds While Photodecomposing and Cleaning. *J. Appl. Electrochem.* 2007, *37*, 1039–1046. 10.1007/s10800-007-9345-x

[305] Sharma, R. K.; Mukhopadhyay, D.; Gupta, P. Microbial Fuel Cell-Mediated Lignin Depolymerization: A Sustainable Approach. *J. Chem. Technol. Biotechnol.* 2019, *94*, 927–932. 10.1002/jctb.5841

[306] Shewa, W. A.; Lalman, J. A.; Chaganti, S. R.; Heath, D. D. Electricity Production from Lignin Photocatalytic Degradation Byproducts. *Energy* 2016, *111*, 774–784. 10.1016/j.energy.2016.05.007

[307] Lam, C. H.; Lowe, C. B.; Li, Z.; Longe, K. N.; Rayburn, J. T.; Caldwell, M. A.; Houdek, C. E.; Maguire, J. B.; Saffron, C. M.; Miller, D. J.; Jackson, J. E. Electrocatalytic Upgrading of Model Lignin Monomers with Earth Abundant Metal Electrodes. *Green Chem.* 2015, *17*, 601–609. 10.1039/c4gc01632g

[308] Zhou, Y.; Gao, Y.; Zhong, X.; Jiang, W.; Liang, Y.; Niu, P.; Li, M.; Zhuang, G.; Li, X.; Wang, J. Electrocatalytic Upgrading of Lignin-Derived Bio-Oil Based on Surface-Engineered PtNiB Nanostructure. *Adv. Funct. Mater.* 2019, *29*, 1–11. 10.1002/adfm.201807651

[309] Fang, Z.; Flynn, M. G.; Jackson, J. E.; Hegg, E. L. Thio-Assisted Reductive Electrolytic Cleavage of Lignin β-O-4 Models and Authentic Lignin. *Green Chem.* 2021, *23*, 412–421. 10.1039/d0gc03597a

# 2 Hydrocracking of Palm Oil Into Biofuel Over Ni-$Al_2O_3$-bentonite (Aluminosilicate) Nanocatalyst

*Karna Wijaya, Resy Norma Annisa, Ani Setyopratiwi, and Akhmad Syoufian*
Department of Chemistry, Faculty of Mathematics and Natural Sciences, Gadjah Mada University, Yogyakarta, Indonesia

*Hasanudin Hasanudin*
Department of Chemistry, Faculty of Mathematics and Natural Sciences, Sriwijaya University, Indralaya, South Sumatra, Indonesia

*Maisari Utami*
Department of Chemistry, Faculty of Mathematics and Natural Sciences, Universitas Islam Indonesia, Yogyakarta, Indonesia

**CONTENTS**



DOI: 10.1201/9781003167693-2

## 2.1 INTRODUCTION

Several different methods and raw materials as alternative energy used in research have been conducted. One of the potential materials is vegetable oil. Vegetable oil can be processed into fuel, which may have similar character to fossil fuel through hydrocracking process. The process may cut off the long hydrocarbon chain of triglyceride in which this chain may act as fuel since it has the chemical characteristics of fuel [1]. Vegetable oil, known as renewable fuel, is proposed to be used as fuel properly as this resource is renewable [2]. Palm oil is known as one of the most potential resources. This resource can be planted and processed on a large scale in the tropical area, such as Indonesia and Malaysia. Therefore, its availability is sufficient, and the cost is relatively low compared to other resources [3].

Palm oil cannot be directly used as fuel. It can be processed as fuel with similar characteristics to gasoline and is known as biogasoline. Palm oil contains several fatty acids; bound on triglyceride structure that has a very long hydrocarbon chain [4]. The most dominant content in palm oil is palmitate acid which has 16 carbon atoms excluding double bond, and oleic acid, which includes 18 carbon atoms with single bonds on its hydrocarbon structure. Fatty acid bonded on triglyceride structure on palm oil can be processed into biogasoline (C12 – C16) by cutting its carbon chain; thus, palm oil can produce the derivative product as fuel, including diesel, kerosene, or gasoline [5].

Biogasoline production from palm oil can be processed through hydrocracking either from a catalytic or non-catalytic process. Hydrocracking is the combination process between cracking and hydrogenation to produce a saturated compound [6]. In this reaction, carbon bonding is decomposing followed by hydrogenation process, a saturated molecule with hydrogen. This reaction is an exothermic process that produces lower weight molecules [7]. Non-catalytic hydrocracking is processed by heating the palm oil at high temperature and pressure without a catalyst. This method consumes a lot of energy and is less effective since hydrocarbon chain cutting is uncontrolled, producing a short hydrocarbon chain in the gas phase rather than the expected biogasoline. The other method is catalytic hydrocracking. This method may be better since the temperature reaction is lower and the product obtained is higher [8].

Catalytic hydrocracking is a cracking method involving a constant supply of both catalyst and hydrogen [9]. There are three catalyst types, which are: (1) metal catalyst: Ni, Pd, Pt [10,11]; (2) non-metal catalyst: $Al_2O_3$, $SiO_2$, zeolite [12,13]; and (3) metal-pillared catalyst: Ni/$\gamma$-$Al_2O_3$, Ni/HZSM-5, Ni/zeolite Y [14–17]. Metal catalyst has limitation, such as being inefficient and unable to work at high temperature; thus, the use of this catalyst in the oil and gas industry is limited. On the other hand, the non-metal catalyst has an acid characteristic in its reaction, particularly in the crude oil cracking process. Whereas metal-pillared catalyst has an acid function in which the cracking reaction and isomerization process come from its pillared material, such as alumina, silica, and zeolite [18–20]. Meanwhile, the hydrogenation function and olefin forming come from its metal. This catalyst is proposed to be the most promising method for hydrocracking process [21].

Indonesia has a high abundance of mineral resources located in several areas. One of the minerals is clay. Soil-based on its mineral contents is divided into montmorillonite, kaolinite, halloysite, chlorite, and illite [22]. Montmorillonite is a phyllosilicate mineral group in which it is the most interesting one because this mineral has the ability for swelling and intercalation process with organic compound forming composite organic-inorganic material and also the high capacity of cation exchange; thus, its interlayer space can accommodate a cation on high abundance [23,24].

Bentonite is a hydrous phyllosilicate with layer-structured material. Each bentonite layer has two tetrahedral layers containing Si(O,OH) and pressed between one octahedral layer containing M(O,OH) with M = Al, Mg, Fe [25]. Between these layers is a space filled with water molecules and other cations. The cations can be exchanged with higher cation content and size, such as polycation or polyhydroxy metal-based, including Al, Zr, Ti, Cr, Fe, and Ga. The cation exchange product produces two dimensional materials such as bentonite, known as pillared clay [26]. After the heating process, pillared polycation will form an oxide cluster producing an open clay layer permanently and producing a dimensional molecular space with a proper porous system. As a result, this pillared clay will be obtained a surface texture that has the acidic base characteristic. This characteristic can be used for several catalytic applications such as crude oil cracking, alkylation, dehydration alcoholic, chemical production, and others [27,28].

Pillared clays can be suitable not only for a variety of analytic applications but also for catalyst supports (an active component) with catalytic characteristics. The benefits of pillared clay for catalyst supports include high surface area, porosity typical form, sufficient thermal stability, and catalytically active surface. Compounding pillared clay with catalytic active components are usually metals transition, including Cu, Ni, Pt, Mn, Mo, Pa, Fe, and Co. The compound will produce a more effective and selective material in accordance with the catalytic properties of metal transition [29,30].

Nickel is a potential active component that can rest on bentonite. As a catalyst, nickel has good activity and selectivity in a reaction. The nickel price is also more affordable than platinum and palladium metal that the use of nickel is more beneficial [31]. Pillared bentonite with alumina was made by Dhahri et al. [32] which reported an increase in physical-chemical properties of bentonite, such as specific surface area of bentonite, pore size, and pore distribution. While the cracking reaction of palm oil using nickel-based mesoporous zeolite Y catalyst has been made by Li et al. [33], who managed to get a fraction biogasoline. Trisunaryanti et al. [34] reported that cooking oil cracking using catalyst Ni-Mo/natural zeolite can produce biogasoline fraction. Many studies have discussed catalytic cracking using pillared clays. However, there are not many research on palm oil hydrocracking using modified bentonite.

Based on the description above, the research that led to the manufacture of fuel biogasoline using Ni-$Al_2O_3$-bentonite catalyst becomes highly potential. In this study, a modified natural bentonite was prepared with the pillared method using a Keggin agent. Next, it is impregnated to use nickel metal before it is used as a catalyst in palm oil hydrocracking biogasoline to examine the catalytic activity. Biogasoline has the properties of biodegradable, renewable, environmentally friendly, less pollutant ($SO_x$, $NO_x$, Pb) production, and others [35]. The purpose of this study is to make a hydrocracking catalyst of modified bentonite with pillared method, as well as assesses the physical-chemistry characteristics. Besides, it also aims to recognize the activity of the modified bentonite as a catalyst in the hydrocracking reaction of palm oil and to compare the properties of the hydrocracking product with fossil fuels.

## 2.2 RESEARCH METHOD

### 2.2.1 MATERIAL

The materials used in this study were: natural bentonite from Bantul Yogyakarta, cooking palm oil, HF, $H_2SO_4$, $AlCl_3.6H_2O$, oxalic acid ($H_2C_2O_4{\cdot}9H_2O$), $AgNO_3$, ammonia, NaOH, ammonia nickel chloride ($NiCl_2{\cdot}6H_2O$), $CaCl_2$, and NaCl which were purchased from e-Merck. $N_2$ and $H_2$ gas were obtained from PT Samator Gas.

### 2.2.2 INSTRUMENTS

The equipment used in this study consisted of a centrifuge (KOKUSAN Ogawa Seiki type H-107), a sieve of 250 mesh, oven, furnace, hotplate stirrer, analytical balance (AND GR-200), a set of refluxes, hydrocracking reactor flow-fixed bed. The analytical instruments used for the characterization of materials catalyst were an IR spectrophotometer (Shimadzu FTIR 8201 PC), X-ray diffraction (Shimadzu models XRD 6000), Brunauer Emmett Teller (BET) (Quantachrome Instruments version 11.0), Transmission Electron Microscopy (TEM) (JEOL JEM-1400), testers acidity with ammonia, Atomic Absorption Spectrometer (AAS) (Perkin Elmer 5100 PC), and X-Ray Fluorescence (XRF) (MiniPal4 PANalytical). The instrument used to characterize the activity and selectivity of catalysts was a GC-MS (Shimadzu QP2010S).

### 2.2.3 Bentonite Material Preparation

Bentonite preparation was analyzed by dispersing bentonite in distilled water with a ratio of 1:2 (w/v). Next, it was deposited for 24 hours, and then its layer of subtlety was taken, which did not contain sand. Afterward, it was dried in an oven at a temperature of 90–110°C, sieved by 250 mesh, and characterized by using AAS.

Bentonite was then washed with HF to remove any impurities such as sand or silica. Refined bentonite was mixed in a solution of HF 1% with a ratio of 1: 2 (w/v) and stirred for 15 minutes. Next, it was washed to neutralize pH using distilled water. After that, bentonite was oven-dried at a temperature of 90–110°C and pulverized to 250 mesh size, resulting in HF-called bentonite. After this preparation stage, the bentonite was ready to be modified.

### 2.2.4 Bentonite Activation Using Sulfuric Acid

HF-bentonite was then activated with sulfuric acid. HF-bentonite was dispersed in 1.2 M sulfuric acid solution at a ratio of 1:4 (w/v), stirred for 1 hour, and then neutralized with distilled water. Afterward, bentonite was dried using an oven at a temperature of 90–110°C and pulverized to 250 mesh size. The result was named H-bentonite. H-bentonite was characterized by XRD, FTIR, acidity test, XRF, and BET.

### 2.2.5 Pillarization with $Al_2O_3$

The bentonite was intercalated and pillared using an $Al_2O_3$ pillared agent. First, the preparation of polycation pillared agent $[Al13]^{7+}$ by mixing slowly 400 mL of 0.4 M $AlCl_3{\cdot}6H_2O$ with 400 mL of 0.88 M NaOH solution (ratio OH/Al 2.2) was initially prepared. Next, it was refluxed for 3 hours at a temperature of 80–90°C. The pillared agent was then mixed with 2% bentonite suspension by dispersing 4 g of H-bentonite in 200 mL of distilled water and mixed evenly. The mixture of the pillared agent with bentonite suspension was stirred for 24 hours. After 24 hours, the mixture was neutralized from Cl ion content using the $AgNO_3$ indicator (until no white precipitate AgCl formed). After that, bentonite-obtained was dried in an oven at a temperature of 70°C and smoothed with a size of 250 mesh. It resulted in $Al_2O_3$-bentonite. $Al_2O_3$-phased bentonite was then calcined at 500°C for 2 hours with an increased temperature of 1°C per 2 minutes. Finally, $Al_2O_3$-bentonite was characterized by XRD, FTIR, acidity test, XRF, and BET.

### 2.2.6 Bentonite Impregnation Using Nickel

Bentonite impregnation started by mixing the $Al_2O_3$-bentonite suspension into a saline solution $NiCl_2{\cdot}6H_2O$ as much as 1% (w/w) of the weight of pillared bentonite. It was then refluxed at the temperature of 80–90°C for 5 hours. Bentonite suspension obtained was dried in an oven at a temperature of 70°C. It resulted in Ni-$Al_2O_3$-bentonite. Next, it was calcined for 5 hours at a temperature of 500°C with $N_2$ gas flowing at a rate of 20 mL/min; and reduced with $H_2$ gas at the same rate for 2 hours at a temperature of 400°C. Finally, Ni-$Al_2O_3$-bentonite was characterized by XRD, FTIR, XRF, BET, acidity test, and TEM.

### 2.2.7 Catalyst Characterization

The analysis was done by destructing 0.04 g of bentonite, which had not been treated with aqua regia in a 10 mL volumetric flask. It was then diluted 100 times. NaCl was used as standard solution for analysis of Na content and $CaCl_2$ for Ca content analysis. Afterwards, the sample and standard solution were analyzed using AAS.

Determination of the total number of bentonite acid sites was analyzed quantitatively and qualitatively. An acidity test was used to quantitatively determine the number of active sites on the catalysts by the gravimetric method using ammonia as its base adsorbate. First, an empty rate that had been put into the oven for 30 minutes was weighed and recorded for its weight. Furthermore, a 1 gram sample rate bentonite was added into the oven at the temperature of 90–100°C for 1 hour. Afterward, the bentonite was weighed and recorded once more time. Then a set of acidity tests was prepared. Bentonite sample was inserted into the testing container, and the container was sealed. The container was vacuumed and subsequently purged using ammonia to reach its saturation. The container was closed tightly, so there was no inflow and outflow of air. The container stayed for 24 hours. After 24 hours, the samples were weighed and recorded. Determination of the number of acid sites qualitatively was specified by FTIR analysis results after its ammonia adsorption get tested. The success was indicated by the change of wavelength vibration on 1,489 $cm^{-1}$ for a Bronsted acid and on 1,635 $cm^{-1}$ for Lewis acidic sites.

### 2.2.8 Palm Oil Hydrocracking

Ni-$Al_2O_3$-bentonite was tested for its catalytic ability. The test has been done by using it as the catalyst for palm oil hydrocracking into biogasoline to examine its catalytic activity. Hydrocracking was carried out using a fixed bed flow reactor with an $H_2$ gas flow rate of 20 mL/min at a temperature of 500°C for approximately 1 hour. The ratio comparison between palm oil and the catalyst in the reaction was 5:1 (w/w). When biogasoline products were obtained, it was then analyzed by GC-MS to identify the compounds in the biogasoline conversion result.

## 2.3 RESULTS AND DISCUSSION

### 2.3.1 Ni-$Al_2O_3$-bentonite Catalyst Preparation

Bentonite is a key material in the manufacture of hydrocracking catalysts for this study. As it is known, there are two types of bentonite-based cation, namely Na-bentonite and Ca-bentonite. Na-bentonite is a type of bentonite that is more suitable when used as a hydrocracking catalyst compared to Ca-bentonite. This is because Na-bentonite has the power to expand 8–15 times better than Ca-bentonite if it is immersed in water. Na-bentonite is also more stable under the influence of acid compared to Ca-bentonite [36,37]. This is important because, in the process of preparing Ni-$Al_2O_3$-bentonite catalyst, there is a procedure in which the bentonite will be activated with sulfuric acid. Because of that reason, it should be certainly known that bentonite has already been modified to be Na-bentonite.

Based on its physical appearance, raw material used in accordance with the characteristics of Na-bentonite was identified. The physical characteristic was cream-colored and well-dispersed in water. However, understanding the physical appearance was not sufficient to determine the Na-bentonite; therefore, an AAS analysis should also be conducted. AAS analysis was used to determine whether the most widely used cations contained in bentonite raw materials shall be used, and these also determine the swelling capacity of bentonite [38].

An atomic absorption spectrophotometer (AAS) is a measurement method based on the amount of radiation absorbed by the atoms when the number of radiations goes through a system containing those atoms. The amount of absorbed radiation depends on the number of atoms to absorb

**TABLE 2.1**
**Elemental Analysis Result**

| Element | Concentration |
|---|---|
| Sodium | 1.93 |
| Calcium | 0.40 |

radiation. The concentration of elements in the sample can be recognized by measuring the intensity of absorbed radiation (absorbance), [39]. The AAS method was selected as it is relatively easy to do, affordable, accurate, and has high selectivity and sensitivity. Since the AAS method is exceptionally specific, metals that form complex mixtures can be analyzed; consequently, the concentration of an element contained in very small concentrations can be specified without any prior separation previously.

AAS analysis was an initial step in making the process of Ni-$Al_2O_3$-bentonite catalyst. The preparation of the analysis was firstly carried out by making aqua regia solution which was a mixture of hydrochloric acid and nitric acid concentrated at a ratio of 1:3. Aqua regia was used to destruct bentonite samples to analyze by dissolving bentonite. Therefore, the contents can be analyzed by AAS. The analyzed bentonite was the initial bentonite without taking any prior treatment. Bentonite was then mixed with aqua regia. After dissolved completely, the mixture was diluted with distilled water into some variation of concentration to be a standard addition of curve parameters. NaCl solution was used as a standard solution for the analysis of cations $Na^+$ and calcium chloride solution for the analysis of $Ca^{2+}$ cation. Various samples and both the standard solution were also analyzed by AAS. In Table 2.1, it can be apparent when the bentonite samples were analyzed, and the the sodium content was greater than the calcium content. It proved that the analyzed bentonite sample was Na-bentonite.

### 2.3.2 Natural Bentonite Modification Into Ni-$Al_2O_3$-bentonite Catalyst

Bentonite is a potential material as cracking catalysts, but unfortunately, bentonite is non-resistant to high temperatures. From the results of studies on bentonite structure, it is known that modifications to the bentonite structure can improve its properties and durability that it can be used as a good cracking catalyst [40].

This study was conducted in several stages of treatment to make modifications for the bentonite hydrocracking catalyst: including the activation process using sulfuric acid, the pillared process using $Al_2O_3$, and impregnation using Ni metal. However, prior to the modification process, bentonite was required to undergo early stages of preparation beforehand. This initial preparation was used to remove impurities contained in bentonite since the bentonite was taken from nature [41].

The first stage of preparation was washing using distilled water. Bentonite was soaked in distilled water at a ratio of 1:2 (w/v) and then left to stand for 24 hours. The water layer is the uppermost layer containing light impurities such as the remains of plants or other. The second layer is a layer of bentonite, and the third layer is a layer of sand. Layers of sand were at the bottom as it had the greatest density. The water and sand layers were considered as impurities; thus, it was separated from the layer of bentonite [42]. Bentonite layers were then centrifuged to further separation for the bentonite from water and fine sand. The obtained bentonite was then washed using HF. It aimed to remove silica impurities apart from the structure of fluosilicate bentonite. Silica ($SiO_2$) was usually derived from silica sand commonly found in soil [21]. The reactions between silica and HF are as follows [43,44]:

$$SiO_{2(s)} + 6\ HF \rightarrow H_2SiF_{6(l)} + 2H_2O$$

$$SiO_{2(s)} + 4\ HF \rightarrow SiF_{4(g)} + 2H_2O$$

Two reactions occurred between silica and HF would eliminate the silica well, as the first reaction will produce a silicon fluoride compound soluble in water that is easily disposed of during drying. The second reaction produced silicon fluoride compound on phase gas to evaporate.

After preparation, bentonite was then modified. The first modification was the activation process using sulfuric acid. Activation aimed to clean the surface of the pores, remove its impurities compounds, and rearrange the exchangeable cations. Activation caused the bentonite to undergo dealumination and decationization [45]. At the time being inside the acid solution, many metals in octahedral and tetrahedral layers can be set apart from the structure, resulting in the charge on the sheet of bentonite becoming more negative. However, load balancing of cation position was immediately replaced by $H^+$ ions from the acid solution. $H^+$ ions were much easier to exchange with aluminum polycation $(Al_{13})^{7+}$ as the charge was much smaller than the polycation charge. It should be noted that after this activation process, it shall require no calcination process. It is because calcination leads to the unavailability of cations in the space between the aluminosilicate sheets or TOT (tetrahedral-octahedral-tetrahedral) that it can be exchanged. Consequently, it did not make them have possible pillarization with polycation aluminum. Besides, the calcination can lead to the collapse of the aluminosilicate's bentonite structure [46].

In the activation process with sulfuric acid, dealumination also occurred. Dealumination is the process of releasing an Al atom of bentonite [47]. Bentonite catalyst which contains high aluminum has low stability at high temperatures that should be dealuminated to optimize the performance of the catalyst in hydrocracking reactions occurring at high temperatures. Dealumination can also increase the Si/Al ratio [48]. Escalating the ratio of Si/Al also increases the acid sites of bentonite. It means that the activity ability of bentonite as catalyst increases. A high comparison between silica and alumina leads to increasing adsorption activity. The high content of Al leads to bentonite frame to have the hydrophilic characteristic. The hydrophilic and polar characteristic of bentonite is its obstacle in terms of adsorption ability.

Subsequent modifications were pillared process. The process went through intercalation on aluminum polycation, acting as pillared agents, into the bentonite. Aluminum polycation with Keggin structure was made by slowly mixing $Al(OH)_3$ with $NaOH_{(aq)}$. It was then refluxed for 3 hours to obtain a clear, homogeneous solution. This mixture caused the forced hydrolysis reaction $Al(OH)_3$ by $NaOH_{(aq)}$. The process continued; consequently, the polymerization reaction happened to produce polycation $[AlO_4Al_{12}(OH)_{24}(OH_2)_{12}]^{7+}$ [49].

At the time of bentonite was intercalated by ions $(Al_{13})^{7+}$, it resulted in the influx of ions into the interlayer space aluminosilicate. The influx of ions $(Al_{13})^{7+}$ made changes to the increased basal spacing of bentonite as ion size was enlarging. Therefore, it can be said that $(Al_{13})^{7+}$ ion is pillared into aluminosilicate bentonite interlayer space. The calcination was then performed to transform ion $(Al_{13})^{7+}$ to this aluminum oxide to be sturdy pillars. Calcination will release water molecules from the ion $(Al_{13})^{7+}$; thus, it becomes the oxide. The following is the calcination reaction ion $((Al_{13})^{7+}$ [50]:

$$[AlO_4Al_{12}(OH)_{24}(OH_2)_{12}]^{7+} \rightarrow 6.5\ Al_2O_3 + 7\ H^+ + 20.5\ H_2O$$

The latest modification stage of bentonite was the coupling process with nickel metal. The process was conducted using the impregnation method by mixing the salt solution $NiCl_2{\cdot}6H_2O$ with bentonite and refluxing it for 5 hours. The process made by nickel metal was coupled evenly on bentonite. The previous pillared bentonite with aluminum allows the nickel to permeate well into the bentonite pores. The presence of nickel in the bentonite will accelerate the hydrogenation reaction and prevent Ni metal sintering at high temperature cracking [31]. The sintering process is a process of catalyst coagulation, which typically occurs in metal catalysts. It is caused by heating or high pressure, resulting in loss of catalyst activity due to the reduced surface area of the catalyst or the changing structure of the catalyst surface [51].

The coupled bentonite was then calcined by flowing $N_2$ gas for 5 hours with the aim to improve the dispersion of the metal. After that, bentonite was reduced with $H_2$ gas for 2 hours to convert the metal ions into metal atoms.

### 2.3.3 Bentonite Structure Analysis Using the X-ray Diffraction Method

X-ray diffraction (XRD) is one of the qualitative methods most often used for the characterization of materials. An XRD method was used to analyze the composition of the phases or compounds in materials and the characterization of crystalline (crystallinity) in the composition of the material. The diffraction pattern of each solid crystalline is very distinctive, that each different solid crystalline have different diffraction patterns. Analysis by XRD diffractogram was conducted by matching a sample material with the material standard diffractogram of data to identify of which components contained in the material. These standards data of diffractogram are usually called as the Joint Committee on Powder Diffraction Standard (JCPDS). Constituent mineral type bentonite 2θ was shown by the data, while the crystallinity of the component structures was indicated by the high and low intensity of the peak [52].

Impurities are the main obstacle in material manufacturing or modification; therefore, the procedure of removing impurities becomes a common part to proceed. Similarly, in the manufacture of bentonite catalyst, impurities usually found in quartz and bentonite are metal ions. Quartz ($SiO_2$) is considered a contaminant as it has hydrophobic properties that can reduce the adsorption of bentonite. It must be very annoying when using bentonite as a catalyst. Quartz in this study was eliminated by HF treatment, while the metal ions were removed by sulfuric acid activation. To achieve the success of the above treatment procedures, analysis using the XRD method is suggested. Reduced impurities cause an increase peak intensity of basal spacing $d_{001}$ and $d_{002}$. The increased peak intensity is shown in Figure 2.1 (b). It points out that the reduction of quartz content inside bentonite makes montmorillonite characteristics more dominant.

Based on diffractograms in Figure 2.1, $2\theta = 5{,}98^\circ$ $d_{001}$ is a typical peak of montmorillonite and $2\theta = 19{,}88^\circ$ $d_{002}$ is a typical peak of montmorillonite (JCPDS No. 12–231). Table 2.2 represents that after treatment with HF, the typical peak of montmorillonite has shifted to a smaller 2θ. This indicates that a large basal spacing $d_{001}$ and $d_{002}$ increase. The increase in basal spacing happened

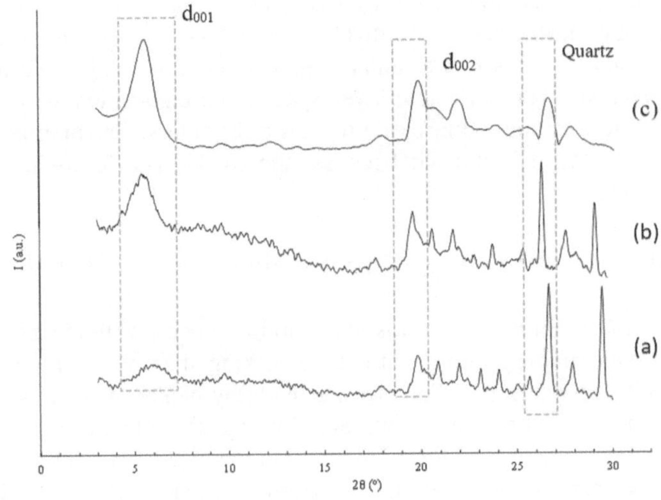


**FIGURE 2.1** Diffractogram of (a) Na-bentonite, (b) HF-bentonite, (c) H-bentonite.

**TABLE 2.2**
**XRD Analysis Results**

| 2θ (°) | | | Materials |
|---|---|---|---|
| Na-bentonite | HF-bentoniet | H-bentonite | |
| 5.94 | 5.58 | 5.46 | Montmorilonite |
| 19.88 | 19.76 | 19.82 | Montmorilonite |
| 26.66 | 26.56 | undetected | $SiO_2$ |

due to the loss of impurities, leading to the distance between the coating and the increased surface area. In the area of 2θ = 26,66°, which is the typical peak of quartz (JCPDS No. 5–490), shows a decrease in the intensity of the peaks. This decrease was due to reduced quartz contents in bentonite [53]. The same thing happened on bentonite activated by using sulfuric acid. The H-bentonite diffractogram showed two peaks, $d_{001}$ and $d_{002}$, which also increased intensity and a shift to smaller 2θ. A decrease in the intensity of quartz also occurs. The ability of sulfuric acid to dissolve metal ions such as $Na^+$, $Mg^{2+}$, $Ca^{2+}$, $Fe^{2+}$, and $K^+$ can also be viewed from the diffractogram, which becomes smooth.

X-ray diffraction analysis was also performed for the pillared bentonite [30]. All five samples of Na-bentonite, HF-bentonite, H-bentonite, bentonite, and A12O3-Ni-A12O3-bentonite are compared based on characters from its basal spacing d001 for each sample. Figure 2.2 shows that d001 peak is shifting after each stage of modification. However, the highlight d001 on-bentonite A12O3 and Ni-A12O3-bentonite do not show the diffractogram. This might be due to the peak shifts towards 2θ smaller than 3. Limitations of the XRD instrument to display data on the 2θ (small angle) make the XRD diffractogram display no complete data, and a 2θ peak shift cannot be observed.

Basal spacing peak shifting towards smaller 2θ after pillarization happened due to dihydroxylation during calcination process changes the bond between the pillar and the clay from ionic to

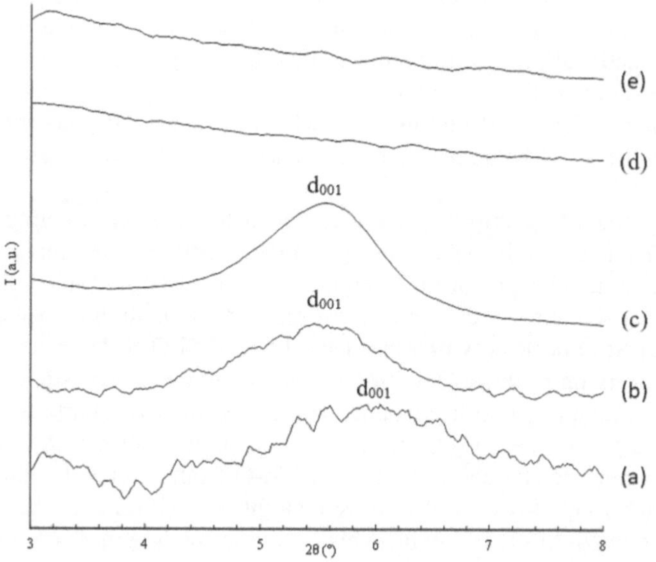


**FIGURE 2.2** Diffractogram and *basal spacing* $d_{001}$ of (a) bentonite, (b) HF-bentonite, (c) H-bentonite, (d) $Al_2O_3$-bentonite, (e) Ni-$Al_2O_3$-bentonite.

**TABLE 2.3**
**Increasing of Basal Spacing of Bentonite**

| Sample | 2θ (°) | $d_{001}$ (Å) | $\Delta d = (d_{001}\text{-}9.6)$ Å |
|---|---|---|---|
| Na-bentonite | 6.04 | 14.62 | 5.02 |
| HF-bentonite | 5.74 | 15.38 | 5.78 |
| H-bentonite | 5.52 | 15.99 | 6.39 |
| $Al_2O_3$-bentonite | – | – | – |
| Ni-$Al_2O_3$-bentonite | – | – | – |

covalent bonds. Therefore, it results in porous network stability (Hand, 2011). This porous network stability generates more surface area that the size of basal spacing increases. The increase in basal spacing $d_{001}$ can be measured quantitatively. Measurement is taken by calculating the difference (Δd) between major $d_{001}$ (Å) peak diffractogram basal spacing on each sample with a standard width of bentonite clay *interlayer layer* (9.6 Å) [54]. The results of the calculations are shown in Table 2.3.

Table 2.3 shows that treatment with HF and sulfuric acid activation on bentonite increases the basal spacing of $d_{001}$ bentonite. Despite the fact that the increase in basal spacing on $Al_2O_3$-bentonite and Ni-Al$_2$O3-bentonite cannot be identified because of the apparent absence of peak $d_{001}$, theoretically, pillarization using $Al_2O_3$ will increase the basal spacing for the replacement of $H^+$ ions on H-bentonite with ions $(Al_{13})^{7+}$ in a bigger size that will make the basal spacing $Al_2O_3$-bentonite. It means the increased pore radius of bentonite. To determine the pore size more accurately on-bentonite $Al_2O_3$ and Ni-$Al_2O_3$-bentonite, it can be analyzed by the BET method.

### 2.3.4 Analysis of the Bentonite Functional Group Using the FTIR Method

An infrared spectrometer (FTIR) analysis was conducted to determine the functional groups contained in bentonite. This method gives the information on the existence of octahedral and tetrahedral unit builders on the aluminosilicate layer of bentonite. This information is obtained due to the adsorption of infrared light by a material that causes vibration and rotation of the bonds between atoms or functional groups in the material. Infrared spectra of clay minerals are usually observed at a wave number of 400–4,000 $cm^{-1}$ [55]. This analysis becomes essential that it is considered to use one of them to determine with which the successful modifications made to the bentonite. FTIR spectra of H-bentonite, bentonite-$Al_2O_3$, and Ni-$Al_2O_3$-bentonite are presented in Figure 2.3.

The analysis was taken by comparing the peak characteristics representing functional groups. Changes in peaks characteristic indicate a change in the structure of the modified bentonite. Data spectra peak characteristic changes of the bentonite are presented in Table 2.4.

Absorption peaks at 470 $cm^{-1}$ are the main absorption peak in all three bentonite samples since it is an indication of Si-O bending vibration of the bonded Si-O-Si layer tetrahedral (TO4) bentonite. While the absorption peaks at 794 $cm^{-1}$ are absorption characteristic of the mineral quartz indicated in all three samples, it still contained impurities quartz. Quartz is difficult to remove because the bond is a part of the $SiO_2$ layer of bentonite aluminosilicate. A peak of 1,041 $cm^{-1}$ is an absorption of asymmetric stretching vibration of Si-O from Si-O-Si bonding contained in the three-sample spectrum [56]. However, they have a slight peak shifting on the spectrum of $Al_2O_3$-bentonite and bentonite Ni-$Al_2O_3$-bentonite. This is caused by the process of pillarization that has formed a covalent bond between the atoms O in the Si-O-Si and alumina from $Al_2O_3$pillared.

Absorption peaks at 1,635 $cm^{-1}$ are indicative of the O-H bending vibration of water molecules [55]. This suggests that there are water molecules in all three samples of bentonite. But the

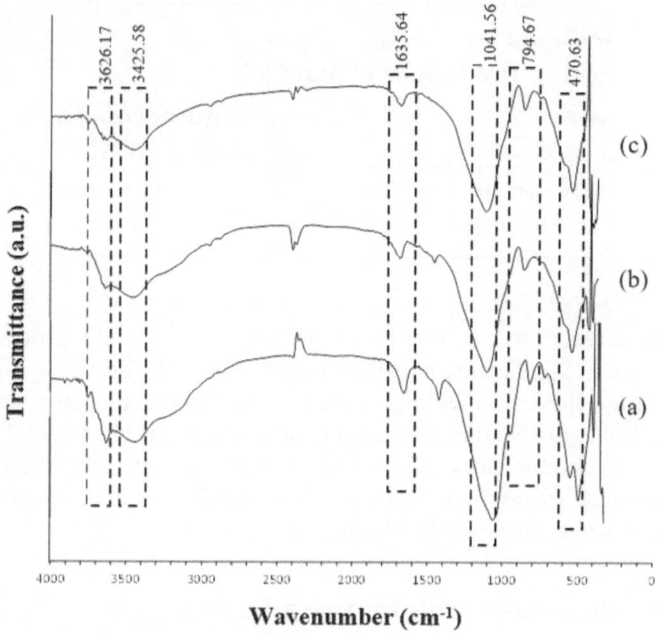


**FIGURE 2.3** Spectrum of FTIR (a) H-bentonite, (b) $Al_2O_3$-bentonite, (c) Ni-$Al_2O_3$-bentonite.

**TABLE 2.4**
**Characteristic Absorptions**

| Wavenumber ($cm^{-1}$) | | | Assignments |
|---|---|---|---|
| H-bentonite | $Al_2O_3$-bentonite | Ni-$Al_2O_3$-bentonite | |
| 470 | 470 | 470 | Bending vibration of Si-O from Si-O-S tetrahedral layer ($TO_4$) |
| 794 | 794 | 794 | Symmetrical stretching vibration of Si-O from Si-O-Al quartz |
| 1041 | 1049 | 1049 | Symmetrical stretching vibration of Si-O from i Si-O-Si |
| 1635 | 1635 | 1635 | Bending vibration of O-H from water |
| 3425 | 3433 | 3448 | Stretching vibration –OH from Si-OH (silanol) |
| 3626 | 3626 | 3626 | Stretching vibration of O-H from Al-OH (aluminol) |

intensity of the spectrum of $Al_2O_3$-bentonite and Ni-$Al_2O_3$-bentonite are lower compared to the spectrum of H-bentonite. This indicates a reduction in the number of water molecules on both bentonites, which may be caused by calcination process at the second stage of the modified bentonite. Absorption peaks at 3,425 $cm^{-1}$ indicate the bond stretching vibration of Si-OH-OH (silanol), and the intensity tends to decrease during the third spectrum. Absorption peaks at 3,626 $cm^{-1}$ indicate the O-H stretching vibration of Al-OH bond (aluminol) [57,58]. But the spectrum of $Al_2O_3$-bentonite and Ni-$Al_2O_3$-bentonite peak intensity decreased. This is due to the high temperature calcination at both stages of bentonite modification caused the release of water molecules on the bentonite structure that Al is idle. Absorption peaks at 3,425 $cm^{-1}$ (silanol) and 3,626 $cm^{-1}$ (aluminol) are part of the octahedral structure ($TO_6$) bentonite. In general, the drop of those three absorption peaks indicated dehydroxylation and dealumination on pillared bentonite and coupled bentonite; thus, Al is in a free state.

**TABLE 2.5**
**Acidity Test Results of Materials**

| Sample | Acidity (mmol/g) |
|---|---|
| H-bentonite | 2.21 |
| $Al_2O_3$-bentonite | 3.22 |
| Ni-$Al_2O_3$-bentonite | 3.25 |

Dealumination is caused by calcination on pillared and coupled process resulting in an absorption peak at 524 $cm^{-1}$, indicating the vibrational bending of Si-O-Al, and an absorption peak at 918 $cm^{-1}$, exhibiting vibration buckling of Al-OH-Al. H-bentonite disappeared on bentonite $Al_2O_3$-bentonite and Ni-$Al_2O_3$-bentonite spectrum. According to Ramos et al. [59], the absorption peak at 600–700 $cm^{-1}$ is an absorption peak of Ni-O bond. However, the absorption peak did not contain Ni-$Al_2O_3$-bentonite spectrum, indicating that Ni metal that falls on bentonite was not in the form of oxides, but in a state of neutral or basic.

### 2.3.5 Acidity Determination on the Bentonite Surface

Acidity is an important parameter for bentonite as a catalyst. The acidity is a parameter on how well a catalyst to accelerate the reaction. The number of acid sites on the catalyst becomes a benchmark of the acidity of a catalyst; a growing number of acid sites indicates the higher the acidity of the catalyst [60]. In this study, the acidity was calculated using ammonia absorption in the vacuum system. Ammonia's small size will fit into the pores of bentonite and bentonite acid bonded to the site. Alkaline nature of ammonia would cause it to naturally bond easily to the acid sites bentonite. Acidity can be calculated from how much the bentonite absorbs it. The results of the bentonite acidity test are presented in Table 2.5.

Acidity in bentonite comes from Lewis and Bronsted acidic sites. The presence of Al in the aluminate and $Al_2O_3$ pillared layer is Lewis acidic sites that accept an electron pair. In contrast, the protons and hydroxyl groups on the layer of bentonite are a Bronsted acid site. From Table 2.5, the increasing bentonite acidity afterward is shown. Increasing acidity in the $Al_2O_3$-bentonite was caused by $Al_2O_3$ pillared, which added Lewis acidic sites on the catalyst. While the Ni-$Al_2O_3$-bentonite, the increasing of acidity occurred because the nickel metal added Bronsted sites on the catalyst. Nickel metal is acidic because its orbital-d was not filled, which makes the reactive react with alkaline ammonia. Bronsted acid reaction sites with ammonia will produce $NH_4^+$ for donating protons to Bronsted acidic sites on ammonia, while the reaction of Lewis acidic sites will generate $NH_2$- for Lewis acidic sites receiving proton donor of ammonia.

Analysis by FTIR has also been done to recognize the outcome of the acidity test qualitatively. From the three samples of bentonite spectrum in Figure 2.4, it can be observed the presence of Lewis and Bronsted acid sites of bentonite. The existence of absorption peaks in the spectrum derived from ammonia species contained in bentonite. Absorption peaks in 1,635 $cm^{-1}$ indicate that the uptake of Bronsted acidic sites is clustered $NH_4^+$, whereas the absorption peak of 1,404 $cm^{-1}$ shows the uptake of Lewis acidic sites [60].

### 2.3.6 Analysis of Compound Content Using Bentonite by XRF Method

Characterization of bentonite with the XRF method aims to determine the content of compounds $Al_2O_3$ and Ni metal added during the modification of bentonite, i.e., on pillarization and impregnation. XRF method was chosen as it can analyze quantitatively and qualitatively. This method can provide data in the form of the element or oxide, and it can analyze multiple elements

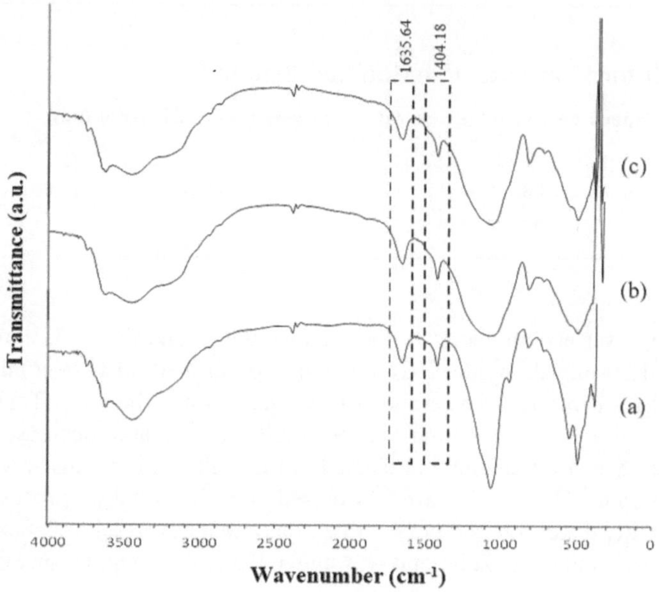


**FIGURE 2.4** The FTIR spectra after acidity test of (a) H-bentonite, (b) $Al_2O_3$-bentonite, (c) Ni-$Al_2O_3$-bentonite.

**TABLE 2.6**
**XRF Analysis Results**

| Sample | Composition (%) | |
|---|---|---|
| | $Al_2O_3$ | Ni |
| H-bentonite | 11.00 ± 0.10 | 0.36 ± 0.005 |
| $Al_2O_3$-bentonite | 17.00 ± 0.20 | 0.38 ± 0.01 |
| Ni-$Al_2O_3$-bentonite | – | 2.28 ± 0.02 |

at once in one analysis [59]. Analysis was performed on H-bentonite, bentonite-$Al_2O_3$, and Ni-$Al_2O_3$-bentonite.

From the data in Table 2.6, it can be viewed that there is a significant rise in pillarization composition $Al_2O_3$ and Ni metal after impregnation. This increase does not determine the success of the specific preparation of the catalyst, but it can be a supporter of the data analysis by XRD and FTIR. The increase of 5% $Al_2O_3$ content after pillarization indicates that at the time of intercalation, ion $(Al_{13})^{7+}$ entered the interlayer layer of bentonite and represents that the process is successful intercalation with $Al_2O_3$.

An increase of 2% composition of Ni metal content in the Ni-$Al_2O_3$-bentonite becomes an indication of the success of Ni metal impregnation into the pores of bentonite. This increase points out that the catalyst Ni-$Al_2O_3$-bentonite is a better catalyst than the H-bentonite. The increasing Ni metal content in $Al_2O_3$-bentonite represents that the acidity of the Ni-$Al_2O_3$-bentonite is also inclining.

### 2.3.7 Analysis of Specific Total Surface Area and Porosity Using the BET Method

Analysis using the BET method can help to determine the specific surface area and porosity owned in bentonite. Knowing the specific surface area and porosity of a catalyst becomes important because

**TABLE 2.7**
**BET Analysis Result for Bentonite and Modified Bentonites**

| Sample | Specific Surface Area ($m^2/g$) | Average Pore Diameter (nm) | Pore Volume (cc/g) |
|---|---|---|---|
| H-bentonite | 109.182 | 7.11 | 0.194 |
| $Al_2O_3$-bentonite | 108.759 | 5.78 | 0.157 |
| Ni-$Al_2O_3$-bentonite | 63.075 | 9.34 | 0.147 |

the greater the area of a surface and pores of the catalyst, the greater the chance of a catalyst to react with the reactants [61]. Analysis by the BET method is performed on H-bentonite catalyst, $Al_2O_3$-bentonite, and Ni-$Al_2O_3$-bentonite. The results of the analysis are shown in Table 2.7.

From the data in Table 2.7, it is known that bentonite surface area decreases after the pillared process. This can be caused by calcination carried out in pillarization stages, which released ions and water in the bentonite. The specific area decreased after the coupling process. This was due to the inclusion of metallic nickel into the pores of bentonite that covered its surface [62]. In addition, the factor on the distribution of nickel metal was uneven, and the coagulation effect may cause the decline of the Ni-$Al_2O_3$-bentonite surface.

Based on the results of this analysis, it is reported that there was a decrease in the average pore of pillared bentonite compared to the previous pillared. This decrease was caused by calcination performed on the pillarization stage. But after the coupling stage, an increase in the average pore took place. This can be caused by the bentonite pore opening during metal coupling with calcination and reduction. Another interesting part that can be observed is the reduction of the pore volume, which occurred after pillarization. Theoretically, pillarization will enlarge as the pillars' $Al_2O_3$ pore size was larger than previously cations contained in the H-bentonite. However, the calcination decreases the pore size due to the occurring dehydro-oxidation. Although the pores decreased, the $Al_2O_3$ pillared makes the bentonite pore more homogeneous and stronger. Afterward a coupling, pore size becomes smaller. This is due to the inclusion of metallic nickel into the pores, making the bentonite filling pores in pore volume in the Ni-$Al_2O_3$-bentonite smaller than $Al_2O_3$-bentonite.

### 2.3.8 Analysis of Morphology Structure Using TEM

Morphological analysis of bentonite structure using TEM allows the identification of an outcome of pillarization and coupling on bentonite catalyst visually. Figure 2.5 (a) shows the arrangement of dark and light layers of the same size. This is the layer or layers of smectite (aluminosilicate layers) [21]. It shows that pillared bentonite with $Al_2O_3$ managed to create a uniformity aluminosilicate layer of bentonite.

Figure 2.5 (b) is bentonite because of nickel metal development after pillarization. Black dots indicate a metal nickel pillared in a bentonite pore. The layers of bentonite also seem obvious. It means that the nickel pillarization does not damage bentonite layer or alumina pillars on bentonite. The presence of nickel metal and pillared aluminosilicate layers seems obvious, indicating that bentonite modification has been pillared and coupled successfully.

### 2.3.9 Activity Test of the Ni-$Al_2O_3$-bentonite Catalyst

The successfully pillared and coupled bentonite was tested by its catalytic activity for the hydrocracking reaction of palm oil. Hydrocracking aims to convert palm oil into biogasoline. The fatty acids in palm oil with its carbon chain of C12-C18 will be converted into hydrocarbon compounds that have similarity, such as C5-C12 carbon chain of petroleum. To compare the

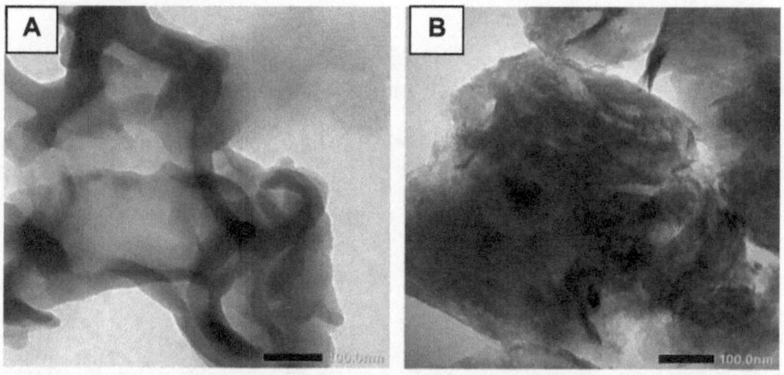

**FIGURE 2.5** TEM analysis result of (a) $Al_2O_3$-bentonite dan (b) Ni-$Al_2O_3$-bentonite.

catalytic activity of the modified bentonite, the test was done three times using different catalysts, namely, H-bentonite, bentonite-$Al_2O_3$, and Ni-$Al_2O_3$–bentonite. Hydrocracking was performed using a fixed-bed flow reactor at temperatures of 500°C for 5 hours. The palm oil used is commercial oil, which is used for cooking purposes.

When the hydrocracking reaction was running, palm oil inside the reactor was initially evaporating because of high temperatures. This steam will rise up and react with the catalyst. At this moment, the hydrocracking reactions took place. Active sites on bentonite and $H_2$ gas that flowed during the hydrocracking process of palm oil reacted with steam to produce hydrocarbon fractions in the form of gas. The fraction of light hydrocarbons ($C_1$-$C_4$) evaporates into the air, while the hydrocarbon fraction in ($C_5$-$C_{12}$) and severe (>$C_{12}$) was condensed into liquid and solid phases in the condenser.

As for a parameter of the catalytic properties of the catalyst, it was starting time when the liquid form of hydrocracking result was running. The temperature in the hydrocracking process increased slowly from 80–500°C, but the product started to get the result before reaching the maximum temperature. The hydrocracking process using H-bentonite started to produce the liquid product at a temperature of 480°C. In the hydrocracking process using $Al_2O_3$-bentonite, it began to form the liquid product at a temperature of 425°C. As for the hydrocracking using Ni-$Al_2O_3$-bentonite, the liquid product began to form at temperatures of 380°C.

### 2.3.10 Catalyst Effect Test Towards the Total Product of Hydrocracking Palm Oil

Modification of bentonite from the beginning aims to increase the activity of bentonite as a catalyst. The use of a catalyst is proof of the increased catalytic activity of modified bentonite. By increasing the catalytic activity of bentonite, hydrocracking produced better products in both quantity and quality. Broadly speaking, the increase in the catalytic activity is attributed to the higher acidity of bentonite as the modifications were made [21]. The process started from activation with sulfuric acid, $Al_2O_3$ pillarization, and metal nickel development. To observe the difference in the catalytic activity of each modification, the product was compared to the three different products of the hydrocracking catalyst.

The liquid product is desired as the main product of the hydrocracking process since biogasoline is in the liquid phase. Based on Table 2.8, it is known that hydrocracking catalyst Ni-$Al_2O_3$-bentonite produces liquid products at most compared to two other catalysts. Meanwhile, hydrocracking $Al_2O_3$-bentonite to produce liquid products was higher than the H-bentonite. The catalyst Ni-$Al_2O_3$-bentonite produced more liquid products due to higher acidity than the other catalysts. The presence of nickel orbitals-d metal that has not yet been fully loaded on this catalyst caused the increased acidity. Meanwhile, the hydrocracking catalyst of $Al_2O_3$-bentonite produces more liquid products from H-bentonite due to $Al_2O_3$ pillared, which makes pores on bentonite enlarged so that the surface area increases.

**TABLE 2.8**
**Palm Oil Hydrocracking Products with Different Catalysts**

| Catalyst | Conversion of Hydrocracking of Palm oil (%) | | |
|---|---|---|---|
| | Solid | Liquid | Coke |
| H-bentonite | 26.16 | 17.37 | 6.77 |
| $Al_2O_3$-bentonite | 58.81 | 22.09 | 15.25 |
| Ni-$Al_2O_3$-bentonite | 11.37 | 46.33 | 12.27 |

The by-product produced in the hydrocracking reaction was a solid product. This solid product was a hydrocarbon fraction with a longer chain, which was in the solid phase. This solid product may also contain traces of palm oil that was not converted into hydrocarbons. The long-chain fatty acid in palm oil was more difficult to break down; eventually, it froze and became a solid product on the hydrocracking reactor condenser. This solid product was observed at most formed when the reactor temperature was still low; this is likely the reason for the non-optimal obtained product. Hydrocracking liquid products were analyzed by GC-MS.

### 2.3.11 Catalyst Effect Test Towards the Liquid Product of the Palm Oil Hydrocracking Reaction

Liquid products obtained from three hydrocracking reactions were analyzed by GC-MS to determine the compounds contained therein. Qualitatively, the analysis of the composition of a sample was analyzed using GC-MS to get chromatogram peaks. A comparison of the hydrocracking product between gasoline and diesel oil chromatogram was also conducted to determine how similar the results of the hydrocracking with two fuels were. Chromatogram of gasoline, diesel oil sourced as well as the liquid product results from hydrocracking catalyst H-bentonite, bentonite-$Al_2O_3$, Ni-$Al_2O_3$-bentonite is presented in Figure 2.6.

In Figure 2.6, the third liquid product resulting in hydrocracking catalyst H-bentonite, $Al_2O_3$-bentonite, Ni-$Al_2O_3$-bentonite has a composition more similar to the composition of diesel compared to gasoline. This is evident from the retention time and peak intensity. Most of the three chromatograms of the liquid products' peaks result from hydrocracking at retention time like peaks in diesel chromatogram, namely the retention time of around 15–40 minutes. This is far different from the gas chromatograms, in which most peaks are at a retention time of around 0–10 minutes. But the hydrocracking products still contained similar components to gasoline fractions, although non-dominant ones.

Meanwhile, to determine the composition of the liquid product of hydrocracking results, it is quantitatively analyzed by calculating the percentage of peak area adapted to the data of the chromatogram of mass spectrometry. From the data chromatogram of GC-MS, it can be identified that the composition of the liquid product was in accordance with the fractions, namely the fraction of gas ($C_1$-$C_4$), the fraction of gasoline ($C_5$-$C_{12}$), the fraction of diesel oil ($C_{13}$-$C_{18}$), and the fraction of heavy oil ($>C_{18}$). The data was then compared to the data belonging to gasoline and diesel oil [62].

Based on Table 2.8, the three liquids of hydrocracking products not only contain gasoline fractions but also diesel oil fractions and heavy fractions. Even more, three containing diesel oil fractions were compared with gasoline. On average, hydrocracking products contain 50–60% diesel oil fractions and 35–40% gasoline fractions. It can be concluded that palm oil hydrocracking using the bentonite catalyst in this study predominantly produce diesel oil fractions. These data show that the liquid hydrocracking product using $Al_2O_3$-bentonite catalyst contains the highest

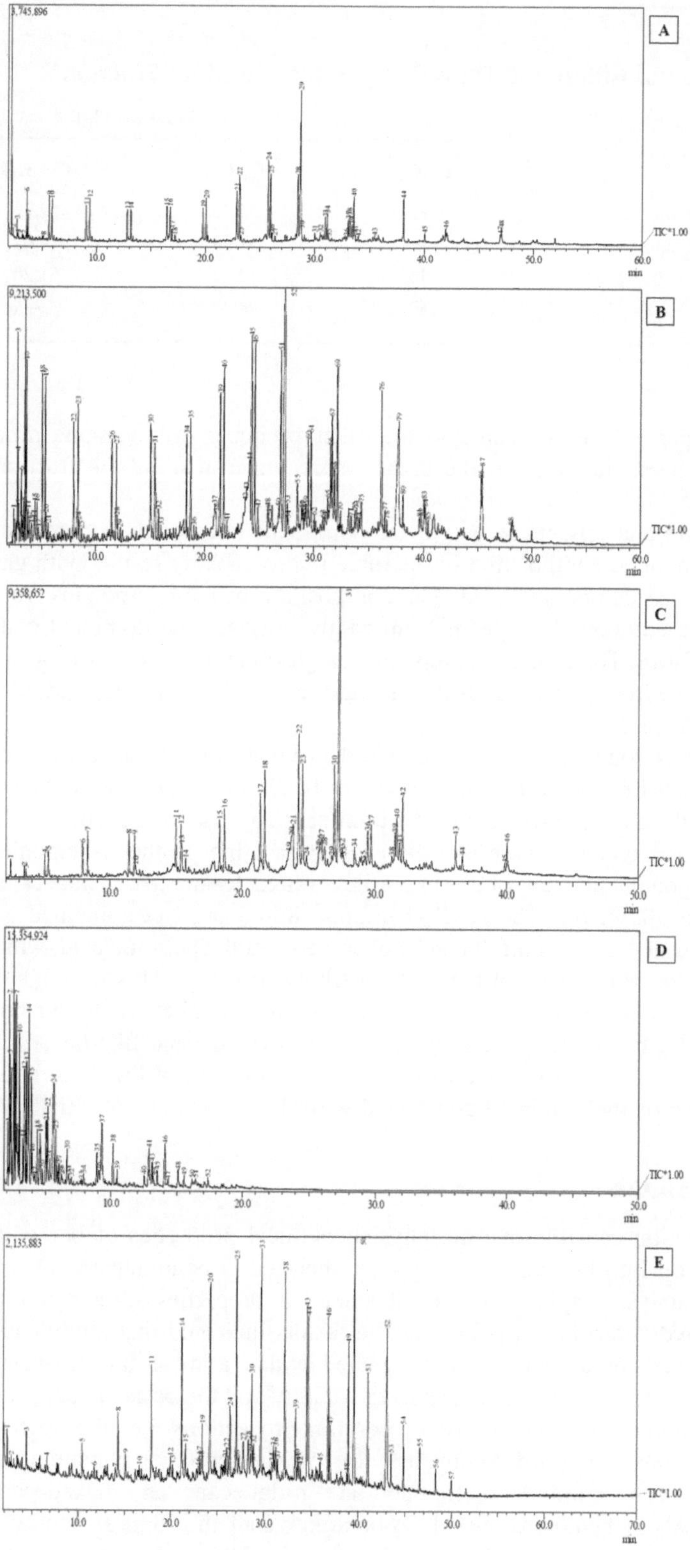


**FIGURE 2.6** Chromatogram of (A) liquid product H-bentonite, (B) liquid product $Al_2O_3$-bentonite, (C) liquid product Ni-$Al_2O_3$-bentonite, (D) gasoline, and (E) diesel oil.

**TABLE 2.9**
**Liquid Product Composition and Data Comparison Based on Fraction**

| Sample | Fraction (%) | | | |
|---|---|---|---|---|
| | $C_1$-$C_4$ | $C_5$-$C_{12}$ | $C_{13}$-$C_{18}$ | $>C_{18}$ |
| Liquid product H-bentonite | 7.93 | 35.83 | 52.73 | 3.51 |
| Liquid product $Al_2O_3$-bentonite | – | 39.58 | 49.62 | 10.80 |
| Liquid product Ni-$Al_2O_3$-bentonite | 0.46 | 30.42 | 61.20 | 7.94 |
| Gasoline | 1.57 | 98.43 | – | – |

gasoline fraction percentage compared to the other products. At the same time, hydrocracking catalyst Ni-$Al_2O_3$-bentonite contains the highest percentage of diesel oil fractions than the other two products.

Based on GC-MS analysis, it can be viewed compound compositions contained in a sample by comparing the compound composition in the three hydrocracking results with gasoline and diesel oil hydrocracking that can be analyzed. The composition of the compounds from each sample is represented by five compounds as being dominantly present. The dominant compound data was obtained from GC-MS. Data on the compound of gasoline and diesel oil, as well as the liquid product results in hydrocracking catalyst H-bentonite, $Al_2O_3$-bentonite, and Ni-$Al_2O_3$-bentonite, are presented in Table 2.9.

In Table 2.9 it is found that the three liquid products of hydrocracking results have some dominant compounds like the dominant compounds of diesel oil. Otherwise, there are no dominant compounds of the third liquid product like the dominant compound in gasoline. This means that the third hydrocracking product is more like diesel than gasoline. In the H-bentonite liquid products are dominant compounds like that of diesel oil, i.e., n-hexadecane like hexadecane and n-dodecane were also found in diesel oil. The $Al_2O_3$-bentonite liquid product contained 3-hexadecane like compound hexadecane on diesel oil. It can be observed that the palmitic acid, which is a compound found in palm oil, is still present in the hydrocracking products. This means hydrocracking with $Al_2O_3$-bentonite does not fully convert the entire palm oil. The last liquid product is a liquid product of Ni-$Al_2O_3$-bentonite, which also contains 3-hexadecane like hexadecane contained in diesel oil. It can be concluded that the results of hydrocracking of the three liquid products have similar composition of the dominant compound with diesel oil is greater than with gasoline.

## 2.4 CONCLUSIONS

Bentonite modification can improve the physicochemical properties of bentonite for the hydrocracking of palm oil into biofuel. This study synthesized H-bentonite, $Al_2O_3$ bentonite, and Ni-$Al_2O_3$-bentonite catalysts, explored the physicochemical properties of the bentonite, and examined its ability in the hydrocracking of palm oil. The results indicated that modifications of bentonite, which were activated, pillarized, and impregnated with Ni, were able to increase the physicochemical properties of the bentonites, such as the acidity and the basal spacing. The hydrocracking process produces liquid products with two predominant fractions, i.e., the fraction of the gasoline and diesel oil fractions. The third dominant hydrocracking products containing diesel oil fraction with a composition of dominant compound are n-dodecane and hexadecane. Compared to H-bentonite and $Al_2O_3$ bentonite, Ni-$Al_2O_3$-bentonite had the highest acidity of 3.25 mmol/g. Pillarization and impregnation with Ni generally decreased the surface area, average pore size, and pore volume of bentonite. The hydrocracking process of palm oil using modified bentonite catalysts represented that the liquid products have similar compositions to that of diesel oil and gasoline fraction. Hydrocracking with Ni-$Al_2O_3$-bentonite resulted in the most liquid product yield of 46.32%, with the highest percentage of diesel oil fraction of 61.20%.

## ACKNOWLEDGMENTS

This research was funded by the Ministry of Research, Technology and Higher Education, Republic of Indonesia, through MP3EI 2015–2016 Research Grant and through PTUPT 2020 (No. 1771/UN1/DITLT/DIT-LIT/PT/2021.

## REFERENCES

[1] D. H. Prajitno, A. Roesyadi, M. Al-Muttaqii, and L. Marlinda, *Bull. Chem. React. Eng. & Catal.* 12, 318 (2017). doi:10.9767/bcrec.12.3.799.318-328
[2] M. Torres-García, J. F. García-Martín, F. J. Jiménez-Espadafor Aguilar, D. F. Barbin, and P. Álvarez-Mateos, *J. Energy Inst.* 93, 953 (2020). doi:10.1016/j.joei.2019.08.006
[3] K. A. Zahan and M. Kano, *Energies* 11, 1 (2018). doi:10.3390/en11082132
[4] R. El-Araby, A. Amin, A. K. El Morsi, N. N. El-Ibiari, and G. I. El-Diwani, *Egypt. J. Pet.* 27, 187 (2018). doi:10.1016/j.ejpe.2017.03.002
[5] I. Istadi, T. Riyanto, L. Buchori, D. D. Anggoro, A. W. S. Pakpahan, and A. J. Pakpahan, *Int. J. Renew. Energy Dev.* 10, 149 (2021). doi:10.14710/ijred.2021.33281
[6] I. G. B. N. Makertihartha, R. B. Fitradi, A. R. Ramadhani, M. Laniwati, O. Muraza, and Subagjo, *Arab. J. Sci. Eng.* 45, 7257 (2020). doi:10.1007/s13369-020-04354-4
[7] C. Peng, Y. Du, X. Feng, Y. Hu, and X. Fang, *Front. Chem. Sci. Eng.* 12, 867 (2018). doi:10.1007/s11705-018-1768-x
[8] R. Sahu, B. J. Song, J. S. Im, Y.-P. Jeon, and C. W. Lee, *J. Ind. Eng. Chem.* 27, 12 (2015). doi:10.1016/j.jiec.2015.01.011
[9] M. Zhang, B. Qin, W. Zhang, J. Zheng, J. Ma, Y. Du, and R. Li, *Catal.* 10, 815 (2020). doi:10.3390/catal10080815
[10] L. Marlinda, M. Al-Muttaqii, I. Gunardi, A. Roesyadi, and D. H. Prajitno, *Bull. Chem. React. Eng. & Catal.* 12, 167 (2017). doi:10.9767/bcrec.12.2.496.167-184
[11] L. A. Dosso, C. R. Vera, and J. M. Grau, *Int. J. Chem. Eng.* 2018, 4972070 (2018). doi:10.1155/2018/4972070
[12] A. Galadima and O. Muraza, *J. Ind. Eng. Chem.* 61, 265 (2018). doi:10.1016/j.jiec.2017.12.024
[13] Q. Cui, Y. Zhou, Q. Wei, X. Tao, G. Yu, Y. Wang, and J. Yang, *Energy & Fuels* 26, 4664 (2012). doi:10.1021/ef300544c
[14] S. M. Ulfa, I. Sari, C. P. Kusumaningsih, and M. F. Rahman, *Procedia Chem.* 16, 616 (2015). doi:10.1016/j.proche.2015.12.100
[15] A. Srifa, N. Viriya-empikul, S. Assabumrungrat, and K. Faungnawakij, *Catal. Sci. Technol.* 5, 3693 (2015). doi:10.1039/C5CY00425J
[16] R. Rasyid, A. Prihartantyo, M. Mahfud, and A. Roesyadi, *Bull. Chem. React. Eng. & Catal.* 10, 61 (2015). doi:10.9767/bcrec.10.1.6597.61-69
[17] L. Mu, W. W. Feng, H. Zhang, X. Hu, and Q. Cui, *RSC Adv.* 9, 20528 (2019). doi:10.1039/c9ra03324f
[18] J. G. Speight, in Refin. Futur. (Second Ed., edited by J. G. B. T.-T. R. of the F. (Second E. Speight (Gulf Professional Publishing, 2020), pp. 303–342. doi:10.1016/B978-0-12-816994-0.00009-9
[19] J. W. Thybaut and G. B. Marin, in Adv. Catal., edited by C. B. T.-A. in C. Song (Academic Press, 2016), pp. 109–238. doi:10.1016/bs.acat.2016.10.001
[20] E. T. C. Vogt, G. T. Whiting, A. Dutta Chowdhury, and B. M. Weckhuysen, in Adv. Catal., edited by F. C. B. T.-A. in C. Jentoft (Academic Press, 2015), pp. 143–314. doi:10.1016/bs.acat.2015.10.001
[21] K. Wijaya, A. D. Ariyanti, I. Tahir, A. Syoufian, A. Rachmat, and Hasanudin, *Nano Hybrids Compos.* 19, 46 (2018). doi:10.4028/www.scientific.net/NHC.19.46
[22] I. Fatimah, *Chemical* 3, 54 (2018). doi:10.20885/ijcr.vol2.iss1.art7
[23] N. M. Alandis, W. Mekhamer, O. Aldayel, J. A. A. Hefne, and M. Alam, *J. Chem.* 2019, 7129014 (2019). doi:10.1155/2019/7129014
[24] N. D. Mu'azu, A. Bukhari, and K. Munef, *J. King Saud Univ. – Sci.* 32, 412 (2020). doi:10.1016/j.jksus.2018.06.003

[25] W. Bleam, in Soil Environ. Chem. (Second Ed., edited by W. B. T.-S. and E. C. (Second E. Bleam (Academic Press, 2017), pp. 87–146. doi:10.1016/B978-0-12-804178-9.00003-3
[26] J. M. B. T.-R. M. in E. S. and E. S. Huggett, in Ref. Modul. Earth Syst. Environ. Sci. (Elsevier, 2015), pp. 358–365. doi:10.1016/B978-0-12-409548-9.09519-1
[27] P. Satwikanitya, I. Prasetyo, M. Fahrurrozi, and T. Ariyanto, *J. Eng. Technol. Sci.* 52, 424 (2020). doi:10.5614/j.eng.technol.sci.2020.52.3.9
[28] Muslimin, A. Darmawan, and R. Lusiana, *J. Sains Dasar* 7, 49 (2018). doi:10.21831/j.%20saind %20dasar.v7i1.22260
[29] M. Marković, S. Marinović, T. Mudrinić, M. Ajduković, N. Jović-Jovičić, Z. Mojović, J. Orlić, A. Milutinović-Nikolić, and P. Banković, *Appl. Clay Sci.* 182, 105276 (2019). doi:10.1016/j.clay. 2019.105276
[30] J. Moma, J. Baloyi, and T. Ntho, *RSC Adv.* 8, 30115 (2018). doi:10.1039/C8RA05825C
[31] Y. Jiang, X. Li, Z. Qin, and H. Ji, *Chinese J. Chem. Eng.* 24, 1195 (2016). doi:10.1016/j.cjche.2016. 04.030
[32] M. Dhahri, M. A. Muñoz, M. P. Yeste, M. A. Cauqui, and N. Frini-Srasra, *React. Kinet. Mech. Catal.* 118, 655 (2016). doi:10.1007/s11144-016-1017-6
[33] T. Li, J. Cheng, R. Huang, J. Zhou, and K. Cen, *Bioresour. Technol.* 197, 289 (2015). doi:10.1016/ j.biortech.2015.08.115
[34] W. Trisunaryanti, I. A. Kartika, R. R. Mukti, H. Hartati, T. Triyono, R. Widyawati, and E. Suarsih, *Biofuels* 1 (2019). doi:10.1080/17597269.2019.1669871
[35] A. Bakhtyari, M. A. Makarem, and M. R. Rahimpour, in Bioenergy Syst. Futur. Prospect. Biofuels Biohydrogen, edited by F. Dalena, A. Basile, and C. B. T.-B. S. for the F. Rossi (Woodhead Publishing, 2017), pp. 87–148. doi:10.1016/B978-0-08-101031-0.00004-1
[36] C. Ruskandi, A. Siswanto, and R. Widodo, *Polimesin* 18, 53 (2020). doi:10.30811/jpl.v18i1.1596
[37] K. Khan, S. A. Khan, M. U. Saleem, and M. Ashraf, *Open Constr. Build. Technol. J.* 11, 274 (2017). doi:10.2174/1874836801711010274
[38] M. B. Amran, S. Aminah, H. Rusli, and B. Buchari, *Heliyon* 6, e04051 (2020). doi:10.1016/ j.heliyon.2020.e04051
[39] S. Arita, M. Naswir, I. Astriana, and Nelson, *Am. J. Eng. Appl. Sci.* 11, 845 (2018). doi:10.3844/ ajeassp.2018.845.851
[40] A. V. Koshelev, E. I. Tikhomirova, and O. V. Atamanova, *Russ. J. Phys. Chem. B* 13, 1051 (2019). doi:10.1134/S199079311906006X
[41] M. Sirait, N. Bukit, and N. Siregar, *AIP Conf. Proc.* 1801, 1 (2017). doi:10.1063/1.4973084
[42] S. Barakan and V. Aghazadeh, *Micro Nano Lett.* 14, 688 (2019). doi:10.1049/mnl.2018.5364
[43] S. R. S. Ganesh, H. Göebel, and P. M. Mathias, *Chem. Eng. Trans.* 69, 325 (2018). doi:10.3303/ CET1869055
[44] S. Y. Yoon, S. E. Choi, and J. S. Lee, *J. Nanomater.* 2013, 1 (2013). doi:10.1155/2013/510524
[45] S. Balci, *Int. J. Chem. React. Eng.* 17, 20180167 (2019). doi:10.1515/ijcre-2018-0167
[46] G. An, Y. Jiang, J. Xi, L. Liu, P. Wang, F. Xiao, and D. Wang, *CrystEngComm* 21, 202 (2019). doi: 10.1039/C8CE01571F
[47] M. Naswir, Y. Gusti Wibowo, S. Arita, W. Hartati, and L. Septiarini, *Int. J. Chem. Sci.* 3, 89 (2019).
[48] F. Yusniyanti, W. Trisunaryanti, and Triyono, *Indones. J. Chem.* 21, 37 (2021). doi:10.22146/ ijc.51496
[49] H. Haerudin, N. Rinaldi, and A. Fisli, *Indones. J. Chem.* 2, 173 (2010). doi:10.22146/ijc.21913
[50] I. Fatimah, N. Narsito, and K. Wijaya, *Indones. J. Chem.* 9, 5 (2010). doi:10.22146/ijc.21557
[51] N. M. Deraz, *Int. J. Nanomater. Nanotechnol. Nanomedicine* 4, 1 (2018). doi:10.17352/2455-3492 .000023
[52] S. L. Abdullahi and A. A. Audu, *ChemSearch J.* 8, 35 (2017).
[53] G. A. P. K. Wardhani, N. Nurlela, and M. Azizah, *Molekul* 12, 174 (2017). doi:10.20884/1.jm.2017. 12.2.382
[54] J. T. Kloprogge, L. V. Duong, and R. L. Frost, *Environ. Geol.* 47, 967 (2005). doi:10.1007/s00254-005-1226-1
[55] N. N. Bukalo, G.-I. E. Ekosse, J. O. Odiyo, and J. S. Ogola, *Open Geosci.* 9, 407 (2017). doi: 10.1515/geo-2017-0031
[56] R. Reddy T, K. S, E. T, and L. Reddy S, *J. Lasers, Opt. Photonics* 04, 1000171 (2017). doi: 10.4172/2469-410x.1000171
[57] A. Tinti, V. Tugnoli, S. Bonora, and O. Francioso, *J. Cent. Eur. Agric.* 16, 1 (2015). doi:10.5513/ JCEA01/16.1.1535

[58] J. M. Ramos, M. T. Mauŕicio, A. C. Costa, O. Versiane, and C. A. T. Soto, *ScienceAsia* 37, 247 (2011). doi:10.2306/scienceasia1513-1874.2011.37.247
[59] B. Sarma and B. C. Goswami, *Pelagia Res. Libr. Asian J. Plant Sci. Res.* 6, 71 (2016).
[60] L. Hauli, K. Wijaya, and R. Armunanto, *Orient. J. Chem.* 34, 1559 (2018). doi:10.13005/ojc/340348
[61] I. Kobayashi, H. Owada, T. Ishii, and A. Iizuka, *Soils Found.* 57, 683 (2017). doi:10.1016/j.sandf.2017.08.001
[62] K. Wijaya, A. Syoufian, and S. Ariantika, *Asian J. Chem.* 26, 70 (2014).

# 3 Optimization-based Development of a Circular Economy Adoption Strategy

*Shubham Sonkusare, Neeraj Hanumante, and Yogendra Shastri*
Department of Chemical Engineering, Indian Institute of Technology Bombay, Mumbai, India

## CONTENTS



## 3.1 INTRODUCTION

Human society's current lifestyle patterns are not sustainable; mainly, the resources we consume and their consumption rate are a matter of concern. With industrialization in the 18th century and the advancement of new technology for ease of life, the Earth's natural resources are being extracted at an unprecedented rate [1]. Rapid urbanization and higher life expectancy have further increased the consumption rate.

DOI: 10.1201/9781003167693-3

Philosopher Adam Smith stated, "Consumption is the sole end and purpose of all production" [2]. There are various examples of consumption increase; in 1950 world produced only 2 million tonnes of plastic per year. Since then, the production of plastic has increased by 200-fold and reached 381 million tons in 2015. Fossil fuel consumption has significantly increased over the past half-century, over eightfold since 1950 and roughly two times since 1980. Over the last five decades, material use has tripled, increasing by a factor of 3.5, from 26.7 billion tonnes in 1970 to 92.0 billion tonnes in 2017 [3]. The exploitation of resources has a negative influence on the ecosystem. It causes deforestation, and a considerable loss of biodiversity has been seen in the last few decades. The United Nations Food and Agriculture Organization (FAO) estimates that the annual rate of deforestation is about 1.3 million square kilometers per decade [4]. Human activities in scientific, political, and social spheres will be affected once this resource pool is depleted. Thus, increasing overall consumption leads to ecosystem collapse [5]. The future state of our ecosystem would face even more hardship if not dealt with aptly.

The critical component of these ill effects is the linear economic model. A linear economic model can be described as one following the extract-produce-use-dispose philosophy. Circular economy is seen as a way of addressing the problems arising out of rising consumption. Ellen MacArthur Foundation describes the circular economy (CE) as "an economy which is restorative and regenerative by design." Furthermore, the European action plan states circular economy is where the value of products, materials, and resources is maintained in the economy for as long as possible, and the generation of waste is minimized [6]. The core of CE is a closed loop of material and use of raw material and energy through multiple phases, which are associated with the 3R principle of CE: reduce, reuse, and recycle [7]. The United Nations adopted the 17 sustainable development goals (SDGs) to ensure a sustainable future for us and the next generations [8]. Circular economy significantly contributes towards SDG 12, focusing on sustainable consumption and production. Within SDG 12 milestones, SDG 12.5 highlights the importance of substantially reducing waste generation through prevention, reduction, recycling, and reuse of produced and used goods. These objectives can be addressed by using a circular economy since CE keeps the material within the economy as long as possible, thereby increasing the efficiency of material use. In addition, it helps to achieve sustainability as materials are being used by conserving resources.

The circular economy concept is aggressively promoted by the European Union, China, Japan, the United Kingdom, France, Canada, and many others. Systematic assessment of the costs and benefits of the adoption of CE is essential. Such studies have been reported for specific sectors or products [9–11]. While these sector specific studies are valuable in developing specific policies and guidelines, they cannot provide a broader view of the adoption of circular economy on global sustainability. Consequently, studies at regional and national scales have also been reported. It is estimated that the global economy will be benefitted by up to 1,000 billion U.S. dollars annually with the implantation of CE [12]. According to the Ellen MacArthur Foundation, the CE model has been estimated to save 630 billion USD for medium-level complex goods and 706 billion USD for fast-moving goods just in terms of material cost [13]. The circular economy can reduce primary material consumption by 32% by 2030 [14].

Europe implemented CE practices in plastic waste management, showing 31.1% recycling of plastic, with 41.6% energy recovery, with the rest going to the landfill [15]. Also, in Europe, implementing circular economy can halve carbon dioxide emissions by 2030. In 2015, around 84.4 million tonnes of raw materials, such as minerals, biomass, and fossil fuels, were extracted worldwide. In comparison, only 8.4 million tonnes (only about 10%) of recycled materials re-entered the economic system. This shows the considerable potential of circularity to shift from extraction to reuse and recycling of used products.

The CE represents an annual material cost-saving opportunity of USD 340 to 380 billion per annum at the EU level for a "transition scenario" [16]. From 2014 to 2019, the European

Investment Bank (EIB) provided €2.5 billion to co-finance circular economy projects in various sectors and aimed to invest at least €10 billion in the circular economy by 2023 [17]. The adoption of a circular economy is also expected to have broader sustainability consequences. It is important to study these complex interlinkages across the human, technological, and ecological sub-systems to better understand the impact of CE adoption. This points towards the use of integrated models for sustainability assessment.

Hanumante et al. [18] reported a study to explore the potential benefits and limitations of global CE adoption. They adapted a global integrated model that captured major components of the global system and incorporated circular economy. Different possible CE adoption trends were simulated, and the impact on all the model compartments was studied. One of the key observations of this work was that the very aggressive adoption of CE was also detrimental to global sustainability. While it ensured increasing prosperity of the human-controlled sectors, it had negative effects on natural sectors. This is not acceptable as per the concept of strong sustainability. It was apparent that there was an optimal pathway of CE adoption that could ensure the sustainability of all sectors.

The main goal of this work is to use systematic methods to identify the optimal pathway for CE adoption within the modeling framework used by Hanumante et al. [18]. This work uses a dynamic optimization formulation to determine the optimal CE adoption trend. The information theory-based metric of Fisher information (FI) is used to quantify sustainability, and the optimization problem is formulated using FI-based objective function.

The chapter is arranged as follows. Section 3.2 provides a brief description of the selected model and details how CE aspects are incorporated. Section 3.3 summarizes the optimization model formulation with Fisher information. Section 3.4 deals with scenario planning. Section 3.5 deals with these scenarios' simulation results and discusses their implications. The last section presents the concluding remarks.

## 3.2 INTEGRATED PLANETARY MODEL

In the various attempts to develop economic and ecological models, the coupling between economic and ecological components is a critical factor [13]. This coupling is modeled through a single link or path in several integrated models with ecological and economic dimensions. However, owing to such coupling, the capability of these models to represent the real world is limited. Furthermore, in some models functioning based on general computational equilibrium, vast amounts of data are required. Thus, in 2006, the U.S. EPA developed a planetary compartmental model [13].

This model has the following features: The model is addressed as a closed mass system; no mass and energy can be exchanged outside the system. Segregating the mass by their property type has been done for incorporation of legal foundation. In addition, the economic sector is integrated with the ecological sector via a price-setting model.

The primary aim of developing the model is to get insight as to which strategies might be desirable or undesirable in an integrated system. Moreover, some understanding could be gained about policy making that can change social and economic behavior to increase the stability of the system. As this model is not calibrated with real-world data and is not tried in any real system, it is abstract in nature. Most importantly, this model permits one to identify solutions to address unsustainable situations.

Shastri et al. [19] used the model to formulate scenarios and study the effect of resource consumption and population growth on system sustainability. Kotecha et al. [20] added the energy sector to it and studied the implications of biofuel on system sustainability. Hanumante et al. [18] worked on the compartmental model and introduced the circular economy. They used this model to study the global implications of adopting the circular economy. They modified the model to incorporate a circular industry functioning based on the principles of circular economy. We will describe this modified model in the following section.

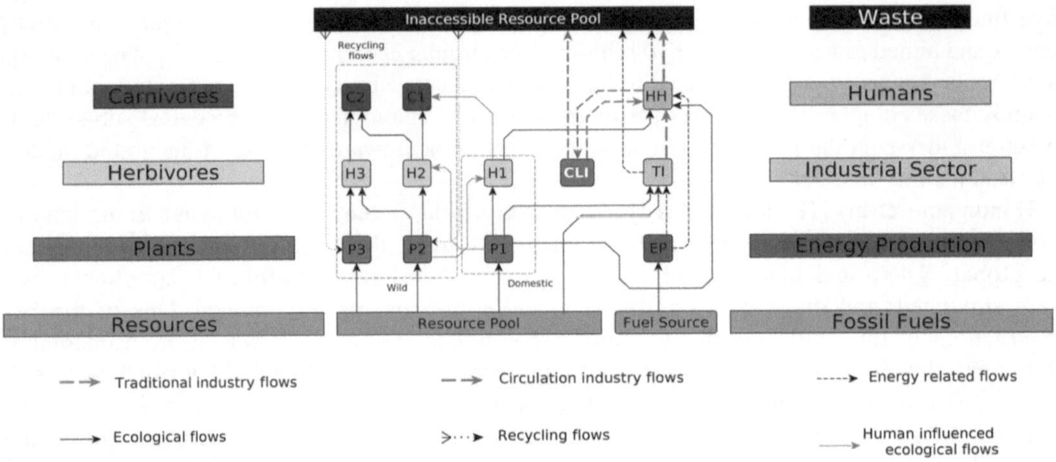


**FIGURE 3.1** Integrated planetary model. A solid black line represents mass transfer between different compartments (e.g., between P3 and H3). The dashed black line represents energy flow (e.g., between EP and HH); the thick dashed line represents a flow of industrial goods (e.g., between TI and HH). The light grey lines are associated with human influenced ecological flows (e.g., between P1 and H2). The black dotted line represents the microbial degradation of waste (i.e., recycling flows).

### 3.2.1 Model Description

The integrated planetary model (Figure 3.1) is a predator-prey model. There are multiple trophic levels, and the total system mass is distributed into various compartments. The higher trophic level feeds upon the lower trophic level for survival. Multiple groups of species are considered at each trophic level; hence, intra-trophic level diversity can be seen in the food web model. In this model, mass is distributed into 14 compartments. The dynamics of mass distribution depend upon economic decisions and biological interactions. The system is closed to mass and energy, except solar energy, which is used for plant growth (representing planetary limits). The stocks and flows of mass for each compartment are modeled with respect to time. Primary producers make the resource available to the rest of the food web by taking resources from the resource pool. The nomenclature of various compartments is shown in Table 3.1.

**TABLE 3.1**
**Description of the Integrated Planetary Model Compartments**

| Trophic level | Compartment | Description |
|---|---|---|
| Resource pools | RP | Resource Pool |
| | IRP | Inaccessible Resource Pool |
| | ES | Energy Source |
| Plants | P2 | Grasslands |
| | P3 | Forests |
| Herbivores | H2, H3 | Feral Herbivores |
| Carnivores | C1, C2 | Feral Carnivores |
| Humans | HH | Human Households |
| Industries | TI | Industrial Sector |
| | EP | Energy Production |
| | P1 | Agriculture |
| | H1 | Livestock |

Various intra-tropic level diversified groups can be seen at the tropic levels mentioned on the left. P1 and H1 constitute the domesticated part of the ecosystem representing agriculture and livestock, respectively. On the other side, P2, P3, H2, H3, C1, and C2 are feral, i.e., no control of humans on these compartments. Flow from the domestic sector to the wild sector represented by solid light grey lines is prevented by fences. P2 represents open access to a plant, such as a grassland consumed by domestic herbivorous H1, again represented by solid light grey lines.

There are four industries in this model: P1, H1, TI, and EP. P1, H1, and TI produce generic goods using resource pool RP. Consumption of products by human households is denoted by the thick dashed line, mass flow from TI to Inaccessible Resource Pool (IRP) through HH. The energy flow is represented by the black dashed line emerging from EP, the energy production industry. Energy is supplied to the industrial sector and human households. Energy flow does not contribute to the mass of the compartment; it is directly transferred to IRP. IRP represents an accumulation of material that primary producers cannot use directly, i.e., plants. Mass from IRP undergoes microbial degradation at a very low rate to P2 and P3, represented by dotted lines. Solid lines represent the mass flow between the compartments.

The flows across the compartments depend not only on their masses but also on the economic situation. The economic variables also depend upon the stocks of the ecological compartments, thus, featuring a complete integration of these dimensions. Conservation of mass is ensured for each flow across the compartments. This model portrays an industrialized world, as the industrial sector dominates the economic component. Equations reflecting these features are discussed in more detail in upcoming sections.

### 3.2.2 Model Governing Equations

This section provides governing equations for different parts of the model, namely the ecological and economic dimensions and the modeling of human society.

#### 3.2.2.1 Ecological Stocks and Flows

The ecological dimension is based on the predator-prey relationship. The species at the higher trophic level feed on the species at the lower trophic level for survival. This intra-trophic level diversity has been created by introducing multiple compartments at each tropic level. As a result of these dependencies, a complex food web model is developed. Equations-3.3 presents the generalized equation of mass of any compartment $j$ connected to a lower trophic level ($i$) and higher trophic level ($j$).

$$M_j^{n+1} = M_j^n + \sum Q_{ij}^n \times \delta_t - \sum Q_{jk}^n \times \delta_t - m_j \times M^n \times \delta_t \tag{3.1}$$

$$Q_{ij}^n = g_{ij} \times M_i^n \times M_j^n \tag{3.2}$$

$$Q_{jk}^n = g_{ik} \times M_j^n \times M_k^n \tag{3.3}$$

where, $M$ represents mass of compartment, $Q$ is mass flow across the compartment, $\delta$ is per capita demand, and $g$ is ecological mass transfer parameter. These equations indicate that the flow between any two ecological compartments is proportional to their masses.

#### 3.2.2.2 Economics-related Flows

The economic dimension is incorporated in the form of a price-setting model (Equation 3.4-3.6). The price is computed based on the system variables, and this price then governs the market dynamics. This is employed in four industries: agriculture P1, animal husbandry H1, energy production EE, and consumer goods TI. Per capita demand for products of these industries is a function of the prices. Production targets for each of these industries are computed based on system variables that determine its raw material demand. Based on availability, these raw material demands are satisfied, and hence corresponding flows are altered [21].

$$P_j^n = a_j^p + b_j^p \times W^n - c_j^p \times \Delta_j^n \tag{3.4}$$

$$Pr_j^n = a_j^{pr} - b_j^{pr} \times W^n - c_j^{pr} \times \Delta_j^n \tag{3.5}$$

$$\Delta_j^n = M_j - \bar{M} + M_{j\ def}^n \tag{3.6}$$

Here, $W^n$ represents the wage rate, $\bar{M}$ represents minimum inventory to be maintained, $Mj\ def^n$ represents the deficit, and $\Delta j^n$ represents the number of units currently in the stock to meet the demand during the next time step. Parameter $a$ represents the base value and $b$ and $c$ capture the sensitivity to the labour availability through wage rate and the availability of the finished goods, respectively.

#### 3.2.2.3 Human Society

Human society is modeled to work in one of the four industries in exchange for wages. The dominant sector fixes this wage rate. In the current work, as an industrialized society is modeled, wage rate, $W^n$, is governed by the consumer goods industry. Population and the wage rate are inversely correlated. It is computed using Equation 3.7, where $a_w$, $c_w$, and $d_w$ are the constant parameters. The second and third terms incorporate the effect of total availability of finished goods and the population [21].

$$W^n = a_w - c_w \times M_{Wa}^n - d_w \times N_{HH}^n \tag{3.7}$$

$$(\lambda_{RP} + \theta_{P1}) \times M_{Wa}^n = M_{IS}^n - \bar{M}_{IS} + E_{IS}^n \tag{3.8}$$

where, $E_{IS}^n$ is the net unsatisfied demand for the consumer goods, which is the difference between the total demand and the total supply.

A population model linked with economic prosperity is used for modeling population dynamics. The per capita birth rate is negatively correlated with the wages indicating that the population is expected to grow at a slower pace with higher prosperity. Equation 3.9 is used to compute the per capita birth rates. $U^n$ represents the weighted price given by Equation 3.10.

$$B^n = \eta_a - \eta_b \times \sqrt{\frac{W^n}{U^n}} \tag{3.9}$$

$$U^n = \frac{P_{P1} \times Q_{P1HH}^n + P_{H1} \times Q_{H1HH}^n + P_{IS} \times Q_{ISHH}^n + P_{EP} \times Q_{EPHH}^n}{Q_{P1HH}^n + Q_{H1HH}^n + Q_{ISHH}^n + Q_{EPHH}^n} \tag{3.10}$$

### 3.2.3 Modeling of Circular Economy

The integrated model described without circulation industry represents a produce-use-discard linear economy. The Industrial Sector (TI) uses raw materials from P1 and RP to produce industrial goods. After usage, humans discard this into the inaccessible resource pool; from there, they return to the ecological food web at a very low rate. In incorporating the circular economy, the stream discarded must be brought back again [18].

The Traditional Industry (TI) uses virgin raw materials and produces goods used by the human household. In addition to the traditional industry (TI), the circular economy has been incorporated through the circulation industry (CLI). This industry uses the product coming out of the human sector and recycles them through a processing sector. The circulation industry consists of two sub-compartments, namely, CLI processing and CLI inventory. The CLI processing sector receives the used goods and makes them fit for use. This is assumed to be done through some generic processing step that uses time and energy. Thus, the material spends some time in this compartment, and some energy is required to complete the processing. Moreover, it is assumed that not all material can be recycled, and there is some material loss during this processing step. Recycled goods are transferred to the CLI inventory compartment from where the human households take them for use. The fraction of post-consumption waste delivered to the circulation industry is called the circulation fraction (*CF*) defined below:

$$CF = \frac{HHCLI}{HHCLI + HHIRP} \tag{3.11}$$

where, *HHCLI* is post consumption flow of industrial goods by human household to circulation industry (CLI) and *HHIRP* is good discarded to inaccessible resource pool from human household. Higher the CF more actively will be circular economy incorporation. This approach of modeling CE assumes that the recycled goods are equivalent to the original goods in terms of quality and service provided. Material loss modeled during the CLI processing step ensures the infinite recycling of material is not considered. Additionally, this work ignores the possible differences in the prices of the original and recycled goods. The other assumptions of the integrated model are not modified.

### 3.2.4 Model Simulations and Observations

Hanumante et al. [18] simulated different growth patterns of the circular economy using this model. Since it is not possible to know the future exactly, Hanumante et al. [18] considered different possible CE adoption trends. Using a combination of different values of maximum circulation fraction, the delay in adaption, as well as the rate of adoption, they simulated 910 possible scenarios. For each scenario, the global sustainability was assessed by analyzing the trends in the masses of different compartments.

Figure 3.2 shows three broad trends that were observed. In collapse mode 1 (Worldview 1), delay in the adoption of CE leads to the extinction of P1, leading to major instability in the system. This is because P1 is an essential resource for humans (food), domestic animals (food), and industry (raw material). In contrast, in collapse mode 2 (Worldview 3), very aggressive adoption of CE leads to the extinction of P2, which can be thought of as representing grasslands in the model. Note that in this collapse mode, the humans dominated sectors flourish, and therefore it can be argued that the system is sustainable if we take a human centric view. The extinction of P2 happens due to a rapid increase in the size of the P1 sector, which deprives P2 of natural resources for survival. From a practical standpoint, this can be viewed as the conversion of grasslands into agricultural land. For the CE adoption trends between these two extremes, none of the model compartments go to extinction, raising the hope for a sustainable system.

These results implied that there is a zone of sustainability as far as CE adoption trends are concerned. Therefore, determining the model's desirable CE adoption trend may not be straightforward and will require more systematic analysis. This idea is the foundation of the work reported subsequently here. We have formulated a dynamic optimization problem to determine the optimal CE adoption trend. One of the challenges in formulating the optimization problem is the identification of

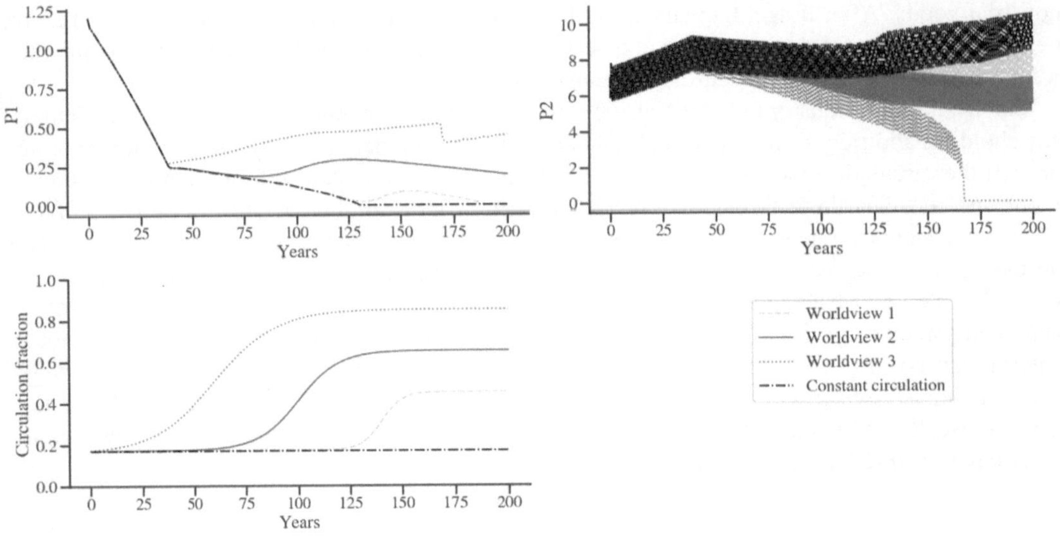


**FIGURE 3.2** Selected circulation fraction trends and the corresponding dynamics of P1 and P2 compartments capturing three distinct regimes of the system.

an appropriate objective function. Given the highly diverse nature of the model, with compartments reflecting very different types of resources or activities, coming up with an objective function is challenging. The next section describes the objective function considered in this work and the formulation of the optimization problem.

### 3.2.5 The Need for a Novel Method

There are various methods and ways of quantifying circular economy, but it gives different units of circularity. There is no standard unit for measuring circularity; also, it is complicated to do so. So there is a need for an index that will account for all factors related to the circular economy and give a single number for calculating sustainability. This can be understood as the same as the GDP of a country, which accounts for all monetary measures of the market value of all final goods and services produced in a specific period and gives it in one quantity.

Information theory, which can mathematically depict sustainability through an information index, can help in such a case. There is information available on concepts such as Shannon's information and Ginni Simpson's information; here, we specifically discuss the Fisher information index. The trajectory of P1 collapse, P2 collapse, and sustainable can be used as a prior indication of the actual collapse to take place.

Furthermore, we must be able to distinguish between the collapse and sustainable behavior. For this purpose, Fisher information is used to find the stability of the system. Fisher information gives the maximum amount of information available from the system of observation. The next part deals with Fisher information calculation and its application on an integrated planetary model with an optimization problem-solving method.

## 3.3 OPTIMIZATION MODEL FORMULATION

In this work, the information theory-based index known as Fisher information (FI) has been used to formulate the objective function. The idea behind an information index is that any data or model variable can be converted to information regardless of its discipline. So, information can act as an interdisciplinary bridge. Information theory is not only used in quantifying sustainability; it has a wide range of applications. For example, the Shannon information index measures system capacity in a biological and ecological system. Biologists use information theory to quantify the organization of genetic and macromolecular structures. Information theory has been used in thermodynamics to measure deviation from equilibrium. With such a broad application, information theory has now gained importance in quantifying sustainability. The Fisher Information index is used for this purpose and successfully in predicting the sustainability in the predator-prey model [22] and complex system [19]. The next section briefly summarizes the concept of Fisher information.

### 3.3.1 Fisher Information

Ronald Fisher introduced a statistical measurement factor known as Fisher information [19]. Fisher information is interpreted as a measure of the ability to estimate a parameter, as the amount of information that can be extracted from a set of measurements, and also as a measure of a state of order or organization of a system or phenomenon. Fisher information is given as

$$I = \int \frac{1}{p(x)} \left( \frac{dp(x)}{dx} \right)^2 dx \tag{3.12}$$

where, $p$ is the probability density function and $x$ is the variable. Fisher information is a local property as it has a derivative of the probability distribution function. This makes it more sensitive as perturbation will affect the probability density function. A highly disordered system is unbiased, or the uniform probability distribution function has low Fisher information. On the other hand, an ordered system has certain systems states that are more probable, which results in higher Fisher information. Fisher information and predictability are directly proportional to each other. Fisher information value of zero implies we cannot predict the system state at all [21]. For the stability of an ecosystem (static or dynamic), the system must not lose or gain species, affecting the system dimensionality and, hence, FI's value. Therefore, the sustainability hypothesis states that a system's time-averaged FI in a persistent regime does not change with time. Any change in the regime will manifest itself through a corresponding change in FI value [22].

Cabezas and Fath [22] stated two additional corollaries to this hypothesis: If the FI of a system increases with time, then the system maintains the state of the organization. Moreover, if the FI of a system decreases with time, then the system loses its state of organization, as shown in Figure 3.3.

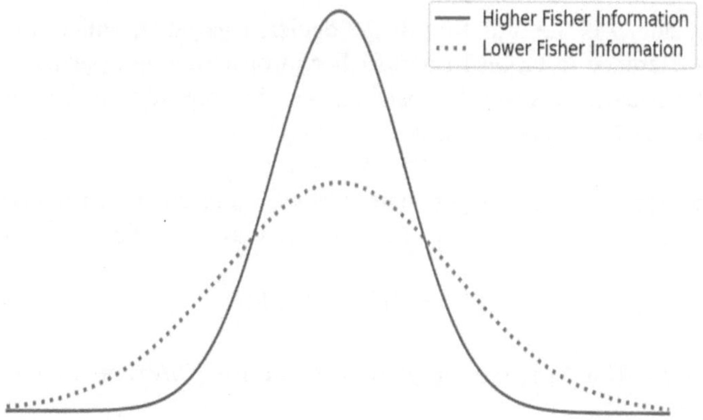


**FIGURE 3.3** Different probability density functions with different values of Fisher information.

Fath et al. [21] have successfully demonstrated this regime shift in the complex food web with a corresponding change in Fisher information. This leads to the minimization of FI variance over time as a possible objective function for control problems. Shastri and Diwekar [23] used an FI-based objective function in an optimal control framework to manage a three-species predator-prey system. They found that the minimization of Fisher information variance around a target value or profile leads to more stable results as compared to the maximization of the Fisher information objective. Based on this learning, Shastri et al. [19] uses the minimization of the FI variance objective function in a more complicated integrated model (the earlier version of the model discussed here). They could show that by using waste discharge fee as a control variable, it is possible to delay or eliminate unsustainability.

### 3.3.2 Optimization Model

An optimization model consists of three parts: objective function, decision variables, and constraints. For the proposed optimization model, constraints are the equations governing the integrated model. These are not reported here for brevity, and interested readers can find the details in Whitmore et al. [24] and Kotecha et al. [20].

Methods to solve optimization problem are generally classified as gradient-based and direct search methods. Since the proposed optimization problem is discontinuous, the direct search method combined with the steepest descent has been used [25,26]. Based on one of the sustainability hypotheses, the minimization of Fisher information variation around a target trajectory has been used as the objective function. The objective function is defined as follows:

$$J = min \int_0^T (I(t) - Ic(t))^2 dt \tag{3.13}$$

where, $I(t)$ is the current FI profile, $Ic(t)$ is the targeted FI profile for a stable system, and T is the total time horizon under consideration. The model presents multiple potential decision variables that can be directly or indirectly modulated. This includes circulation fraction, rate of resource consumption by humans, and population. Among these, the rate of resource consumption and population are very difficult to control since they involve social and behavioral aspects. In contrast, the circulation fraction is relatively easier to control through policy interventions such as regulations, subsidies, and taxes. Therefore, it is selected as the time-dependent decision variable and is modified as per the following function:

$$CF[k+1] = CF[k] + \alpha \times p \tag{3.14}$$

where, $\alpha$ is step size and $p$ is the direction of the decision variable, which is circulation fraction, and is equal to the difference in target FI and Fisher information at a particular time ($Ic - It$). In addition to the model equations, several constraints can be imposed on the circulation fraction to make the formulation realistic. These can be:

- **Growth in circulation fraction per year:** Growth in circulation fraction per year (G) is the change in circulation fraction in one year and can be modeled as follows:

$$G = cf[i+1] - cf[i] \tag{3.15}$$

  where, $cf[i]$ and $cf[i+1]$ is the circulation fraction in $'i'th$ time step and $'i+1'th$ time step, respectively.

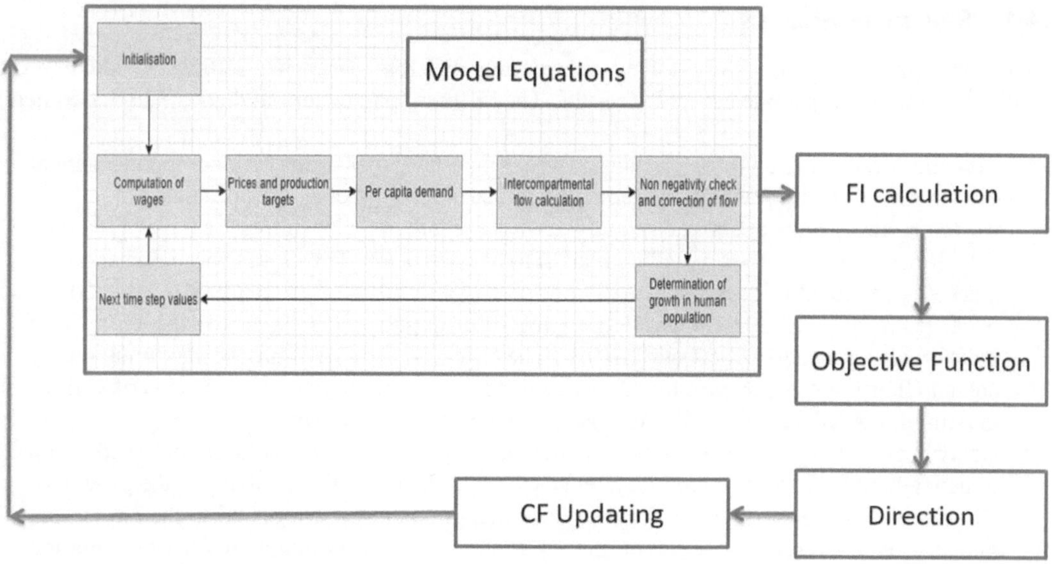


**FIGURE 3.4** Sequence of steps depicting the optimization procedure.

- **Maximum and minimum achievable circulation fraction:** It is essential to put an upper and lower limit to CF to make the problem realistic. This is to ensure the fact that 100% circulation of goods is not possible. In addition to it, non-negativity constraints should be satisfied.
- **Delay in the adoption of circulation fraction:** Delay in growth is one of the possible constraints that ensure the circulation fraction does not increase for a certain period. This constraint is to provide the policymakers time that is required for the adoption of new technology.

Figure 3.4 summarizes the solution procedure of the optimization problem. For an initial guess of CF, the current global circulation of industrial goods is considered. The following section will explain the scenarios studied in this work.

## 3.4 SCENARIO PLANNING

The first thing before solving the optimization problem is the calculation of Fisher information, as the objective function is calculated using FI. FI must give some indication prior to P1 collapse, P2 collapse, and sustainable case. In this case, input to the Fisher information calculation is state variable time series data. The FI profile comprises a stepwise calculation in FI, where each step represents the time average FI for one year.

One of the challenges while calculating FI is what state variables must be chosen to calculate FI. There are multiple options to calculate FI; these options were specified according to their role in the model. They were ecological, economic, and mixed variables. For studying the behavior of FI, we identified scenarios for which analysis is carried out. By exploring these three options, the ecological option of calculating FI is able to predict the collapse and track the system behavior, hence is selected for solving an optimization problem.

The optimization model results are compared with those reported by Hanumante et al. [18] based on heuristics. It is expected that the optimization model will provide better results leading to improved sustainability of the system.

### 3.4.1 Sensitivity Analysis

Several optimization models were solved considering different parameter settings to identify the sensitivity of particular parameters on the results. The different parameters studied are discussed here:

- Maximum rate of growth of circulation fraction: The paper industry's compound annual growth rate (CAGR) was 2.97% from 1961 to 2018. Redling et al. [27] predicted the global waste market to grow at about 6%, and Bizwit Research and Consulting LLP [28] predicted CAGR for the waste paper recycling market to be over 3.4%. The global waste management market is expected to grow from $285 billion to $435 billion by 2016 to 2023, CAGR of 6.2% from 2017 to 2023. Considering this information, the CAGR for the circulation industry is assumed to be between 1–10% . The base case considers that the maximum rate of annual CF growth is 5% over the CF value during the current year. The impacts of various maximum CAGR between 1% and 10% in a step of 1% are quantified.
- Delay in growth of CF: Delay in the growth of CF captures the time required for the adoption of new technology. As a dynamic system requires time to stabilize, a delay in the growth of CF will give a proper time gap for the consumed goods to get circulated to the circulation industry. Moreover, the adoption of new technology (i.e., delay in the growth of circulation fraction) will give time for policymaking and implementation. In this work, the delay is simulated by enforcing the time at which CF can start to increase. This time is varied between 10 years and 120 years from the simulation start time in an interval of 10 years.
- Maximum achievable circulation fraction: The maximum circulation fraction cannot be 100% given that recycling/remanufacturing always leads to some wastes. According to the Aluminum Association, 90% of aluminum in building and automotive parts is recycled at the end of life [29]. Based on limited available data, the highest recycling rates for plastic in 2014 were in Europe (30%) and China (25%), whereas in the United States, plastic recycling has remained steady at 9% since 2012 [30]. According to a report published in 2015 steel recycling rate in North America was found to be 81% [31]. Therefore, the optimization model is solved while varying the maximum CF achievable between 17% and 90%.

The two primary components defining the case are population growth and consumption increase. The projected population growth is based on the UN projection of the human population from the year 2000 to the year 2300 United Nations [32]. The human population is projected to peak around 2050 and become reasonably stable for the medium population growth projection. The consumption increase is modeled as a demand coefficient relative to consumption at time t = 0. The consumption increase of 'x' represents that total consumption by humans of goods has increased by 'x' time by the end of the simulation horizon (2300) as compared to t = 0. The consumption growth pattern is shown in Figure 3.5.

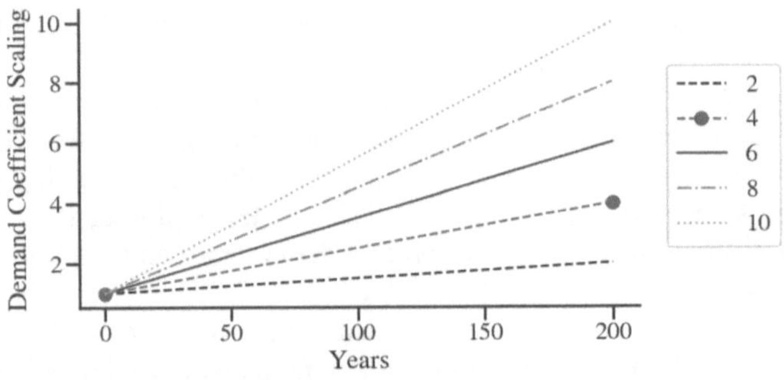


**FIGURE 3.5** Different consumption level increases modeled in the analysis along with their increase pattern. Legends in the plot indicate the level of consumption increase.

A consumption level increase of 6 has been considered for simulation with reference from Hanumante et al. [18]. Simulations have been carried out over 200 years with a time step of 1 week. Calculation of FI has been made as an average time value of 1 year as the rolling window. For illustrative purposes, the reported results are for consumption level 6. With the above parameters taken into consideration now, we will deal with the results and discussion in the next section.

## 3.5 RESULTS AND DISCUSSION

The optimization problem has been solved for the scenarios stated in Section 4. The following section will deal with the result and discussion part.

### 3.5.1 Optimization Problem Solution

Simulations for the optimization problem are carried out for consumption level 6 and 5% growth of CF per year. The optimization results show that the circulation fraction rises rapidly from its initial value of 0.17 and reaches 0.40 at around year 25. After that, the CF is maintained at that value, rising again towards the end of the simulation horizon after 150 years. As can be seen, the optimal solution has a much better FI profile with lesser fluctuations.

The P1 and P2 profiles reported in Figure 3.6 also confirm that the system is sustainable and does not lead to the extinction of any of the two compartments. Since this is the main requirement for sustainability, it can be argued that the optimal solution has achieved its goal.

Qualitatively, the CF profile is very different from the ones considered by Hanumante et al. [18]. Based on the typical adoption of new technology, the CF profile followed an S-shaped behavior. However, the optimization results indicate that the profile should be such that a rapid increase in CF is followed by a stable value for a long period of time. It must, of course, be acknowledged that this is a non-convex problem, and multiple optimal solutions or multiple solutions with optima very close to the one reported may exist.

### 3.5.2 Sensitivity Analysis

Sensitivity analysis has been done for three cases, namely, the growth of circulation fraction per year, delay in growth, and maximum circulation fraction.

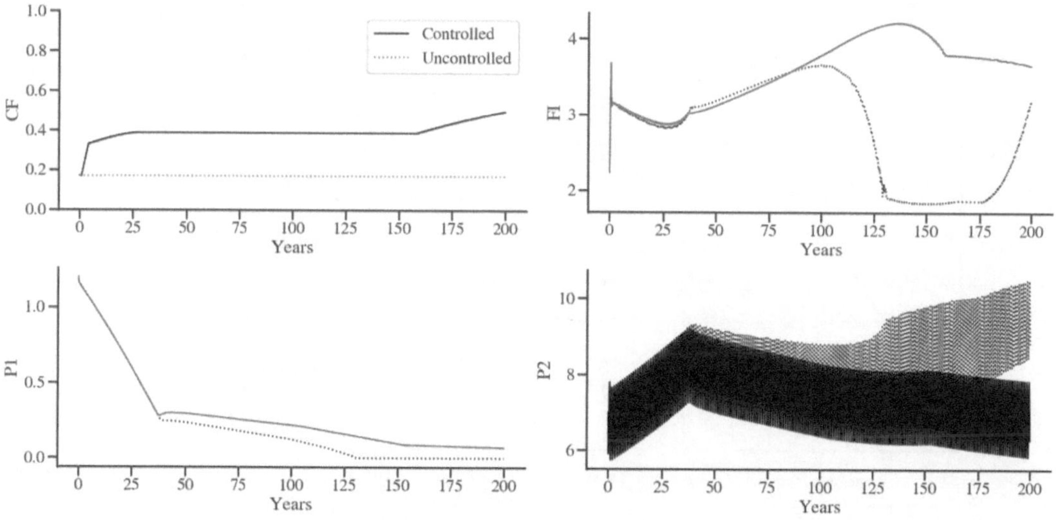


**FIGURE 3.6** Trends for circular fraction, Fisher information, P1 and P2 for optimal problem solution for the consumption level increase of 6 with an upper limit of 5% growth in circular fraction per year.

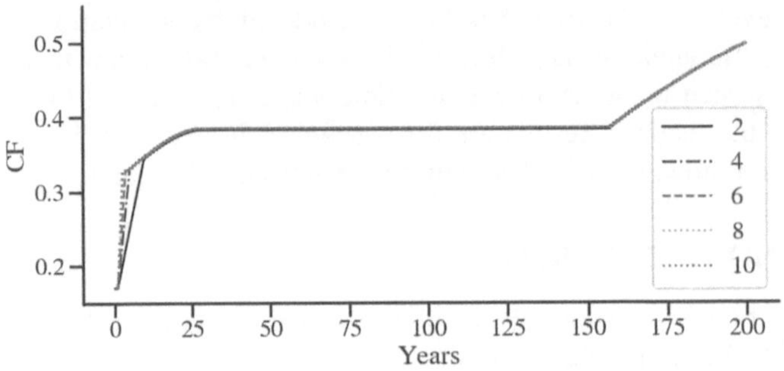


**FIGURE 3.7** Optimal circulation fraction corresponding to optimized solution for different upper limits on the annual increase in circular fraction. Legend in the plot shows the limit on the growth of circulation fraction per year.

#### 3.5.2.1 Annual Rate of Growth in CF

The optimization problem has been solved for different growth per year in CF to get the optimal trend of CF. The optimal trend of CF for maximum growth of CF can be seen in Figure 3.7.

At the growth rate of 2% per year, we are getting minima, and we have CF trends corresponding to that, as shown in Figure 3.7. So, for consumption level 6 growth per year of 2%, CF is an optimized solution. A similar trend can be applied to all other consumption levels to get the optimized solution. The circulation of goods from the human compartment is plotted in Figure 3.8.

The flow of industrial goods from human households to the circulation industry is increasing with time. As consumption increases with time, the quantity of material coming to the human compartment increases, so more circulation needs to be done to reduce the extraction load from primary producers. The population gets stabilized at around 100 years of simulation time, as shown in Figure 3.8. The increase in capacity of processing of circulation industry wage rate increases as the population is stabilized. And as the circulation quantity increases, the domesticated compartment is stabilized, as shown in Figure 3.9.

#### 3.5.2.2 Delay in Growth in CF

Sensitivity analysis has been done for the delay in growth of CF from 10 years to 120 years for consumption level 6. Results show that delay in the growth of the circulation fraction is beneficial,

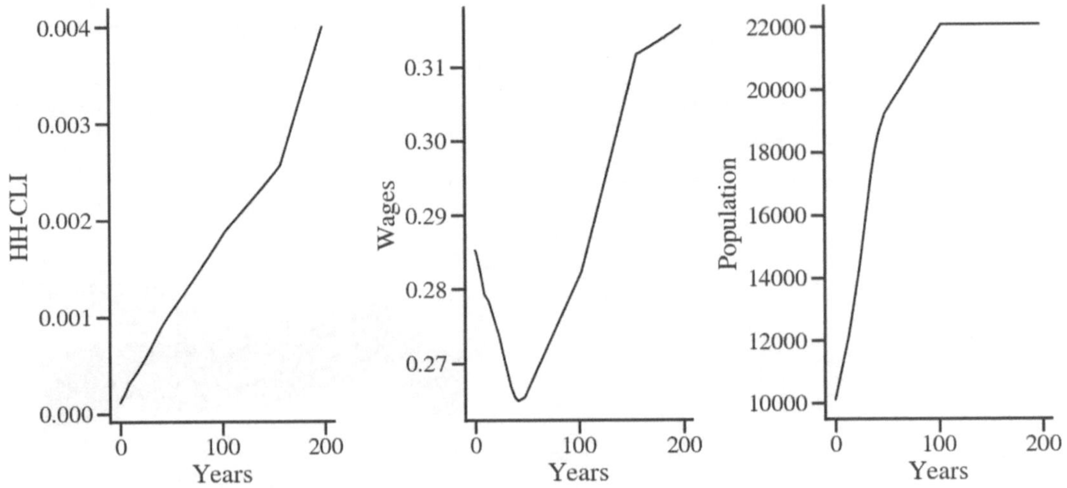


**FIGURE 3.8** Optimal HH-CLI, wage rate, and population for the optimal solution for consumption level increase of 6.

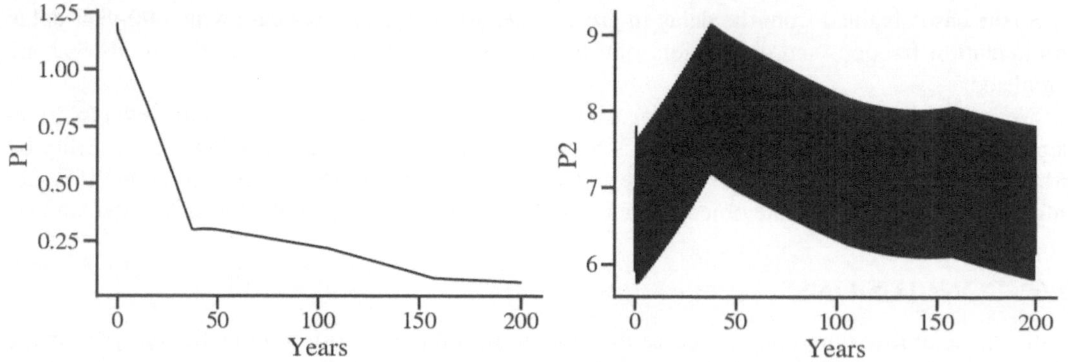


**FIGURE 3.9** Optimal profiles for P1 and P2 mass for consumption increase level 6.

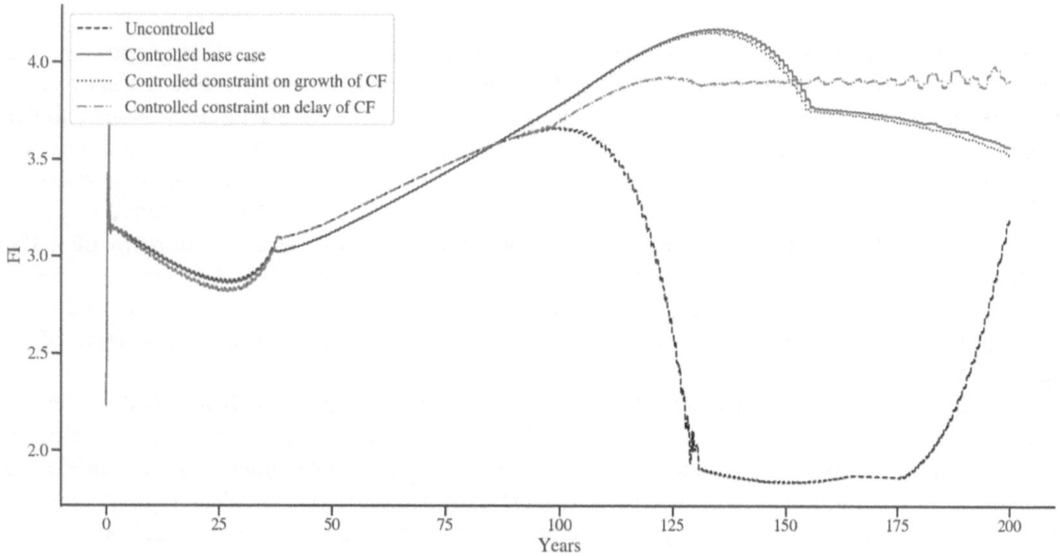


**FIGURE 3.10** Fisher information profile for uncontrolled and different controlled cases.

and 90 years of delay in the growth of CF gives the optimal case. It does not mean that the amount of goods circulated is constant. As consumption increases, the fraction is constant; however, the amount of goods circulated increases.

Figure 3.10 compares the uncontrolled FI with the optimal control FI of the base case, a case with a maximum growth of CF, and a case with a delay in the growth of CF. The results show that FI is more stable for a controlled system than an uncontrolled one. Moreover, FI decreases with time for uncontrolled systems, indicating the system collapse. On the other hand, FI is increasing with time in controlled cases showing the self-sustaining state of the system. The system is more stable when the growth of CF is delayed by 90 years which goes in hand with the analogy of FI; that is, the FI of the stable system does not change with time.

### 3.5.2.3 Maximum Limit of CF

The maximum limit of CF has been limited—to 0.55, 0.5, 0.45, and 0.4—for analyzing the effect. The objective function value decreases with the decrease in the circulation fraction maximum value. For CF limits of 0.45 and 0.4, the system is unable to get a sustainable regime.

So the case obtained from the delay in growth scenario is the optimal case with a 90-year delay in circulation fraction, and circulation industry is growing without increasing the goods that are circulated.

We have chosen Fisher information to solve the optimization problem and develop time-dependent policies. Fisher information has been a promising tool for quantifying sustainability by properly indicating P1 and P2 collapse, and sustainable mode. So optimization with Fisher information used to formulate objective function is a proper tool to quantify systems sustainability.

## 3.6 CONCLUSIONS

With sustainability emerging as the central theme in today's world, systematic analysis of the complex system is essential. This work aims to develop time-dependent management policies for sustainability in the integrated planetary model. An index that can measure sustainability in a time-dependent function has been identified as the Fisher information (FI) index. Fisher information has been used for objective function formulation, and dynamic optimization has been solved.

Optimization problem formulation with minimization of variance of FI can optimize the solution with constraint as circulation fraction. Comparison of results with predefined trends gives the verification of optimal trends. Sensitivity analysis shows the rate of 2% circulation growth per year is optimal for different growth patterns, which is feasible in reality. Analysis concerning the delay in growth gives an optimal scenario for 90 years of delay in the growth of circulation fraction. The flow of human households to circular industry (HH-CLI) has a non-negative slope for a sustainable regime. With the decrease in the maximum value of the circulation fraction, the system moves towards a less optimal solution. As a concluding remark, HH-CLI needs to increase or be constant as the increasing amount of goods consumed is increasing, hence, circular industry. The optimal scenario is the growth of the circulation fraction of 2% per year with a delay of 90 years in growth.

Limitations of the model used for CE studies are that the consumer goods are modeled in the aggregate form as a single product. This study does not consider severe events such as floods, economic crises, and pandemics. The goods produced by the circulation industry are considered to be identical to the traditional industry products. Perfect substitution is difficult in reality. There is always a down cycling in the process of circulation of goods except for some exceptions.

## REFERENCES


[1] Kleczkowski, A., Ellis, C., Hanley, N. and Goulson, D., 2017. Pesticides and bees: Ecological-economic modelling of bee populations on farmland. *Ecological Modelling*, *360*, pp. 53–62. Smith, Adam. 1776. *An Inquiry into the Nature and Causes of the Wealth of Nations*. London: Methuen & Co (Book IV, chapter 8, 49).

[2] Smith, A., 2010. *The Wealth of Nations: An Inquiry Into the Nature and Causes of the Wealth of Nations*. Harriman House Limited.

[3] De Wit, M., Hoogzaad, J. and Von Daniels, C., 2020. *The Circularity Gap Report 2020*. Ruparo: Amsterdam, The Netherlands.

[4] Pinstrup-Andersen, P. and Pandya-Lorch, R., 1998. Food security and sustainable use of natural resources: a 2020 vision. *Ecological Economics*, *26*(1), pp. 1–10.

[5] Shastri, Y., Diwekar, U. and Cabezas, H., 2008. Optimal control theory for sustainable environmental management. *Environmental Science & Technology*, *42*(14), pp. 5322–5328.

[6] Haupt, M. and Hellweg, S., 2019. Measuring the environmental sustainability of a circular economy. *Environmental and Sustainability Indicators*, *1*, p. 100005.

[7] Lieder, M. and Rashid, A., 2016. Towards circular economy implementation: a comprehensive review in context of manufacturing industry. *Journal of Cleaner Production*, *115*, pp. 36–51.

[8] Desa, U.N., 2016. Transforming our world: The 2030 agenda for sustainable development.

[9] Grossmann, H., Handke, T. and Brenner, T., 2014. Paper recycling. In *Handbook of Recycling* (pp. 165–178). Elsevier.

[10] Jacobsen, N.B., 2006. Industrial symbiosis in Kalundborg, Denmark: a quantitative assessment of economic and environmental aspects. *Journal of Industrial Ecology*, *10*(1-2), pp. 239–255.

[11] Rizos, V., Tuokko, K. and Behrens, A., 2017. The Circular Economy: A review of definitions, processes and impacts. *CEPS Papers*, (12440).

[12] Korhonen, J., Honkasalo, A. and Seppälä, J., 2018. Circular economy: the concept and its limitations. *Ecological Economics*, *143*, pp. 37–46.

[13] MacArthur, E., 2015. Circularity indicators: An approach to measuring circularity. *Methodology*, *23*.

[14] *Ellen macarthur foundation* 'The circular economy in detail'. Last access date 19-11-2022.

[15] Horodytska, O., Kiritsis, D. and Fullana, A., 2020. Upcycling of printed plastic films: LCA analysis and effects on the circular economy. *Journal of Cleaner Production*, *268*, p. 122138.

[16] MacArthur, E., 2013. Towards the circular economy. *Journal of Industrial Ecology*, *2*(1), pp. 23–44.

[17] Goovaerts, L., Schempp, C., Busato, L., Smits, A., Žutelija, L. and Piechocki, R., 2018. Financing Innovation and Circular Economy. In *Designing Sustainable Technologies, Products and Policies* (pp. 427–432). Cham: Springer.

[18] Hanumante, N.C., Shastri, Y. and Hoadley, A., 2019. Assessment of circular economy for global sustainability using an integrated model. *Resources, Conservation and Recycling*, *151*, p. 104460.

[19] Shastri, Y., Diwekar, U., Cabezas, H. and Williamson, J., 2008. Is sustainability achievable? Exploring the limits of sustainability with model systems. *Environmental Science & Technology*, 42(17), pp. 6710–6716.

[20] Kotecha, P., Diwekar, U. and Cabezas, H., 2013. Model-based approach to study the impact of biofuels on the sustainability of an ecological system. *Clean Technologies and Environmental Policy*, *15*(1), pp. 21–33.

[21] Fath, B.D., Cabezas, H. and Pawlowski, C.W., 2003. Regime changes in ecological systems: an information theory approach. *Journal of Theoretical Biology*, *222*(4), pp. 517–530.

[22] Cabezas, H. and Fath, B.D., 2002. Towards a theory of sustainable systems. *Fluid Phase Equilibria*, *194*, pp. 3–14.

[23] Shastri, Y. and Diwekar, U., 2006. Sustainable ecosystem management using optimal control theory: Part 2 (stochastic systems). *Journal of Theoretical Biology*, *241*(3), pp. 522–532.

[24] Whitmore, H.W., 2006. *Integration of an Economy under Imperfect Competition with A Twelve-cell Ecological Model.* National Risk Management Research Laboratory, Office of Research and Development, US Environmental Protection Agency.

[25] Kirk, D.E., 2004. *Optimal Control Theory: An Introduction.* Courier Corporation.

[26] Nocedal, J. and Wright, S., 2006. *Numerical Optimization.* Springer Science & Business Media.

[27] Redling, A., 2018. Rapid growth projected for global waste market in coming years. *Recycling Today.*

[28] Bizwit Research and Consulting LLP, 2018. Global waste paper recycling market to reach usd 54.9 billion by 2025. https://www.marketwatch.com/press-release/global-waste-paper-recycling-market-to-reach-usd-549-billion-by-2025-2018-09-24

[29] Brough, D. and Jouhara, H., 2020. The aluminium industry: A review on state-of-the-art technologies, environmental impacts and possibilities for waste heat recovery. *International Journal of Thermofluids*, *1*, p. 100007.

[30] Geyer, R., Jambeck, J.R. and Law, K.L., 2017. Production, use, and fate of all plastics ever made. *Science Advances*, *3*(7), p. e1700782.

[31] Bowyer, J., Bratkovich, S.T.E.V.E., Fernholz, K.A.T.H.R.Y.N., Frank, M.A.T.T., Groot, H.A.R.R.Y., Howe, J.E.F.F. and Pepke, E., 2015. Understanding steel recovery and recycling rates and limitations to recycling. *Dovetail Partners Inc.: Minneapolis, MN, USA*, pp. 1–12.

[32] Nations, U., 2004. The United Nations on world population in 2300. *Population and Development Review*, *30*, pp. 181–187.

# 4 "Waste"-to-energy for Decarbonization

## *Transforming Nut Shells Into Carbon-negative Electricity*

*Daniel Carpenter, Eric C.D. Tan, Abhijit Dutta, Reinhard Seiser, and Stephen M. Tifft*
National Renewable Energy Laboratory, Golden, Colorado, USA

*Michael S. O'Banion*
The Wonderful Company, Los Angeles, California, USA

*Greg Campbell and Rick Becker*
V-Grid Energy Systems, Camarillo, California, USA

*Carrie Hartford*
Jenike & Johanson, inc., San Luis Obispo, California, USA

*Jayant Khambekar*
Jenike & Johanson, Inc., Houston, Texas, USA

*Neal Yancey*
Idaho National Laboratory, Idaho Falls, Idaho, USA

**CONTENTS**



DOI: 10.1201/9781003167693-4

## 4.1 INTRODUCTION AND BACKGROUND

Agricultural wastes are generated each day as a result of the increasing demands of the fast-growing population and have become a growing challenge for various crops, including the nut industry [1]. For instance, pistachio production in the United States has grown from 222,000 tons in 2011 to 577,500 tons in 2021, a 60% increase [2]. Agricultural wastes are organic waste, and they are mostly landfilled [3]. However, when organic waste is landfilled, it releases methane, a potent greenhouse gas (GHG) that has global warming potential (GWP) about 30 times greater than carbon dioxide over a 100-year time horizon. Besides GHG emissions that cause global warming, landfills also emit a number of hazardous air pollutants and volatile organics, which can cause adverse health effects. Therefore, it is desirable to divert organic waste from landfills for significant environmental and public health benefits. Furthermore, agricultural waste valorization can also bring economic benefits.

To reduce landfill-related emissions and convert waste for a new income stream, farms and agricultural companies across the country are seeking alternative methods for dealing with waste while lowering their environmental impacts. The Wonderful Company (TWC) in Los Angeles, CA, is the world's largest almond and pistachio grower, generating over 250,000 dry tons of waste material annually, including hulls, shells, and wood. Industry wide, this amounts to well over 5 million tons/year of highly aggregated waste available in the United States, presenting an opportunity to both mitigate a disposal problem for the industry and decarbonize local electricity production.

Supported by the Feedstock-Conversion Interface Consortium (FCIC) to "develop science-based knowledge and tools to understand biomass feedstock and process variability, improving overall operational reliability, conversion performance, and product quality across the biomass value chain [4]," National Renewable Energy Laboratory (NREL), in partnership with TWC, Idaho National Laboratory (INL), and Jenike & Johanson (Tyngsboro, MA) developed a general methodology for designing integrated biorefinery solids preprocessing, handling, and feeding systems based on the physical, mechanical, and chemical attributes of the starting biomass material. Using detailed feedstock attributes, preprocessing optimization, controlled conversion testing, and bulk flow measurements, the goal of the project was to design an integrated preprocessing, material handling, and reactor in-feed system for a commercial gasifier using TWC's pistachio shell waste material.

Consistent and reliable preprocessing, conveyance, and reactor in-feed systems, particularly for low-cost waste biomass feedstocks with variable chemical and physical properties, remain critical technical and economic risk factors for the emerging bioeconomy. Solids feeding and handling systems for these feedstocks have often been adapted from other industries without fully understanding the potential inconsistency of commercially available materials. This can have major impacts on handling systems and conversion processes. The result is that most systems have been unable to achieve economically viable throughputs, often reaching less than half of design capacity due to feeding difficulties [5]. Cost-effective solutions for achieving high biorefinery on-stream factors will require co-optimizing conversion processes with an acceptance window designed to handle reasonable feeding upsets and off-spec material, and active management of critical feedstock attributes to ensure reasonably consistent feedstock behavior, including mechanical flowability, grindability, and chemical composition necessary to ensure reliable feeding and conversion performance.

This study aimed to (1) understand the impacts of pistachio waste physical, mechanical, and chemical attributes and the variability in these attributes on the performance of preprocessing, conveying, and reactor feeding systems; (2) use this information to design a feed train and high-temperature reactor feeding system for this material, including specifying equipment for storage, preprocessing, conveyance, and introduction into a gasifier; (3) demonstrate consistent and reliable processing and reactor feeding of pistachio waste (wood, hulls, shells) into a gasification process;

and (4) develop a generalized methodology for achieving the first three goals that are applicable to other biomass feedstocks.

V-Grid Energy Systems (Camarillo, CA) (https://vgridenergy.com/company/) was brought on as a partner to conduct a 500-hour field trial. V-Grid offers a trailer-mounted, 100 kW downdraft gasifier/genset system with a nominal biomass throughput of 200 lb/h. The results indicated that modeled electricity production costs are largely sensitive to staffing requirements, biochar selling price, and operational uptime. This chapter presents results from bench-scale preprocessing and conversion tests, bulk handling system and gasifier design recommendations, and V-Grid's extended gasifier field trials, followed by details of the techno-economic analysis (TEA) and life cycle assessment (LCA) for this process. Finally, a generalization of the methodology applied to other feedstocks and conversion processes is discussed.

## 4.2 PISTACHIO WASTE CRITICAL MATERIAL ATTRIBUTES

This section presents the key results and findings of the feedstock characterization, preprocessing, conversion, and handling system designs. Table 4.1 shows the typical chemical properties of raw pistachio waste. The chemical composition results for the selected pistachio waste show similar results among years and types of treatments for the processing of the waste. The largest variation in the composition occurs with % total ash as high as 13.65%, which may be largely due to the high amount of fines and contaminants observed in the raw feedstock. From a thermal conversion standpoint, the critical attributes of most of the pistachio shell wastes are within the values typically seen in biomass like straws and other agricultural residues [6,7]. However, some inorganic constituents known to cause slagging issues in gasifier operations, particularly phosphorous and potassium, are quite high in a few samples [8]. Accordingly, the sulfur content of all samples is relatively low, but the nitrogen and chlorine are high enough to be of potential concern from emissions (NOx) and corrosivity (HCl) perspectives.

The highest priority waste streams ultimately considered in this project were "harvest waste" and "process waste" from TWC's pistachio processing operations (see Figure 4.1). Harvest waste, which includes "floaters and blanks," is quite heterogeneous and consists of blank shells, inedible meats, sticks, leaves, dirt clots, adhered hulls, and other debris incidentally collected during harvest. Process waste originating inside the processing operation represents a cleaner stream and is primarily comprised of half shells and residual meats. Both waste streams are generally stockpiled together at TWC's processing facilities. These two streams accounted for most of the gasification tests at V-Grid and the combination of the two is denoted as mixed pistachio waste or abbreviated as "P-Shells". Several preprocessing options of varying cost ($2–$37/ton) and complexity (between 2- to 7-unit operations) were considered for this material, including mechanical sieving, air-classification, size reduction, densification, and washing.

Once a downstream conversion system was identified for this application, several feedstock material attributes (MAs) potentially impacting the gasification step were presumed critical, including particle size distribution, moisture content, inorganics content/speciation, and protein content. Each of these is discussed below.

### 4.2.1 Particle Size Distribution

Particle size requirements vary depending on the specific gasifier design, but for a downdraft system such as V-Grid's, "fines" are known to cause potential plugging in the gasifier grate, and oversized material will cause blockages in conveyance systems and, depending on the residence time, may not convert completely. For V-Grid's system, particles smaller than ¼" generally need to be removed. The proportion of fines was found to be quite variable. Sieve analyses showed that this material could reach 40 wt.% in some samples, and 25 wt.% was indicated anecdotally in the field for the mixed waste primarily used for V-Grid's extended tests (see Section 4.4 below). At the

**TABLE 4.1**
**Chemical Properties of Raw Pistachio Waste**

| Feedstock Storage Pile | Pile #1 Super Sack 1 | Pile #1 Super Sack 2 | Pile #5 Super Sack 1 | Pile #5 Super Sack 2 | Roasted/Salted Pile #5 Super Sack 1 | Roasted/Salted Pile #5 Super Sack 2 | Pile #8 Super Sack 1 | Pile #8 Super Sack 2 |
|---|---|---|---|---|---|---|---|---|
| Year | 2016 | 2016 | 2017 | 2017 | 2017 | 2017 | 2017 | 2017 |
| ***Proximate analysis (wt.% dry basis)*** | | | | | | | | |
| Ash | 5.18 | 7.51 | 11.1 | 11.91 | 4.44 | 13.65 | 7.47 | 10.89 |
| Volatile matter | 76.81 | 76.26 | 75.16 | 72.09 | 78.43 | 74.18 | 76.26 | 73.53 |
| Fixed carbon | 18.01 | 16.23 | 13.74 | 16 | 17.13 | 12.17 | 16.27 | 15.58 |
| HHV (MJ/kg) | 18.31 | 17.47 | 17.61 | 18.34 | 21.63 | 18.85 | 17.84 | 17.17 |
| LHV (MJ/kg) | 16.96 | 16.09 | 16.23 | 17.09 | 20.18 | 17.42 | 16.59 | 15.84 |
| ***Ultimate analysis (wt.% dry basis)*** | | | | | | | | |
| C | 47.45 | 45.38 | 45.16 | 45.22 | 50.66 | 46.58 | 45.87 | 44.62 |
| H | 6.3 | 6.43 | 6.39 | 5.8 | 6.68 | 6.59 | 5.8 | 6.15 |
| N | 0.94 | 1.07 | 1.28 | 1.32 | 2.04 | 1.7 | 1.2 | 1.24 |
| S | 0.072 | 0.075 | 0.091 | 0.114 | 0.167 | 0.155 | 0.073 | 0.092 |
| O (by diff.) | 40.05 | 39.55 | 35.99 | 35.64 | 36.01 | 31.32 | 39.59 | 37.02 |
| ***Elemental analysis (wt.% in total ash)*** | | | | | | | | |
| Si | 38.79 | 51.73 | 56.85 | 58.26 | 15.49 | 46.58 | 52.77 | 56.76 |
| Al | 9.48 | 12.3 | 13.24 | 11.81 | 5.04 | 12.12 | 11.09 | 12.86 |
| Ti | 0.4 | 0.45 | 0.51 | 0.46 | 0.14 | 0.55 | 0.4 | 0.45 |
| Fe | 3.06 | 3.91 | 4.1 | 3.78 | 2.36 | 4.13 | 5.2 | 3.67 |
| Ca | 5.81 | 4.49 | 4.79 | 4.95 | 3.68 | 3.27 | 4.59 | 4.59 |
| Mg | 2.76 | 2.61 | 2.48 | 2.57 | 2.32 | 2.59 | 2.31 | 2.44 |
| Na | 1.24 | 1.63 | 2.43 | 2.34 | 24 | 10.3 | 2.18 | 3.52 |
| K | 20.75 | 12.9 | 9.3 | 7.75 | 17 | 6.72 | 13 | 9.33 |
| P | 5.63 | 3.9 | 3.6 | 3.92 | 10.89 | 3.89 | 4.3 | 2.92 |
| S | 1.74 | 0.82 | 0.71 | 0.69 | 0.17 | 0.57 | 0.11 | 0.63 |
| Cl | 0.06 | 0.15 | 0.61 | 0.49 | 24.91 | 9.9 | 0.91 | <0.01 |

| | | | | | | | | |
|---|---|---|---|---|---|---|---|---|
| ***Compositional analysis (wt.% dry basis)*** | | | | | | | | |
| Total Ash | 5.54 | 10.38 | 11.79 | 11.70 | 4.77 | 14.84 | 7.25 | 12.47 |
| Structural Inorganics | 2.91 | 5.58 | 7.74 | 8.54 | 1.21 | 9.48 | 3.58 | 8.48 |
| Non-structural inorganics | 2.63 | 4.80 | 4.05 | 3.16 | 3.56 | 5.36 | 3.67 | 3.99 |
| Sucrose | 0.02 | 0.09 | 0.76 | 0.56 | 1.36 | 2.53 | 0.15 | 0.75 |
| Free Glucose | 0.00 | 0.14 | 0.31 | 0.17 | 0.54 | 0.51 | 0.02 | 0.25 |
| Free Fructose | 0.09 | 0.00 | 0.00 | 0.38 | 0.88 | 0.00 | 0.13 | 0.00 |
| Water Extractable Others | 5.58 | 5.54 | 5.75 | 6.62 | 8.67 | 9.24 | 3.89 | 3.75 |
| Ethanol Extractives | 2.04 | 2.74 | 7.44 | 8.85 | 15.22 | 9.99 | 3.88 | 5.41 |
| Lignin | 29.41 | 25.54 | 23.35 | 25.43 | 25.13 | 23.72 | 29.80 | 24.43 |
| Glucan | 15.78 | 15.25 | 13.37 | 11.29 | 9.76 | 9.47 | 12.79 | 14.49 |
| Xylan | 28.35 | 26.86 | 22.80 | 20.58 | 18.34 | 14.55 | 26.98 | 25.11 |
| Galactan | 1.19 | 1.11 | 0.95 | 0.99 | 0.96 | 0.81 | 1.14 | 0.99 |
| Arabinan | 0.46 | 0.95 | 0.84 | 0.59 | 0.70 | 1.09 | 0.52 | 1.26 |
| Mannan | 0.00 | 0.42 | 0.49 | 0.00 | 0.00 | 0.00 | 0.00 | 0.43 |
| Acetyl | 2.97 | 3.44 | 3.52 | 3.12 | 2.97 | 2.49 | 3.90 | 3.80 |
| Total % | 91.43 | 92.47 | 91.38 | 90.28 | 89.29 | 89.25 | 90.45 | 93.13 |

**FIGURE 4.1** Waste stream images showing the two main types of wastes employed in the study: a) harvest waste including full empty shells. b) Process waste including residual meats.

time of the extended trials, V-Grid had added a screen to remove this material, and a figure of 15 wt.% was used as a consistent basis to adjust experimentally measured mass flows. Co-feeding pelletized fines to the gasifier was considered a potential strategy to increase waste utilization, produce additional electricity and biochar, and improve overall economics. Pelletized fines were used for a portion of the extended trials, and this was considered as a separate case in the TEA and LCA (see Sections 4.5 and 4.6 below).

To determine whether size reduction of larger material (e.g., whole or half shells) would be needed, devolatilization tests were conducted with pistachio shells of various sizes (including intact half-shells) in a flow tube reactor. In this test, the biomass sample is inserted into a preheated furnace with a simulated gasifier atmosphere, and the volatilized products are monitored in real time. Figure 4.2 shows the extent of conversion as a function of time at 700°C, indicating full

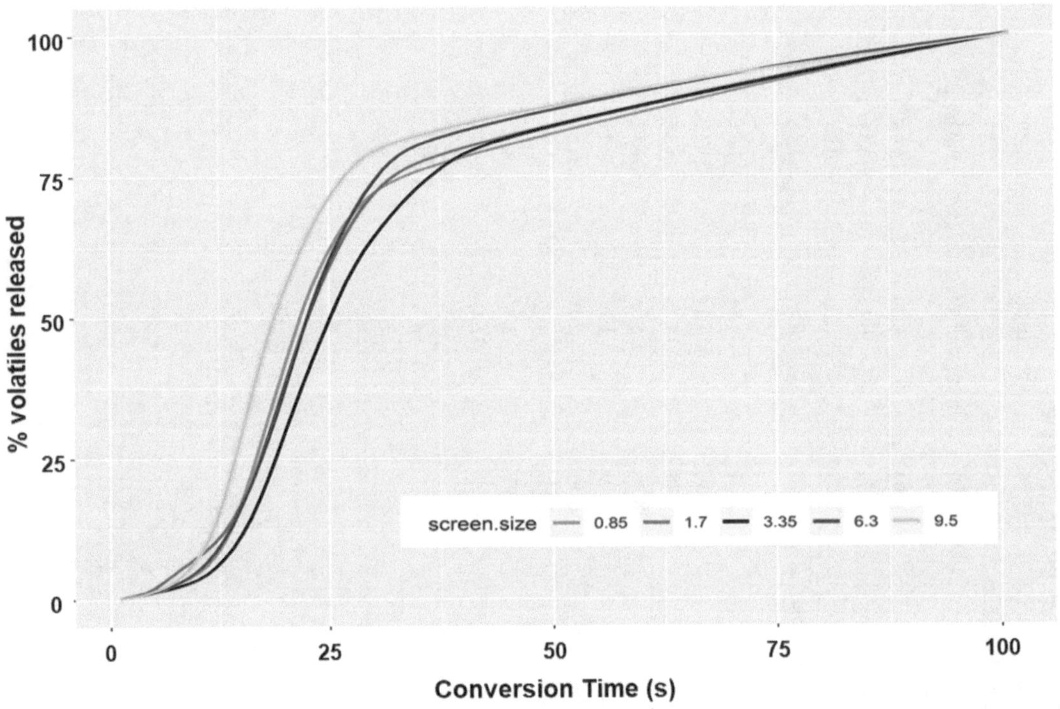


**FIGURE 4.2** Pistachio shell conversion vs. time for various particle sizes (in mm) at 700°C.

conversion occurs in under 2 minutes. This is well under the expected residence time of >10 minutes in a downdraft gasifier system, and it was deemed no further size reduction was needed.

### 4.2.2 Moisture Content

Moisture content is critical for solids flowability and thermal efficiency for any thermal conversion process. For gasification, a moisture content of less than 30 wt.% is generally desired [9]. The moisture content of all bulk samples analyzed was less than 10 wt.%, despite having been rained on just prior to collection, and all samples were found to be free-flowing. Moisture content is thus unlikely to be problematic for either flowability or gasification, and a drying step should not be required. One exception may be the separated fines, which are likely more susceptible to flowability issues with higher moisture.

### 4.2.3 Inorganic Content and Composition (Ash)

High-ash feeds, particularly those high in alkali metals, can form low-melting eutectics that cause slagging issues in the gasifier. Additionally, high sulfur and nitrogen can lead to increased $SO_x$ and $NO_x$ emissions during electricity production. The inorganic composition of the pistachio waste samples was typical of other agricultural residues and grasses (total inorganics = 2.3%–13.7%; K = 0.7%–1.1%; P = 0.3%–0.5%; N = 0.9%–2.0%; S = 75–1,770 ppm). While the sulfur content was relatively low, the nitrogen content could be problematic from an emissions standpoint. The genset operating on producer gas employs a three-way catalytic converter for emissions reduction of hydrocarbons, carbon monoxide, and nitrogen oxides. Nitrogen compounds in the producer gas increase the production of nitrogen oxides during combustion but are a relatively smaller contribution compared to thermal NOx. Sulfur may poison palladium-type automotive catalysts if above certain concentrations. The phosphorous and potassium are relatively high in a few samples, which could lead to solids agglomeration in the gasifier bed, particularly if there are hot spots. This, in fact, has been observed occasionally in V-Grid's system but is currently mitigated by daily cleaning of the gasifier vessel. The ash melting behavior of these samples was not investigated further.

### 4.2.4 Protein Content

High-protein feeds such as pistachio meats and hulls can generate higher levels of tar in the producer gas, making cleanup prior to heat exchangers and power generation equipment more difficult and costly. Air classification [10] was investigated as a potential strategy to separate residual meats from the shells, and an air-classified sample was used during the extended trials. For this particular sample, a significant difference was not achieved in either the feed nitrogen content (an indicator of protein) or the measured tar levels, though substantial variability was noted in both of those measurements.

## 4.3 BULK MATERIAL HANDLING SYSTEM DESIGN

Jenike & Johanson is the world's leading company in powder and bulk solids handling, processing, and storage technology. As a bulk material engineering specialist, the company provided scaled engineering designs, equipment specifications, and cost estimates for bulk handling of TWC's mixed pistachio waste and the separated fine material (see Figure 4.3). The design basis was 20 tons per day (TPD) total throughput, which was chosen to provide flexibility to accommodate either a larger stationary downdraft gasifier or several modular units connected in parallel. The design could also be scaled down using the same concept and smaller components. The cost estimate (± 30%) for the complete 20 ton/day system was $680,000. It should be noted that this

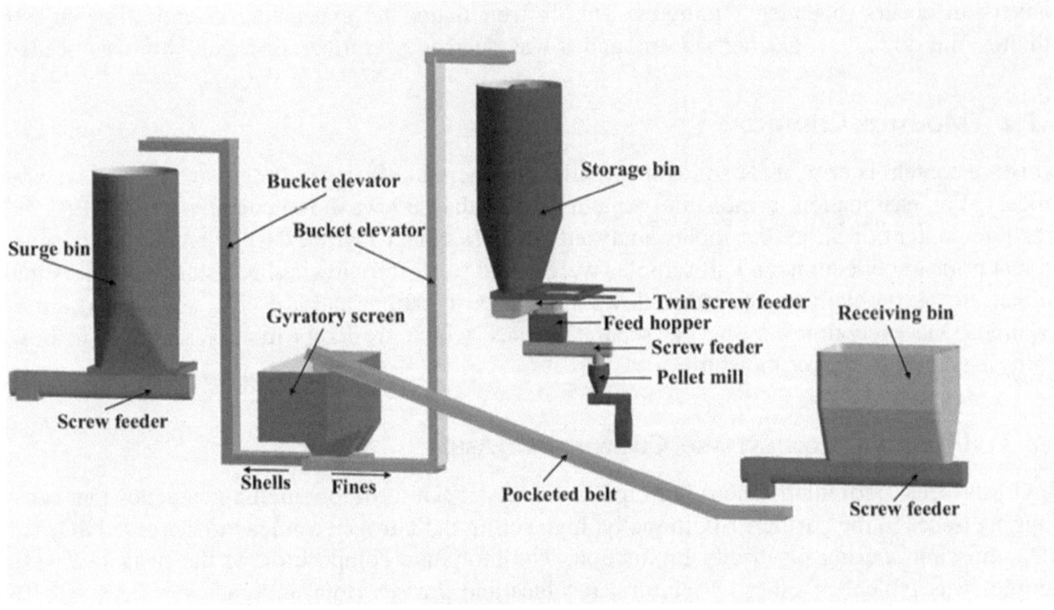


**FIGURE 4.3** Solids handling system design by Jenike & Johanson for a future 20 TPD operation. The shells (green components) are conveyed directly to the gasifier bank and the fines (blue components) are diverted to a pellet mill to be inserted back into the downstream feed train with the shells (not shown).

specific capital cost was not included in the techno-economic analysis below as it represents a future scenario with costs amortized across multiple units, which was not considered here. Instead, the TEA equipment costs were estimated for a single gasifier/genset unit based on publicly available information that reasonably aligns with installed equipment costs at V-Grid, as explained in Section 4.5.

## 4.4 V-GRID GASIFIER EXTENDED TRIALS AND DESIGN RECOMMENDATIONS

Extended field trials were carried out by V-Grid Energy Systems using TWC's pistachio shell waste material, including operation with various feedstocks and two different gasifiers (a 6-inch diameter and a scaled-up 8-inch version). The goal of these tests was to demonstrate at least 500 hours of operation while collecting data on system uptime and issues resulting in downtime and collecting samples for analysis.

System uptime, operational data, and downtime/issues were documented, and samples were collected for offline characterization. Over the entire testing period of 753 hours, which included six different feedstocks and two gasifier sizes, the uptime accounted for 577 hours, regeneration for 99 hours, scheduled maintenance for 51 hours, and unscheduled maintenance for 26 hours. The overall on-stream factor using the original gasifier with mixed pistachio waste, which accounted for 450 of the uptime hours, was 80% (Figure 4.4). Approximately 280 samples of feed, char, and condensate were collected and shipped to NREL, and a subset of these were characterized for particle size distribution, proximate, ultimate, and ash composition. Operational and analytical data from these extended trials were used for the techno-economic analysis and life-cycle assessment of the process (see the following sections). Additional details for all feedstocks are shown in Figures 4.5 and 4.6.

Furthermore, the test study identified areas that can potentially improve the reliability of V-Grid's feed handling and gasification systems and included an engineering design review of a scaled-up gasifier vessel. From a preprocessing standpoint, starting with consistent material and

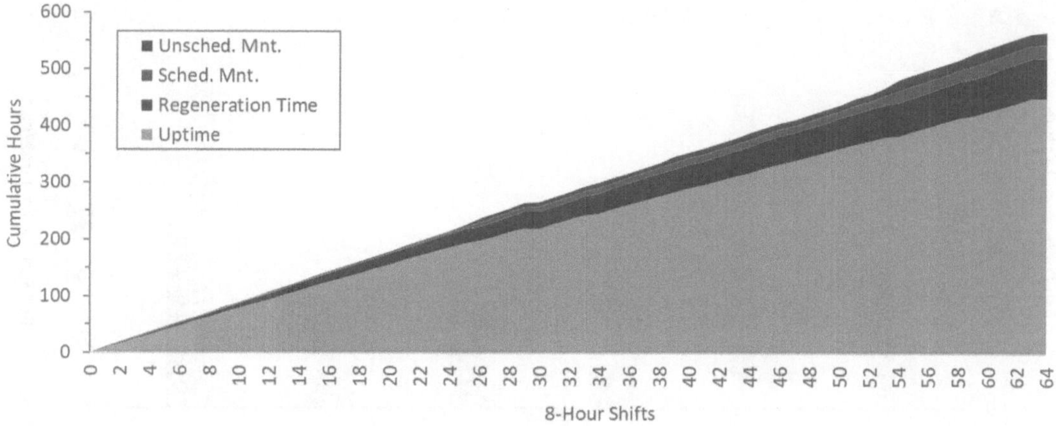


**FIGURE 4.4** Cumulative time-on-stream and maintenance cycles for gasifier tests using mixed pistachio waste. Uptime accounted for 450 hours, regeneration for 72 hours, scheduled maintenance for 23 hours, and unscheduled maintenance for 22 hours.

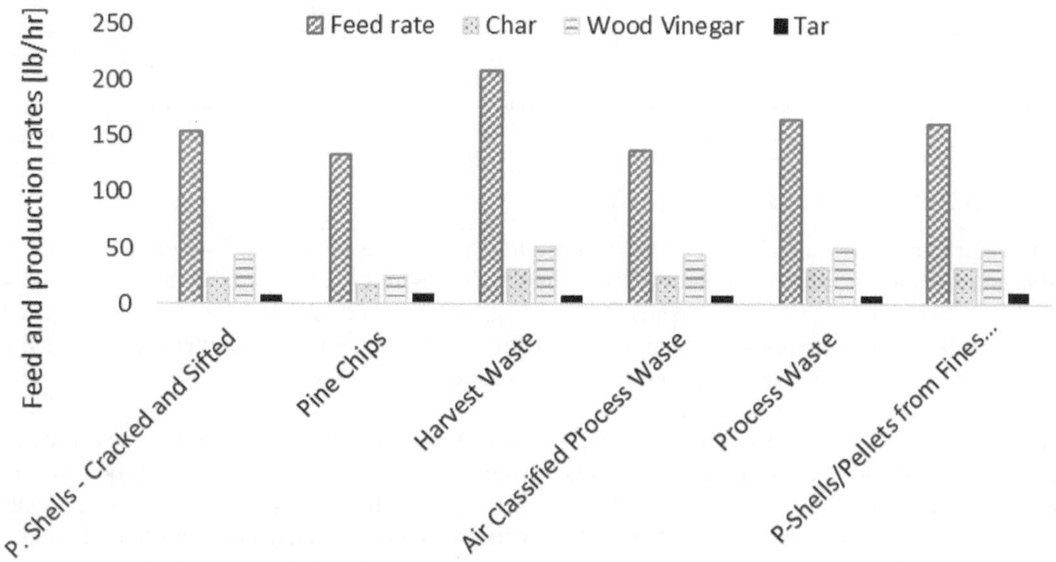


**FIGURE 4.5** Feed and production rates for various feedstocks. P-shells – cracked and sifted (mixed pistachio waste) and P-Shells/Pellets from Fines (80/20) were used for the techno-economic study. "Wood vinegar" represents the aqueous-phase by-product from the process.

loading methods when transferring pistachio reject material from TWC operations to the conversion system is important. For example, minimizing the amount of dirt will prevent excessive wear in preprocessing, conveyance, and gasifier systems. V-Grid's system works well with relatively clean pistachio half-shells, so size reduction does not appear to be necessary. However, the removal of large material (e.g., sticks) and fines (<1/4") will be needed, as well as determining a use for the significant proportion of fines observed (typically ~10%, but up to 40% by mass in some cases). Recommendations for gasifier system modifications, primarily around the reduction of fine char particles and tar, were made based on observations of V-Grid's system, and preliminary gasification tests and collective experience at NREL.

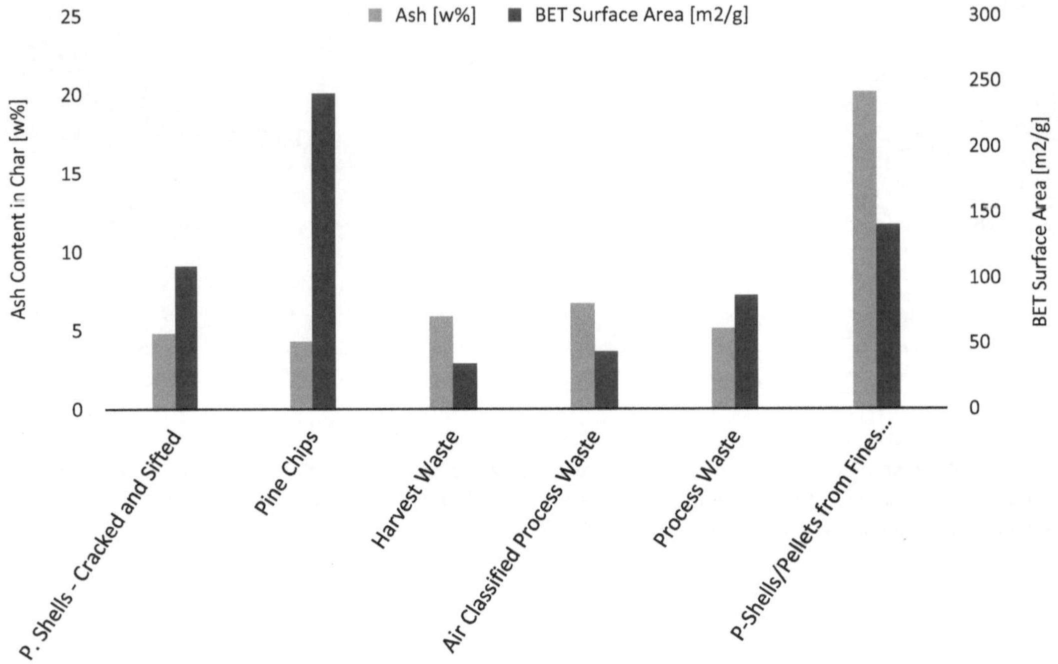


**FIGURE 4.6** Ash content and BET surface area of char produced from various feedstocks.

## 4.5 TECHNO-ECONOMIC ANALYSIS

This techno-economic analysis (TEA) determined the minimum electricity selling price (MESP) of biopower produced from agricultural waste materials based on experimental and operational results for two cases: (A) mixed pistachio waste and (B) 80% pistachio shells + 20% pelletized fines. The MESP value represents the minimum selling price of electricity that meets the economic parameters assumed for the economic analysis. In addition to electricity, the process also co-produces biochar, tar (as fuel), and wood vinegar. For the valuations of co-products, (a) biochar value was treated as a sensitivity variable, with sensitivity analysis for other parameters performed around a $1,000/ton biochar price, (b) tar was treated as a zero-cost waste, assuming there would be some credit from its heating value with a net zero cost for disposal, and (c) wood vinegar was assumed to be treated as wastewater in this TEA with treatment costs derived from other biomass TEA reports [11]. The current study did not explore alternate applications of tar and wood vinegar that may potentially provide additional economic benefits.

### 4.5.1 Process Model Description and Assumptions

Figure 4.7 depicts the Aspen Plus process model architecture; the key unit operations include gasification (GASIFIER), syngas electrical generator (GENSET), and separators for biochar (CHARSEP), tar (TARSEP), and wood vinegar (VINEGSEP). The model was based on measured compositions of feedstocks and products (char, tar, wood vinegar, product gas), as well as operational process flows. The analytical results were averaged when multiple data points were available, as shown in Table 4.2.

Anomalies in mass/atomic balances after the integration of all the experimental data were assumed to be from offsets in measurements, e.g., the calibration of the air flow meter or additional air intrusion through the lock hopper. The experimental and analytical results needed to be harmonized to achieve complete atomic balance and avoid related errors in the Aspen Plus model; the harmonization was executed in Excel by matching inlet and outlet quantities of atoms/species. The

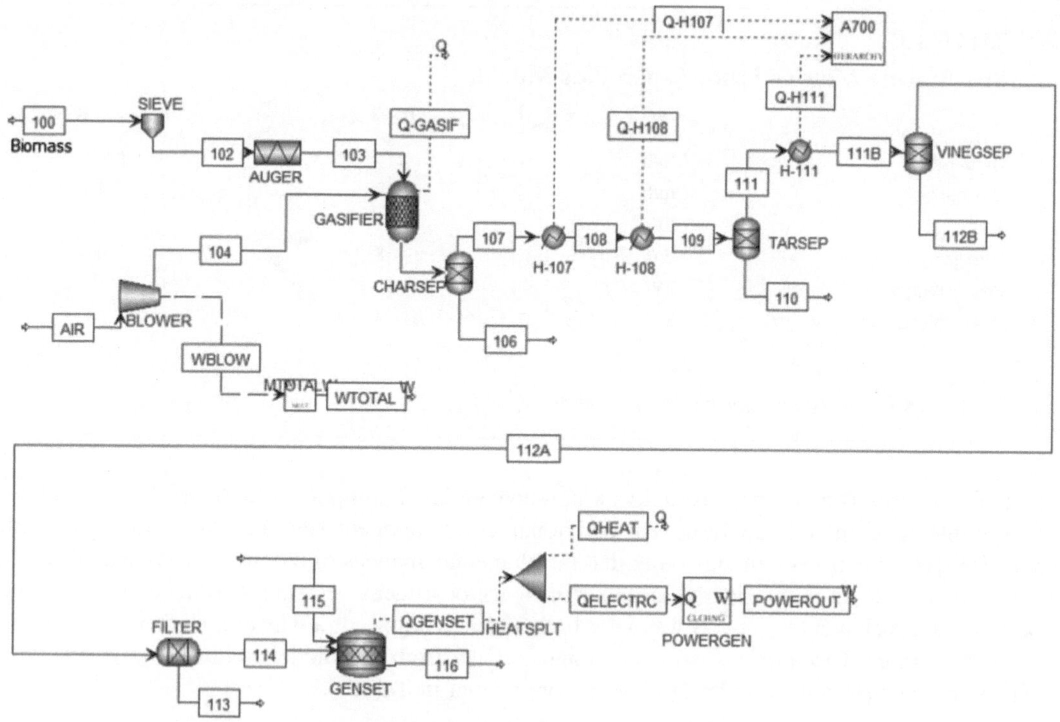


**FIGURE 4.7** Simple Aspen Plus model flowsheet for TEA.

**TABLE 4.2**
**Elemental Analyses Used as Starting Basis for Process Model (Averaged Values When Multiple Data Points Were Available)**

| | Case A: Mixed Pistachio Waste | | | | Case B: 80% Pistachio Shells, 20% Fines | | | |
|---|---|---|---|---|---|---|---|---|
| | Feed (dry) | Biochar (dry) | Tar (wet) | Vinegar (wet) | Feed (dry) | Biochar (dry) | Tar (wet) | Vinegar (wet) |
| **Carbon** | 48.45 | 86.61 | 62.77 | 7.73 | 47.76 | 67.91 | 60.03 | 8.30 |
| **Hydrogen** | 6.36 | 1.53 | 7.42 | 10.36 | 6.36 | 1.15 | 7.93 | 10.37 |
| **Oxygen** | 44.05 | 6.04 | 27.19 | 81.51 | 38.75 | 5.87 | 28.72 | 80.70 |
| **Nitrogen** | 0.37 | 0.72 | 2.40 | 0.34 | 1.79 | 1.13 | 2.98 | 0.48 |
| **Sulfur** | 0.03 | 0.05 | 0.10 | 0.01 | 0.13 | 0.14 | 0.20 | 0.01 |
| **Ash** | 0.73 | 5.04 | 0.11 | 0.04 | 5.21* | 23.81 | 0.14 | 0.14 |

*Note*

* *Measured ash was 13.39%. This was deemed high after some scrutiny and inspection of previous analysis. The ash value was adjusted down to 5.21% to reflect parity with measured ash in char.*

order of matching (and related manipulations) of relevant components was ash, sulfur, carbon, hydrogen, and final oxygen and nitrogen. Surrogate compounds used to represent tar were naphthalene ($C_{10}H_8$), eugenol ($C_{10}H_{12}O_2$), benzyl-mercaptan ($C_7H_8S$), and water. Surrogate compounds used to represent wood vinegar were acetic acid ($C_2H_4O_2$), phenol ($C_6H_6O$), and water. Proportions of the surrogate compounds used for tar and wood vinegar were manipulated to match the measured

**TABLE 4.3**
**Key Process Metrics From Aspen Plus Model**

| | Unit | Case A | Case B |
|---|---|---|---|
| Biomass | lb/h | 215 | 200* |
| Biochar | lb/h | 27 | 43 |
| Tar | lb/h | 11 | 16 |
| Vinegar | lb/h | 68 | 67 |
| Power Output | kW | 57 | 39 |
| Power Need | kW | 4 | 4 |

*Note*
* *Lower flow rate assumed for higher ash material.*

elemental analysis. Biochar was treated as a non-conventional component in Aspen Plus, matching the experimental elemental analyses. An additional ash component was introduced to capture ash imbalances from the measurements (note that the elemental analysis of biochar already includes ash, and the additional ash is considered an adjustment factor to achieve species balance closure). The measured feedstock ash for Case B was very high and was adjusted downward, as mentioned in the Table 4.2 footnote. Key process parameters and outputs from the models, after using the derived inputs with complete atomic balance closures, are shown in Table 4.3.

### 4.5.2 Economic Assumptions

The TEA reported here uses nth-plant economics, which is associated with a successful industry, has been established with many operating plants using similar process technologies. The financial assumptions were similar to those provided in [11].

The minimum electricity selling price (MESP) values were calculated using a discounted cash flow rate of return (DCFROR) analysis. MESP is a function of cash inflows and outflows from capital costs, product sales, operating costs, and financial assumptions. If not otherwise mentioned, all currency is in 2016 U.S. dollars (2016$). The MESP value assumes profit to the current "waste"-to-power plant of 10% internal rate of return (IRR) over a 30-year period. Furthermore, it assumes 60/40 debt/equity financing on capital, self-sustained utilities, and greenfield/standalone construction for an average of 215 lb/hr of nut shells input. Given the relatively small scale of this operation compared to most BETO-sponsored design reports (i.e., 2,000 dry ton/day), some additional assumptions were made for the current TEA.

The two cases included in the TEA reflect (A) Pathway #1 and (B) 20% Pistachio Fines cases, as shown in Table 4.4. Given the consistently low moisture values found in pistachio storage piles (<10%), the low cost of mechanical separations, and size reduction being unnecessary in V-Grid's

**TABLE 4.4**
**Feedstock Costs for Case A and Case B**

| | Case A | | Case B |
|---|---|---|---|
| Feedstock Pathway → | Pathway #1 - PS (Baseline) | Pistachio Fines | 80% PS + 20% Fines |
| Conveyor | $0.70 | $0.70 | $0.70 |
| Mechanical Separation | $1.00 | | $0.80 |
| Palletization | | $9.38 | $1.88 |
| **Cost per Ton** | **$1.70** | **$10.08** | **$3.38** |

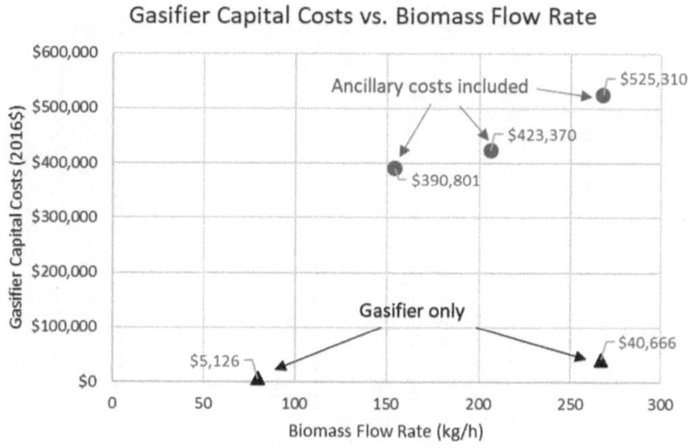


**FIGURE 4.8** Cost estimation for gasifier and ancillaries.

system, the total modeled cost for preprocessing the fines, using palletization and/or briquetting, was estimated to be $9.38 to $10.93 per ton. Conveyor costs were added for both feedstocks, while the $1.00/ton mechanical separation cost was added to allow the separation of fines after the size reduction of pistachio shells.

Other assumptions are listed below.

- The capital expenditure (CAPEX) was estimated from publicly available information that included ancillary costs. The estimated cost was $283,000 using a scaling exponent of 0.6 for scaling down from larger equipment specified in the literature (Figure 4.8, created with data from Table 4.5). A small installation factor of 1.2 was used because of the modular nature of the equipment. These assumptions were compared with information from V-Grid and the information aligned reasonably well; we did not feel it was necessary to change the baseline assumptions.
- The genset cost was based on public information. The 50 kW system cost was $14,950 with an additional installation factor of 1.2 applied.
- Cooling water cost for industrial use was 5¢/1,000 gallon from previous TEA [11]. This cost contribution was negligible, so a sensitivity analysis was not conducted for this.
- Biochar selling price had a significant impact on the MESP (minimum electricity selling price). Current prices can be as high as ~$2,500/ton. However, prices will come down with greater availability. For the base case, we assumed $1,000 per U.S. short ton.
- The number of employees necessary for operations and maintenance has a significant impact on the economics. Given the clustered approach to these installations, it was assumed that personnel will be shared among multiple units. 0.25 of a maintenance worker (one worker per four units) was allocated for the 50 kW unit for the base case.
- Financial assumptions from other BETO projects were generally maintained here. Co-product prices for tar (zero-cost) and wood vinegar (sent to wastewater treatment) were different from the V-GRID business model (as available to us); this has been discussed earlier.
- As actual operations involve two 8-hour shifts of gasifier operation, the current model assumes 16 hours of operation per day.

### 4.5.3 TEA Results for Case A and Case B

Char production has an overwhelming impact on the economics of the process because of an assumed selling price of $1,000/ton. At this price point, the minimum electricity selling price (MESP) was determined to be 39.9¢/kWh for Case A and 35.5¢/kWh for Case B.

**TABLE 4.5**
**Capital Expenditure (CAPEX) from Publicly Available Information that Included Ancillary Costs**

| Year Published [Source] | Biomass Rate | Reported Energy Output | Reported Capital Cost, in Base Currency | Year Reported | Exchange Rate during Year | Cost ($US) | Cost (2016 $US) | Other Assumptions |
|---|---|---|---|---|---|---|---|---|
| 2008 [12] | 266.7 kg/h | N/A | RMB 30.0 Ten Thousand | 2008 | RMB 6.9451/$1 | $43,195.92 | $40,666.03 | Just gasifier cost, does not include ancillary systems |
| 2013 [13] | 206.3 kg/h | 164 kW | £289,140 | 2012 | £1.5802/$1 | $456,899.03 | $423,370.17 | Ancillary costs included, using a willow chip feedstock (requires extensive drying), CHP system |
| 2013 [13] | 154.6 kg/h | 162 kW | £266,897 | 2012 | £1.5802/$1 | $421,750.64 | $390,801.10 | Ancillary costs included, using a miscanthus feedstock (no drying needed), CHP system |
| 2015 [14] | 267.6 kg/h | 190 kW | € 438,000 | 2014 | €1.2755/$1 | $558,669.00 | $525,309.84 | Some ancillary costs included, using a wet olive pomace feedstock (requires extensive drying), CHP system |
| 2017 [15] | 80 kg/h | 45 kW | IDR 52,800,000 | 2008 | IDR 9698.0793/$1 | $5,444.38 | $5,125.51 | Just Gasifier Cost, does not include ancillary systems |
| 2017 [16] | Unknown | 500 m^3/day syngas | E£ 30,000 | 2017 | E£ 17.8351/$1 | $1,682.08 | $1,605.60 | Runs in two 3-hr batches/day, no assumption listed for amount or type of biomass, cost includes gasifier with control system, no installation, or ancillary costs |

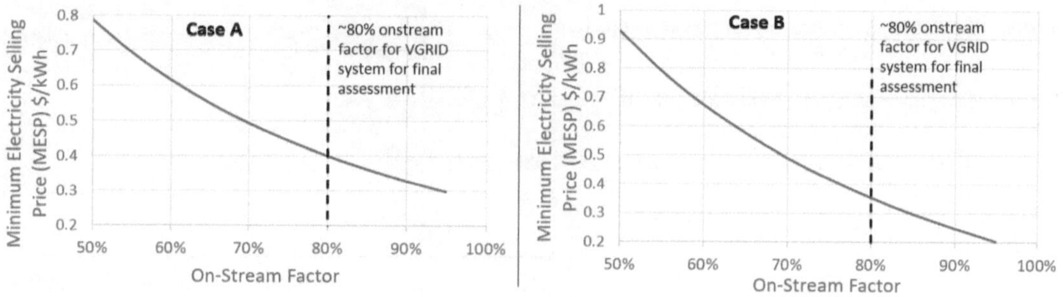


**FIGURE 4.9** Modeled impact of on-stream factor on the minimum electricity selling price.

Note that the MESP is sensitive to the onstream factor. For example, an increase of the onstream time from the current 16 hours of operation per day to 24 decreases the MESP to 18.3¢/kWh for Case A (using an 80% onstream factor), the primary reason being higher capital utilization, and therefore lower MESP.

Processing the fines did not add any significant cost to the feed handling process (Table 4.3). The modeled total electricity production was also lower in Case B because of the lower carbon content in the feedstock (Table 4.1) and a lower throughput assumed for the higher ash material (Table 4.2). However, the Case B MESP is still lower than Case A because of the higher production of high-value biochar (Table 4.2) from the higher ash feedstock that included fines pellets, resulting in an overall 11% reduction in MESP.

### 4.5.4 TEA Sensitivity Analysis and Discussion

The impacts of uncertainties on the MESP were analyzed via sensitivity analyses. The effect of on-stream factor (this is applied on top of a 16 hours per day operation) is shown in Figure 4.9. Since the biochar price is the predominant driver for economics, its impact is captured in Figure 4.10.

Sensitivities to changes in key variables were evaluated for the baseline Case A and Case B scenarios (Figure 4.11).

An additional sensitivity related to feedstock processing costs based on a scenario where an *optional* size reduction cost of $17.21/ton of pistachio shells was added showed in increase in the MESP for the base Case A from 39.9¢/kWh to 43.5 ¢/kWh, and for the base Case B from 35.5¢/kWh to 39.8¢/kWh.

The TEA shows that uptime and biochar prices have significant impacts on the economics of the process. Additional revenue streams from tar and wood-vinegar can also be beneficial but they were not considered here. Comparisons between Case A and Case B show that Case B can have

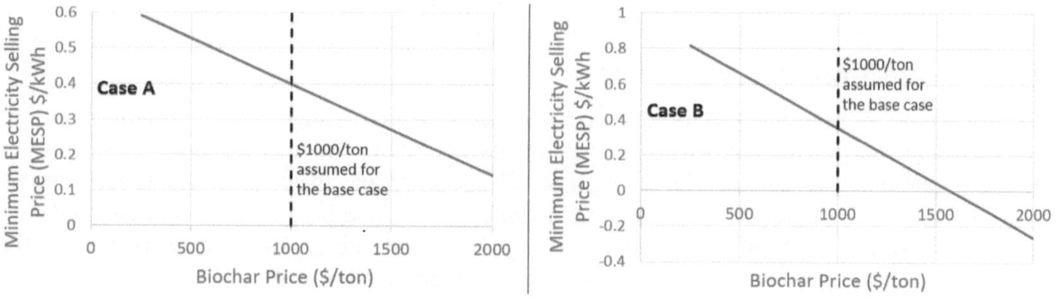


**FIGURE 4.10** Modeled impact of biochar price on the minimum electricity selling price.

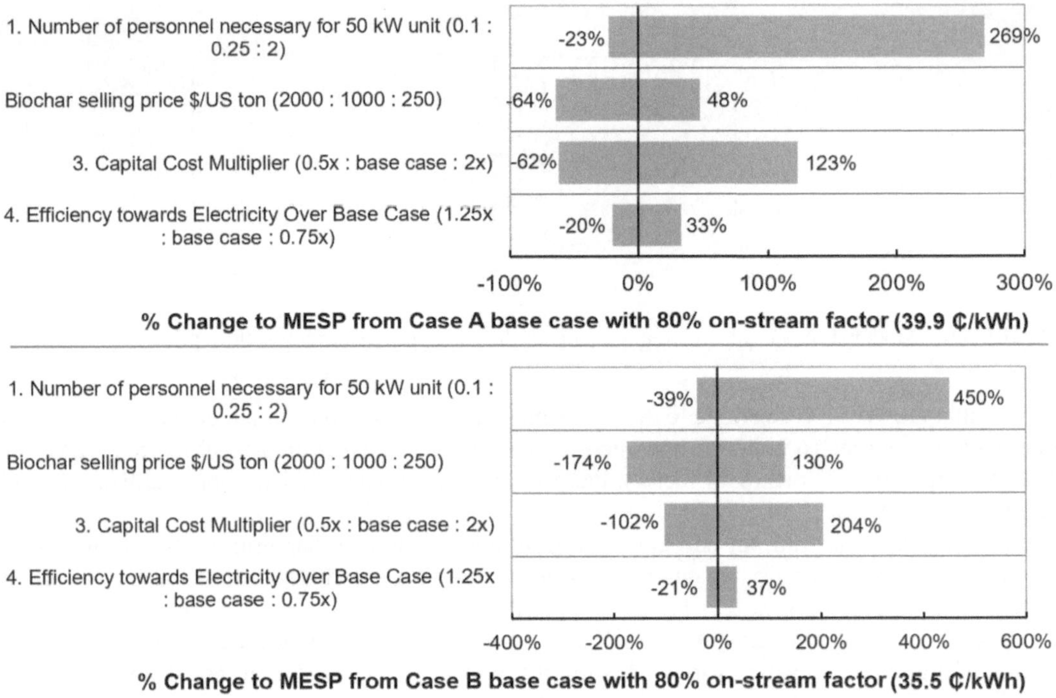


**FIGURE 4.11** Sensitivity analysis showing the effect of key factors and known uncertainties.

significant benefits with the higher biochar price assumptions; addition of fines in Case B does not add significantly to the unit price of feedstock but reduces electricity production while increasing char production. Additional efforts to increase capital utilization by reducing maintenance time and increase operations beyond 16 hours per day can also have significant benefits. As identified earlier, temperature control to minimize the least profitable products will be important; in this context, tar and wood vinegar minimization or maximization should be considered based on their economic values.

## 4.6 LIFE-CYCLE ASSESSMENT

In conjunction with the TEA study, this study also estimated the life-cycle greenhouse gases (GHG) for biopower derived from (A) mixed pistachio waste and (B) 80% pistachio shells + 20% pelletized fines. Assumptions around the use and avoided decay and disposal of pistachio waste, along with the life-cycle assessment (LCA) methodology, results, and discussion, are presented below.

### 4.6.1 LCA Method and Assumptions

This LCA study focused on determining the carbon intensity (CI) for electricity production from pistachio shells via gasification. The system boundary for this LCA study (Figure 4.12) was from the field to the power plant gate, including embodied energy and material flows. The functional unit for CI is 1 kWh of electricity. Life-cycle inventory data were based on the process modeling outputs from Aspen Plus process models developed for the TEA. Table 4.6 summarizes the material and energy inputs and outputs to the electricity production system.

The biomass feedstock (pistachio shells) is an agricultural solid waste, and there is no additional share of environmental burdens associated with the pistachio shells production. This is different

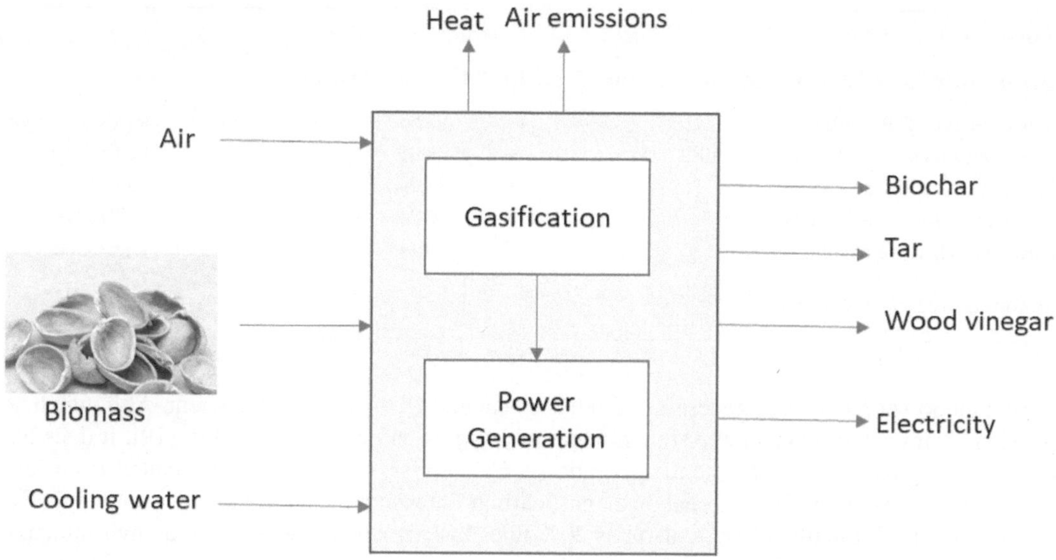


**FIGURE 4.12** System boundary of valorization of pistachio shells for electricity generation.

**TABLE 4.6**
**Material Inputs/Outputs for Life-cycle Inventory**

| | | Case A | Case B |
|---|---|---|---|
| **Material and energy inputs** | **Unit** | **Flow** | **Flow** |
| Biomass feedstock (P-Shells) | lb/h | 214.88 | 0.00 |
| Biomass feedstock (P-Shells 80, Fines 20) | lb/h | 0.00 | 199.90 |
| Gasifier air flow | lb/h | 249.12 | 249.12 |
| Genset air flow | lb/h | 551.16 | 504.73 |
| Cooling water | gal/h | 30 | 30 |
| **Releases to environment** | **Unit** | **Flow** | **Flow** |
| Heat | btu/h | 578679 | 402425 |
| $CO_2$ exhaust | lb/h | 226.50 | 155.99 |
| 20% fines lost for P-Shells | lb/h | 42.98 | 0.00 |
| **Main Products** | **Unit** | **Flow** | **Flow** |
| Electricity produced (net) | hp | 70.27 | 47.18 |
| **Co-products** | **Unit** | **Flow** | **Flow** |
| Biochar (for carbon sequestration) | lb/h | 26.82 | 43.43 |
| Wood vinegar (otherwise as wastewater) | lb/h | 68.32 | 67.48 |
| Tar as fuel (otherwise as solid waste) | lb/h | 10.90 | 15.69 |

than for herbaceous feedstocks such as corn stover, in which burdens required to harvest stover (e.g., additional nutrient replacement) have to be considered using a "purpose-driven" approach [17]. Furthermore, the thermoelectricity facility is in the vicinity of the pistachio shells collection facility, suggesting that the transportation distance is negligible. Moreover, the biomass can be fed directly to the gasifier, and size reduction is not required. Consequently, the carbon intensity associated with the feedstock production and logistics only encompasses the storage and handling, estimated to be 3.05 kg carbon dioxide equivalent ($CO_2e$)/dry ton, based on Hartley et al. [18].

**TABLE 4.7**
**Carbon Intensity Factors Estimated and Used for LCA Evaluation**

| | | |
|---|---|---|
| Biomass feedstock (P-Shells) | 3.05 | kg $CO_2$e/dry ton |
| Biomass feedstock (P-Shells 80, Fines 20) | 3.05 | kg $CO_2$e/dry ton |
| Biochar (for carbon sequestration) | −1.24 | kg $CO_2$e/kg |
| Wood vinegar (otherwise as wastewater) | −2.24 | g $CO_2$e/ton |
| Tar as fuel (otherwise as solid waste) | −0.11 | kg $CO_2$e/kg |

*Negative values indicate "credits."*

Both biopower plant cases generate and release heat and $CO_2$ to the environment. Although heat emissions from power plants have shown negative impacts on ecosystem quality [19], it does not contribute to global warming potential. Similarly, $CO_2$ emission is entirely originated from biomass and is exempted from greenhouse gas carbon intensity accounting according to IPCC methodology [20]. Finally, Case A also has 20% fines lost, in which the associated environmental burden for its handling was assumed negligible.

In this system, electricity is the main product, and the analysis has accounted for the electric power consumption for the process (i.e., net electricity product = total electricity production − power plant electricity consumption). There are three co-products: biochar, tar, and wood vinegar. All co-products were treated as avoided products using the product displacement method, based on the concept of displacing the existing product with the new product [21]. Biochar can potentially sequester carbon when applied to soils, and reduce $N_2O$ and $CH_4$ emissions from soil, increase fertilizer efficiency, and increase soil organic carbon [22,23]. The Carbon Stability Factor (CSF) is defined as the proportion of the total carbon in freshly produced biochar that remains fixed as recalcitrant carbon over a defined time period (10 years, 100 years, etc., as defined). For example, fresh biochar with a CSF of 0.75 indicates that 75% of the biochar carbon will remain as stable carbon, and 25% of the biochar carbon will be converted into $CO_2$ and escape to the atmosphere over the defined time horizon [24]. In this study, we adopt a carbon stability factor of 50%, and with a carbon content of 67.9 wt% in the biochar, the carbon intensity credit for the biochar was determined to be 0.56 kg $CO_2$e/lb biochar.

Wood vinegar is an extra liquid product generated and is primarily composed of acetic acid and water. Wood vinegar is considered in this study as a co-product. However, it retains a high water content (87.2%); we assumed its co-product carbon intensity credit is equivalent to that of the wastewater treatment, to be conservative. Note that the carbon intensity for the wastewater is 2.24 g $CO_2$e/ton [25]. The lower heating values for the tar co-product streams are similar for Case A and Case B, 10,694 Btu/lb and 10,726 Btu/lb, respectively. The credit for the tar is assumed to be the carbon intensity of natural gas (4.74 kg $CO_2$e/MMBtu) [26] and was determined to correspond to 0.051 kg $CO_2$e/lb of tar. The key carbon intensity factors used in the LCA evaluation are summarized in Table 4.7.

### 4.6.2 Carbon Intensity Results and Discussion

Figure 4.13 shows the carbon intensity (CI) associated with the production of electricity from pistachio shells. The carbon intensities for Case A and Case B (without any co-product credits) were determined to be 5.77 g $CO_2$e/kWh and 7.99 g $CO_2$e/kWh, respectively, and are essentially attributed to biomass feedstock storage and handling. The lower power output for Case B is responsible for the nearly 40% higher carbon intensity when compared to Case A.

The co-product credits are predominantly from biochar, −0.29 kg $CO_2$e/kWh for Case A and −0.70 kg $CO_2$e/kWh for Case B, suggesting that biochar carbon sequestration can be a

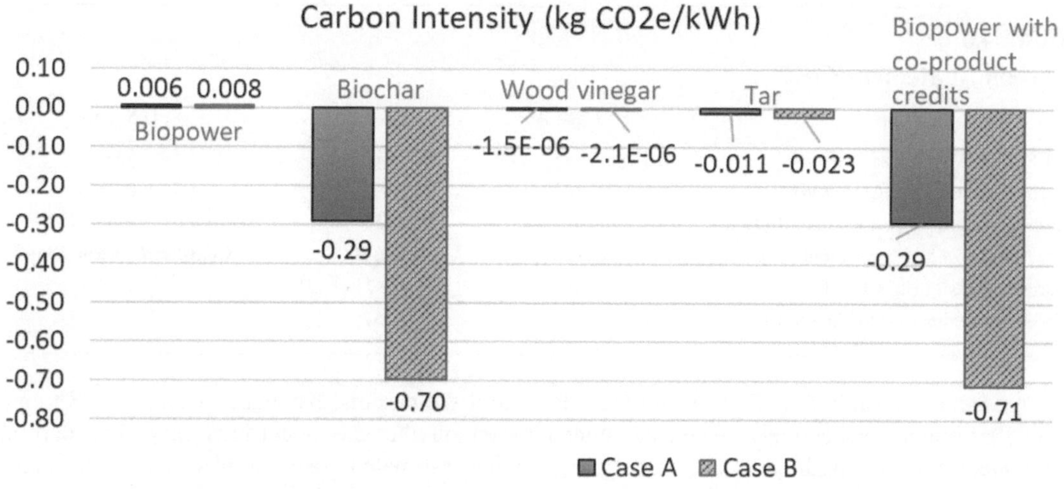


**FIGURE 4.13** Carbon intensity associated with the electricity production and co-product credits per 1 kWh.

complementary and preventive measure to lower greenhouse gas emissions for climate change mitigation [27]. Case B exhibits a higher biochar credit due to higher biochar yield (1.23 lb/kWh compared to 0.51 lb/kWh for Case A). The co-product contributions from wood vinegar and tar are relatively insignificant. With co-product credits, the specific carbon intensity for Case B (–0.71 kg $CO_2$e/kWh) is significantly lower than Case A (-0.29 kg $CO_2$e/kWh), and both are considerably lower than the U.S. electricity mix (0.45 kg $CO_2$e/kWh) [26]. Therefore, it is clear that the valorization of pistachio shells for biopower generation can help decarbonize the electricity grid. This study did not consider any greenhouse gas (GHG) emissions associated with the disposal and handling of this solid waste. Future studies can define the disposal system and quantify any related GHG emissions. However, any GHG emission from the waste disposal will be treated as credits for biopower, further strengthening the value proposition of valorization of pistachio shells for biopower.

From a climate change mitigation perspective, higher biochar yield with lower electricity generation (Case B) is more favorable than Case A. This can be further examined by quantifying the marginal carbon abatement cost with the integration of the TEA and LCA results. Marginal abatement costs for avoided GHG emissions were calculated as the difference between the GHG for the average U.S. mix and electricity alternatives from this study divided by the difference between the costs of the U.S. mix and electricity alternatives. The case with 100% electricity from coal was also evaluated as a sensitivity case for comparison. The marginal carbon abatement costs were determined using the base case TEA results and the carbon intensity results according to Equation (4.1).

$$Carbon\ Abatement\ Cost = \frac{MESP_{Biomass} - MESP_{US\ Mix\ or\ Coal}}{CI_{US\ Mix\ or\ Coal} - CI_{Biomass}} \tag{4.1}$$

The carbon abatement cost is in \$/metric ton $CO_2$e emissions and represents the marginal increase in the electricity price for a given reduction in carbon intensity when switching from fossil/mixed fuels to biomass. The base case MESPs are 39.9 and 35.5 ¢/kWh for Case A and Case B, respectively. The electricity prices for both the U.S. electricity mix and electricity from the coal-fired power plant were assumed to be the same at 5.9¢/kWh. The respective carbon intensity is 0.45 kg $CO_2$e/kWh for the U.S. mix and 1.17 kg $CO_2$e/kWh for electricity from coal-fired power plants [26]. Electricity from coal-fired power plants was considered here as it is regarded as the "dirtiest" electricity and still dominates in the electricity mix in many states, for example, 72% in Kentucky, 84% in Wyoming, and 91% in West Virginia [28]. The carbon abatement cost results

**TABLE 4.8**
**Carbon Abatement Cost**

| U.S. Mix as Baseline | Case A | Case B | U.S. Mix |
|---|---|---|---|
| MESP (¢/kWh) - Base case | 39.9 | 35.5 | 5.90 |
| Carbon intensity (kg $CO_2e$/kWh) | −0.29 | −0.71 | 0.45 |
| Carbon abatement cost ($/tonne $CO_2e$) | 458 | 255 | -- |
| *Coal-fired plant as a sensitivity case* | **Case A** | **Case B** | **Coal-fired Power Plant** |
| Carbon intensity (kg $CO_2e$/kWh) | −0.29 | −0.71 | 1.17 |
| Carbon abatement cost ($/tonne $CO_2e$) | 233 | 157 | -- |

are presented in Table 4.8. Case A and Case B exhibit the carbon abatement costs of $458/tonne and $255/tonne, respectively, when calculated based on the U.S. electricity mix. The carbon abatement costs for both cases are sharply lower when estimated based on electricity from coal-fired power plants, $233 and $157 per metric ton $CO_2e$ emissions for Case A and Case B. As a reference, the U.S. Department of Energy's Office of Fossil Energy and Carbon Management has targeted to achieve less than $100/tonne $CO_2e$ for both capture and storage within a decade [29]. The current study shows that producing biopower from waste feedstock offers a feasible alternative to mitigate climate change at a cost comparable to or better than other emerging carbon removal technologies, such as direct air capture of $CO_2$ from air [30].

## 4.7 GENERAL METHODOLOGY TO OPTIMIZE PREPROCESSING, CONVEYANCE, AND CONVERSION

The feasibility of transforming pistachio waste for biopower also lies in the success of the critical processing steps. A major focus of this work was to establish the best general approach to co-optimize feedstock preprocessing, conveyance behavior, and conversion performance to maximize process reliability and productivity. Our approach in this project was to systematically evaluate several potential options in parallel pertaining to preprocessing methods and convertibility of the material under different conditions, then simultaneously determine the combined set of desired feedstock attributes, considering overall system performance and cost:

1. Characterize pistachio shell attributes and variability (chemical, physical, flow) presumed to be important for gasification
2. Explore preprocessing methods potentially beneficial for conveyance and thermal conversion
3. Perform conversion testing vs. feedstock attributes and process conditions
4. Determine ranges of attributes for acceptable conversion and choose the minimum preprocessing to achieve this
5. Measure solid flow characteristics of preprocessed material to determine commercial equipment design specifications
6. Demonstrate long-term process reliability with preprocessed feedstock

Co-optimization of multiple unit operations simultaneously proved difficult, especially without a specific conversion process determined at the outset of the project. This was because, while testing of preprocessing and conversion options provided useful data on outcomes and costs, the variable space for these unit operations was too large and needed to be constrained.

Each conversion process has its own set of requirements for process conditions and feedstock. For example, some processes may require feedstock particles to be smaller than a certain size or a moisture content of less than a certain threshold. There may also be requirements from a temperature and pressure standpoint. Thus, as a first step, the conversion process should determine appropriate specifications for the feedstock. This was a sentiment echoed by project partner Jenike, as well as Forest Concepts, a leading biomass feedstock processing technology provider. Forest Concepts has developed a detailed protocol and interview structure to define technical specifications that are tied to functional specifications (the "why") for a variety of conversion processes [31]. An example functional specification might be: "the residence time distribution in a reactor should be limited to 3–5 s," which corresponds to technical specifications around the feedstock's hydrodynamic properties such as particle size/shape distribution, particle envelope density, etc.

This is also consistent with the Quality by Design (QbD) approach recently adopted by the FCIC as a guiding principle [32]. With QbD, the starting point for determining preprocessing and handling requirements is knowing the final product quality specifications. However, a complete set of specifications may or may not be known to the conversion technology provider. A wide range of processes are currently under development across the nascent bioenergy industry, with variations in reactor design, feedstocks, process temperatures, process gas composition, feeding characteristics, and product targets. As such, the specific concerns (and level of concern) among process developers regarding feedstock quality vary significantly and appear largely proportional to the scale and stage of development. Most bioenergy development projects have tested multiple feedstock types and qualities, and a few have developed feedstock technical specifications for some critical material attributes (CMAs) for their processes. Many potential CMAs and levels still need to be specified, representing a potential gap in large-scale bioenergy development.

Once the CMAs are known, specifications are set, and the process conditions are determined, the next step should be testing a representative sample of the feedstock at representative process and handling conditions. This testing should focus on characterizing the flow behavior of such sample. Jenike has developed shear testing methods for effectively characterizing the flow behavior of bulk solids [33]. The test data obtained by conducting such testing then provides the basis for designing appropriate hoppers, feeders, and transfer systems. This information can also be used to select appropriate handling equipment. As this is an empirical approach, it is critical to include truly representative samples that capture the full range of anticipated flow attributes – e.g., from multiple sources, with varying moisture contents, preprocessed to a "reactor ready" state, etc.

Regarding operation with forest residues in comparison to the pistachio shells used in this project, typical biomass feedstocks, such as woodchips, corn stover, and switch grass, have much more complex flow behavior compared to pistachio shells. Pistachio shells are of a more consistent particle size distribution with a low aspect ratio, allowing this material to move more freely than particles of woodchips and corn stover. This also decreases the potential of interlocking for pistachio shells. Furthermore, compared to woodchips and corn stover, the moisture content (and moisture holding capacity) of pistachio shells is much lower. As a result, one can expect pistachio shells to have a more manageable flow behavior compared to typical biomass feedstocks. Even then, arching is still possible with pistachio shells. Pistachio shells can form ratholes, especially the pistachio shell fines, if the hopper wall angles are not steep enough.

In V-Grid's case, the material handling system and gasifier had already to some extent been optimized for using pistachio shells. When testing forest residues, the material needed to be chipped to a ½" minus screen size, and the feed auger needed to be changed to a different configuration. Still, neither the gasifier nor the genset was modified. This is consistent with recent industry feedback and the broader QbD approach of FCIC—that, due the existing investments, primary conversion processes and downstream unit operations (genset, catalytic synthesis reactor, pyrolysis vapor upgrader, etc.) are likely to remain mostly fixed. In the near term, rather than starting with a broad acceptance envelope for the process, more of the burden will likely be placed on preprocessing operations to produce a feedstock with the appropriate specifications. Longer term, as these processes scale up and lower volumes of ideal feedstocks are available for low cost, it may be more cost effective to evolve a process and/or employ secondary mitigation steps rather than meet stringent feedstock specifications.

## 4.8 CONCLUSIONS AND LESSONS LEARNED

This study demonstrated that waste pistachio nut shells could be used as a renewable feedstock for climate-friendly electricity generation via industrial gasification technology. For processing TWC's pistachio waste material, among critical material attributes, fines content in the biomass (<1/4") had the largest potential to reduce the operating time of the gasifiers due to plugging. Pelletizing fines and co-feeding them with the mixed pistachio waste increased the average feed density and feed rate, and increased biochar production. Compared to pine wood chips, mixed pistachio waste yielded higher biochar quantity but slightly reduced quality as determined by BET surface area measurements.

A scaled-up gasifier was able to increase throughput and thereby improve the economics without negative impacts (higher throughput for the same personnel). During the extended 500-h trial, unplanned maintenance was lower than planned maintenance, and both were lower than the regeneration time required to clean the system of carbon deposits.

Biochar production and selling price were major economic and carbon drivers for the process; the number of personnel required to operate the process was the most prominent economic sensitivity. Biochar production rate and selling price had an overwhelming impact on the modeled Minimum Electricity Selling Price (MESP), which ranged from 35.5 to 39.9¢/kWh (based on 16 h/day operation). A modest improvement in the on-stream factor and utilization of pelletized fines could result in a noticeable reduction in MESP.

LCA results show that valorization of pistachio shells for biopower generation is a "carbon negative" process that can help decarbonize the U.S. electricity grid; the specific carbon intensity was up to –0.71 kg $CO_2$e/kWh, compared to 0.45 kg $CO_2$e/kWh for the average U.S. electricity mix. Pistachio trees absorb carbon dioxide during growth via photosynthesis, offsetting a percentage of the greenhouse gases released during the gasification of pistachio shells and syngas combustion. Moreover, highly stable biochar permanently sequesters a significant fraction of biochar carbon in the ground, more than offsetting the remaining emissions. Carbon sequestration of biochar from pistachio waste can be a complementary climate change mitigation strategy.

In general, a systematic Quality by Design type of methodology is the preferred approach for designing preprocessing and material conveyance systems, where a downstream technology (end user) for the produced intermediate is specified at the outset. A commercial gasifier partner was not identified until the project was substantially underway, so knowledge of a full feedstock specifications was not immediately available.

Finally, pistachio nut shells are no longer an agricultural waste once it is transformed to biopower. The present utilization of pistachio "waste" for carbon-negative electricity production is a bio-based circular carbon economy that stresses capturing atmospheric carbon via photosynthesis and exploiting this unique feature to the fullest extent possible, as well as closing the carbon cycle by utilizing biogenic carbon [34].

## 4.9 DISCLAIMER

The work performed in part by the employees of the National Renewable Energy Laboratory, operated by Alliance for Sustainable Energy, LLC, was supported by the U.S. Department of Energy (DOE) under Contract No. DE-AC36-08GO28308. The views expressed in the article do not necessarily represent the views of the DOE or the U.S. Government. The U.S. Government retains and the publisher, by accepting the article for publication, acknowledges that the U.S. Government retains a nonexclusive, paid-up, irrevocable, worldwide license to publish or reproduce the published form of this work, or allow others to do so, for the U.S. Government purposes.

## REFERENCES

[1] Koul B, Yakoob M, Shah MP. Agricultural waste management strategies for environmental sustainability. *Environmental Research*. 206, 112285 (2022).

[2] USDA/NASS QuickStats Ad-hoc Query Tool [Internet]. Available from: https://quickstats.nass.usda.gov/

[3] U.S. EPA. Downstream Management of Organic Waste in the United States: Strategies for MEthane Mitigation [Internet]. (2022). Available from: https://www.epa.gov/system/files/documents/2022-01/organic_waste_management_january2022.pdf

[4] Feedstock-Conversion Interface Consortium [Internet]. *Energy.gov*. Available from: https://www.energy.gov/eere/bioenergy/feedstock-conversion-interface-consortium

[5] Merrow E, Phillips K, Meyers C. Understanding Cost Growth and Performance Shortfalls in Pioneer Process Plants [Internet]. *RAND Corporation*. Available from: https://www.rand.org/pubs/reports/R2569.html

[6] Howe D, Westover T, Carpenter D, *et al.* Field-to-Fuel Performance Testing of Lignocellulosic Feedstocks: An Integrated Study of the Fast Pyrolysis–Hydrotreating Pathway. *Energy Fuels*. 29(5), 3188–3197 (2015).

[7] Carpenter D, Westover T, Howe D, *et al.* Catalytic Hydroprocessing of Fast Pyrolysis Oils: Impact of Biomass Feedstock on Process Efficiency. *Biomass and Bioenergy*. 96, 142–151 (2017).

[8] Du S, Yang H, Qian K, Wang X, Chen H. Fusion and Transformation Properties of the Inorganic Components in Biomass Ash. *Fuel*. 117, 1281–1287 (2014).

[9] Susastriawan AAP, Saptoadi H, Purnomo. Small-scale Downdraft Gasifiers for Biomass Gasification: A Review. *Renewable and Sustainable Energy Reviews [Internet]*. 76, 989–1003 (2017). Available from: https://linkinghub.elsevier.com/retrieve/pii/S1364032117304471

[10] Lacey JA, Aston JE, Westover TL, Cherry RS, Thompson DN. Removal of Introduced Inorganic Content from Chipped Forest Residues Via Air Classification. *Fuel [Internet]*. 160, 265–273 (2015). Available from: https://linkinghub.elsevier.com/retrieve/pii/S0016236115007917

[11] Dutta A, Iisa M, Talmadge M, *et al.* Ex Situ Catalytic Fast Pyrolysis of Lignocellulosic Biomass to Hydrocarbon Fuels: 2019 State of Technology and Future Research [Internet]. Available from: https://www.osti.gov/servlets/purl/1605092/

[12] Lv P, Wu C, Ma L, Yuan Z. A Study on the Economic Efficiency of Hydrogen Production from Biomass Residues in China. *Renewable Energy*. 33(8), 1874–1879 (2008).

[13] Huang Y, McIlveen-Wright DR, Rezvani S, *et al.* Comparative Techno-economic Analysis of Biomass Fuelled Combined Heat and Power for Commercial Buildings. *Applied Energy*. 112, 518–525 (2013).

[14] Borello D, De Caprariis B, De Filippis P, *et al.* Thermo-Economic Assessment of a Olive Pomace Gasifier for Cogeneration Applications. *Energy Procedia*. 75, 252–258 (2015).

[15] Susanto H, Suria T, Pranolo SH. Economic Analysis of Biomass Gasification for Generating Electricity in Rural Areas in Indonesia. *IOP Conf. Ser.: Mater. Sci. Eng*. 334(1), 012012 (2018).

[16] Hamad MAF, Radwan AM, Amin A. Review of Biomass Thermal Gasification [Internet]. In: *Biomass Volume Estimation and Valorization for Energy*. Tumuluru JS (Ed.), InTech (2017) [cited 2022 Oct 29]. Available from: http://www.intechopen.com/books/biomass-volume-estimation-and-valorization-for-energy/review-of-biomass-thermal-gasification

[17] Tao L, Tan ECD, McCormick R, *et al.* Techno-economic Analysis and Life-cycle Assessment of Cellulosic Isobutanol and Comparison with Cellulosic Ethanol and N-Butanol. *Biofuels, Bioproducts and Biorefining [Internet]*. 8(1), 30–48 (2014). Available from: https://onlinelibrary.wiley.com/doi/abs/10.1002/bbb.1431

[18] Hartley DS, Thompson DN, Cai H. Woody Feedstocks 2019 State of Technology Report [Internet]. Idaho National Lab. (INL), Idaho Falls, ID (United States). Available from: https://www.osti.gov/biblio/1607741

[19] Raptis CE, Boucher JM, Pfister S. Assessing the Environmental Impacts of Freshwater Thermal Pollution from Global Power Generation in LCA. *Sci Total Environ*. 580, 1014–1026 (2017).

[20] AR5 Climate Change 2014: Impacts, Adaptation, and Vulnerability — IPCC [Internet]. Available from: https://www.ipcc.ch/report/ar5/wg2/

[21] Dupuis DP, Grim RG, Nelson E, *et al.* High-Octane Gasoline from Biomass: Experimental, Economic, and Environmental Assessment. *Applied Energy [Internet]*. 241, 25–33 (2019). Available from: https://linkinghub.elsevier.com/retrieve/pii/S0306261919303617

[22] Biochar for Environmental Management: Science, Technology and Implementation [Internet]. *Routledge & CRC Press*. Available from: https://www.routledge.com/Biochar-for-Environmental-Management-Science-Technology-and-Implementation/Lehmann-Joseph/p/book/9780367779184

[23] Zhang C, Zeng G, Huang D, et al. Biochar for Environmental Management: Mitigating Greenhouse Gas Emissions, Contaminant Treatment, and Potential Negative Impacts. *Chemical Engineering*

*Journal [Internet]*. 373, 902–922 (2019). Available from: https://www.sciencedirect.com/science/article/pii/S1385894719311635

[24] Biochar SS. Tool for Climate Change Mitigation and Soil Management [Internet]. In: *Encyclopedia of Sustainability Science and Technology*, Springer New York, 845–893 (2012) [cited 2022 Feb 28]. Available from: https://www.academia.edu/21089920/Biochar_Tool_for_Climate_Change_Mitigation_and_Soil_Management

[25] LTS. DATASMART LCI Package [Internet]. Available from: https://ltsexperts.com/services/software/datasmart-life-cycle-inventory/

[26] Argonne National Laboratory. Greenhouse Gases, Regulated Emissions and Energy use in Transportation (GREET) model. [Internet]. (2020). Available from: https://greet.es.anl.gov/

[27] Enríquez-de-Salamanca Á. Climate Change Mitigation in Forestry: Paying for Carbon Stock or for Sequestration? *Atmosphere*. 13(10), 1611 (2022).

[28] US EPA O. Emissions & Generation Resource Integrated Database (eGRID) [Internet]. (2020). Available from: https://www.epa.gov/egrid

[29] Office of Fossil Energy and Carbon Management [Internet]. Energy.gov. Available from: https://www.energy.gov/fecm/office-fossil-energy-and-carbon-management

[30] Keith DW, Holmes G, St., Angelo D, Heidel K. A Process for Capturing CO2 from the Atmosphere. *Joule [Internet]*. 2(8), 1573–1594 (2018). Available from: https://www.sciencedirect.com/science/article/pii/S2542435118302253

[31] Dooley JH, Lanning DN, Lanning CJ. Structured Interview Guide and Template for Specification of Woody Biomass Fuel and Feedstocks [Internet]. In: *2011 Louisville, Kentucky, August 7 - August 10, 2011*, American Society of Agricultural and Biological Engineers (2011) [cited 2022 Jan 27]. Available from: http://elibrary.asabe.org/abstract.asp?JID=5&AID=37417&CID=loui2011&T=1

[32] Davis B, Schlindwein WS. Introduction to Quality by Design (QbD) [Internet]. In: *Pharmaceutical Quality by Design*. Schlindwein WS, Gibson M (Eds.), John Wiley & Sons, Ltd, Chichester, UK, 1–9 (2018) [cited 2022 Oct 31]. Available from: https://onlinelibrary.wiley.com/doi/10.1002/9781118895238.ch1

[33] The Jenike Direct Shear Tester [Internet]. Jenike & Johanson (2015). Available from: https://jenike.com/time-tests-and-the-test-of-time-the-jenike-direct-shear-tester/

[34] Tan ECD, Lamers P. Circular Bioeconomy Concepts—A Perspective. *Frontiers in Sustainability [Internet]*. 2 (2021). Available from: https://www.frontiersin.org/articles/10.3389/frsus.2021.701509

# 5 Carbon Recycling

## *Waste Plastics to Hydrocarbon Fuels*

*Halima Abu Ali, Peter Eyinnaya Nwankwor, David John, Hassan Tajudeen, and Jean Baptiste Habyarimana*
Department of Petroleum Chemistry, American University of Nigeria, Yola, Nigeria

*Wan Jin Jahng*
Department of Petroleum Chemistry, American University of Nigeria, Yola, Nigeria

Department of Energy, Department of Ophthalmology, Julia Laboratory, Suwon, Korea

## CONTENTS



## 5.1 INTRODUCTION AND BACKGROUND

Plastic polymers are among the most prominent products and the most profitable molecules of the chemical industry, fetching the top 50 global chemical companies with more than $950 billion in annual revenues [1–3]. Many plastics are resistant to heat and water, and, therefore, degrade slowly. The versatile and inexpensive nature of plastics, coupled with their non-biodegradability, makes them exist for a long time in the oxidative environment. Chemical companies produce an estimated 8.3 billion tons of plastics, and 6.3 billion tons become waste with a recycling efficiency of 9%. The current study seeks to present a novel means of converting these waste excesses to energy products. The study also provides potential ways through which the proposed methodology

DOI: 10.1201/9781003167693-5

could become a means of achieving a net carbon balance in the environment, thereby lessening the risk of contributing to climate change.

### 5.1.1 Current Plastic Waste Disposal Method and Environmental Impacts

The alarming rate of plastic waste accumulation requires new methods of disposal that would provide adequacy and efficiency in waste management and sustainability. Current plastic waste disposal methods have been adopted in many countries as a viable solution to waste accumulation; however, it presents environmental drawbacks. Plastic waste takes hundreds of years to degrade. Besides, the degradation in landfills generates toxic molecules, including phthalates, chlorine, bisphenol A, furans, antimony trioxide, and dioxins that leach into the ground, potentially sipping into groundwater or even end up in the ocean. Greenhouse gases, including methane and carbon dioxide, are released into the atmosphere in large amounts due to photo-degradation that occurs when the waste is exposed to sunlight over a long period.

Incineration is more hazardous than landfilling due to the hazardous molecules and small particles released into the air, settling on vegetation and the human body. As a significant contributor to air pollution, the developed countries ensured that specific emission standards were set for the release of exhaust gas to be permissible; however, in developing countries, no such standards exist, and exhaust from incineration is released directly into the air without any form of filtering, desulfurization, denitrification, or carbon capture. In addition to $CO_2$, $NO_x$, and $SO_x$, other toxic molecules, including dioxins, furans, halogens, carcinogenic aromatic compounds, and hazardous metals such as mercury, are released into the air. The harmful effects of waste plastics on the environment, including vegetation, animal, and human lives, are unparalleled compared to the effects of other waste disposals.

The halogens and carbon emissions increase the carbon budget and speed up climate change. Dioxins and furans are responsible for causing neurological disorders and cancer. Countries including the United Kingdom, the United States, and China introduced a program known as Energy from Waste, which generates electricity from the exhaust stack in the incineration process. Incorporating energy from waste material transforms incineration into a sustainable mechanism for environmental, social, and economic benefits. A drawback in the process lies in retrofitting existing incineration plants with turbines and generators and the high cost of connecting the plant network to the grid.

The majority of the used plastics are accumulated in water and land, polluting the environment, and posing a health threat to wildlife and humans. There will be more plastics than live fish in the ocean by 2050. Disposal of plastics through landfills and incineration is inefficient, expensive, and hazardous, therefore more environmentally friendly methods are required.

### 5.1.2 Catalytic Pyrolysis

Pyrolysis of waste polymers showed a promising alternative approach among the current methods for treating contaminated wastes. Pyrolysis is a cracking reaction under high temperatures in the range of 400°C to 500°C and pressure into helpful carbon energy as solid, liquid, or gas.

Catalytic pyrolysis involves the degradation of the polymeric materials under high temperatures in the absence of oxygen and the presence of a Lewis acid catalyst. The catalyst affects specific parameters, including reaction temperature, kinetics, gas formation, solid residue, by-product impurities, aromaticity, paraffin formation, reactivity, and molecular weight distribution. Catalytic pyrolysis demonstrated shorter reaction time, increased short- to medium-chain distillate concentration, less solid product yield, removal of impurities via adsorption, and aromatics formation with better octane rating. Zeolite is a hydrated crystalline aluminosilicate mineral widely used as a cracking catalyst [4]. Zeolites are microporous, allowing ions to fit for the cracking reaction, and the Si/Al ratio determines the acidity as a versatile and suitable catalyst for bond-breaking reactions. The high Si/Al ratio in zeolite demonstrates a high Lewis acidity optimized for hydrocarbon cracking. Titanium dioxide ($TiO_2$) is a naturally occurring ilmenite, rutile, and anatase metal oxide applicable in the catalyzed pyrolysis of waste plastics [5]. Due to its high refractive index, titanium dioxide is used for the pigment to provide white color and high opacity to paint, plastics, and paper for thin-film coating and UV blocking.

### 5.1.3 Aims and Objectives

The main aims of the research conducted are:

1. To determine the catalytic efficiency of specific metal oxides, including titanium dioxide, zeolite, and potash, as novel catalysts.
2. To generate methane, gasoline, and diesel products.
3. To determine the molecular mechanism of catalytic cracking using a kinetics model.

Furthermore, we assessed the effects of the different catalysts on the yield and composition of liquid hydrocarbon products, reaction temperatures, and the molecular analysis of the collected products compared to positive controls.

## 5.2 HYDROCARBON COMPOSITION OF LIQUID PRODUCTS FROM WASTE PLASTICS

Low-density polyethylene (LDPE) is one of the most mass-produced plastics, with a density of 0.910–0.940 g/cm$^3$ of short-chain branches. A schematic diagram of the one-stage fixed bed pyrolysis reaction system is shown in Figure 5.1. Its increased ductility and low tensile strength make it a suitable substrate for catalytic cracking. In a fixed-bed reactor system, LDPE has been cracked over ZSM-5 zeolite to yield $C_4$–$C_8$ aliphatic and $C_7$–$C_{12}$ aromatic compounds. Reaction kinetics (conversion yield vs. reaction time) using metal oxide are presented in Figure 5.2 [6]. Polyvinyl chloride (PVC) is produced via the radical addition polymerization of chloroethene. PVC is heat-, fire-, and water-resistant. The various chemical methods for recycling PVC are pyrolysis, catalytic de-chlorination, and hydrothermal treatment using a transition metal oxide catalyst. PVC is an excellent substrate for pyrolysis having a high laminar burning velocity of 178.6 cm/s.

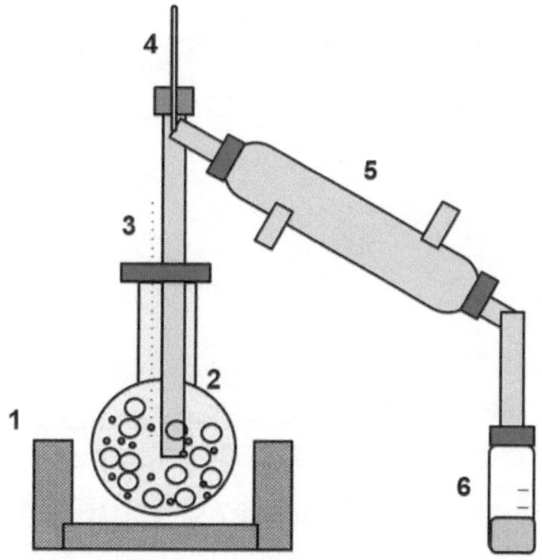


**Schematic diagram of the one-stage fixed bed pyrolysis reaction system.**
1. Heating mantle; 2. Waste plastics and catalyst; 3. Reactor; 4. Thermometer; 5. Condenser; 6. Collection flask

**FIGURE 5.1** Schematic diagram of the one-stage fixed bed pyrolysis reaction system.

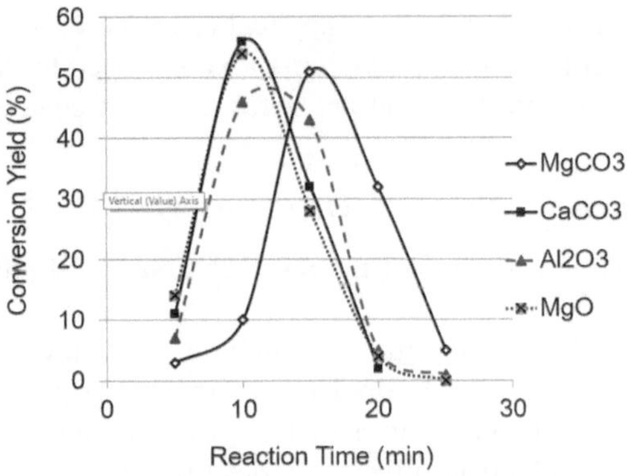


**FIGURE 5.2** Reaction kinetics of metal oxide-catalyzed pyrolysis. Each catalyst showed a specific reaction time and conversion yield of pyrolysis.

Polystyrene (PS) is an inexpensive plastic popularly known as Styrofoam. PS is a vinyl polymer synthesized by free radical vinyl polymerization of the styrene monomer. Polystyrene pyrolysis under 400–600°C in the presence of nitrogen-generated $C_2$–$C_{15}$ products.

The GC-MS analysis of liquid products from the catalytic pyrolysis of LDPE over zeolite and titanium dioxide showed various hydrocarbon structures and fragments (Figure 5.3). The LDPE products cracked over zeolite indicate the m/z of an unknown molecule at RT 1.182 mins showing m/z = 55.1 (base peak) and m/z = 84.1 ($M^+$). The NIST database search confirmed the above product as 2-methyl-1-pentene with m/z = 56.0 (base peak) and m/z = 84.0 ($M^+$). The GC analysis of an unknown molecule at RT 1.478 mins showed m/z = 56.1 (base peak) and m/z = 100 ($M^+$). The NIST database search determined the product as 1,3-dimethylcyclopentene with m/z = 56.0 (base peak) and m/z = 100 ($M^+$). The LDPE products cracked over $TiO_2$ showed m/z = 55 (base peak) and m/z = 71 ($M^+$) of an unknown molecule at RT 1.027 mins. The NIST database search demonstrates 1,2-dimethylcyclopropane with m/z = 55 (base peak) and m/z = 71 ($M^+$). The

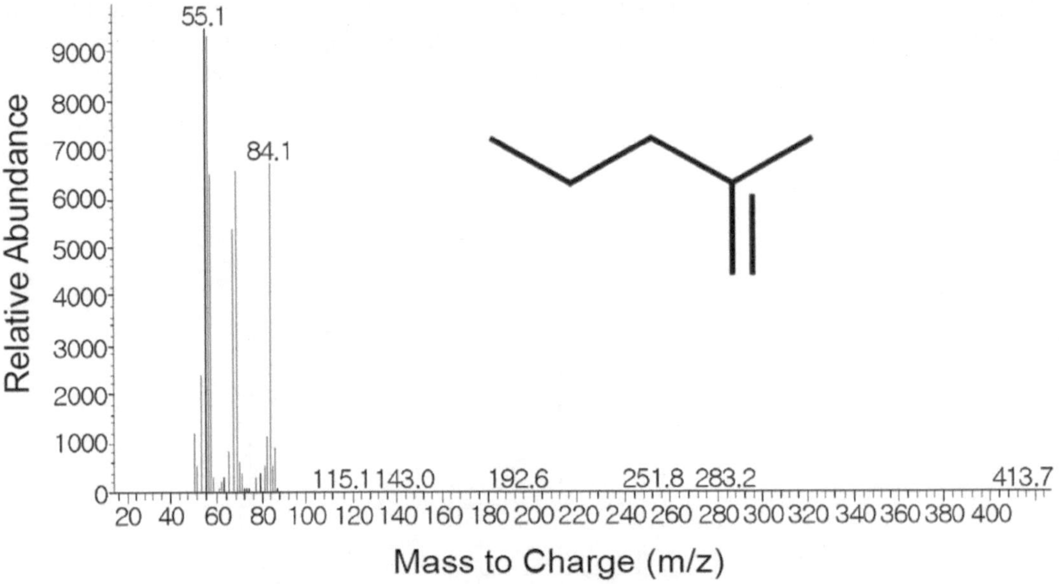


**FIGURE 5.3** GCMS analysis. Pyrolysis products were analyzed using gas chromatography-mass spectrometry (GCMS) to determine the molecular weight and the fragmentation pattern to identify each product. LDPE pyrolysis using zeolite and titanium oxide generated C1–C16 hydrocarbon gas/liquid products, including 2-methyl-1-pentene (m/z = 56). GCMS spectrum demonstrated the base peak (55.1 m/z)as the most abundant ion and the molecular ion peak (84.1 m/z).

liquid products were mainly $C_5$ (mainly 1,2-dimethylcyclopropane), $C_6$ (mainly 2-methylpentane), $C_7$ (mainly 1,3-dimethylcyclopentane), and $C_8$ (2-octene, 4-octene, octane, 3-ethylhexane), in addition to $C_9$ and $C_{10}$ with less amount. The liquid products from catalytic pyrolysis using zeolite and titanium catalysts were mainly aliphatic compounds. The production of a high amount of gasoline range hydrocarbon fractions can be attributed to the use of metal oxide catalysts compared to non-catalytic pyrolysis.

The current experiments demonstrate that the liquid products from waste over selected catalysts mainly consisted of hydrocarbons in the gasoline-range hydrocarbon fractions of $C_5$–$C_8$. Our data indicate that the one-stage pyrolysis system effectively synthesized gasoline-range hydrocarbons from LDPE using zeolite and titanium dioxide. GC-MS data indicate that a specific catalyst may determine the chemical composition of the liquid products. Our previous experiments using other metal oxide catalysts have typically yielded kerosene and diesel-range distillates within the range of $C_8$–$C_{22}$.

The impact of catalysts on the hydrocarbon composition was reported on the cracking reaction of waste HDPE plastics over sulfate-modified zirconium [7]. The products were mainly hydrocarbon gas and light liquids within the range of $C_1$–$C_7$, with methane and ethane appearing in negligible amounts. At the same time, the major components were $C_3$ (propane and propene), $C_4$ (n-butane, 1-butene, iso-butane), and $C_5$ (n-pentane, pentene, and iso-pentane). Pyrolysis of LDPE over zinc-loaded ZSM-5 was reported to produce monocyclic-ring aromatic structures with a liquid yield of 51 % at the reaction temperature of 380°C [8].

The physical appearance of the liquid products from LDPE and PS was pale-yellow transparent liquid; however, the liquid extraction without catalyst turned into wax after two minutes of extraction, suggesting higher molecular weight products. The wax production was correlated with low residence times and high heating rates that reduced secondary reactions. For PVC, the dark-green opaque liquid was extracted using zeolite and titanium dioxide catalysts, while a dark-brownish liquid was extracted from the pyrolysis of PVC without catalysts.

## 5.3 EFFECTS OF CATALYST ON PYROLYSIS TEMPERATURE

The temperature profile and corresponding yields for different reactants in the experiments are as follows; the liquid-producing temperatures for LDPE and PS were lower over zeolite (LDPE = 90–178°C; PS= 62–164°C) and no catalyst (LDPE = 82–140°C; PS = 40–140°C) conditions than over titanium dioxide (LDPE = 99–198°C; PS = 68–168°C). In comparison, the reverse was the case for PVC, which was lower over titanium dioxide (68–168°C) compared to zeolite (70–172°C) and no catalyst conditions (80–154°C) [9–11].

However, the liquid yield (wt%) varied according to the starting materials. Despite the use of a catalyst, polystyrene produced the highest average liquid yield of 82.4% and at lower temperatures than LDPE and PVC. Using the zeolite catalyst, the gas expulsion temperature of PS (62°C) was 8°C and 28°C lower than PVC and LDPE, respectively. Our experiments suggest that the high stability of the phenyl ring as a critical role in stabilizing the thermal cracking intermediates formed during the pyrolysis reaction [9–11].

LDPE generally favored higher cracking temperatures in comparison to PVC and PS. Under the zeolite catalyst condition, the gas expulsion temperature was 90°C compared to 70°C and 62°C for PVC and PS, respectively. A similar trend is also observed using titanium dioxide, where the gas expulsion temperature for LDPE (99°C) was 31°C higher than that of PVC and PS. Higher pyrolysis product collection temperature for LDPE at 425°C against PS products expelled at a lower temperature of 350°C.

## 5.4 PRODUCT VISCOSITY AND DENSITY ANALYSIS

The liquid products from the LDPE pyrolysis were analyzed by petroleum fuel's physicochemical properties, including density and kinematic viscosity [10]. The density of liquid products from LDPE,

PVC, and PS at 25°C range from 0.783–0.803 g/mL, 0.63–1.142 g/mL, and 0.876–0.906 g/mL, respectively, and it is compatible with commercial gasoline fuel with a range of 0.728–0.746 g/mL.

The kinematic viscosity of the liquid fractions from LDPE over different catalysts at 75°C is shown in centistokes (cSt) 3.1749–3.5143 cSt. Even though there is no direct relationship between density and viscosity, low API is associated with higher viscosity. Several studies support that temperature has the most considerable influence on product yield, as high as 54%, 83.15%, and 97% from 400–600°C. However, one of the specific aims of our current study is energy conservation, targeting the maximum yield at a lower temperature. Previously, a higher liquid yield (71–81%) was reported using zeolite at 450°C [4,6,9–11]. A discrepancy is reported in the high boiling point range of oil products from ZSM-5 compared to the zeolite-Y catalyst, suggesting composition variation. ZSM-5 could have an aromatic forming capacity leading to stronger intermolecular interactions, causing a shift in boiling range to a higher temperature. Furthermore, zeolite enhances liquid yield at a higher temperature, possibly due to interactions for higher boiling fractions.

Catalytic pyrolysis of HDPE showed that liquid yield is not correlated linearly with temperature as an optimum yield of 91.2% was reached at 450°C despite the temperature varied from 420–510°C [12–14]. However, despite the decline after 450°C, the olefin composition positively correlated with temperature, while the paraffin content had an inverse correlation. Similarly, despite the temperature range of 400–550°C, the maximum product yield of 23.9% was obtained at 450°C. The reaction rate of HDPE cracking peaked at 467°C, while the first liquid fraction was collected at 450°C.

Our results indicate that potash and zeolite catalysts generate liquid fuels at a lower temperature (100–200°C) than at higher temperatures (400–500°C), thus conserving energy to achieve waste-fuel conversion [9–11].

## 5.5 EFFECT OF CATALYST-POLYMER RATIO ON PRODUCT YIELD

We examined the influence of the catalyst/polymer ratio on the liquid product yield. We observed that the increased catalyst/polymer ratio (0.4–1) inhibited the liquid yield. However, zeolite increased the liquid yield by 50%. Potash could have an affinity toward gas and char products, especially at a higher catalyst ratio.

Our results suggest that a low potash ratio could improve liquid conversion yield; however, for zeolite, the liquid conversion is favored at a higher zeolite/polymer ratio of at least 400°C. The role of the catalyst-polymer ratio in converting HDPE to fuels has been evaluated. It is reported that using a fluidized catalytic cracking (FCC) catalyst at an optimum catalyst-polymer ratio of 20 wt%, a liquid conversion of 91.2% was achieved at a temperature of 450°C. Our study, with a pyrolysis temperature of less than 200°C, is significantly different from the commonly observed high temperatures for pyrolysis reactions as high as 400°C to 600°C, implying the considerable difference in yield regardless of the catalyst-polymer ratio. Also, it was reported that increasing the catalyst-polymer ratio beyond 20 wt% boosted gasification and coke formation [15–17].

Our results support the idea that the liquid yield could be assigned to a combination of temperature and the catalyst/polymer ratio. Increased temperature has been well established to lead to higher liquid yield. As shown in the current study, there is a mixture of HDPE with various phases contacting the catalyst as temperature increases for the one-stage thermal catalytic cracking. Our data show that the catalyst/polymer ratio impacts the product yield, with low potash concentration leading to a high yield. At the same time, zeolite appears to require higher catalyst loading to favor high liquid yields.

## 5.6 MINERAL DISTRIBUTION IN CATALYST

Next, we determined the molecular composition of potash using various analytical techniques, including an inductively coupled plasma-optical emission spectrophotometer (ICP-OES) and X-ray diffractometer (XRD). XRD pattern of the potash catalyst demonstrated the presence of minerals, including quartz (COD: 96-500-0036), trona (COD: 96-900-7658), hanksite (COD: 96-900-0375), and halite (COD: 96-900-6374). With the intensity of XRD peaks roughly equivalent to the abundance of the mineral, it can be deduced that prominent peaks for the minerals are seen at peak positions ($2\theta$) of 26.647° (011) for quartz, 33.957° ($51\bar{1}$) for trona, 34.247° (220) for hanksite, and 31.843° (002) for halite. From the data, it is evident that carbides, oxides, sulfides, and chlorides of potassium, silicon, and sodium provide the catalytic sites for HDPE degradation. Oxides and hydroxides of K, Ca, Fe, Co, Zn, Mn, and Ni are potent catalyst materials in plastic to fuel processes. Metal oxides, including $SiO_2$, $TiO_2$, and modified $\gamma$-$Al_2O_3$, have shown potential in the catalytic process. The potential of catalytic centers available in potash was further analyzed using ICP-OES for a total of 27 metals, including Al, Sb, As, Ba, Be, B, Cd, Ca, Cr, Co, Cu, Fe, Pb, Mg, Mn, Mo, Ni, K, Se, Si, Ag, Sr, Na, Tl, Ti, V, and Zn. Nine of the metals are predominant in the order K> Na> Fe > Si > Mg > Al> Cu > Ca> Ni. With potash having the highest concentration of ~5,000 mg/Kg (72.2% composition) of the sample, our data showed that potassium in various combinations with non-metals could provide the predominant active catalytic sites.

Similarities exist between potash and zeolite in terms of their components. The acidity of solid catalysts plays a vital role in their catalytic function, and the $SiO_2$:$Al_2O_3$ ratio influences it. The ICP-OES and the XRD pattern demonstrate that potash has aluminum and silica with an average concentration of silicon (245 mg/Kg) and aluminum (143 mg/Kg).

## 5.7 POTENTIAL CRACKING MECHANISM

Our data demonstrated that the $CaCO_3$ catalyst has the highest percentage yield (90.5%), followed by $Al_2O_3$, and MgO, while $MgCO_3$ has the lowest percentage yield (73.5%) for the condensed liquid [4,9–11]. $MgCO_3$ has the highest yield (5.2%) of solid products, followed by $Al_2O_3$ (3.8%) and $CaCO_3$ (1.9%). MgO showed the lowest percentage yield (0.3%) of solid residues. $MgCO_3$ has the highest gas product yield, while $Al_2O_3$ gives the lowest percentage yield of the gas products.

Based on the reaction rate, MgO demonstrated the highest catalyst efficiency to convert a large amount of feedstock polypropylene within a short period compared to the other catalysts that include $MgCO_3$, $CaCO_3$, and $Al_2O_3$ (Figure 5.2). However, $MgCO_3$ showed the lowest catalytic ability for polypropylene cracking into liquid products than MgO, $CaCO_3$, and $Al_2O_3$. $MgCO_3$ was proved to be an active catalyst for higher gas production (20.3%) during degradation. Our data demonstrated that the catalytic efficiencies were in the order of $MgO>CaCO_3>Al_2O_3>MgCO_3$.

The kinetic study, including liquid yield for every 5 minutes using the various catalysts, showed the molecular mechanism as a time and ratio-dependent first-order reaction in polypropylene cracking. $CaCO_3$ has the highest liquid conversion rate between 10–15 minutes of reaction time. The liquid hydrocarbon products were characterized using FTIR for functional group analysis.

A potential mechanism was derived based on GC-MS results for the thermal cracking of polypropylene, consisting of three possible processes: C—C bond breaking, C—H bond breaking, and hydrogenation. C–C bond cleavage may occur either on each carbon-carbon bond (C—C, 82.94 kcal/mol) or on carbon-hydrogen ($CH_3$—H, 104 kcal/mol, $CH_2$—H, 98 kcal/mol, CH—H, 95 kcal/mol and C—H, 93.00 kcal/mol. The average C—H bonding energy is 98.09 kcal/mol. The cleavage of these bonds may result in various products; due to the higher bonding energy of the C—H bond, the possibility of breaking the C—H bond is lower than that of the C—C bond breakage. High temperatures lead to the formation of free radicals, which react with hydrocarbons, producing new hydrocarbons and new free radicals, as shown in the following mechanism.

$$R—CH_2—CH_2—CH_2—CH_3 + \bullet CH_3 \rightarrow R—CH_2—CH_2—CH_2—CH_2\bullet + CH_4$$

Then free radicals could be decomposed to generate olefins and new radicals.

$$CH_2—CH_2—CH_2—CH\bullet—CH_2—CH_2— \rightarrow R—CH = CH_2 + CH_3—CH_2\bullet$$

Polypropylene cracking occurs through the end-chain scission or depolymerization, producing monomers, dimers, or oligomers with random chain scission breaking down into smaller fragments of various lengths of C—C bond. The C—C bond cleavage may occur at any position; however, $CH_3—CH_2$- bond-breaking shows higher possibilities than $—CH_2—CH_2—$ breaking.

The slightest possibility is the breaking of the C—H bond. Hydrogen generated from this mechanism may react with alkenes to generate alkanes. Unsaturated longer-chain hydrocarbons could be more easily hydrogenated than shorter-chain ones because longer-chain hydrocarbons are mostly saturated alkanes. Our data reveal that thermal cracking of polypropylene may occur within 150–300°C temperature ranges. A kinetic model of polypropylene cracking reaction was determined, and the rate constants for the reactions were studied under the assumption that no mass transfer resistance happens inside and outside of catalyst powder and no catalyst deactivation occurs inside the rotating basket. The kinetic model was used as there are hundreds of components involved in the intermediates/products. The kinetic experiments show that every reaction is the first order to the reactant. The polypropylene degradation with 10% of the catalyst diminishes the activation energy in an iso-conversion of 95%, confirming the influence of the catalyst on the activation energy. Finally, we noted that the degradation time diminished significantly when the reaction temperature increased. The activation energy of the catalytic cracking reaction was calculated in the range of 13.2–24.3 kcal/mol.

Our cracking reaction required high activation energy to break down the polypropylene molecules using a $CaCO_3$ catalyst; however, $Al_2O_3$ cleaved polypropylene at the lowest activation energy compared to other catalysts. The GC-MS showed that the liquid yield has a carbon range from $C_8$ to $C_{18}$, which is in the range for naphtha, petrol, kerosene (paraffin), and diesel. The various fractions could be used as fuel for car engines, solvents in petrol, fuel for aircraft and stoves, and road vehicles and trains. The liquid products were a mixture of aliphatic and cyclic liquid hydrocarbons. FTIR analysis demonstrated that most molecules in the liquid products were unsaturated alkenes, partly because polypropylene cracking reactions include C—H bond cleavage, introducing unsaturation, especially with metal oxide catalysts. The calorimetric analysis showed that the liquid has a heating value of 9.8–10.8 Kcal/g. This energy could be increased further at an optimum set temperature and heating rate.

Our study introduced different catalysts for converting waste plastics into desired fuels [4,9–11]. It was demonstrated that the $CaCO_3$ catalyst has the highest liquid percentage yield of 90%, while the $MgCO_3$ catalyst has the highest yield of 5.2% of solid and gas product yield. Moreover, the MgO catalyst has the highest degradation conversion rate; $Al_2O_3$ initiated a cracking reaction with the lowest activation energy but a lower yield. The cracked liquid products might be used in industrial boilers, burners, power generators, automobiles, and jet fuel. This process will bring potential solutions for protecting and preserving our environment and the ecosystem. Furthermore, our technology will create new opportunities for the thermal catalytic process to reduce polymer waste from the environment and enhance the carbon/energy recycling strategy [18,19].

## 5.8 RESULTS SUMMARY

In line with our aims, the study demonstrated the catalytic cracking of waste LDPE, HDPE, PP, PVC, and PS over metal oxide catalysts, including zeolite, titanium (IV) oxide, and new catalysts potash were investigated in the one-stage pyrolysis-catalysis reactor [4,9–11]. In addition, the liquid hydrocarbon yield in the catalytic degradation of waste plastic over the catalyst was determined [4,9–11]. With an average liquid yield of 12.8%, PVC produced more gas products,

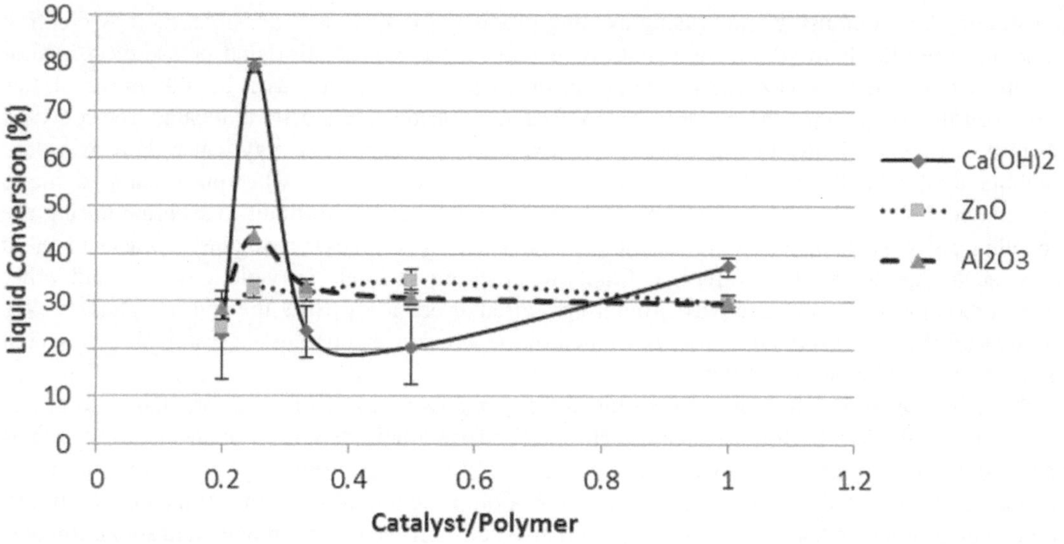


**FIGURE 5.4** Liquid conversion as a function of catalyst-to-polymer ratio. Various ratios were examined to optimize the highest liquid conversion yield.

while PS, with an average liquid yield of 87.7%, generated mostly liquid products. Among the reactants, the yield of gas products is PVC > LDPE > PS, and the polymer structures and the stability of the intermediate may determine the liquid yield.

Our data demonstrate that the cracking yield increased with reactant structure stability. Similar results were reported on the catalytic degradation of HDPE, LDPE, polypropylene, and PS over an H-Y zeolite in a two-stage batch reactor at 500°C over a reaction time of 43.75 minutes. The PS feedstock yielded the highest liquid products (71%), while LDPE showed a 42% liquid yield.

Our previous experiments examined catalytic cracking of waste plastics and metal oxide catalysts, including 1. low-density polyethylene (LDPE) over $Ca(OH)_2$, $Al_2O_3$, and ZnO, 2. high-density polyethylene (HDPE) over zeolite and a novel potash catalyst, and 3. polypropylene (PP) over $MgCO_3$, $CaCO_3$, $Al_2O_3$, and MgO at a catalyst loading of 40 wt%. The liquid yields were 73.5% ($MgCO_3$), 90.5% ($CaCO_3$), 90.1% ($Al_2O_3$), and 86.3% (MgO) on PP; and 34.7% (potash) and 32.4% (zeolite) on HDPE, respectively. For the catalysts, including ZnO, $Ca(OH)_2$, and $Al_2O_3$, we also tested the reusability of these various metal oxide catalysts. The re-used catalysts generated lower liquid products and required higher temperatures for the reaction (data not shown).

It is noteworthy that the catalytic pyrolysis using zeolite consistently gave higher yields for all reactants than the titanium dioxide catalyst. Zeolite and titanium dioxide are among the most commonly used metal oxide catalysts with several advantages, including Lewis acidity, large surface area, and non-toxicity. The surface of titanium dioxide is suitable for pyrolysis and can give different structures for diverse applications. The numerous acid sites determined by the Si/Al ratio enhance zeolites' C-C bond scission catalytic activities.

In addition, our data demonstrate that the catalyst/polymer ratio determines the liquid conversion yield, as shown in Figure 5.4.

## 5.9 CIRCULAR ECONOMY AND SUSTAINABILITY

Landfilling and incineration have been adopted in many countries as viable solution to plastic waste accumulation; however, it presents a lot of environmental drawbacks, as elaborated in the Introduction section. Meanwhile, recycling can be a more sustainable option for dealing with waste plastics. Recycling has been crucial in the promotion of a circular economy. The circular economy is a strategy based on using and reusing existing products for as long as possible, thereby creating value and keeping them within the economy to reduce waste generation and pressure on natural resources. Recycling is an excellent method for the disposal of plastic waste, but the drawback lies in the fact that only so much can be recycled. Residual plastics are plastic that cannot be recycled because of their difficulty in the process. For example, some plastics may cause damage to the machine involved in the recycling process, such as plastic wraps that may become entangled due to their thinness and flexibility. Others are made up of layers that are difficult to separate during the recycling process. With emerging technologies like catalytic cracking of plastic waste to yield fuels, such threats can be overcome. Catalytic cracking of fuels to yield gasoline and other hydrocarbon products is in line with promoting a circular economy. Still, the end-use phase makes one wonder if it is a sustainable process. To determine whether this process is sustainable, it is vital to begin by defining sustainability.

There are several definitions for sustainability, but two key definitions are the Brundtland definition and the Triple Bottom Line definition. The Brundtland definition of sustainability states that sustainable development meets current needs without compromising the ability of future generations to meet their own needs. The Triple Bottom Line definition focuses on a people, profit, and planet approach whereby a business should also consider the environmental and social impacts rendered along with economic returns. Both definitions are in line with the goals of the proposed process, but since the triple bottom definition takes a more holistic approach to defining sustainability, we shall explore how the catalytic cracking of plastics satisfies this definition.

Economically, the process will help with waste reduction while being profitable due to the products being generated. For example, there would be relatively low to zero fuel costs because plastic waste is all around us and in abundance; therefore, low operating costs. Existing refineries can be retrofitted to accommodate this process; therefore, the required capital cost would be insignificant.

Socially, collecting consumer waste and transforming it into valuable products counts as boosting public convenience. In addition, job creation would result from an additional and relatively novel residual plastic recycling method, thereby boosting employment rates and positively impacting the economy.

Immense environmental benefits are to be gained from adopting this method of dealing with plastic waste, mainly if applied to residual plastic waste. Catalytic cracking of plastics would be most beneficial in presenting a solution to the difficulty of placing residual plastic waste in the circular framework. Residual plastics that generally end up being sent to landfills, incinerated, or dumped in oceans can now be transformed in a way that does not cause harm to the environment because the transformation process itself does not lead to the release of toxic chemicals or particulate matter. However, a drastic reduction in the rate of marine habitat degradation can be attained due to reduced dumping of plastic waste in oceans. The rate at which terrestrial and aquatic animals die from getting entangled in plastics or choking from plastics would also reduce drastically with this additional means of dealing with them. Overall, there would be a positive impact on human lives, terrestrial lives, marine lives, oceans, air quality, and the environment at large because of a reduction in toxicity and emissions associated with current processes.

From the above, it can be stated that the catalytic cracking of plastic waste satisfies the economic, social, and environmental aspects of the Triple Bottom Line definition of sustainability. However, to fully claim that it is sustainable, a solution must be presented to tackle emissions associated with the end-use phase of the products.

### 5.9.1 Carbon Recycling

The most sustainable option for waste materials lies in carbon recycling. The only drawback associated with the environmental impact of the catalytic cracking of plastic waste lies in the end-use phase of the products since hydrocarbons are responsible for the majority of the $CO_2$

emissions. The world is currently transitioning away from fossil-fuel-based energy sources to cleaner alternatives. Various technologies exist to aid with fuel switching and energy efficiency improvements; however, fossil fuels still have a significant role in the rising energy demand at a cost-effective rate. A key solution for the decarbonization of the end-use of fossil fuels lies in carbon fixation and direct carbon capture.

### 5.9.2 Carbon Fixation

*Carbon fixation* is a naturally occurring process in which carbon dioxide is incorporated into organic compounds during the transformation of sunlight to chemical energy [20]. On average, it is estimated that about 450 gigatons of carbon dioxide are absorbed by nature every year. The main enzyme that actively transforms carbon dioxide in plants is ribulose-1,5-bisphosphate carboxylase/oxygenase (RuBisCo); however, a drawback is relatively slow and unspecific. Scientists are currently developing ways to improve the RuBisCos fixation mechanism via metabolic engineering and synthetic carbon fixation pathways.

The agricultural industry is already a significant contributor to carbon fixation; however, with the increasing energy demand, the rate of carbon emissions is bound to go up, along with the increased need for fossil fuels, before a complete transition to clean energy is attained. There is a need for increased expansion of plant population through afforestation schemes and decreased deforestation. With more sources of RuBisCo in place, natural carbon fixation can keep up with increased carbon emissions.

### 5.9.3 Carbon Capture

Another option is direct carbon capture and storage. One such technology relevant to the end-use of fossil fuel products is direct air capture. Direct air capture involves the permanent removal of carbon dioxide from the earth's atmosphere, after which it is transported and stored in deep geological formations such as aquifers [21]. Some of the captured carbon dioxides can also be transported in parts of the chemical industry. However, unlike the pre-combustion and post-combustion carbon capture processes that have been proposed for use in industry, direct air capture is not limited by factors such as the number of emission point sources, the concentration of $CO_2$ in flue gas, or even location [21–23]. It can be set up anywhere to filter ambient air regardless of the concentration of $CO_2$. This is most suitable and ideal for the increasing emissions regarding fossil fuel use. Researchers and engineers are also looking into integrating carbon capture and storage in internal combustion engines to capture and store emissions before the exhaust is released into the atmosphere [20]. Since the significant catalytic cracking products are gasoline and diesel, which are mainly used in the transport industry, this technology would prove to be highly integral to ensuring that the catalytic cracking of plastics is a fully sustainable process.

Direct air capture is at the demonstration stage in terms of technological maturity, with only about 15 such plants operating globally. However, for commercialization and wide-scale adoption to occur, there is a need for drastic cost reductions due to effective policy incentives and business models. Carbon capture from internal combustion engines is still in the research and development phase, with no demonstrations yet in place.

With these proposed mechanisms for dealing with the end-use phase of the products from catalytic cracking fully incorporated and adopted globally, the claim that the catalytic cracking of plastics is a fully sustainable process can be made.

## 5.10 CONCLUSIONS

The current study tested the hypothesis of the carbon recycling application of used polymers. Thermal catalytic cracking reactions were carried out in a fixed bed to synthesize natural gas

(C1–C3, methane-rich gas), gasoline, kerosene, and diesel range hydrocarbon fuels from waste plastics. Specific metal oxides, including zinc oxide (ZnO), calcium carbonate ($CaCO_3$), aluminum oxide ($Al_2O_3$), titanium (IV) dioxide ($TiO_2$), zeolite, and potash were tested as pyrolysis catalysts using low-density polyethylene (LDPE), high-density polyethylene (HDPE), polypropylene, polyvinyl chloride (PVC), and polystyrene (PS) reactants. In addition to the catalyzed pyrolysis reactions, we also investigated the non-catalyzed thermal degradation of the plastic substrates for the negative control. The liquid yield, reaction temperature profile, and physical appearance of the synthesized liquid products were determined. The pyrolysis reactions demonstrated that the optimum catalyst-polymer ratio is 1:2.5. The distillate collection temperatures ranged between 82–198°C (LDPE), 68–172°C (PVC), and 40–168°C (PS). Our experiments showed that LDPE, PVC, and PS could readily be pyrolyzed to produce 44% (LDPE), 13% (PVC), and 89% (PS) hydrocarbon liquid products using a zeolite catalyst. Gas chromatography-mass spectrometry (GC-MS) analysis demonstrated that the structure and chemical composition of the products have the $C_5$ (1,2-dimethylcyclopropane), $C_6$ (2-methylpentane), $C_7$ (1,3-dimethylcyclopentene, 1-heptene), and $C_8$ (2-octene, 4-octene, octane, 3-ethylhexane), indicating methane, gasoline to diesel range hydrocarbon ranges.

Further, we tested the low-temperature catalytic cracking of waste plastics, including high-density polyethylene (HDPE) polymer using potash as a novel catalyst and zeolite for carbon recycling. We determined the effects of the potash and zeolite catalysts, temperature, and catalyst/polymer ratio on the liquid pyrolysis yield and hydrocarbon contents. Potash catalysts produced an average liquid conversion of 34.7% at a catalyst ratio of 30 wt% over a distillate temperature range of 76 to 140°C. In contrast, zeolite generated 19.5% at the same catalyst ratio over a distillate temperature range of 90 to 120°C. In addition, a higher potash ratio promoted a smaller carbon range of products with purer compositions. Our experiments demonstrated that specific catalysts, including potash, could convert waste plastics into a $C_1$–$C_{16}$ range of valuable and recyclable products as potential renewable fuel sources as a carbon recycling strategy.

## REFERENCES


[1] Suhrhoff, T.J., Scholz-Bottcher, B.M.: Qualitative Impact of Salinity, Uv Radiation and Turbulence on Leaching of Organic Plastic Additives from Four Common Plastics: A Lab Experiment. *Mar. Pollut. Bull.* **102**, 84–94 (2016).

[2] Idumah, C.I., Nwuzor, I.C.: Novel Trends in Plastic Waste Management. *SN Appl. Sci.* **1**, 1402 (2019).

[3] Miandad, R., Barakat, M.A., Aburiazaiza, A.S., Rehan, M., Nizami, A.S.: Catalytic Pyrolysis of Plastic Waste: A Review. *Process Saf. Environ. Prot.* **102**, 822–838 (2016).

[4] Nwankwor, P.E., Onuigbo, I.O., Chukwuneke, C.E., Yahaya, M.F., Agboola, B.O., Jahng, W.J.: Synthesis of Gasoline Range Fuels by the Catalytic Cracking of Waste Plastics Using Titanium Dioxide and Zeolite. *Int. J. Energy Environ. Eng.* (2020). 10.1007/s40095-020-00359-9.

[5] Liu, M., Inde, R., Nishikawa, M., Qiu, X., Atarashi, D., Sakai, E., Nosaka, Y., Hashimoto, K., Miyauchi, M.: Enhanced Photoactivity with Nanocluster-grafted Titanium Dioxide Photocatalysts. *ACS Nano*. **8**(7), 7229–7238 (2014). 10.1021/nn502247x.

[6] Rahimi, N., Karimzadeh, R.: Catalytic Cracking of Hydrocarbons over Modified ZSM-5 Zeolites to Produce Light Olefins: A Review. *Appl. Catal. A Gen.* **398**, 1–17 (2011). 10.1016/j.apcata.2011.03.009.

[7] Almustapha, M.N., Farooq, M., Andresen, J.M.: Sulphated Zirconia Catalysed Conversion of High Density Polyethylene to Value-added Products Using A Fixed-bed Reactor. *Journal of Analytical and Applied Pyrolysis*. **125**, 296–303 (2017). 10.1016/j.jaap.2017.03.013.

[8] Wang, Y., Cheng, L., Gu, J., Zhang, Y., Wu, J., Yuan, H., Chen, Y.: Catalytic Pyrolysis of Polyethylene for the Selective Production of Monocyclic Aromatics over the Zinc-Loaded ZSM-5 Catalyst. *ACS Omega*. **3**, 2752–2765 (2022) Jan 13;7(3):2752-2765. doi: 10.1021/acsomega.1c05401. PMID: 35097272; PMCID: PMC8793055.

[9] John, D., Chukwuneke, C.E., Onuigbo, I.O., Yahaya, M.F., Agboola, B.O., Jahng, W.J.: Low-Temperature Synthesis of Kerosene- and Diesel-range Fuels from Waste Plastics Using Natural Potash Catalyst. *Int. J. Energy Environ. Eng.* **12**, 531–541 (2021). 10.1007/s40095-021-00387-z.

[10] Habyarimana, J.B., Njiemon, M., Abdulnasir, R., Neksumi, M., Yahaya, M., Sylvester, O., Joseph, I., Okoro, L., Agboola, B., Uche, O., Jahng, W.J.: Synthesis of Hydrocarbon Fuel by Thermal Catalytic Cracking of Polypropylene. *Int. J. Sci. Eng. Res.* **8**, 1193–1203 (2017). 10.14299/ijser.2017.01.014.

[11] Chukwuneke, C., Sylvester, O., Kubor, K., Lagre, S., Siebert, J., Uche, O., Agboola, B., Okoro, L., Jahng, W.J.: Synthesis of C5-C22 Hydrocarbon Fuel From Ethylene-Based Polymers. *Int. J. Sci. Eng. Res.* **5**, 805–809 (2014). 10.14299/ijser.2014.08.008.

[12] Ahmad, I., Khan, M., Khan, H., Ishaq, M., Tariq, R., Gul, K., Ahmad, W.: Pyrolysis Study of Polypropylene and Polyethylene Into Premium Oil Products. *Int. J. Green Energy.* **12**, 140303064405005 (2014). 10.1080/15435075.2014.880146.

[13] Kumbar, V., Dostal, P.: Temperature Dependence Density and Kinematic Viscosity of Petrol, Bioethanol and Their Blends. *Pakistan J. Agric. Sci.* **51**, 175–179 (2014).

[14] Graça, I., Lopes, J.M., Ribeiro, M.F., Ramôa Ribeiro, F., Cerqueira, H.S., de Almeida, M.B.B.: Catalytic Cracking in the Presence of Guaiacol. *Appl. Catal. B Environ.* **101**(3–4), 613–621 (2011). 10.1016/j.apcatb.2010.11.002.

[15] Shah, J., Rasul, M., Adnan, A.: (2014). Conversion of Waste Polystyrene through Catalytic Degradation Into Valuable Products. *Korean J. Chem. Engineering.* **31**, 1389–1398 (2014). 10.1007/s11814-014-0016-4.

[16] Wu, C., Williams, P.T.: Effects of Gasification Temperature and Catalyst Ratio on Hydrogen Production from Catalytic Steam Pyrolysis-Gasification of Polypropylene. *Energy & Fuels.* **22**(6), 4125–4132 (2008). DOI: 10.1021/ef800574w

[17] Tarifa, P., Reina, T.R., González-Castaño, M., Arellano-García, H.: Catalytic Upgrading of Biomass-Gasification Mixtures Using Ni-Fe/$MgAl_2O_4$ as a Bifunctional Catalyst. Energy & Fuels, **36**(15), 8267–8273 (2008). DOI: 10.1021/acs.energyfuels.2c01452.

[18] Jan, M.R., Shah, J., Gulab, H.: Catalytic Degradation of Waste High-density Polyethylene Into Fuel Products Using $BaCO_3$ as a Catalyst. *Fuel Process. Technol.* **91**, 1428–1437 (2010). 10.1016/j.fuproc.2010.05.017.

[19] Normile, D.: Round and Round: A Guide To The Carbon Cycle. *Science* **325**, 1642–1643 (2009).

[20] Weigmann, K. Fixing Carbon. *EMBO reports.* **20**, 2 (2019). doi: 10.15252/embr.201847580.

[21] IEA: Direct Air Capture, IEA, Paris (2020). https://www.iea.org/reports/direct-air-capture.

[22] Sanz-Pérez, E.S., Murdock, C.R., Didas, S.A., Jones, C.W.: Direct Capture of $CO_2$ from Ambient Air. *Chemical Reviews.* **116**(19), 11840–11876 (2016). doi: 10.1021/acs.chemrev.6b00173.

[23] Sharma, S., Maréchal, F.: Carbon Dioxide Capture From Internal Combustion Engine Exhaust Using Temperature Swing Adsorption. *Frontiers in Energy Research.* **7**, 143 (2019). doi: 10.3389/fenrg.2019.00143.

# 6 Recycling Plastic Waste to Produce Chemicals

## *A Techno-economic Analysis and Life-cycle Assessment*

*Robert M. Baldwin, Eric C.D. Tan, Avantika Singh, Kylee Harris, and Geetanjali Yadav*
National Renewable Energy Laboratory, Golden, Colorado, USA

**CONTENTS**



## 6.1 INTRODUCTION AND BACKGROUND

The problem of plastic waste is well known;[1] some pertinent data on the plastic waste in MSW managed in the United States in 2019 for a variety of polymers and their management pathways is shown in Table 6.1. As can be seen, the only polymer currently recycled in significant quantities is polyethylene terephthalate or PET. For this material, mechanical recycling technologies exist that

DOI: 10.1201/9781003167693-6

**TABLE 6.1**
**Plastic Polymer Types and Their Waste in MSW Managed in the United States in 2019**

| Different Type of Plastic Waste | | | | Plastic Waste Handling Distribution | | |
|---|---|---|---|---|---|---|
| Resin Identification Code | Polymer Name | Common Products (kt, %) | Total Plastic Waste Managed, in kt (%) | Plastic Landfilled, in kt (%) | Plastic Waste Combusted, in kt (%) | Plastic Waste Recycled, in kt (%) |
| 1 PETE | Polyethylene terephthalate (also PET) | Bottles, jars, containers, trays, carpet | 5,986 (14%) | 4,554 (76%) | 533 (9%) | 899 (15%) |
| 2 HDPE | High-density polyethylene | Bottles, milk jugs, bags, containers, toys | 7,910 (18%) | 6,448 (82%) | 693 (9%) | 768 (10%) |
| 3 PVC | Polyvinyl chloride | Pipes, siding, pool liners, bags, shoes, tile | 8,189 (19%) | 7,202 (88%) | 716 (9%) | 271 (3%) |
| 4 LDPE | Low-density polyethylene (another form is linear low-density polyethylene, or LLDPE) | Bags, wrap, squeezable bottles, flexible container lids, agricultural film, cable coating | 15,139 (34%) | 13,290 (88%) | 1,524 (10%) | 325 (2%) |

| | | | | | | |
|---|---|---|---|---|---|---|
| 5 PP | Polypropylene | Tupperware plastics, yogurt tubs, hangers, diapers, straws | 699 (2%) | 614 (88%) | 66 (9%) | 18 (3%) |
| 6 PS | Polystyrene or expanded polystyrene (PS/EPS) | Disposable cups and plates, take-out containers, packing peanuts | 3,094 (7%) | 2,815 (91%) | 263 (9%) | 16 (1%) |
| 7 Other | Polycarbonate, nylon, acrylonitrile butadiene styrene (ABS), acrylic, polylactide (PLA), etc. | CDs, safety glasses, medical storage, baby bottles | 3,115 (7%) | 2,796 (90%) | 278 (9%) | 41 (1%) |
| **Total plastic waste managed** | | | **(">44,131 (100%)** | **(">37,720 (86%)** | **(">4,073 (9%)** | **(">2,339 (5%)** |

Note: Created using data from literature [6].

produce a degraded recycled polymer (r-PET); the technology is essentially limited to a single cycle due to properties degradation associated with the recycling technology. The net result of this mismatch between recycling and production is that waste plastics are either diverted to landfills or combusted, or worst, are discarded to the ecosphere creating environmental disasters such as the 'great garbage patch' in the Pacific Ocean [2]. Some projections suggested that the annual mass of mismanaged plastic waste could be more than double by 2050 if plastic production and waste generation continue to grow [3]. Thus, it is critical to strengthen the transition from the current linear economy to a circular one by recycling plastic waste to increase the time these materials spend within the technosphere through alternate use cycles to balance our use of finite natural resources while making our economic system more resilient [4].

Recycling technologies are under development now that expand greatly the options for recycling polymers, including chemical recycling [5], where polymers are deconstructed via pyrolysis back to monomers or liquids suitable for co-processing in conventional steam cracking reactors, or to synthesis gas (a mixture of CO and $H_2$) that can be used to make polymer precursors. These process pathways are important first steps to an eventual circular economy for polymers as opposed to the current linear economy where plastics are discarded after a single use.

This chapter focuses on two near-term chemical deconstruction technologies: 1) pyrolysis and 2) gasification. We developed the conceptual processes for these two conversion technologies using data from the literature and performed a detailed techno-economic analysis (TEA) and life cycle assessment (LCA) of five pathways encompassing chemical recycling by pyrolysis and gasification.

## 6.2 RECYCLING OF POLYMERS FOR THE PRODUCTION OF CHEMICALS VIA PYROLYSIS

There are two generic pyrolysis pathways for recycling waste plastics – principally polyolefins – to produce olefins. Basic data and information required to build process flowsheets for two near-term commercial pathways for producing olefins from waste plastics were gleaned from the public literature. The two generic pathways are 1) direct production of olefins by thermal or catalytic pyrolysis (i.e., the one-step process) and 2) thermal or catalytic pyrolysis to produce naphtha followed by co-processing in an existing steam cracking reactor to produce olefins (i.e., the two-step process).

The first pathway is the production of olefins directly from polymer via pyrolysis mainly for polystyrene. In the case of pyrolysis applied to other polyolefins, however, scant information exists in the public literature that can be used to conduct either a technical or economic analysis – the technology is simply too immature at this time. For the second pathway, the basic process schematically consists of two steps: 1) plastics pyrolysis at low severity conditions to produce a pyrolysis oil, and 2) co-processing the pyrolysis oil with fossil naphtha in a conventional steam cracker to produce ethylene and propylene. This study focuses on techno-economic analysis (TEA) and life-cycle assessment (LCA) of these two near-commercial technologies that represent this general two-step methodology, as well as a novel one-step process for direct production of C2 and C3 olefins using *ex situ* catalytic fast pyrolysis. In all cases, the pathways involve either thermal fast pyrolysis or *in-situ* or *ex-situ* catalytic fast pyrolysis for the first step. The basic information required to build the conceptual process flowsheets was mainly extracted from the public literature.

### 6.2.1 Process Description

Pyrolysis is essentially thermal devolatilization in the absence of air. In general, pyrolysis of materials such as polymers takes place at temperatures in the range of 400–650°C. Pressures employed for pyrolysis reactions are typically ambient to just over ambient (<2 atm.); some pyrolysis processes employ a high vacuum. High heating rates are important for pyrolysis to minimize the amount of char that is formed; most pyrolysis processes operate under fast pyrolysis (FP) conditions with heating rates greater than 1,000 K/s. For this study, only fast pyrolysis reaction conditions were assumed giving rise to reactor residence times of 1 second or less.

Fast pyrolysis can be either non-catalytic (thermal) or can employ catalysts in the *in-situ* or *ex-situ* mode. The major difference between the *in-situ* and *ex-situ* mode lies in where the catalyst is employed. For the *in-situ* mode, the catalyst is in the main pyrolysis reactor in contact with the devolatilizing plastic. In the *ex-situ* mode, the catalyst is in a separate reactor directly downstream and close-coupled to the pyrolysis reactor and the catalyst only 'sees' pyrolysis vapors. For *ex-situ* catalytic fast pyrolysis (CFP), the vapors from thermal pyrolysis pass directly into the catalyst bed without condensation. However, there can be intervening unit operations like hot gas filtration between the pyrolysis reactor and catalytic upgrading reactor to remove catalyst poisons and entrained solids (char fines, for example). The *in-situ* mode has the advantage of simplicity (a one-pot system). Still, it has the major disadvantage that it is impossible to protect the catalyst from poisoning by contaminants in the waste plastic feedstock. It is also not very flexible, as the pyrolysis and catalysis must take place at exactly the same reaction conditions.

The *ex-situ* mode is more complex (a separate catalytic upgrading reactor results in higher CAPEX), but it allows the catalyst to be protected from contaminants and offers a greater degree of flexibility since the pyrolysis reactor and catalytic upgrading reactors don't have to be operated at the same conditions (temperature, space time, etc.).

Process, economic, and sustainability analysis were conducted on four conceptual processes where pyrolysis is used as the primary conversion step for the plastic waste. Both of these pathways are examples of the two-step methodology; details on each of these pathways are outlined below.

The information for the non-catalytic thermal pyrolysis (Pathway 1) has primarily been developed from websites and open literature. The integrated process is intended to produce olefins by co-processing naphtha from plastics pyrolysis with fossil naphtha in an existing steam cracker. The feedstock type was unknown, but it likely will be mixed plastic waste containing primarily polyolefins. The pyrolysis technology is thermal pyrolysis (i.e., non-catalytic). Details on reaction conditions (temperature, pressure, space time) are not available in the public domain. The reactor technology is also unknown. It is unlikely that this dramatically impacts the process pathway development as yields are not likely to be strongly influenced by reactor type for thermal pyrolysis. The downstream processing (DSP) is distillation to remove light ($C5^-$) and heavy ends ($C12^+$), producing a pyrolysis naphtha cut which is the feedstock to the steam cracking reactor.

The information for the *in-situ* CFP (Pathway 2) was also gleaned from open literature [7]. We anticipated the overall objective for this process is to produce olefins from waste plastics by co-processing with fossil naphtha in an existing steam cracker. The feedstock is unknown but will likely be mixed plastic waste containing primarily polyolefins. The catalyst for the *in-situ* CFP is unknown. Details on reaction conditions (i.e., temperature, pressure, space time) were also unknown, although the website cited above reports a pyrolysis temperature <400°C [7]. The reactor technology is unknown and might have a substantial impact on process economics. The downstream processing (DSP) was not reported. Some separations technology is likely to be employed to remove light ($C5^-$) and heavy compounds ($C12^+$) from the catalytic fast pyrolysis (CFP) liquids prior to blending with fossil naphtha and to produce a pyrolysis naphtha 'cut'.

As a summary, the pyrolysis pathways evaluated here are 1) non-catalytic thermal pyrolysis, 2) *in-situ* catalytic fast pyrolysis, 3) *ex-situ* catalytic fast pyrolysis (Scenario I, with pyrolysis and catalytic upgrading taking place in two separate reactor systems), and 4) *ex-situ* catalytic fast pyrolysis (Scenario II, with direct production of aromatics and olefins).

### 6.2.2 Steam Cracker Mass Balance

A basic assumption is that the mass balance for an industrial steam naphtha cracker will not be influenced by co-processing with pyrolysis oil derived from waste plastics. This is likely true at

**TABLE 6.2**
**Typical Product Distribution from Naphtha Cracker**

| Product | wt% (based on feed) |
|---|---|
| Ethylene | 35 |
| Propylene | 15 |
| C4 | 8.5 |
| C5 | 25.5 |
| Gas ($H_2$, $CH_4$) | 16 |

*Note:* Created using data from literature [9].

low (<10 vol%) blending ratios of pyrolysis oil to fossil naphtha. At higher blend ratios, the chemistry of the steam cracking reaction could be impacted due to the abundance of unsaturates (olefins) in the pyrolysis oil. The impact of hydrocarbon types present in pyrolysis oil could be especially critical if significant amounts of aromatics (benzene, toluene, xylenes, or BTX) are present in addition to olefins. BTX is undesirable in the feed to a naphtha steam cracker as they are largely unreactive (e.g., do not contribute to the formation of low MW olefins) and can serve as potent precursors to coke and tar, resulting in yield loss and poor carbon efficiency in the cracker. The presence of aromatics in the products from the steam cracker adds to the cost of separation and purification of the desired products (low MW olefins), and economics would dictate that they be removed from the feed to the cracker rather than the products if present in significantly high amounts [8]. For the process pathways under investigation, we assumed that aromatics are not produced in levels that would require additional separations. A typical product distribution from an industrial naphtha steam cracker is shown in Table 6.2. These data will be used to establish a baseline for the mass balance for the steam cracking reactor.

### 6.2.3 Feedstock Considerations

The feedstock for this technology will consist of waste plastics pre-sorted to remove materials that can currently be recycled and hence have some monetary value. This represents primarily polyethylene terephthalate (PET) and polystyrene (PS). PET is currently recycled either mechanically or chemically to produce r-PET – a material of some commercial value (≈ 40¢/lb) but lower than that of virgin PET (≈ 90¢/lb). Accordingly, we assumed that the feedstock for this analysis is free from polyesters. Polystyrene is efficiently 'unzipped' by thermal pyrolysis at temperatures of c.a. 500°C to produce styrene in high yields [10]. Commercial technology exists for recycling PS to the monomer (styrene), notably by Agilyx (Tigard, OR). Therefore, we assumed that PS would likely be removed for sale to a recycler due to the high price of the monomer relative to ethylene and propylene.

Further, styrene is an aromatic molecule that is not desirable in the feedstock to a steam cracker. This could be removed in a downstream process (DSP); however, it is not clear if the same technology used to recover BTX would be effective for styrene. For this analysis, we assumed that PS had been removed from the feedstock in the materials recovery facility (MRF). A final feedstock consideration is polyvinyl chloride (PVC). Our survey indicates that the concentration of PVC in waste plastic streams in post-consumer waste is generally <1 wt% and that problems begin to become evident when concentrations rise above 3 wt%. Accordingly, we assumed that the concentration of PVC is low enough (<3 wt%) so as to not pose a significant problem or impact the operations of the integrated process.

**TABLE 6.3**
**Yields from Fluidized Bed Thermal Pyrolysis of Plastics and Mixed Plastics**

| Columns | 2 | 3 | 4 | 5 | 6 | 7 |
|---|---|---|---|---|---|---|
| Temperature (°C) | 530 | 510 | 510 | 510 | 510 | 450 |
| Material | HDPE | HDPE | LLDPE | PP | **Mix** | LLDPE |
| Input (kg) | 1.6 | 2.7 | 5.5 | 4.9 | 4.6 | 1.8 |
| Throughput (kg/h) | 0.9 | 0.9 | 1.1 | 1.0 | 0.9 | 1.0 |
| Methane | 0.8 | 0.3 | 0.4 | 0.6 | 0.4 | 0.1 |
| Ethylene | 2.0 | 0.6 | 0.8 | 0.5 | 0.6 | 0.2 |
| Ethane | 0.8 | 0.3 | 0.6 | 1.0 | 0.5 | 0.2 |
| Propene | 1.8 | 0.7 | 0.8 | 3.0 | 1.5 | 0.2 |
| Propane | 0.7 | 0.3 | 0.3 | 0.4 | 0.3 | 0.2 |
| Butenes | I.I | 0.3 | 0.5 | 0.7 | 1.0 | 0.2 |
| **Total gas** | **7.6** | **2.6** | **3.4** | **6.3** | **4.7** | **1.1** |
| Pentenes | 0.1 | 0.2 | 0.3 | 0.5 | 1.0 | 0.1 |
| Pentanes | 0.03 | 0.1 | 0.1 | 1.2 | 1.7 | 0.1 |
| Pentadienes | 0.1 | 0.1 | 0.1 | 0.1 | 0.2 | 0.0-1 |
| Hexenes | 0.9 | 1.0 | 1.0 | 1.7 | 2.3 | 0.4 |
| Hexanes | 0.2 | 0.2 | 0.2 | 0.2 | 0.-1 | 0.1 |
| Hexadienes | 0.1 | 0.1 | 0.1 | 0.5 | 0.-1 | 0.03 |
| Heptenes | 0.8 | 0.6 | 0.6 | 0.6 | 1.0 | 0.3 |
| Heptanes | 0.3 | 0.4 | 0.3 | + | 0.3 | 0.2 |
| Heptadiencs | 0.01 | + | + | 0.5 | 0.2 | 0.01 |
| Octenes | 0.6 | 0.4 | 0.5 | 0.3 | 0.-1 | 0.3 |
| Octanes | 0.2 | 0.2 | 0.2 | 0.3 | 0.4 | 0.2 |
| Octadienes | 0.03 | 0.01 | 0.02 | 0.2 | 0.01 | 0.01 |
| Nonenes | 0.7 | 0.4 | 0.4 | 7.8 | 4.2 | 0.2 |
| Nonanes | 0.2 | 0.2 | 0.2 | 0.3 | 0.-1 | 0.2 |
| Nonadienes | 0.1 | 0.1 | 0.04 | 0.1 | 0.1 | 0.02 |
| Decenes | 1.0 | 0.7 | 0.7 | 0.8 | 0.8 | 0.4 |
| Decanes | 0.2 | 0.2 | 0.2 | + | 0.1 | 0.2 |
| Decadienes | 0.1 | 0.1 | 0.1 | + | 0.03 | 0.04 |
| Total $C_5$-$C_{10}$ | 6.9 | 5.2 | 5.6 | 15 | 14 | 3.4 |
| Total $C_{11}$-$C_{20}$ | 9.9 | 6.9 | 6.7 | 13 | 5.7 | 6.1 |
| Total > $C_{20}$ | 34 | 2-1 | 24 | 35 | 37 | 18 |
| Total BTX-aromatics | 0.3 | 0.05 | 0.05 | 0.02 | 0.1 | 0.03 |
| Total waxes bp < 500°C | 51 | 36 | 37 | 64 | 57 | 28 |
| Total waxes bp > 500°C | 42 | 61 | 60 | 30 | 38 | 71 |

Mix PP/LLDPE/HDPE: 40/15/45, wt%; + detected, but not quantified
*Note:* Created using data from literature [11].

The above discussion on feedstock issues leads us to a fundamental assumption regarding feedstock composition; this is a critical assumption that will guide our TEA and LCA studies. This study assumed that the feedstock consists of a mixture of only PE (all varieties) and PP with a minor contribution from PVC. These two polyolefins represent approximately the non-recycled (e.g., waste) plastics in the United States alone (Table 6.1). Data for yields from thermal fast pyrolysis of a suite of polyolefins (HDPE, LLDPE, PP) and a mixture of these same materials are shown in Table 6.3. Comparing the yields at 510°C, the following two observations can be made.

First, the two polyethylene samples' yields of gas and liquid products are similar – the differences shown are likely not statistically or experimentally significant. Secondly, yields of gas and liquid from polypropylene are very different from PE, with much higher (2X) gas yields, much higher (approximately 2X) yields of low MW liquids (boiling points < 500°C), and much lower yields (approximately half) of high MW waxes.

Also shown in the table are data from pyrolysis of a 40/15/45 mixture of PP/LLDPE/HDPE, which is characteristic of a feedstock that could be obtained from a materials recovery facility (MRF) [11]. As shown, the yield structure for pyrolysis of this feedstock is intermediate between PE and PP; this yield structure was used as the baseline for the development of the conceptual process pathways.

### 6.2.4 Reactor Considerations

For this analysis, we assumed that a recirculating bed process is the reactor technology for both the thermal pyrolysis reactor and the *in situ* catalytic pyrolysis reactor. Recirculating beds are likely required due to the build-up of char from the plastics pyrolysis reaction, with one bed for the pyrolysis reaction and the second for char combustion, as depicted in Figure 6.1. The thermal pyrolysis reactor could be a single fluid bed (low char-make) or a system of dual recirculating beds where an inert heat transfer medium is circulated between the reactor and a char burner. In the latter case, char combustion would be used to generate heat for the endothermic pyrolysis reactions, and – if feasible – the system would be run in heat balance. In the single-bed configuration, an external heat source will be needed for the pyrolysis reactor.

For the *ex-situ* catalytic upgrading reactor, we assumed it would use a fluidized catalytic cracking (FCC)-type configuration with the catalytic pyrolysis reactor consisting of a short-contact-time entrained-flow riser (about 1 second residence time) coupled with a fluid bed catalyst regenerator (see Figure 6.2). In general, these systems are operated in heat balance with the exothermic heat of coke burning in the regenerator, balancing the endothermic heat of reaction in the riser using sensible heat in the catalyst as the energy carrier. Details from the literature on coke-make are scarce, with only one reference reporting coke-on-catalyst of 2% after the reaction [12]. This value is similar to the coke-make value for commercial FCC units; typical values for FCC catalyst entering and leaving the regenerator are 1–2 wt% coke on spent catalyst entering the

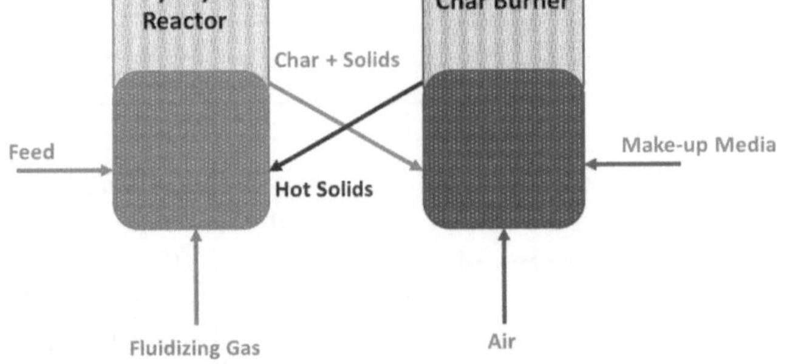


**FIGURE 6.1** Circulating dual fluid bed configuration.

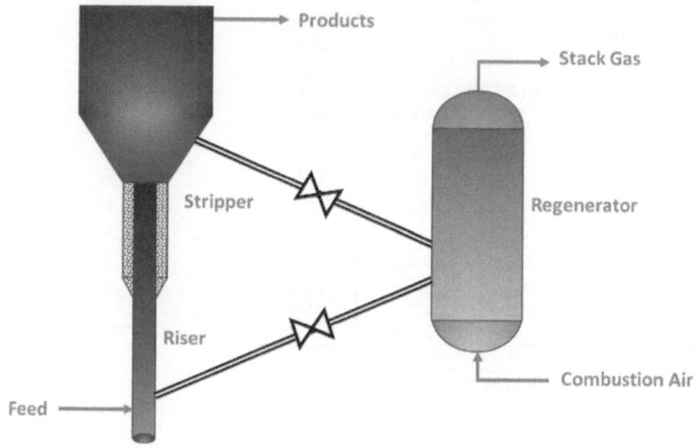


**FIGURE 6.2** FCC-style recirculating reactor system with regenerator.

regenerator and <0.1 wt% coke on the regenerated catalyst. These values were used in the flowsheets; however, it should be recognized that this represents a significant assumption that will impact subsequent economic and sustainability analyses.

### 6.2.5 Detailed Process Modeling

As is normally the case, the literature is quite scattered and divergent in some critical technical areas. For this study, information that was derived from batch reactor studies was not utilized as this information is not necessarily relevant to processing in continuous systems due to uncertainties in reaction time and the impact of heating rate on pyrolysis yield. It is well understood that rapid heating promotes the production of pyrolysis liquids. In contrast, slow heating promotes the formation of carbonaceous residues (char) – since the processes under study here aim to make pyrolysis naphtha from waste plastics, processes that employ slow heating rates – such as would be the case for batch reactors – are not relevant. Accordingly, and where possible, data used to build process flowsheets were obtained only from the literature and studies at NREL where continuous flow experiments have been performed. This section provides results of the detailed process modeling for the four pyrolysis pathways using Aspen Plus. The information is presented in two different forms: 1) process flowsheets from Aspen Plus that show the individual unit operations and material flows for each unit; 2) stream summary tables that give detailed information on the flowrate and composition of each stream in the flowsheet. In addition to showing the unit operations, the flowsheets present information on yields (normal font, in %) and carbon efficiencies (bold italic font, in%) for each stream. The total mass closure for each of these flowsheets is 100%, while carbon balances are 99%$^+$. The data in these simulations is essential in the economic and sustainability analyses that are detailed in the next section.

#### 6.2.5.1 Thermal Pyrolysis Process

Figure 6.3 shows the process flowsheet for a simulated thermal pyrolysis pathway. The stream summary information is shown in Table 6.4. The data for the thermal pyrolysis pathway was taken from Table 6.3 (Column 6), which shows a homologous series of normal alkanes, alkenes, and alkynes, with the olefins representing the dominant peaks. As shown, the full-range pyrolysis oil contains materials that have carbon numbers well in excess of C12, representing heavy hydrocarbons that would not be suitable for steam cracking feedstocks due to the tendency to make tars and coke. In contrast, the naphtha (i.e., gasoline) fraction of this pyrolysis oil is relatively small – about 8%. However, it is well documented that steam crackers can accept heavier feedstocks – up to as high as C15 in some cases. Under these conditions, a larger fraction of the plastic pyrolysis oil could be considered as feedstock to the cracker. Information in the literature in the form of a BASF

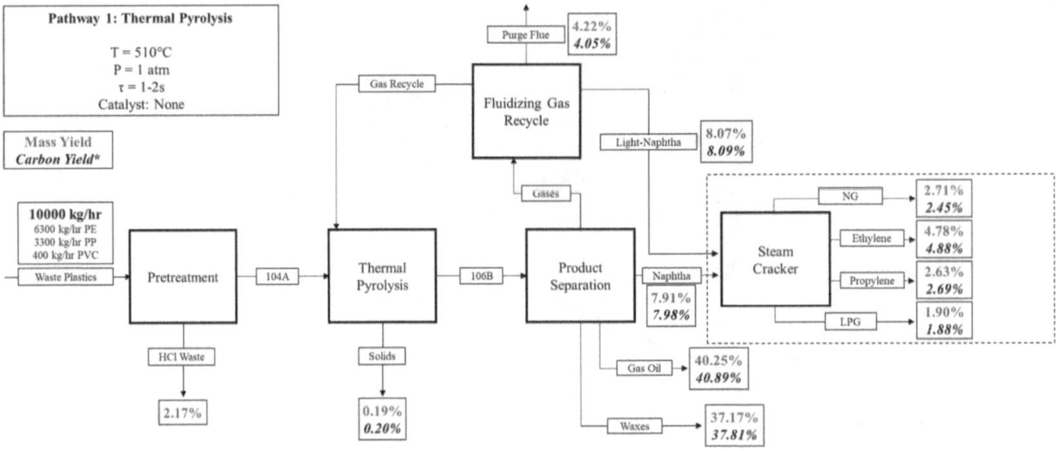


**FIGURE 6.3** Process flowsheet for thermal pyrolysis process.

patent [13] suggests that the plastic naphtha could consist of material from "180° to 280°C, preferably from 220° to 260°C, in particular from 230° to 250°C, and said fraction is fed as a feed material to a steam cracker." However, data that could be used to quantitatively determine this fraction are not available. For this analysis, we assumed that the C5–C10 fraction from Table 6.3 serves as a reasonable estimate of this value.

Recognizing the uncertainty associated with the conditions of the distillation step, a sensitivity analysis on this parameter was included in the techno-economic analysis (TEA). Additionally, the technology for the thermal pyrolysis reactor will depend on the degree of char-make in the pyrolysis reactor. Since polyolefins have very low fixed carbon, it is anticipated that the char-make will correspondingly be very low, in which case the char burner would not be used, and process heat would be supplied by an external source (natural gas); Figure 6.4 shows an example of this type of reactor system. In this configuration, char buildup is controlled by the elutriation of char fines from the fluid bed and/or by utilizing a purge on the recycle stream. The term 'media' in this figure could represent an inert heat carrier such as sand or a disposable catalyst/reactant.

### 6.2.5.2 *In-situ* and *Ex-situ* Catalytic Fast Pyrolysis (CFP)

Figure 6.5 and Table 6.5 present the process flowsheet and stream summary for the *in-situ* catalytic fast pyrolysis (Pathway 2), where the pyrolysis and catalytic upgrading reactions are taking place in a single vessel. This configuration is essentially identical to that shown in Figure 6.2 for the non-catalytic pyrolysis pathway except for an additional stream for catalyst rejection. The exact nature and composition of this 'catalyst' is not known but is described in the literature as a "… consumable in a closed system."[7] It is not clear what this means.

When compared to the *ex-situ* mode of operation, this configuration has the advantage of simplicity associated with fewer process vessels but has the disadvantage of forcing the pyrolysis and catalytic upgrading reactions to take place at exactly the same conditions. This prevents optimization of the reaction system and also exposes the catalyst to any poisons that might be present in the feed. Important distinctions between this pathway and that shown in Table 6.3 are: 1) the pyrolysis temperature was lower; 2) the fraction of pyrolysis liquids suitable for blending with fossil naphtha was assumed to be higher (pyrolysis naphtha yield of 38% vs. 8%) due to the use of a catalyst in the pyrolysis step (e.g., the pyrolysis liquids are lower molecular weight) and the CFP oil is sufficiently upgraded in the catalytic pyrolysis step to reduce heavy ends. However – as was the case for the PE/SABIC pathway – the actual naphtha yield is unknown. Since it is expected that this single parameter will have a significant impact on the process economics, we treated this yield as an unknown parameter that was the subject of a sensitivity analysis. Recenso reported a yield of

**TABLE 6.4**
**Stream Table for Thermal Pyrolysis Process**

| Streams | Units | Waste Plastics | HCl Waste | 104A | Gas Recycle | Solids | 106B | Gases | Gas Oil | Waxes | Light-Naphtha | Naphtha | Purge Flue |
|---|---|---|---|---|---|---|---|---|---|---|---|---|---|
| **Mass Flows** | **kg/hr** | **10,000.0** | **217.3** | **9,782.7** | **7,600.0** | **19.5** | **17,363.3** | **8,829.6** | **4,025.0** | **3,717.1** | **807.3** | **791.5** | **422.3** |
| **Mass Flow of Carbon Atoms** | **kg/hr** | **8,376.54** | **–** | **8,376.5** | **6,108.4** | **16.7** | **14,385.9** | **7,125.1** | **3,425.0** | **3,167.2** | **677.3** | **668.6** | **339.4** |
| PE | kg/hr | 6,300.00 | – | 6,300.0 | – | 9.7 | – | – | – | – | – | – | – |
| PP | kg/hr | 3,300.0 | – | 3,300.0 | – | 9.7 | – | – | – | – | – | – | – |
| H2 | kg/hr | – | – | – | – | – | – | – | – | – | – | – | – |
| CH4 | kg/hr | – | – | – | 627.0 | – | 663.1 | 662.8 | 0.1 | 0.1 | 0.9 | 0.1 | 34.8 |
| C2H6 | kg/hr | – | – | – | 2,081.6 | – | 2,218.1 | 2,215.0 | 0.8 | 0.5 | 17.7 | 1.8 | 115.7 |
| C2H4 | kg/hr | – | – | – | 872.5 | – | 926.9 | 925.9 | 0.3 | 0.2 | 4.9 | 0.5 | 48.5 |
| C3H8 | kg/hr | – | – | – | 732.6 | – | 797.4 | 794.3 | 0.4 | 0.2 | 21.0 | 2.4 | 40.7 |
| C3H6 | kg/hr | – | – | – | 1,729.2 | – | 1,877.1 | 1,870.8 | 0.9 | 0.6 | 45.4 | 4.8 | 96.1 |
| C4H10 | kg/hr | – | – | – | 545.6 | – | 640.4 | 632.9 | 0.5 | 0.3 | 57.0 | 6.7 | 30.3 |
| N-PENTAN | kg/hr | – | – | – | 596.4 | – | 876.0 | 843.7 | 1.0 | 0.5 | 214.2 | 30.9 | 33.1 |
| N-C6H14 | kg/hr | – | – | – | 196.4 | – | 497.4 | 443.2 | 0.8 | 0.4 | 235.9 | 53.0 | 10.9 |
| BENZENE | kg/hr | – | – | 163.3 | 25.8 | – | 64.6 | 57.6 | 0.1 | 0.1 | 30.3 | 6.9 | 1.4 |
| TOLUENE | kg/hr | – | – | – | 3.8 | – | 33.0 | 22.6 | 0.1 | 0.0 | 18.5 | 10.3 | 0.2 |
| XYLENE | kg/hr | – | – | – | – | – | – | – | – | – | – | – | – |
| N-C8H18 | kg/hr | – | – | – | 12.4 | – | 353.1 | 158.1 | 1.1 | 0.5 | 145.0 | 193.5 | 0.7 |
| N-C10H22 | kg/hr | – | – | – | 0.1 | – | 146.2 | 11.9 | 0.9 | 0.3 | 11.8 | 133.0 | 0.0 |
| N-C12H26 | kg/hr | – | – | – | 0.0 | – | 292.1 | 3.3 | 3.7 | 1.1 | 3.3 | 284.0 | 0.0 |
| N-C15H32 | kg/hr | – | – | – | – | – | – | – | – | – | – | – | – |
| N-C20H42 | kg/hr | – | – | – | 0.0 | – | 1,168.5 | 0.0 | 1,069.4 | 35.6 | 0.0 | 63.5 | 0.0 |
| C24H50 | kg/hr | – | – | – | 0.0 | – | 3,116.0 | 0.0 | 2,770.9 | 345.1 | 0.0 | 0.0 | 0.0 |
| C28H58 | kg/hr | – | – | – | – | – | – | – | – | – | – | – | – |
| N-C30H62 | kg/hr | – | – | – | – | – | 3,505.5 | – | 173.9 | 3,331.6 | – | – | – |
| HCL | kg/hr | – | 217.3 | 11.4 | 176.4 | – | 187.9 | 187.6 | 0.1 | 0.0 | 1.4 | 0.1 | 9.8 |
| CACL2 | kg/hr | – | – | – | – | – | – | – | – | – | – | – | – |
| PVC | kg/hr | 400.0 | – | 8.0 | – | – | – | – | – | – | – | – | – |

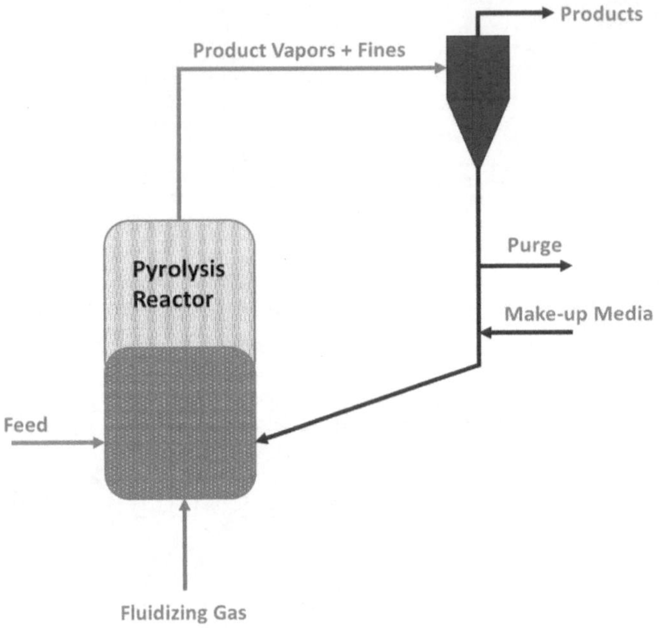


**FIGURE 6.4** Single fluidized bed reactor configuration.

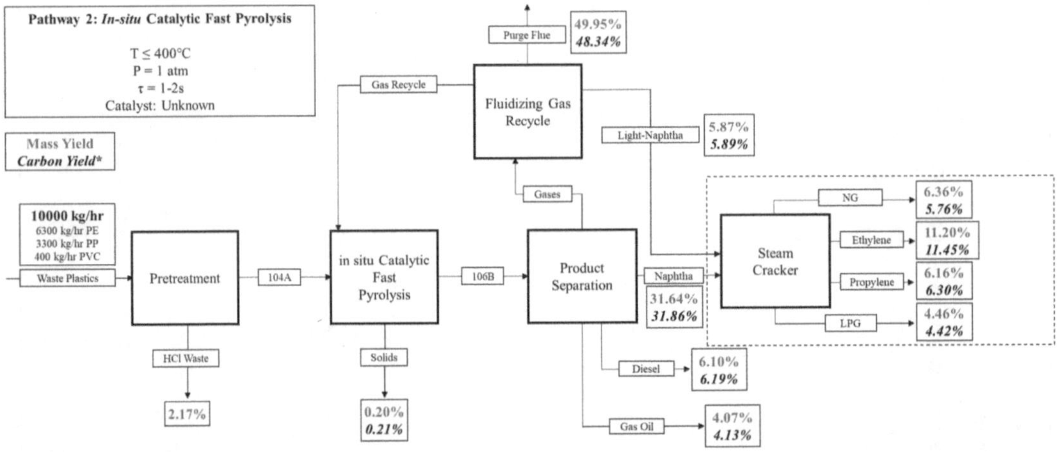


**FIGURE 6.5** In-situ catalytic fast pyrolysis process.

53% pyrolysis oil from their process [13], which correlates well with the sum of approximately 48% (naphtha + light naphtha + diesel + gas oil) from our simulations. The remainder of the product was assumed to be light gases. The composition of this gaseous stream was unknown but was assumed to be heavy in hydrocarbons and low molecular weight olefins (Table 6.5, Purge Flue stream). This stream is being used as a fuel gas to improve the energy integration of the process.

Figure 6.6 and Table 6.6 present the process flowsheet and stream summary for Pathway 3—one scenario of the *ex-situ* catalytic fast pyrolysis variant where the catalyst and choice of upgrading conditions in the catalytic reactor dictate the yield of olefins and aromatics is small (Scenario I, low aromatics). Here pyrolysis and catalytic upgrading take place in two separate reactor systems.

The first reactor is thermal, where an inert fluidization and/or heat transfer medium would be likely used in a fluid bed system to carry out pyrolysis (similar to Figure 6.1), and a recirculating

**TABLE 6.5**
**Stream Table for *In-situ* Catalytic Fast Pyrolysis Process**

| Streams | Units | Waste Plastics | HCl Waste | 104A | Gas Recycle | Solids | 106B | Gases | Diesel | Gas Oil | Light-Naphtha | Naphtha | Purge Flue |
|---|---|---|---|---|---|---|---|---|---|---|---|---|---|
| **Mass Flows** | **kg/hr** | **10,000.0** | **217.3** | **9,782.7** | **7,600.0** | **20.2** | **17,362.5** | **13,181.8** | **610.2** | **406.8** | **586.9** | **3,163.7** | **4,994.9** |
| **Mass Flow of Carbon Atoms** | **kg/hr** | **8,376.54** | **–** | **8,376.5** | **6,160.6** | **17.3** | **14,235.9** | **10,702.5** | **518.1** | **346.3** | **493.0** | **2,669.1** | **4,048.9** |
| PE | kg/hr | 6,300.0 | – | 6,300.0 | – | 10.1 | – | – | – | – | – | – | – |
| PP | kg/hr | 3,300.0 | – | 3,300.0 | – | 10.1 | – | – | – | – | – | – | – |
| H2 | kg/hr | – | – | – | – | – | – | – | – | – | – | – | – |
| CH4 | kg/hr | – | – | – | 1,298.5 | – | 2,153.6 | 2,152.9 | 0.1 | 0.0 | 1.0 | 0.7 | 853.4 |
| C2H6 | kg/hr | – | – | – | 1,606.0 | – | 2,674.3 | 2,669.1 | 0.2 | 0.0 | 7.7 | 5.0 | 1,055.5 |
| C2H4 | kg/hr | – | – | – | 967.5 | – | 1,608.6 | 1,606.5 | 0.1 | 0.0 | 3.2 | 2.0 | 635.9 |
| C3H8 | kg/hr | – | – | – | 2,175.1 | – | 3,668.2 | 3,641.6 | 0.4 | 0.0 | 37.0 | 26.2 | 1,429.5 |
| C3H6 | kg/hr | – | – | – | 781.6 | – | 1,315.0 | 1,306.8 | 0.1 | 0.0 | 11.6 | 8.0 | 513.7 |
| C4H10 | kg/hr | – | – | – | 139.4 | – | 245.5 | 239.5 | 0.0 | 0.0 | 8.5 | 6.0 | 91.6 |
| N-PENTAN | kg/hr | – | – | – | 297.4 | – | 601.4 | 552.7 | 0.1 | 0.0 | 59.8 | 48.5 | 195.5 |
| N-C6H14 | kg/hr | – | – | – | 262.4 | – | 787.8 | 611.9 | 0.3 | 0.0 | 177.0 | 175.6 | 172.4 |
| BENZENE | kg/hr | – | – | 163.3 | 20.4 | – | 61.6 | 47.6 | 0.0 | 0.0 | 13.8 | 14.0 | 13.4 |
| TOLUENE | kg/hr | – | – | – | 3.7 | – | 34.2 | 16.1 | 0.0 | 0.0 | 10.0 | 18.0 | 2.4 |
| XYLENE | kg/hr | – | – | – | – | – | – | – | – | – | – | – | – |
| N-C8H18 | kg/hr | – | – | – | 30.1 | – | 992.8 | 251.2 | 0.9 | 0.0 | 201.3 | 740.8 | 19.8 |
| N-C10H22 | kg/hr | – | – | – | 0.9 | – | 1,519.5 | 54.3 | 3.5 | 0.0 | 52.8 | 1,461.7 | 0.6 |
| N-C12H26 | kg/hr | – | – | – | 0.0 | – | 658.0 | 3.1 | 3.9 | 0.0 | 3.1 | 651.0 | 0.0 |
| N-C15H32 | kg/hr | – | – | – | 0.0 | – | 303.7 | 0.0 | 297.6 | 0.0 | 0.0 | 6.1 | 0.0 |
| N-C20H42 | kg/hr | – | – | – | – | – | 303.7 | 0.0 | 288.7 | 15.0 | – | 0.0 | – |
| C24H50 | kg/hr | – | – | – | – | – | 404.9 | – | 14.2 | 390.8 | – | – | – |
| C28H58 | kg/hr | – | – | – | – | – | – | – | – | – | – | – | – |
| N-C30H62 | kg/hr | – | – | – | – | – | 1.0 | – | 0.0 | 1.0 | – | – | – |
| HCL | kg/hr | – | 217.3 | 11.4 | 17.2 | – | 28.6 | 28.6 | 0.0 | 0.0 | 0.1 | 0.1 | 11.3 |
| CACL2 | kg/hr | – | – | – | – | – | – | – | – | – | – | – | – |
| PVC | kg/hr | 400.0 | – | 8.0 | – | – | – | – | – | – | – | – | – |

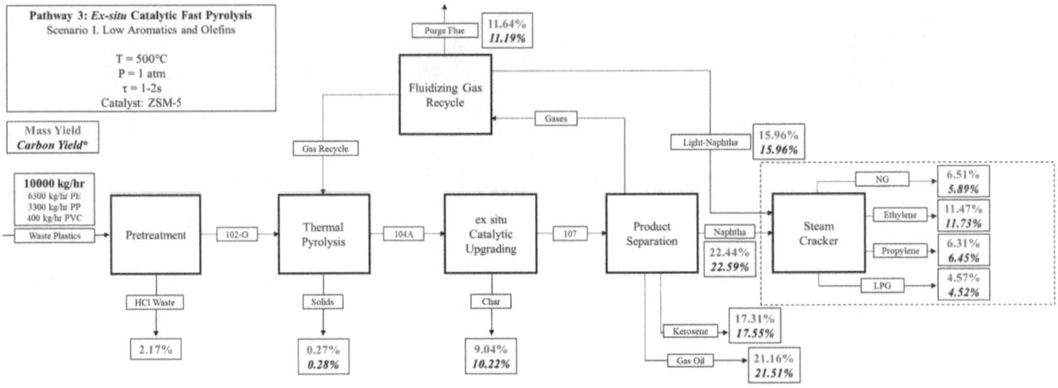


**FIGURE 6.6** *Ex-situ* catalytic fast pyrolysis process (Scenario I).

reaction system (similar to Figure 6.2) would be used for the catalytic upgrading step. Note that the configuration of the thermal pyrolysis reactor will depend on the char-make. If sufficiently high, a char burner would be included to provide process heat. If low, the char burner would not be included, and a reactor configuration similar to that shown in Figure 6.4 would likely be used for the thermal pyrolysis reactor. The carbon range for the upgraded pyrolysis oil was not reported [14], but it is reasonable to assume that the upgraded pyrolysis oil consists of a much higher distribution of hydrocarbons from C5–C12 (naphtha range). As shown in Figure 6.3, after the separation of $C12^+$ and $C5^-$, the non-catalytic pyrolysis naphtha (including light naphtha) yield is small (16%). In comparison, the yield is much higher (38%) for the catalytic pathway, as expected. These data were from a semi-batch experiment, with a continuous flow of pyrolysis vapors but without continuous feeding of polymer. As such, the data must be treated with a degree of skepticism and the pyrolysis naphtha yield was treated as an adjustable parameter in the economic analysis.

Figure 6.7 and Table 6.7 present a different data set for *ex-situ* CFP from experiments performed recently at NREL on a feedstock consisting of pure LLDPE. There are two remarkable features to this pathway: 1) direct path to low molecular weight (MW) olefins without co-processing (e.g., the 'one-step' approach), and 2) pathway to high aromatics production (i.e., Scenario II, high aromatics).

These experiments were conducted in a bench-scale continuous-flow fluidized bed pyrolyzer coupled to a fixed-bed catalytic upgrading reactor; the temperature in both reactors was set at 550°C. No attempt was made to optimize the system for the production of pyrolysis liquids. The results suggest that it is possible to produce low MW olefins directly from the catalytic upgrading step; a yield of over 40% combined $C2^= + C3^=$ was obtained. The composition of the upgraded pyrolysis oil is light pyrolysis oil (C5–C12, 39 wt% yield) and heavy pyrolysis oil (C12–C18, <0.5 wt% yield) obtained in ex-situ mode (550°C, ZSM-5 catalyst). The total pyrolysis oil liquid product (Stream 107, Figure 6.7) was approximately 90% aromatics; the distribution of aromatics from four experiments is shown in Figure 6.8. This system is highly 'tunable,' and the product distribution can be altered by simply changing the catalyst and reaction conditions. In the TEA and LCA, we treated the gas/liquid/aromatics split as an adjustable parameter to investigate the impact of different upgrading strategies on process economics and sustainability.

## 6.2.6 Process Economics

### 6.2.6.1 Feedstock Cost

Before discussing the results of our economic modeling of these four pyrolysis pathways, it is important to detail how feedstock cost was estimated for this project. We assumed the feed to a

**TABLE 6.6**
**Stream Table for *Ex-situ* Catalytic Fast Pyrolysis Process (Scenario I)**

| Streams | Units | Waste Plastics | HCl Waste | 102-O | Gas Recycle | Solids | 104A | Char | 107 | Gases | Kerosene | Gas Oil | Light-Naphtha | Naphtha | Purge Flue |
|---|---|---|---|---|---|---|---|---|---|---|---|---|---|---|---|
| **Mass Flows** | **kg/hr** | **10,000.0** | **10,000.0** | **9,782.7** | **7,600.0** | **27.2** | **17,355.6** | **904.4** | **16,451.2** | **10,360.3** | **1,730.9** | **2,115.6** | **1,596.4** | **2,244.4** | **1,163.8** |
| **Mass Flow of Carbon Atoms** | **kg/hr** | **8,376.5** | – | **8,376.54** | **6,120.3** | **23.3** | **14,432.3** | **856.5** | **13,558.8** | **8,394.3** | **1,470.3** | **1,802.0** | **1,336.7** | **1,892.2** | **937.3** |
| PE | kg/hr | 6,300.0 | 6,300.0 | 6,300.0 | – | 9.6 | – | – | – | – | – | – | – | – | – |
| PP | kg/hr | 3,300.0 | 3,300.0 | 3,300.0 | – | 9.6 | – | – | – | – | – | – | – | – | – |
| H2 | kg/hr | – | – | – | – | – | – | – | – | – | – | – | – | – | – |
| CH4 | kg/hr | – | – | – | 1,641.4 | – | 1,646.9 | – | 1,897.4 | 1,896.6 | 0.2 | 0.0 | 3.8 | 0.6 | 251.4 |
| C2H6 | kg/hr | – | – | – | 1,421.8 | – | 1,537.2 | – | 1,663.8 | 1,660.2 | 0.4 | 0.0 | 20.6 | 3.3 | 217.7 |
| C2H4 | kg/hr | – | – | – | 965.5 | – | 1,003.8 | – | 1,124.4 | 1,122.7 | 0.2 | 0.0 | 9.3 | 1.5 | 147.9 |
| C3H8 | kg/hr | – | – | – | 834.9 | – | 962.1 | – | 1,011.1 | 1,003.2 | 0.4 | 0.0 | 40.4 | 7.5 | 127.9 |
| C3H6 | kg/hr | – | – | – | 1,061.4 | – | 1,193.6 | – | 1,279.7 | 1,271.1 | 0.4 | 0.0 | 47.2 | 8.2 | 162.5 |
| C4H10 | kg/hr | – | – | – | 514.6 | – | – | – | 703.4 | 684.9 | 0.4 | 0.0 | 91.5 | 18.1 | 78.8 |
| N-PENTAN | kg/hr | – | – | – | 777.8 | – | 1.0 | – | 1,508.4 | 1,379.7 | 1.3 | 0.0 | 482.8 | 127.3 | 119.1 |
| N-C6H14 | kg/hr | – | – | – | 284.5 | – | 576.0 | – | 1,196.3 | 917.6 | 1.6 | 0.0 | 589.6 | 277.0 | 43.6 |
| BENZENE | kg/hr | – | 163.3 | 163.3 | 16.0 | – | 54.0 | – | 65.7 | 50.2 | 0.1 | 0.0 | 31.7 | 15.4 | 2.4 |
| TOLUENE | kg/hr | – | – | – | 3.0 | – | 31.7 | – | 62.8 | 28.5 | 0.1 | 0.0 | 25.1 | 34.1 | 0.5 |
| XYLENE | kg/hr | – | – | – | – | – | – | – | – | – | – | – | – | – | – |
| N-C8H18 | kg/hr | – | – | – | 11.1 | – | 346.4 | – | 989.9 | 240.9 | 3.0 | 0.0 | 228.1 | 746.0 | 1.7 |
| N-C10H22 | kg/hr | – | – | – | 0.1 | – | 143.9 | – | 711.0 | 24.2 | 4.9 | 0.0 | 24.1 | 681.9 | 0.0 |
| N-C12H26 | kg/hr | – | – | – | 0.0 | – | 287.6 | – | 296.3 | 1.3 | 4.9 | 0.0 | 1.3 | 290.0 | 0.0 |
| N-C15H32 | kg/hr | – | – | – | 0.0 | – | 1.0 | – | 654.2 | 0.0 | 617.0 | 4.0 | 0.0 | 33.2 | 0.0 |
| N-C20H42 | kg/hr | – | – | – | – | – | 1,150.5 | – | 1,185.1 | 0.0 | 1,044.3 | 140.8 | – | 0.0 | – |
| C28H58 | kg/hr | – | – | – | – | – | 3,067.9 | – | 2,022.5 | – | 51.7 | 1,970.8 | – | – | – |

*(Continued)*

**TABLE 6.6** *(Continued)*

**Stream Table for *Ex-situ* Catalytic Fast Pyrolysis Process (Scenario I)**

| Streams | Units | Waste Plastics | HCl Waste | 102-O | Gas Recycle | Solids | 104A | Char | 107 | Gases | Kerosene | Gas Oil | Light-Naphtha | Naphtha | Purge Flue |
|---|---|---|---|---|---|---|---|---|---|---|---|---|---|---|---|
| N-C30H62 | kg/hr | – | – | – | – | – | 5,272.9 | – | – | – | – | – | – | – | – |
| HCL | kg/hr | – | 228.7 | 11.4 | 67.7 | – | 79.2 | – | 79.2 | 79.0 | 0.0 | 0.0 | 0.9 | 0.1 | 10.4 |
| CACL2 | kg/hr | – | – | – | – | – | – | – | – | – | – | – | – | – | – |
| PVC | kg/hr | 400.0 | 8.0 | 8.0 | – | 8.0 | – | – | – | – | – | – | – | – | – |
| BUTDIENE | kg/hr | – | – | – | – | – | – | – | – | – | – | – | – | – | – |
| 1-BUTENE | kg/hr | – | – | – | – | – | – | – | – | – | – | – | – | – | – |
| COKE | kg/hr | – | – | – | – | – | – | 904.4 | – | – | – | – | – | – | – |

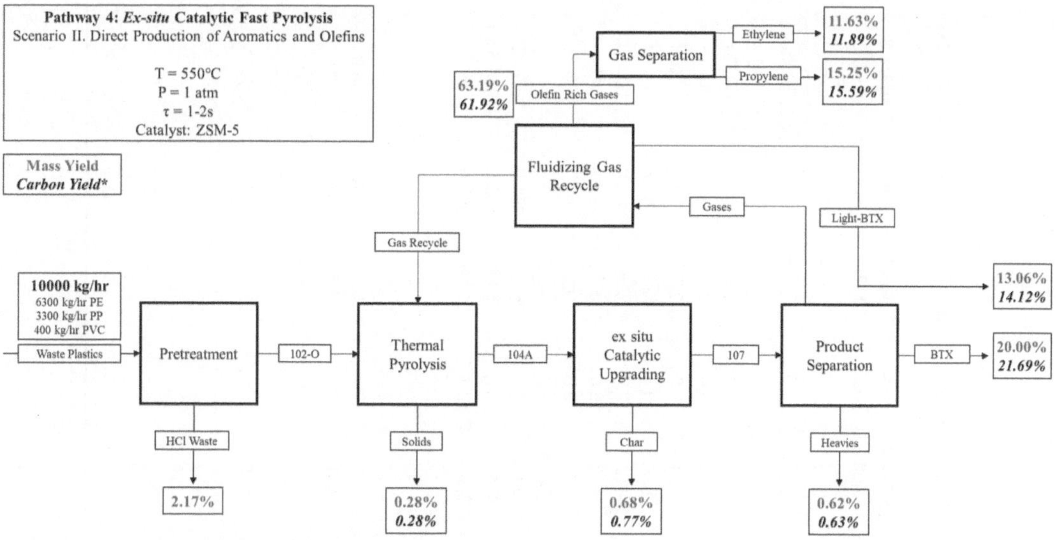


**FIGURE 6.7** *Ex-situ* catalytic fast pyrolysis process (Scenario II).

pyrolysis plant making plastic naphtha consists of a mixture of 63% PE (all types), 33% PP, and 4% PVC. To estimate the cost of this feedstock, we have consulted public data [15] on the cost of reclaimed plastics, which shows in Table 6.8 for post-consumer recycled PET.

We assumed these data are representative of all post-consumer plastics (not just PET) and that the bale price is zero (i.e., the wastes as the feed to the pyrolysis processes have no value), resulting in the weighted average cost of the mixed plastic waste approximately $0.30/lb or $600/tonnes delivered to the processing facility. This value was used as the baseline cost for the feedstock. Tipping fees could vary widely by region in the USA – from a high of $73 in the Pacific region to a low of $41 in the South Central region, with an average of $55/tonnes[6]. Hence, an additional 2 to 3 cents per pound could be used to reduce the total cost shown above. Because of the uncertainty in the mixed plastic waste cost, sensitivity cases were run for each processing scenarios using 15, 30, and 45¢/lb feedstock cost.

### 6.2.7 Techno-economic Analysis Results

Techno-economic analysis (TEA) results are shown in Table 6.9 through Table 6.12 for the four pyrolysis pathways considered for this study. For each pathway, a base case was calculated, and sensitivity to the calculated minimum fuel selling price (MFSP) was determined for five different parameters: naphtha yield, total capital investment (TCI), feedstock cost, internal rate of return (IRR), and income tax rate.

In order to simplify the interpretation of these results, Figure 6.9 to Figure 6.11 present 'tornado plots' that show how each of these parameters impacts the selling price for plastic naphtha compared to fossil naphtha for the first three pathways.

Several conclusions can be drawn from these results. First, the feedstock cost is the dominant economic driver for the production of naphtha from plastics. Second, the naphtha yield is a factor, but it is overwhelmed by the impact of feedstock cost. Lastly, only one case (thermal pyrolysis with low feedstock cost) appears to be economically feasible but only at a feedstock cost that is half the estimated current value for waste polyolefins. Capital investment (represented by TCI or total capital investment) is not a factor, and the impact of the tax rate is essentially zero.

For the fourth pyrolysis pathway, a direct comparison to the production of olefins can be made as this technology does not involve the co-processing of plastic naphtha. Figure 6.12 presents

**TABLE 6.7**
**Stream Table for *Ex-situ* Catalytic Fast Pyrolysis Process (Scenario II)**

| Streams | Units | Waste Plastics | HCl Waste | 102-O | Gas Recycle | Solids | 104A | Char | 107 | Gases | Heavies | Light-BTX | BTX | Olefin Rich Gases |
|---|---|---|---|---|---|---|---|---|---|---|---|---|---|---|
| **Mass Flows** | **kg/hr** | **10,000.0** | **217.3** | **9,782.7** | **7,600.0** | **27.8** | **17,355.0** | **68.4** | **17,286.5** | **15,224.4** | **61.9** | **1,305.7** | **2,000.2** | **6,318.8** |
| **Mass Flow of Carbon Atoms** | **kg/hr** | **8,376.5** | – | **8,376.54** | **6,238.8** | **23.8** | **14,538.1** | **64.8** | **14,477.9** | **12,608.3** | **52.6** | **1,182.4** | **1,817.0** | **5,187.1** |
| PE | kg/hr | 6,300.0 | – | 6,300.0 | – | 9.9 | – | – | – | – | – | – | – | – |
| PP | kg/hr | 3,300.0 | – | 3,300.0 | – | 9.9 | – | – | – | – | – | – | – | – |
| H2 | kg/hr | – | – | – | – | – | – | – | – | – | – | – | – | – |
| CH4 | kg/hr | – | – | – | 1,480.1 | – | 1,536.5 | – | 2,712.7 | 2,712.3 | 0.0 | 1.6 | 0.4 | 1,230.6 |
| C2H6 | kg/hr | – | – | – | 533.6 | – | 688.0 | – | 982.7 | 981.6 | 0.0 | 4.4 | 1.1 | 443.6 |
| C2H4 | kg/hr | – | – | – | 1,399.4 | – | 1,474.6 | – | 2,572.5 | 2,570.6 | 0.0 | 7.7 | 2.0 | 1,163.5 |
| C3H8 | kg/hr | – | – | – | 2,056.8 | – | 2,179.0 | – | 3,845.4 | 3,828.8 | 0.0 | 62.1 | 16.5 | 1,710.0 |
| C3H6 | kg/hr | – | – | – | 1,834.4 | – | 2,003.7 | – | 3,425.2 | 3,412.0 | 0.0 | 52.3 | 13.3 | 1,525.2 |
| C4H10 | kg/hr | – | – | – | 49.4 | – | – | – | 97.8 | 96.2 | 0.0 | 5.7 | 1.6 | 41.1 |
| N-PENTAN | kg/hr | – | – | – | 0.4 | – | 1.0 | – | 1.0 | 0.9 | 0.0 | 0.1 | 0.0 | 0.4 |
| N-C6H14 | kg/hr | – | – | – | – | – | 306.7 | – | – | – | – | – | – | – |
| BENZENE | kg/hr | – | – | 163.3 | 155.4 | – | 256.1 | – | 843.1 | 641.3 | 0.0 | 356.7 | 201.9 | 129.2 |
| TOLUENE | kg/hr | – | – | – | 66.8 | – | 265.4 | – | 1,439.0 | 675.1 | 0.0 | 552.9 | 763.8 | 55.5 |
| XYLENE | kg/hr | – | – | – | 10.2 | – | 109.3 | – | 1,282.5 | 280.7 | 2.3 | 262.0 | 999.6 | 8.5 |
| N-C8H18 | kg/hr | – | – | – | – | – | 247.3 | – | – | – | – | – | – | – |
| N-C10H22 | kg/hr | – | – | – | – | – | 49.5 | – | – | – | – | – | – | – |
| N-C12H26 | kg/hr | – | – | – | – | – | 98.9 | – | – | – | – | – | – | – |
| N-C15H32 | kg/hr | – | – | – | 0.0 | – | 1.0 | – | 58.7 | 0.1 | 58.6 | 0.1 | 0.0 | 0.0 |
| N-C20H42 | kg/hr | – | – | – | 0.0 | – | 1,484.1 | – | 1.0 | 0.0 | 1.0 | 0.0 | 0.0 | 0.0 |
| C28H58 | kg/hr | – | – | – | – | – | 2,473.4 | – | – | – | – | – | – | – |
| N-C30H62 | kg/hr | – | – | – | – | – | 4,155.4 | – | – | – | – | – | – | – |
| HCL | kg/hr | – | 217.3 | 11.4 | 13.6 | – | 25.0 | – | 25.0 | 25.0 | 0.0 | 0.1 | 0.0 | 11.3 |
| CACL2 | kg/hr | – | – | – | – | – | – | – | – | – | – | – | – | – |
| PVC | kg/hr | 400.0 | – | 8.0 | – | 8.0 | – | – | – | – | – | – | – | – |
| BUTDIENE | kg/hr | – | – | – | – | – | – | – | – | – | – | – | – | – |
| 1-BUTENE | kg/hr | – | – | – | – | – | – | – | – | – | – | – | – | – |
| COKE | kg/hr | – | – | – | – | – | – | 68.4 | – | – | – | – | – | – |

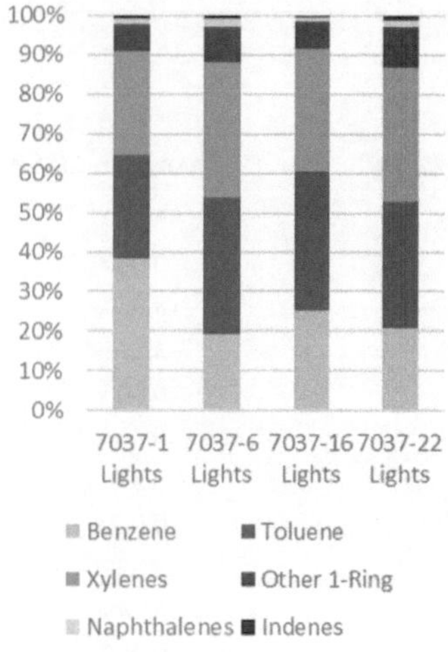


**FIGURE 6.8** Distribution of aromatics in upgraded pyrolysis liquids.

**TABLE 6.8**
**Cost of Reclaimed Plastics**

| Cost Component | Average Cost ($/lb) |
|---|---|
| Bale price, PET | $0.19 |
| Transportation | $0.02 |
| Yield cost | $0.09 |
| Bale price adjusted for yield | $0.31 |
| Conversion to flake | $0.19 |
| TOTAL COST | $0.49 |

results for the production of ethylene and propylene via *ex-situ* catalytic fast pyrolysis using the results shown in Table 6.12. As above, feedstock cost is the dominant economic driver. Estimated selling prices for ethylene and propylene are 42% and 40% higher respectively than fossil prices (5-year average).

## 6.3 RECYCLING OF POLYMERS FOR THE PRODUCTION OF CHEMICALS VIA GASIFICATION

This section presents the results of analyzing process pathways for recycling waste plastics to produce olefins or formaldehyde via gasification using methanol as the central intermediate. Basic data and information required to build process flowsheets for near-term commercial pathways for producing methanol from waste plastics were gleaned from the public literature and, where possible, supplemented with information from NREL and companies actively engaged in attempting to commercialize these pathways.

Unlike the case for pyrolysis, the conversion of waste plastics by gasification produces essentially only one product – synthesis gas (syngas) – a mixture of hydrogen and carbon monoxide with other minor components (principally carbon dioxide) present depending on the nature of the gasification technology being employed. A second major difference, compared to pyrolysis, is that gasification will be much less dependent on the composition of the feedstock and – to some extent

**TABLE 6.9**
**Economic Sensitivity Analysis for Thermal Pyrolysis**

| Thermal Pyrolysis | Value | MFSP ($/kg) | ΔMFSP ($/kg) | % Change in MFSP |
|---|---|---|---|---|
| -5% | 1,108.70 | 3.47 | 0.86 | 33.23% |
| Naphtha Yield (kg/h) | 1,598.84 | 2.60 | 0.00 | 0.00% |
| +5% | 2,100.32 | 2.13 | −0.47 | −18.05% |
| −15% | $37,297,681 | 2.52 | −0.08 | −3.23% |
| TCI (USD) | $43,879,624 | 2.60 | 0.00 | 0.00% |
| +30% | $57,043,511 | 2.77 | 0.17 | 6.46% |
| −50% | 15 | 0.53 | −2.07 | −79.52% |
| Feedstock Cost (cents/lb) | 30 | 2.60 | 0.00 | 0.00% |
| +50% | 45 | 4.67 | 2.07 | 79.52% |
| 5% | 5 | 2.45 | −0.15 | −5.74% |
| IRR (%) | 10 | 2.60 | 0.00 | 0.00% |
| 15% | 15 | 2.76 | 0.15 | 5.92% |
| 15% | 15 | 2.59 | −0.01 | −0.41% |
| Income Tax Rate (%) | 21 | 2.60 | 0.00 | 0.00% |
| 35% | 35 | 2.63 | 0.03 | 1.16% |

**TABLE 6.10**
**Economic Sensitivity Analysis for *In-situ* Catalytic Fast Pyrolysis**

| *In-situ* CFP | Value | MFSP ($/kg) | ΔMFSP ($/kg) | % Change in MFSP |
|---|---|---|---|---|
| −5% | 3,247.50 | 2.26 | 0.34 | 17.52% |
| Naphtha Yield | 3,750.57 | 1.93 | 0.00 | 0.00% |
| +5% | 4,247.25 | 1.70 | −0.23 | −11.74% |
| −15% | $49,259,742 | 1.88 | −0.05 | −2.45% |
| TCI (USD) | $57,952,638 | 1.93 | 0.00 | 0.00% |
| +30% | $75,338,429 | 2.02 | 0.09 | 4.91% |
| −50% | 15 | 1.04 | −0.88 | −45.77% |
| Feedstock Cost (cents/lb) | 30 | 1.93 | 0.00 | 0.00% |
| +50% | 45 | 2.81 | 0.88 | 45.77% |
| 5% | 5 | 1.84 | −0.08 | −4.34% |
| IRR (%) | 10 | 1.93 | 0.00 | 0.00% |
| 15% | 15 | 2.01 | 0.09 | 4.47% |
| 15% | 15 | 1.92 | −0.01 | −0.31% |
| Income Tax Rate (%) | 21 | 1.93 | 0.00 | 0.00% |
| 35% | 35 | 1.94 | 0.02 | 0.87% |

– is a feedstock-agnostic pathway. Hence, unlike pyrolysis, gasification is likely to readily apply to highly heterogenous feedstocks of mixed plastic wastes. Complexities arise, however, on the downstream pathways used to upgrade syngas to olefins and other polymer precursors. For purposes of this report, two downstream pathways – both of which use gasification to produce methanol – will be analyzed: methanol-to-olefins (MTO) and methanol-to-formaldehyde.

**TABLE 6.11**
**Economic Sensitivity Analysis for *Ex-situ* Catalytic Fast Pyrolysis (Scenario I)**

| *Ex-situ* CFP I | Value | MFSP ($/kg) | ΔMFSP ($/kg) | % Change in MFSP |
|---|---|---|---|---|
| -5% | 3,367.49 | 2.01 | 0.13 | 7.11% |
| Naphtha Yield | 3,840.81 | 1.88 | 0.00 | 0.00% |
| +5% | 4,316.48 | 1.74 | −0.14 | −7.37% |
| −15% | $84,497,682 | 1.80 | −0.08 | −4.25% |
| TCI (USD) | $99,409,038 | 1.88 | 0.00 | 0.00% |
| +30% | $129,231,749 | 2.04 | 0.16 | 8.50% |
| −50% | 15 | 1.02 | −0.86 | −45.89% |
| Feedstock Cost (cents/lb) | 30 | 1.88 | 0.00 | 0.00% |
| +50% | 45 | 2.74 | 0.86 | 45.89% |
| 5% | 5 | 1.74 | −0.14 | −7.45% |
| IRR (%) | 10 | 1.88 | 0.00 | 0.00% |
| 15% | 15 | 2.02 | 0.14 | 7.67% |
| 15% | 15 | 1.87 | −0.01 | −0.53% |
| Income Tax Rate (%) | 21 | 1.88 | 0.00 | 0.00% |
| 35% | 35 | 1.90 | 0.03 | 1.50% |

### 6.3.1 Process Description

Gasification takes place at reaction conditions that are very different than pyrolysis. Typical reaction parameters include temperature: 800–1,500°C (higher temperatures (plasma) are used occasionally), pressure (ambient to over 40 atm), catalytic and non-catalytic, steam or reactive gases (oxygen, hydrogen) often used, and reaction times (very fast (seconds) with recirculating fluid bed/entrained flow reactors).

To date, the gasification of 'pure' waste plastics has not been practiced on a commercial scale. Enerkem in Canada uses sorted municipal solid waste from the city of Edmonton as a feedstock, which is reported to be high in plastics [16]. Data on the gasification of pure mixed plastic waste for the production of syngas is limited.

This study focuses on a single gasification pathway that produces methanol from syngas followed by syngas upgrading to either olefins or formaldehyde. As outlined below, only commercially proven technologies were used for the syngas and methanol conversion steps.

#### 6.3.1.1 Syngas-to-Methanol

Production of methanol from syngas is a commercially proven technology 17,18. Catalysts, reaction conditions, and reactor types are well developed, and the system is optimized for clean feedstocks like natural gas. This study assumed that conventional methanol production technology could be utilized for the conversion of syngas produced by the gasification of waste plastics.

#### 6.3.1.2 Methanol-to-Olefins

Technology currently exists for the catalytic conversion of methanol to low molecular weight (LMW) olefins, principally ethylene ($C2^=$) and propylene ($C3^=$). Two different schemes have been developed and demonstrated at a large scale; both can be considered to represent commercial-ready technologies:

1. Methanol-to-olefins (MTO):[19] this process has been developed by UOP to manufacture ethylene and propylene from methanol. The ratio of $C2^=$ to $C3^=$ can be adjusted

**TABLE 6.12**
**Economic Sensitivity Analysis, *Ex-situ* Catalytic Fast Pyrolysis (Scenario II). Gray Rows Are for Ethylene, and White Rows Are for Propylene**

| *Ex-situ* CFP II | Value | MFSP ($/kg) | ΔMFSP ($/kg) | % Change in MFSP |
|---|---|---|---|---|
| −2.5% | 917.45 | 2.46 | 0.34 | 15.88% |
| −2.5% | 1,280.24 | 3.35 | 0.46 | 15.88% |
| Ethylene Yield | 1,163.47 | 2.12 | 0.00 | 0.00% |
| Propylene Yield | 1,525.18 | 2.89 | 0.00 | 0.00% |
| +2.5% | 1,409.23 | 1.89 | −0.23 | −10.93% |
| +2.5% | 1,769.99 | 2.58 | −0.32 | −10.93% |
| −15% | $78,128,430 | 2.03 | −0.09 | −4.09% |
| −15% | $78,128,430 | 2.78 | −0.12 | −4.09% |
| TCI (USD) | $91,915,800 | 2.12 | 0.00 | 0.00% |
| TCI (USD) | $91,915,800 | 2.89 | 0.00 | 0.00% |
| +30% | $119,490,540 | 2.29 | 0.17 | 8.18% |
| +30% | $119,490,540 | 3.13 | 0.24 | 8.18% |
| −50% | 15 | 1.10 | −1.02 | −48.09% |
| −50% | 15 | 1.50 | −1.39 | −48.09% |
| Feedstock Cost (cents/lb) | 30 | 2.12 | 0.00 | 0.00% |
| Feedstock Cost (cents/lb) | 30 | 2.89 | 0.00 | 0.00% |
| +50% | 45 | 3.14 | 1.02 | 48.09% |
| +50% | 45 | 4.29 | 1.39 | 48.09% |
| 5% | 5 | 1.97 | −0.15 | −7.19% |
| 5% | 5 | 2.69 | −0.21 | −7.19% |
| IRR (%) | 10 | 2.12 | 0.00 | 0.00% |
| IRR (%) | 10 | 2.89 | 0.00 | 0.00% |
| 15% | 15 | 2.28 | 0.16 | 7.41% |
| 15% | 15 | 3.11 | 0.21 | 7.41% |
| 15% | 15 | 2.11 | −0.01 | −0.51% |
| 15% | 15 | 2.88 | −0.01 | −0.51% |
| Income Tax Rate (%) | 21 | 2.12 | 0.00 | 0.00% |
| Income Tax Rate (%) | 21 | 2.89 | 0.00 | 0.00% |
| 35% | 35 | 2.15 | 0.03 | 1.45% |
| 35% | 35 | 2.94 | 0.04 | 1.45% |

by choice of operating conditions to give relatively high yields of propylene relative to ethylene (c.a. 1.8/1). A detailed process flow diagram (PFD) for this process is shown in Figure 6.13.

2. Methanol-to-propylene (MTP): this process has been developed by Lurgi to manufacture primarily propylene (Figure 6.14);[20] the other product is a gasoline-range liquid hydrocarbon. Ethylene yields from the MTP process are very low (essentially nil).

The Lurgi MTP process is much more selective than MTO for C3 olefins. Carbon efficiency (mass of carbon in $C3^=$ product per unit mass carbon in methanol) is 53% for MTO and 67% for MTP, reflecting the higher selectivity of the MTP process. The "advanced" MTO process can be designed to run at a variety of $C3^=/C2^=$ ratios, as illustrated in Figure 6.15.

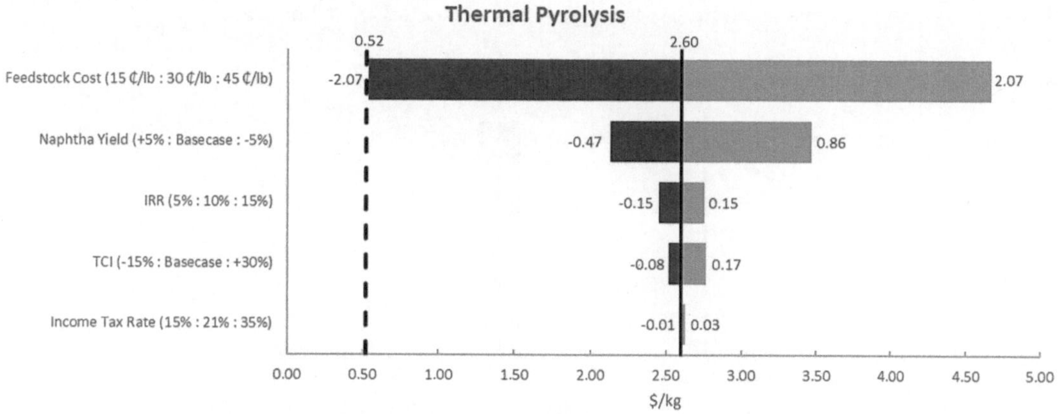


**FIGURE 6.9** Sensitivity analysis for the production of naphtha from plastic waste via thermal pyrolysis. Solid line represents base case ($2.60/kg). Dashed line is 5-year average selling price for fossil naphtha (52¢/kg).

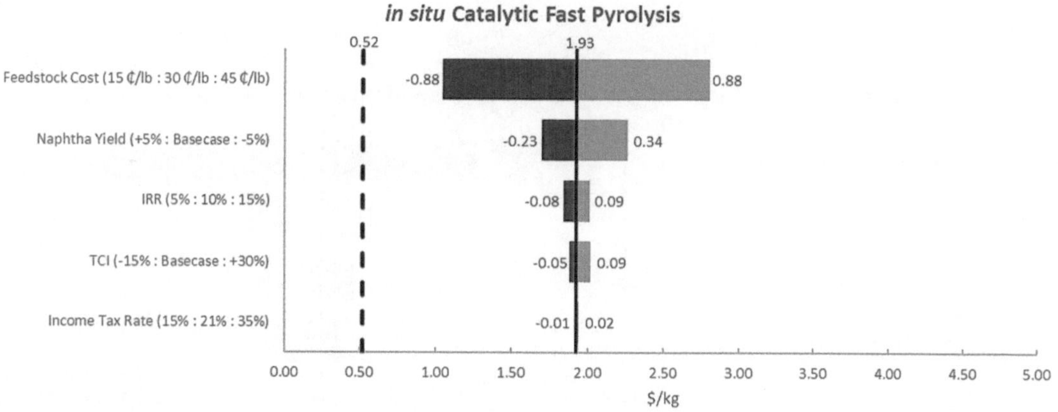


**FIGURE 6.10** Sensitivity analysis for the production of naphtha from plastic waste via *in-situ* catalytic fast pyrolysis. Solid line represents base case ($1.93/kg). Dashed line is 5-year average selling price for fossil naphtha (52¢/kg).

### 6.3.1.3 Methanol-to-Formaldehyde

Formaldehyde ($CH_2O$) is an important industrial chemical that is a precursor to the synthesis of many polymeric materials, including urea-, phenol-, and melamine-formaldehyde resins; approximately 30% of the world's production of methanol in 2017 went to the production of formaldehyde.

In the Formox process, vaporized methanol is mixed with air in the reactor, and formaldehyde is formed. The overall reaction is as follows:

$$CH_3OH + 1/2\ O_2 \rightleftharpoons CH_2O + H_2O$$

There are four side reactions in the Formox process producing dimethyl ether and formic acid as by-products:

$$CH_2O + 1/2\ O_2 \rightleftharpoons CO + H_2O$$

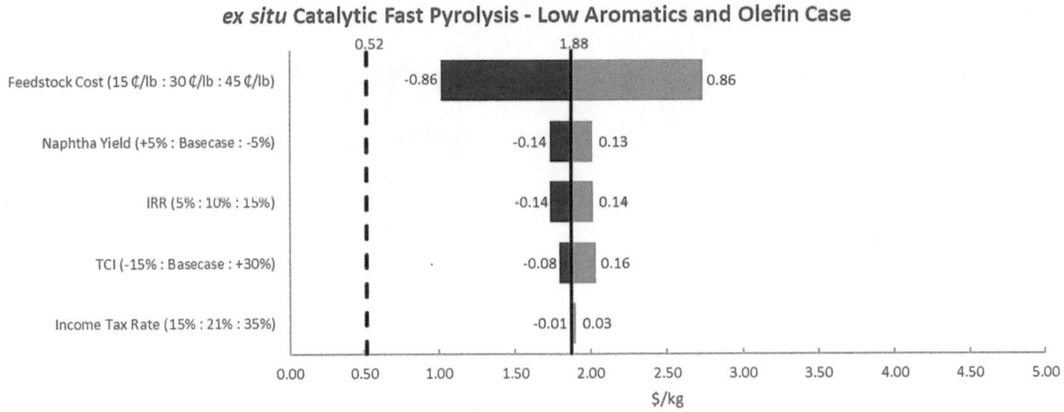


**FIGURE 6.11** Sensitivity analysis for the production of naphtha from plastic waste via *ex-situ* catalytic fast pyrolysis. Solid line represents base case ($1.88/kg). Dashed line is 5-year average selling price for fossil naphtha (52¢/kg).

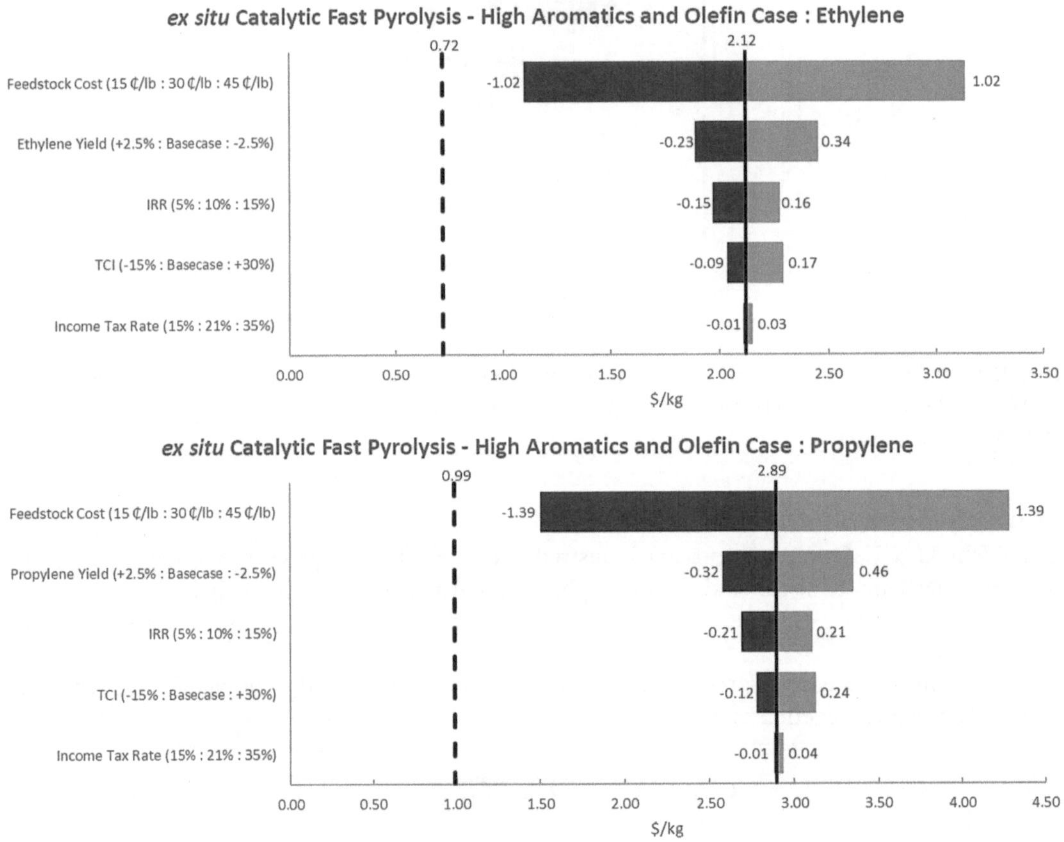


**FIGURE 6.12** Sensitivity analysis for the direct production of ethylene and propylene via *ex-situ* CFP. Dashed line is 5-year average selling price for fossil olefin (72¢/kg for ethylene and 99¢/kg for propylene)

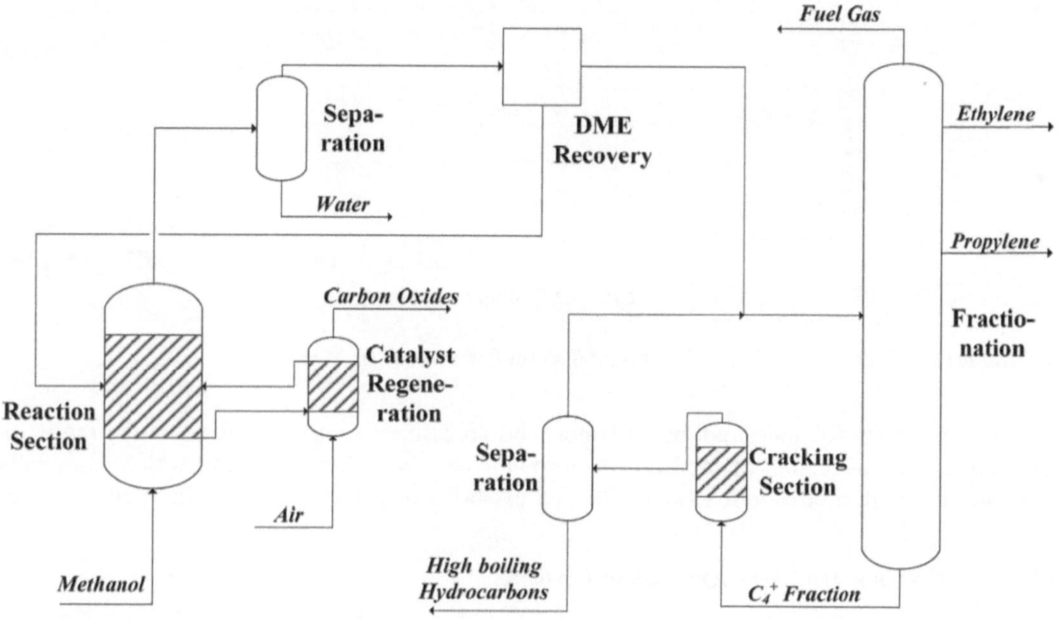


**FIGURE 6.13** UOP methanol-to-olefins (MTO) process.

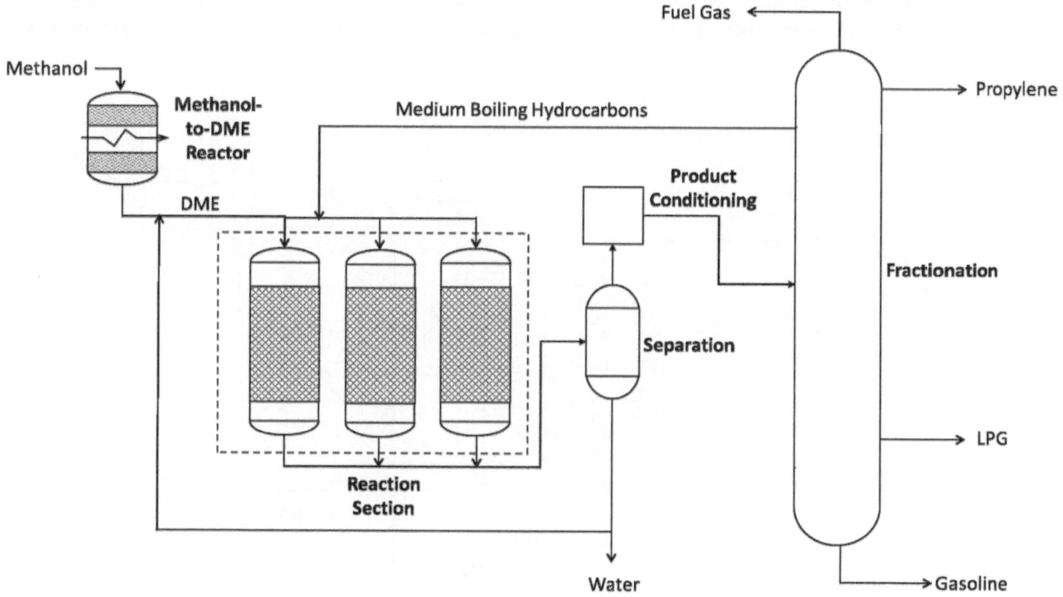


**FIGURE 6.14** Lurgi methanol-to-propylene (MTP) process.

$$2CH_3OH \rightleftharpoons CH_3OCH_3 + H_2O$$

$$CH_2O + 1/2\ O_2 \rightleftharpoons HCOOH$$

$$CH_2O + O_2 \rightleftharpoons CO_2 + H_2O$$

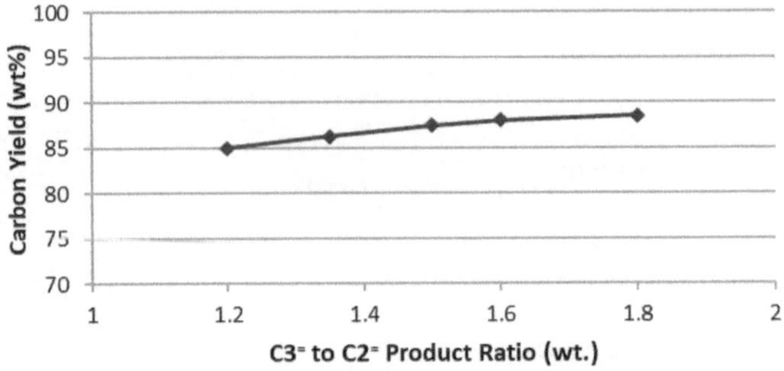


**FIGURE 6.15** Variation in C3 to C2 olefin production from advanced MTO.

All the by-products are undesired, but at atmospheric pressure and between temperature ranges of 270–400°C, the conversion of methanol to formaldehyde is almost complete, which essentially eliminates the undesired side reactions. The process flowsheet is shown in Figure 6.16.

### 6.3.2 Feedstock and Reactor Considerations

The feedstock for the gasification pathways was assumed to be identical to that used for the pyrolysis pathways. Because the primary conversion technology is gasification, the overall process will not be particularly sensitive to the composition of the feed, and restrictions and assumptions that were made for the pyrolysis pathways based on feed composition are largely irrelevant for gasification pathways. Reactors used for methanol synthesis represent commercial packed-bed

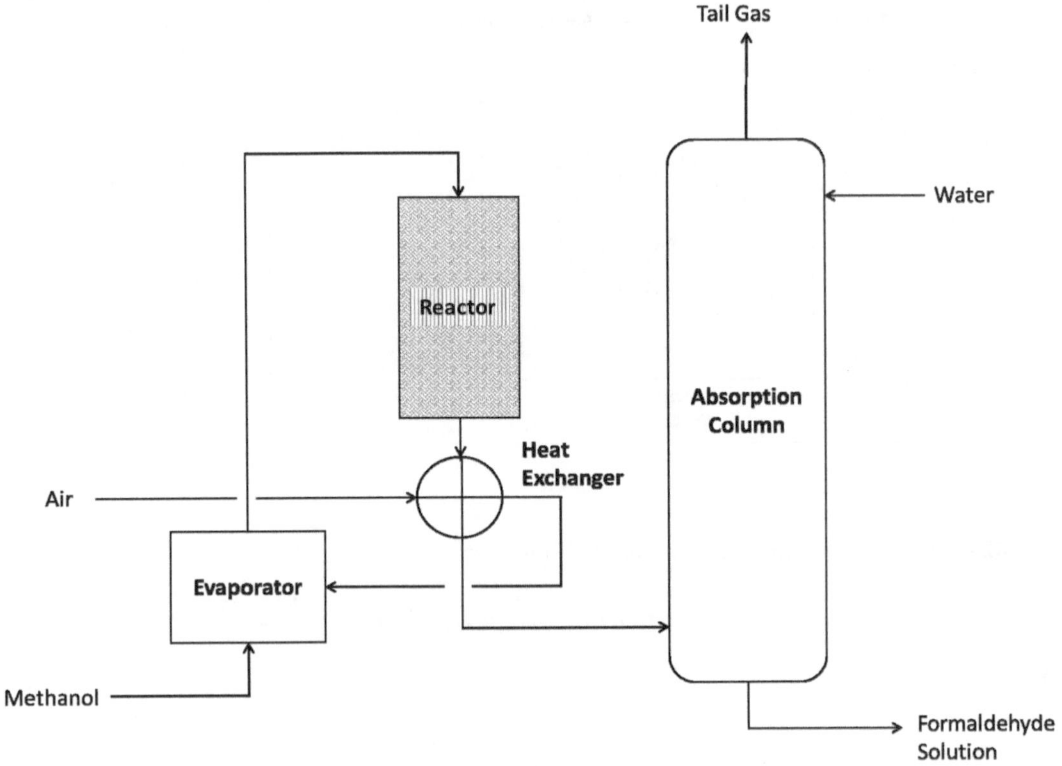


**FIGURE 6.16** Johnson-Matthey Formox process.

shell-in-tube configurations. The gasification reactor is assumed to be a direct heated fluid bed similar in configuration to that shown in Figure 6.4. Since waste polyolefins will contain essentially zero organic oxygen, making syngas requires a source of oxygen in the reactor. This is supplied by using partial oxidation chemistry in the primary conversion reactor to make a hydrogen-rich synthesis gas (≈1:1 $CO:H_2$ on a molar basis). The partial oxidation reaction for hydrocarbons like waste plastics generally proceeds as described by this stoichiometric relationship:

$$C_nH_m + n/2\ O_2 = >nCO + m/2\ H_2$$

The use of oxygen in the primary conversion reactor necessitates the addition of an air separation plant to the overall process flowsheet to provide the needed oxygen. While air could be used, the diluting effect of nitrogen is highly undesirable for the downstream catalytic conversion steps. The use of oxygen has the advantage of producing very clean syngas with low tar content, thus minimizing the need to invest in costly syngas cleaning unit operations.

### 6.3.3 Methanol from Waste Plastics Detailed Process Modeling

The process flowsheet is shown in Figure 6.17. The first reactor is a fluid bed oxygen-blown gasifier operating at elevated pressure (up to 30 atmospheres) and temperatures in the range of 1,000°C. Gas cleanup is utilized largely to remove any residual tars from the gasification step. However, the use of a pressurized oxygen-blown gasifier will minimize tars in the product gases. Gas conditioning mostly removes carbon dioxide, which is produced during the partial oxidation of waste plastics in the gasifier. One advantage of partial oxidation for the production of syngas is that a hydrogen-rich product is made; however, a water-gas shift step (i.e., $CO + H_2O <=> CO_2 + H_2$) to adjust the $H_2/CO$ molar ratio to 2:1 prior to methanol synthesis will likely be required (i.e., $2H_2 + CO <=> CH_3OH$).

Methanol synthesis was assumed to take place at 250°C and 50 atmospheres (735 psi) total pressure; an industry-standard copper-zinc oxide on alumina catalyst is also assumed. Final cleanup of the methanol to remove by-products from the methanol synthesis reaction, such as dimethyl ether, was assumed to produce the final product at a purity necessary for subsequent use in the production of olefins or formaldehyde (generally on the order of $99\%^+$ purity). For purposes of this project, an existing NREL biomass gasification model[17] was used as the starting point for the development of the process model. The Aspen Plus model for the gasification and methanol

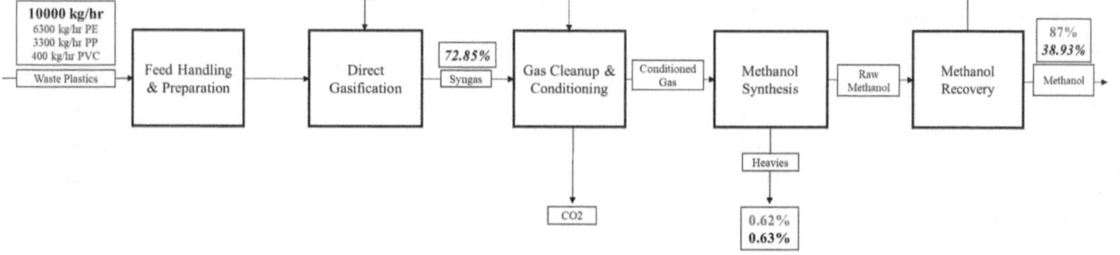


**FIGURE 6.17** Gasification and syngas conversion to methanol.

**TABLE 6.13**
**Waste Plastics Gasification Simulation Results**

| Streams | Units | Waste Plastics | Oxygen | Syngas | Conditioned Gas | Recycled Off-Gases | Raw Methanol | Methanol |
|---|---|---|---|---|---|---|---|---|
| Mass Flows | kg/hr | 10,000.0 | 10,331.1 | 20,331.1 | 19,058.1 | 17,355.6 | 9,421.8 | 8,727.0 |
| Mass Flow of Carbon Atoms | kg/hr | 8,376.5 | - | 6,102.30 | | | | 3,261.0 |

synthesis processes was amended by the addition of an air separation plant (as described above) and modified to use waste plastics as the feedstock. Product flow rates were computed from the Aspen Plus simulation and are shown in Table 6.13. Yields of olefins and formaldehyde from methanol were calculated using published data on MTO and Formox.

### 6.3.4 Techno-Economic Analysis Results

Results from techno-economic analysis for the production of methanol via gasification of waste plastics are shown in Table 6.14. The mass yield of methanol is very high – almost 90% (Figure 6.17), which shows the mass efficiency of the overall process. Carbon efficiency (39%) is in the range normally found for gasification processes for other feedstocks such as biomass. Unfortunately, the cost of producing methanol by gasification of waste plastics is extremely high, $1.37/kg for the base case compared to the 5-year average methanol selling price of 34¢/kg. The primary driver, in this case, is, again, feedstock cost. A 50% reduction in feedstock price will reduce the MFSP to 95¢/kg, which is still a factor of 2.8 higher than the current selling price. The reasons for this very unfavorable economic scenario are twofold: 1) CAPEX required for

**TABLE 6.14**
**Economics for the Production of Methanol from Waste Plastics via Gasification**

| Gasification Model Results | Methanol Price ($/lb) | Methanol Price ($/kg) | ΔMeOH Price ($/kg) | % Change in MFSP |
|---|---|---|---|---|
| –5% | 0.65 | 1.43 | –0.07 | –4.84% |
| Yield | 0.62 | 1.37 | 0.00 | 0.00% |
| +5% | 0.59 | 1.30 | 0.07 | 4.84% |
| –15% | 0.6 | 1.32 | 0.04 | 3.23% |
| TCI (USD) | 0.62 | 1.37 | 0.00 | 0.00% |
| +30% | 0.66 | 1.46 | –0.09 | –6.45% |
| –50% | 0.43 | 0.95 | 0.42 | 30.65% |
| Feedstock Cost (cents/lb) | 0.62 | 1.37 | 0.00 | 0.00% |
| +50% | 0.8 | 1.76 | –0.40 | –29.03% |
| 5% | 0.57 | 1.26 | 0.11 | 8.06% |
| IRR (%) | 0.62 | 1.37 | 0.00 | 0.00% |
| 15% | 0.67 | 1.48 | –0.11 | –8.06% |
| 15% | 0.62 | 1.37 | 0.00 | 0.00% |
| Income Tax Rate (%) | 0.62 | 1.37 | 0.00 | 0.00% |
| 35% | 0.63 | 1.39 | –0.02 | –1.61% |

**TABLE 6.15**
**Sensitivity Analysis for the Production of Ethylene and Propylene from Waste Plastics Derived Methanol via MTO Process**

| MTO | MeOH Price from Gasification ($/lb) | MFSP ($/kg) Ethylene | ΔMFSP ($/kg) Ethylene | MFSP ($/kg) Propylene | ΔMFSP ($/kg) Propylene |
|---|---|---|---|---|---|
| −5% | 0.65 | 3.79 | 0.17 | 5.17 | 0.23 |
| Yield | 0.62 | 3.62 | 0.00 | 4.94 | 0.00 |
| +5% | 0.59 | 3.45 | −0.17 | 4.72 | −0.23 |
| −15% | 0.60 | 3.49 | −0.12 | 4.77 | −0.17 |
| TCI (USD) | 0.62 | 3.62 | 0.00 | 4.94 | 0.00 |
| +30% | 0.66 | 3.87 | 0.25 | 5.28 | 0.34 |
| −50% | 0.43 | 2.57 | −1.05 | 3.50 | −1.44 |
| Feedstock Cost (cents/lb) | 0.62 | 3.62 | 0.00 | 4.94 | 0.00 |
| +50% | 0.80 | 4.62 | 1.00 | 6.31 | 1.36 |
| 5% | 0.57 | 3.32 | −0.30 | 4.53 | −0.41 |
| IRR (%) | 0.62 | 3.62 | 0.00 | 4.94 | 0.00 |
| 15% | 0.67 | 3.92 | 0.30 | 5.35 | 0.41 |
| 15% | 0.62 | 3.62 | 0.00 | 4.94 | 0.00 |
| Income Tax Rate (%) | 0.62 | 3.62 | 0.00 | 4.94 | 0.00 |
| 35% | 0.63 | 3.68 | 0.06 | 5.02 | 0.08 |

pressurized oxygen-blow gasification, and 2) the lack of economy of scale for a plant of this size. Oxygen-blown gasification has significantly higher CAPEX than pyrolysis due primarily to the high cost of the air separation unit (ASU). Additional cost is entailed in gas cleanup to remove inorganics (such as chlorides from PVC) that will foul downstream catalysts and are likely to survive the gasification step. Cheaper gasification technologies could be employed, such as low-pressure steam gasification, but these processes produce very dirty syngas that is high in tars, especially for the waste plastics in the feed [21]. Accordingly, pressurized oxygen-blown gasification will likely be the preferred technology for waste plastics gasification. Because the plant size is very small (240 metric tonnes per day feedstock), economies of scale are not present that can help defray the high capital costs associated with this pathway. Similar conclusions have been reached for biomass gasification, where plant sizes are limited to 2,000 metric tonnes per day, which is nearly ten times higher than the case here.

Because the gasification pathways do not involve co-processing, the economics of the production of olefins from waste plastics can be determined and compared directly to fossil production. Results for the production of ethylene and propylene via MTO are shown in Table 6.15. Tornado plots that present the results of sensitivity analysis for the production of ethylene and propylene by the gasification pathway are shown in Figure 6.18 and Figure 6.19, respectively.

It is evident that the cost of production of C2 and C3 olefins by waste plastics gasification is approximately a factor of 5 greater than fossil; $3.62/kg and $4.94/kg for ethylene and propylene, compared to the selling price (5-year averages) of 0.72¢/kg and 0.99¢/kg, respectively. As was the case for pyrolysis, the cost of production is dominated by feedstock cost. Reduction of feedstock cost by 50% reduces the selling price of ethylene to $2.57/kg and propylene to $3.50/kg. However, even with dramatically lower (zero or negative) feedstock costs, the margin for production by this pathway is likely to still prevent this from being cost competitive. These very unfavorable economics are largely due to the high cost of production of methanol by gasification of waste plastics ($1.37/kg vs. 34¢/kg) and the high cost of additional installed capital associated with MTO.

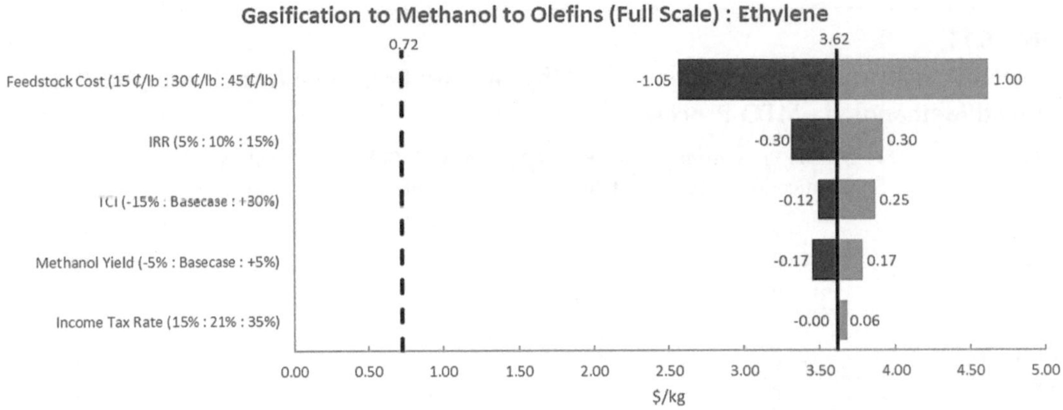


**FIGURE 6.18** Sensitivity analysis for the production of ethylene via direct gasification of waste plastics and methanol-to-olefins (MTO). Dashed line is 5-year average selling price for fossil ethylene (72¢/kg).

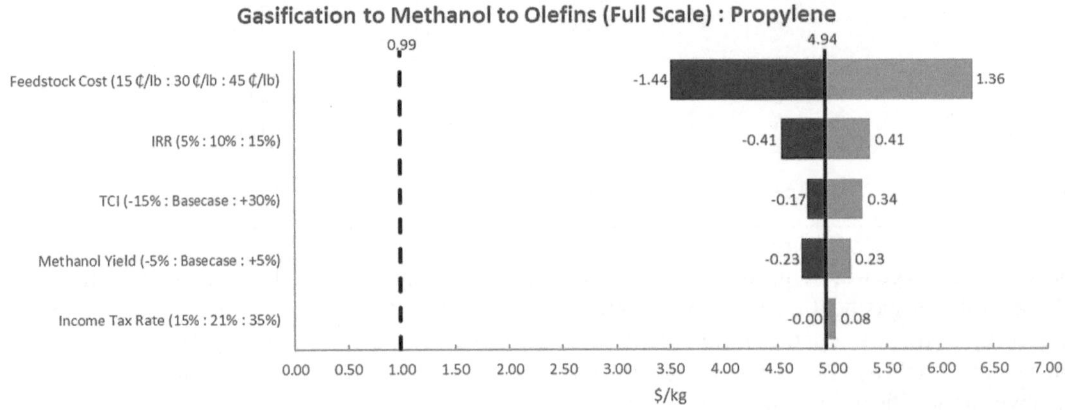


**FIGURE 6.19** Sensitivity analysis for the production of propylene via direct gasification of waste plastics and methanol-to-propylene (MTP). Dashed line is 5-year average selling price for fossil propylene (99¢/kg).

The very high cost of methanol will negatively impact the production of formaldehyde from methanol. Methanol conversion is about 95–99% with a formaldehyde selectivity of 91–94%, resulting in a range of mass yield of formaldehyde from 86% to 93% [22]. The 5-year average selling price for formaldehyde is 32¢/kg. With methanol production at $1.37/kg, the economics for formaldehyde from this pathway will be very challenging; accordingly, no further economic analysis was carried out.

## 6.4 LIFE-CYCLE ASSESSMENT

This attributional life-cycle assessment (LCA) study focuses on the life-cycle greenhouse gas (GHG) emissions and fossil energy consumption (FEC) for the production of ethylene and propylene, and chemicals (formaldehyde). The system boundary for the "cradle-to-gate" LCA study is depicted in Figure 6.20. We account for the stages in the life cycle of the olefins, including plastic waste feedstock production and logistics (transportation), production of intermediates (either naphtha or methanol), and olefin production from the intermediates. The functional unit is one kilogram of the product produced. The greenhouse gas (GHG) emissions are represented in

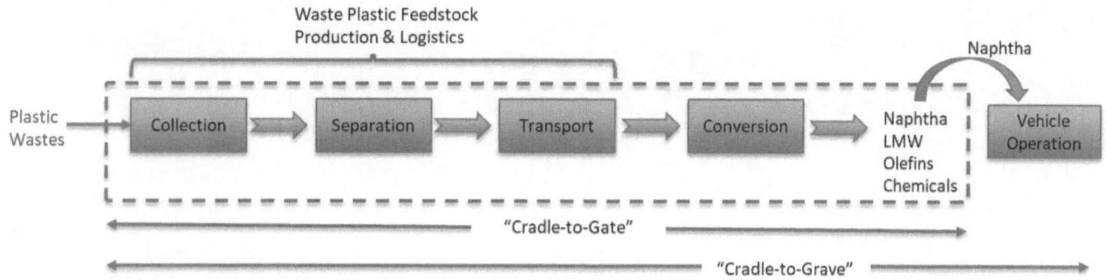


**FIGURE 6.20** Life-cycle system boundary for LCA evaluation.

**TABLE 6.16**
**Transportation of Waste Plastics from MRF to a Conversion Facility**

| Logistics | Value | Unit |
|---|---|---|
| Distance traveled | 160 | km |
| Carried payload | 8.9 | tonnes |
| Payload-distance | 1422.5 | tkm |
| Fuel type | Diesel | |
| Method of transportation | Refuse truck | |

**TABLE 6.17**
**LCA Sensitivity Study Parameter Assumptions**

| | Base Case | Sensitivity 1 | Sensitivity 2 |
|---|---|---|---|
| Electricity | U.S. Avg | EU Avg | U.S. Avg |
| MRF-to-conversion facility distance | 160 km | 160 km | 240 km |

kilograms of carbon dioxide equivalent ($CO_{2eq}$), using a 100-year GHG emission factor. The FEC is represented in megajoules (MJ).

Table 6.16 summarizes the logistics (transportation) of waste plastics from a materials recovery facility (MRF) to a conversion facility. Note that the environmental burden associated with the transportation of plastic waste collection to the MRF is not included, as the transport would take place regardless of whether the waste is used as a feedstock or is landfilled. Additionally, the electricity consumption associated with the plastic waste sorting step was assumed to be 5.7 MJ per metric tonnes.

For each pathway, three scenarios are considered: 1) base case, which uses the average U.S. electricity mix and assumes the MRF-to-conversion facility distance of 160 km, 2) same as scenario 1) except using the EU electricity mix instead of the U.S. electricity mix, and 3) same as scenario 1) except the MRF-to-conversion transportation distance is assumed to be 240 km, or 1.5 times of the base case, as summarized in Table 6.17.

The LCA models are established in SimaPro [23] to link units quantifying the life-cycle impacts based on the process material and energy flows (hereafter referred to as the life-cycle inventory or LCI), as summarized in (Table 6.18). The LCI was derived from process models or literature for parameters such as chemical inputs, power demands, process recycles or co-products, and other input and output mass and energy balance information. Ecoinvent database [24] was used to fill the

**TABLE 6.18**
**Life-cycle Inventory for the LCA Study**

| | Pyrolysis Pathways | | | | | Gasification Pathways | | |
|---|---|---|---|---|---|---|---|---|
| | *Naphtha Production* | | | | *Downstream Process* | *Methanol Production* | *Downstream Process* | |
| | **Thermal pyrolysis** | ***In-situ* pyrolysis** | ***Ex-situ* pyrolysis (Scenario I, low aromatics)** | ***Ex-situ* pyrolysis (Scenario II, high aromatics)** | **Naphtha to LMW olefins** | **Methanol production** | **Methanol to formaldehyde** | **Methanol to LMW olefins** |
| **Products (kg/hr)** | | | | | | | | |
| Naphtha | 1,599 | 3,751 | 3,841 | | | | | |
| Ethylene | | | | 1,163 | 113,398 | | | 35,135 |
| Propylene | | | | 1,525 | 62,375 | | | 35,078 |
| Methanol | | | | | | 8,727 | | |
| Formaldehyde (37%) | | | | | | | 5,592 | |
| Methane rich gas | | | | | 52,466 | | | 124 |
| Ethane rich gas | | | | | 11,921 | | | 1,101 |
| Propane rich gas | | | | | 1,784 | | | 582 |
| C4/C5 mixture | | | | | 43,393 | | | 15,344 |
| **By-products (kg/hr)** | | | | | | | | |
| Kerosene | | | 1,731 | | | | | |
| Gas oil | 4,025 | 407 | 2,116 | | | | | |
| Diesel | | 610 | | | | | | |
| Waxes | 3,717 | | | | | | | |
| Xylene | | | | 1,262 | | | | |
| Toluene | | | | 1,317 | | | | |
| Benzene | | | | 559 | | | | |
| Excess steam (300 psi) | | 46,517 | | | | | | |
| Excess steam (250 psi) | | | | | | | 4,362 | |

| Resource Consumption (kg/hr) | | | | | | | | |
|---|---|---|---|---|---|---|---|---|
| Naphtha | | | | | 379,638 | | | |
| Methanol | | | | | | | 2,405 | 208,737 |
| PE | 6,300 | 6,300 | 6,300 | 6,300 | | 6,300 | | |
| PP | 3,300 | 3,300 | 3,300 | 3,300 | | 3,300 | | |
| PVC | 400 | 400 | 400 | 400 | | 400 | | |
| Total Feedstock | 10,000 | 10,000 | 10,000 | 10,000 | | 10,000 | | |
| Olivine makeup | 8 | | 8 | 8 | | 50 | | |
| Catalysts | | | | | | | 3 | |
| Magnesium oxide (MgO) | | | | | | 2 | | |
| Tar reformer catalyst | | | | | | 1 | | |
| Methanol synthesis catalyst | | | | | | 3 | | |
| Catalyst (SAPO-34 Catalyst) | | | | | | | | 0.02 |
| Dimethyl Disulfide (DMDS) | | | | | | 1 | | |
| Amine (MDEA) makeup | | | | | | 0.4 | | |
| Caustic Soda (50% aq.) | | | | | | | | 220 |
| Boiler feed water makeup | | | | | | 13,357 | | |
| Boiler feed water chemicals | | | | | | 0.2 | | |
| Process water makeup | 507 | 507 | 507 | 507 | | | 3,523 | 165,342 |
| Cooling water makeup | 31,628 | 54,068 | 48,418 | 53,000 | 1,632,931 | | 15,642 | 1,096,219 |
| Cooling tower chemicals | | | | | | 0.03 | | |
| Chilled water makeup | 9,332 | 15,052 | 14,026 | 16,630 | | | | |
| Steam (300 psi) | 694 | | 191 | 1,799 | | | | |

(Continued)

**TABLE 6.18** *(Continued)*
**Life-cycle Inventory for the LCA Study**

| | Pyrolysis Pathways | | | | | Gasification Pathways | | |
|---|---|---|---|---|---|---|---|---|
| | *Naphtha Production* | | | | *Downstream Process* | *Methanol Production* | *Downstream Process* | |
| | Thermal pyrolysis | *In-situ* pyrolysis | *Ex-situ* pyrolysis (Scenario I, low aromatics) | *Ex-situ* pyrolysis (Scenario II, high aromatics) | Naphtha to LMW olefins | Methanol production | Methanol to formaldehyde | Methanol to LMW olefins |
| Steam (1,700 psi) | | | | | 415,490 | | | |
| Steam (1500 psi) | | | | | | | | 189,730 |
| No. 2 diesel fuel | | | | | | 6 | | |
| Natural gas | 599 | | | 1,072 | | | | |
| Electricity (kWh) | 1,599 | 2,756 | 2,023 | 2,048 | | 2,696 | 280 | 3,741 |
| **Waste Streams (kg/hr)** | | | | | | | | |
| Residual solids (coke, spent catalyst, etc.) | 19 | 20 | 27 | | | 2 | | 1,901 |
| Wastewater | 724 | 724 | 724 | 724 | | 1409 | | 165,523 |
| Carbon dioxide (CO2) | 2,880 | 14,672 | 6,595 | 3,352 | 148,778 | 9,836 | | 6,530 |

**TABLE 6.19**
**Market Prices for Main Products and Co-products**

| Products/Co-products | Price ($/kg) |
|---|---|
| Naphtha | 0.521 |
| Ethylene | 0.724 |
| Propylene | 0.988 |
| Methanol | 0.585 |
| Formaldehyde | 0.511 |
| Methane rich gas | 0.169 |
| Ethane rich gas | 0.169 |
| Propane rich gas | 0.364 |
| C4/C5 mixture | 0.364 |
| Kerosene | 0.561 |
| Gas oil | 0.481 |
| Diesel | 0.544 |
| Waxes | 0.547 |
| Xylene | 0.726 |
| Toluene | 0.701 |
| Benzene | 0.811 |
| Excess steam | 0.027 |

*Note*: Obtained from PEP Yearbook [25] using 5-year average, except methanol [26] and steam.

data gaps related to the underlying processes. The GHG emission and FEC burdens were allocated among all products in proportion to the economic values (Table 6.19).

### 6.4.1 Life-Cycle Assessment Results

A "cradle-to-grave" life-cycle assessment (LCA) was first performed on the three pyrolysis pathways that proceed via the manufacture of pyrolysis naphtha; results are shown in Figure 6.21. In this case, the vehicle operation is the same as combustion, so these results compare the greenhouse gas (GHG) emissions for these pathways to that of fossil naphtha combustion in terms of energy intensity (per unit energy).

What is readily apparent is that the carbon intensity of the pyrolysis processes – as measured by $CO_2$ equivalents ($CO_2$-eq) – is a very strong function of the technology used to manufacture pyrolysis naphtha. The thermal and *ex-situ* (I) (i.e., Scenario I, low aromatics) routes all have life-cycle GHG emissions that are approximately the same as that for fossil naphtha. The very large difference noted for *in-situ* CFP is due to the yield from the process, which is approximately 50% pyrolysis liquids. Although not explicitly stated, it can be assumed that the remainder is gaseous. The composition of this stream is unknown but can be assumed to consist mostly of hydrocarbons. These gases are used as fuel gases (e.g., burned) in the plant to raise energy, but due to the low heating value of this stream, a large amount is needed for energy production. The resulting $CO_2$ emissions are high. One other difference that is apparent in all comparisons is the very large difference in GHG emissions for production of the naphtha. In the case of a fossil feedstock, naphtha production is done by simple fractional distillation in a single unit operation (the crude still).

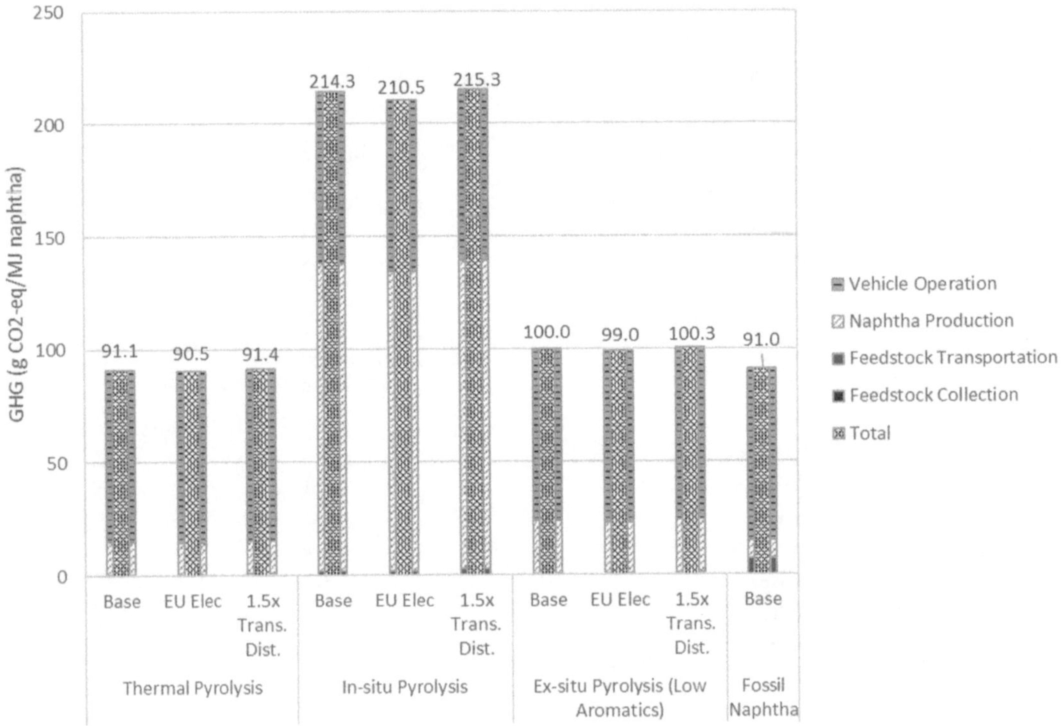


**FIGURE 6.21** Life-cycle GHG emissions for plastic wastes derived naphtha (in g $CO_2$-eq/MJ naphtha).

In contrast, plastic naphtha production uses two basic unit operations that are energy intensive. The energy required for the endothermic depolymerization reactions in the pyrolysis reactor must be supplied by additional process heat as pyrolysis char formation is negligible and cannot be used to meet demands for energy generation. These factors alone and the additional process complexity associated with the recovery and recycling of process gases used for fluidization impart a considerable difference in processing complexity for the plastics case compared to simple distillation for the fossil case. Figure 6.22 shows the life-cycle fossil energy consumption (FEC) comparison, represented in MJ/MJ. The FEC profiles resemble those of the GHG emissions.

Following are the LCA results for the production of ethylene and propylene by all pathways in this study that proceed through either pyrolysis naphtha or methanol as the central intermediate. The life-cycle inventory (LCI) in the form of material and energy flows for all pathways is presented in Table 6.18. The LCA results for the production of ethylene from plastic wastes from all pathways are shown in Figure 6.23 and Figure 6.24.

As shown, thermal pyrolysis and *ex-situ* (II) pathways (i.e., Scenario II, high aromatics) have GHG footprints that are approximately the same or slightly smaller as production from fossil naphtha. The lower GHG footprint for the *ex-situ* (II) pathway is perhaps not a direct comparison since olefins are produced as primary pyrolysis products, and a steam cracker is not used, but it is interesting to note the potential advantage of such a pathway. This pathway was also found to be potentially economically attractive (see Figure 6.12). Comparisons of life cycle FEC for the production of ethylene for all pathways are shown in Figure 6.24.

Similar results were obtained for the production of propylene; GHG emissions and fossil energy consumptions are shown in Figure 6.25 and Figure 6.26. While the numbers are different, as expected, the GHG trends noted above for ethylene production were observed for propylene production.

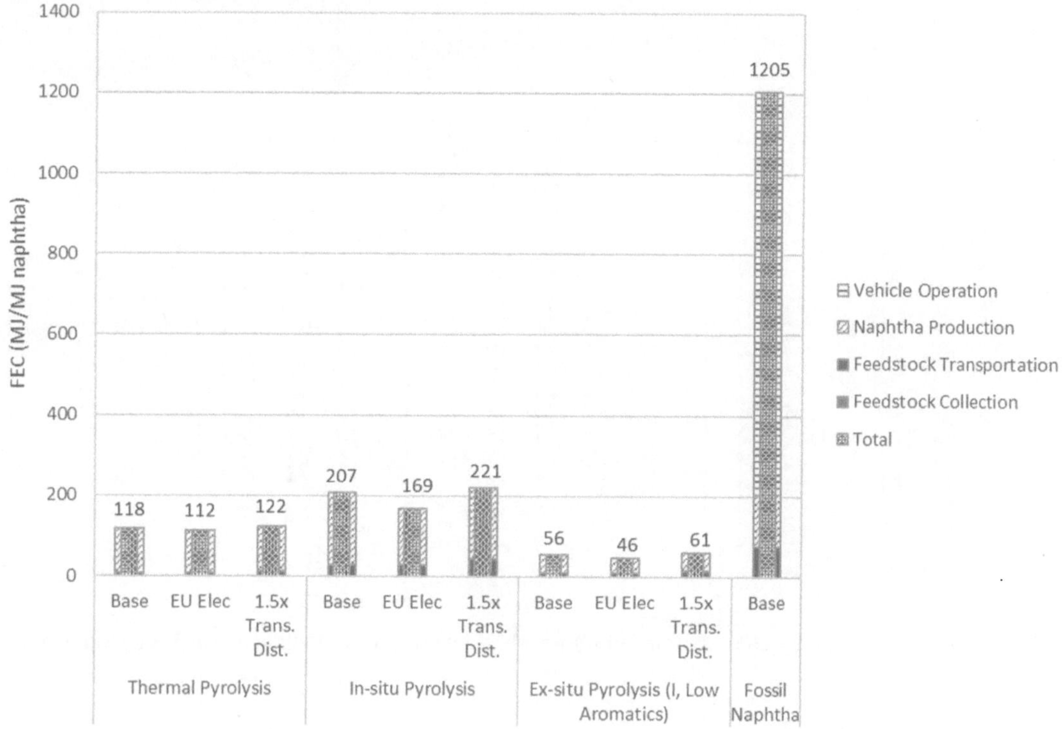


**FIGURE 6.22** Life-cycle FEC for plastic waste-derived naphtha (in MJ/MJ naphtha).

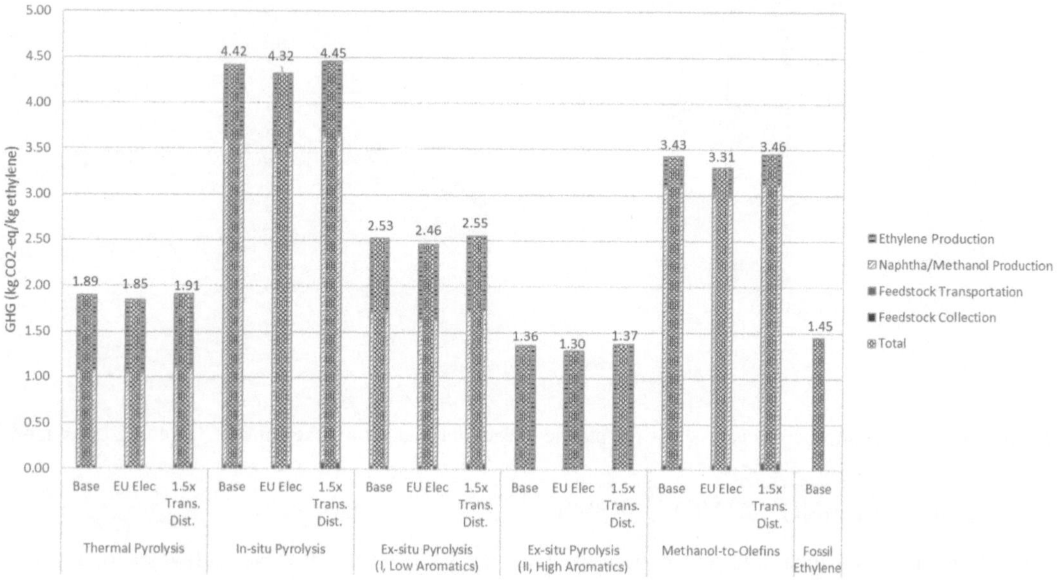


**FIGURE 6.23** GHG emissions for plastic waste–derived ethylene (in kg $CO_2$-eq/kg ethylene).

Figure 6.27 and Figure 6.28 show LCA results for the production of formaldehyde from waste plastics via gasification. Although the economics for these conversion pathways of making formaldehyde from waste plastics are not particularly attractive, it would appear that significant advantages are present in terms of a reduction in fossil energy consumption as shown in Figure 6.28.

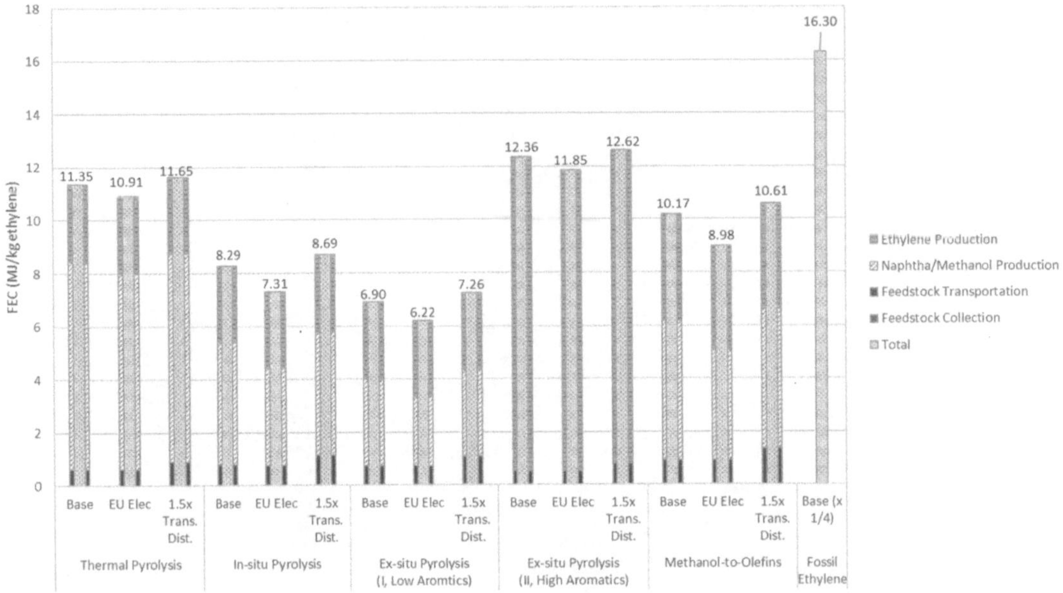


**FIGURE 6.24** Fossil energy consumption (FEC) for plastic waste–derived ethylene (in MJ/kg ethylene).

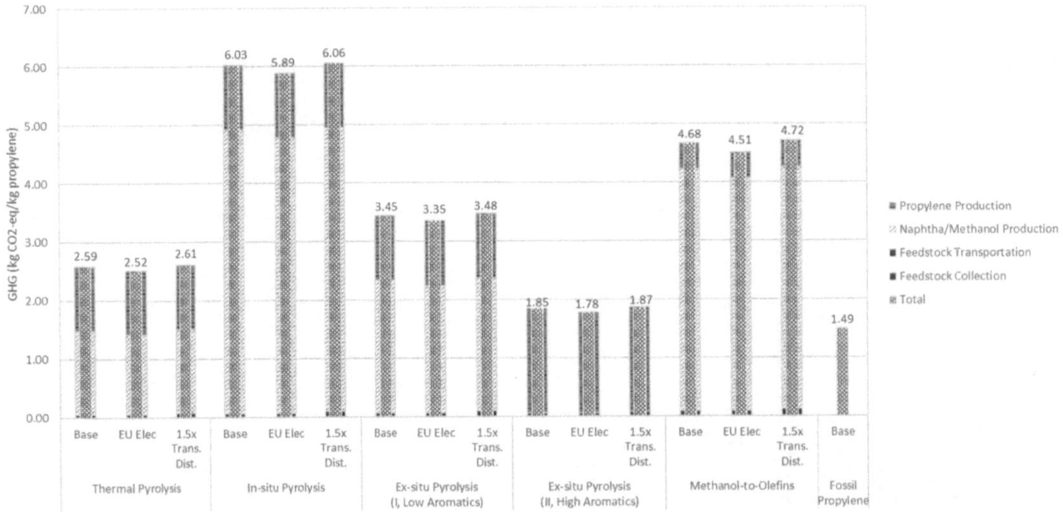


**FIGURE 6.25** Greenhouse gas emissions for plastic waste–derived propylene (in kg $CO_2$-eq/kg propylene).

## 6.5 SUMMARY AND CONCLUSIONS

This study presents a technical and economic overview of selected process pathways for the production of certain polymer precursors from waste plastics. The technologies evaluated all use either pyrolysis or gasification as the primary conversion step to produce either pyrolysis naphtha or syngas. In the former case, this naphtha is co-processed in existing naphtha cracking reactors to make C2 and C3 olefins. In the case of gasification, existing technology was used to take syngas to C2 and C3 olefins or formaldehyde. Data required for techno-economic analysis (TEA) and life-cycle assessment (LCA) were taken primarily from the open literature. Where necessary, reasonable assumptions were made, and sensitivity analyses were performed to test the impact of these assumptions.

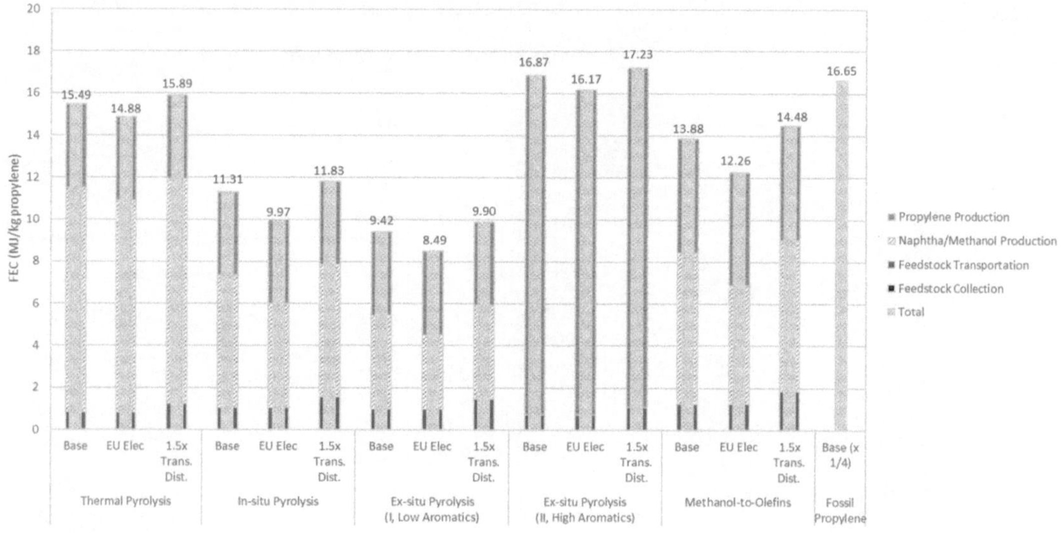


**FIGURE 6.26** Fossil energy consumption for plastic waste-derived propylene (in MJ/kg propylene).

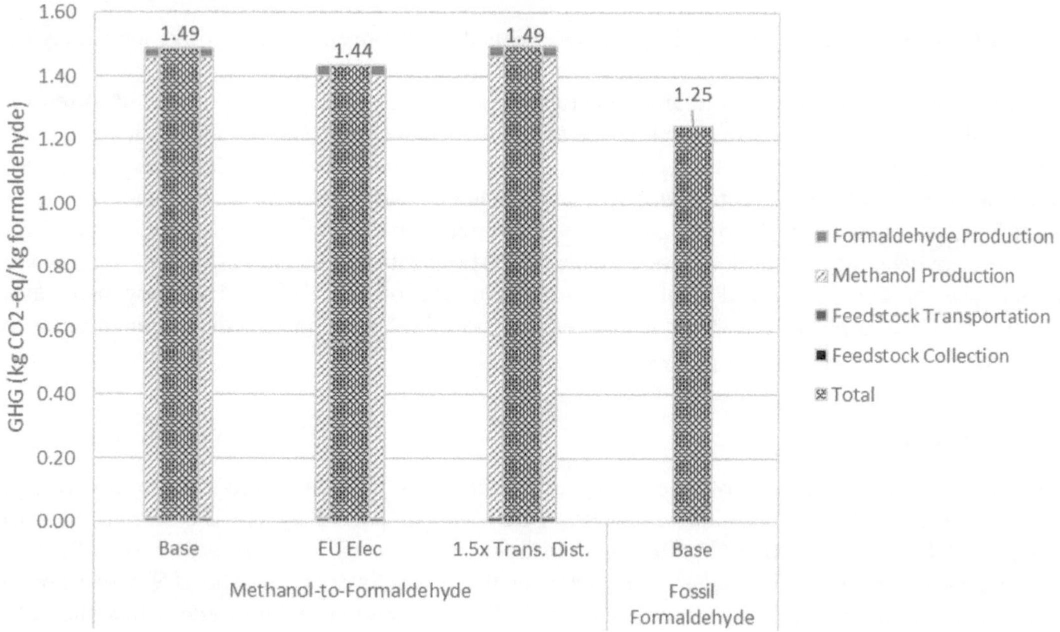


**FIGURE 6.27** Greenhouse gas emissions for plastic waste-derived formaldehyde (in kg $CO_2$-eq/kg formaldehyde).

TEA and LCA results show that all pathways are economically challenged and that the economics are dominated by feedstock cost. There is evidence that a direct route to C2 and C3 olefins via *ex-situ* catalytic fast pyrolysis or low-temperature steam gasification could be cost competitive with conventional petroleum-based C2 and C3, but these results are based on low technology readiness level (TRL) process data that require confirmation and scale-up. LCA modeling of the production of C2 and C3 olefins did not show any significant reductions in GHG emissions for any of the pathways. This is also true for formaldehyde produced by gasification.

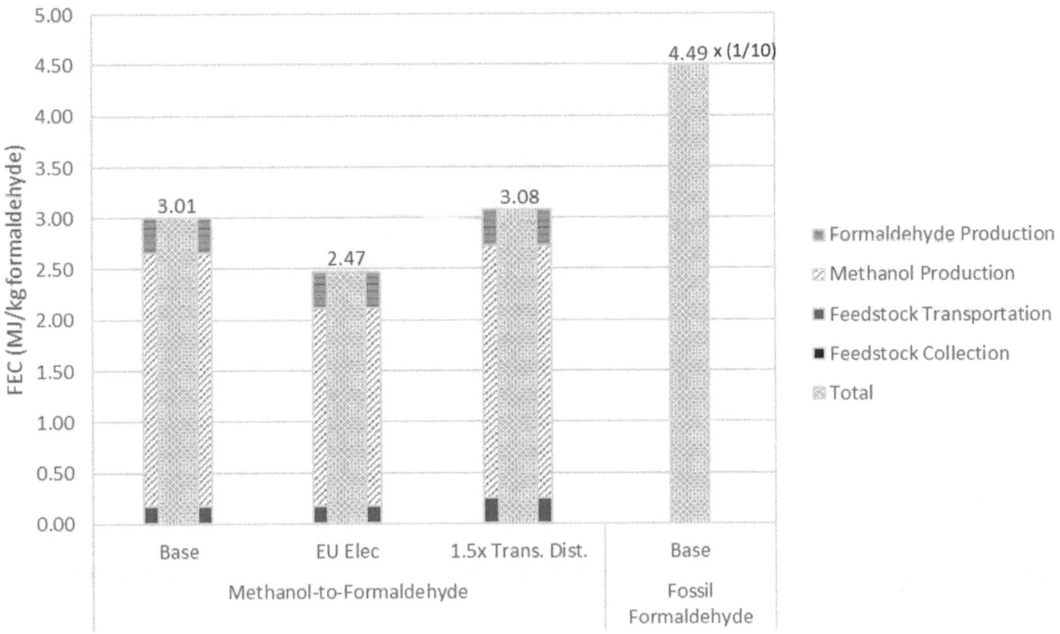


**FIGURE 6.28** Fossil energy consumption for plastic waste-derived formaldehyde (in MJ/kg formaldehyde).

Firm conclusions are difficult to draw due to the large degree of uncertainty in the data used for these analyses. However, it is clear that in the absence of financial incentives such as a carbon tax, economic and sustainability metrics may not be the overarching driving force for using thermochemical pathways to produce polymer precursors. Other factors, such as waste minimization or waste mitigation, may prove to be greater incentives in the short term until carbon credits for recycling are generally available. In the absence of these credits and incentives, it is difficult to see a rationale for the widespread deployment of these and other similar technology pathways. Nevertheless, the results of this study can serve as the baseline for comparing the current process to other plastic waste valorization processes.

## DISCLAIMER

The work performed in part by the employees of the National Renewable Energy Laboratory, operated by Alliance for Sustainable Energy, LLC, was supported by the U.S. Department of Energy (DOE) under Contract No. DE-AC36-08GO28308. The views expressed in the article do not necessarily represent the views of the DOE or the U.S. Government. The U.S. Government retains and the publisher, by accepting the article for publication, acknowledges that the U.S. Government retains a nonexclusive, paid-up, irrevocable, worldwide license to publish or reproduce the published form of this work, or allow others to do so, for the U.S. Government purposes.

## REFERENCES

[1] Tullo, A. H. Plastic Has a Problem; Is Chemical Recycling the Solution? *Chemical & Engineering News*. October 6, 2019.

[2] National Geographic. *Great Pacific Garbage Patch | National Geographic Society*. https://education.nationalgeographic.org/resource/great-pacific-garbage-patch (accessed 2022-10-03).

[3] Lau, W. W. Y.; Shiran, Y.; Bailey, R. M.; Cook, E.; Stuchtey, M. R.; Koskella, J.; Velis, C. A.; Godfrey, L.; Boucher, J.; Murphy, M. B.; Thompson, R. C.; Jankowska, E.; Castillo Castillo, A.; Pilditch, T. D.; Dixon, B.; Koerselman, L.; Kosior, E.; Favoino, E.; Gutberlet, J.; Baulch, S.; Atreya,

M. E.; Fischer, D.; He, K. K.; Petit, M. M.; Sumaila, U. R.; Neil, E.; Bernhofen, M. V.; Lawrence, K.; Palardy, J. E. Evaluating Scenarios toward Zero Plastic Pollution. *Science* 2020, *369* (6510), 1455–1461. 10.1126/science.aba9475.

[4] Tan, E. C. D.; Lamers, P. Circular Bioeconomy Concepts—A Perspective. *Frontiers in Sustainability* 2021, *2*.

[5] Garcia, J. M.; Robertson, M. L. The Future of Plastics Recycling. *Science* 2017, *358* (6365), 870–872. 10.1126/science.aaq0324.

[6] Milbrandt, A.; Coney, K.; Badgett, A.; Beckham, G. T. Quantification and Evaluation of Plastic Waste in the United States. *Resources, Conservation and Recycling* 2022, *183*, 106363. 10.1016/j.resconrec.2022.106363.

[7] Ask-EU. *Waste to Resources 2017*. https://www.ask-eu.de/default.asp?Menue=149&Jahrgang=1142&Ausgabe=2855&ArtikelPPV=29224&AnbieterID=31 (accessed 2022-10-05).

[8] Lichtarowicz, M. *Cracking and Related Refinery*. https://www.essentialchemicalindustry.org/processes/cracking-isomerisation-and-reforming.html (accessed 2022-10-05).

[9] Arpe, H.-J. *Industrial Organic Chemistry, 5Th Edition*, 5th ed.; John Wiley: Weinheim, 2010.

[10] Anuar Sharuddin, S. D.; Abnisa, F.; Wan Daud, W. M. A.; Aroua, M. K. A Review on Pyrolysis of Plastic Wastes. *Energy Convers. Manage.* 2016, *115*, 308–326. 10.1016/j.enconman.2016.02.037.

[11] *Feedstock Recycling and Pyrolysis of Waste Plastics: Converting Waste Plastics into Diesel and Other Fuels*, 1st edition; Scheirs, J., Kaminsky, W., Eds.; Wiley: Chichester, 2006.

[12] Sharratt, P. N.; Lin, Y.-H.; Garforth, A. A.; Dwyer, J. Investigation of the Catalytic Pyrolysis of High-Density Polyethylene over a HZSM-5 Catalyst in a Laboratory Fluidized-Bed Reactor. *Ind. Eng. Chem. Res.* 1997, *36* (12), 5118–5124. 10.1021/ie970348b.

[13] Stabel, U.; Woerz, H.; Kotkamp, R.; Fried, A. Recycling of Plastics in a Steam Cracker. US5731483A, March 24, 1998.

[14] Williams, E. A.; Williams, P. T. The Pyrolysis of Individual Plastics and a Plastic Mixture in a Fixed Bed Reactor. *J. Chem. Technol. Biotechnol.* 1997, *70* (1), 9–20. 10.1002/(SICI)1097-4660(199709)70:1<9::AID-JCTB700>3.0.CO;2-E.

[15] *A Landscape Mapping of the Molecular Plastics Recycling Market*. Closed Loop Partners. https://www.closedlooppartners.com/research/advancing-circular-systems-for-plastics/ (accessed 2022-10-07).

[16] *Plastics*. American Chemistry Council. https://www.americanchemistry.com/chemistry-in-america/chemistry-in-everyday-products/plastics (accessed 2022-10-07).

[17] Tan, E. C. D.; Talmadge, M.; Dutta, A.; Hensley, J.; Schaidle, J.; Biddy, M.; Humbird, D.; Snowden-Swan, L. J.; Ross, J.; Sexton, D.; Yap, R.; Lukas, J. *Process Design and Economics for the Conversion of Lignocellulosic Biomass to Hydrocarbons via Indirect Liquefaction: Thermochemical Research Pathway to High-Octane Gasoline Blendstock Through Methanol/Dimethyl Ether Intermediates*; NREL/TP-5100-62402; National Renewable Energy Laboratory: Golden, CO, 2015.

[18] Tan, E. C.; Talmadge, M.; Dutta, A.; Hensley, J.; Snowden-Swan, L. J.; Humbird, D.; Schaidle, J.; Biddy, M. Conceptual Process Design and Economics for the Production of High-Octane Gasoline Blendstock via Indirect Liquefaction of Biomass through Methanol/Dimethyl Ether Intermediates. *Biofuels, Bioproducts Biorefining 10* (1), 17–35. 10.1002/bbb.1611.

[19] DuBose, B. *UOP Sees Methanol-to-olefins as Solution to Rising Global Propylene Gap*. https://www.hydrocarbonprocessing.com/conference-news/2015/02/uop-sees-methanol-to-olefins-as-solution-to-rising-global-propylene-gap (accessed 2022-10-07).

[20] *Lurgi MTPTM - Methanol-to-Propylene*. Air Liquide. https://www.engineering-airliquide.com/lurgi-mtp-methanol-propylene (accessed 2022-10-07).

[21] Devi, L.; Ptasinski, K. J.; Janssen, F. J. J. G. A Review of the Primary Measures for Tar Elimination in Biomass Gasification Processes. *Biomass and Bioenergy* 2003, *24* (2), 125–140. 10.1016/S0961-9534(02)00102-2.

[22] *Ullmann's Encyclopedia of Industrial Chemistry*, 5th ed.; Elvers, B., Hawklins, S., Schulz, G., Eds.; Wiley-VCH Verlag GmbH & Co.: Hoboken, NJ, 1991.

[23] PRé Consultants. *SimaPro*; PRé Sustainability: Amersfoort, the Netherlands, 2019.

[24] Wernet, G.; Bauer, C.; Steubing, B.; Reinhard, J.; Moreno-Ruiz, E.; Weidema, B. The Ecoinvent Database Version 3 (Part I): Overview and Methodology. *Int J Life Cycle Assess* 2016, *21* (9), 1218–1230. 10.1007/s11367-016-1087-8.

[25] *PEP Yearbook*. IHS Markit. https://ihsmarkit.com/products/chemical-process-economics-program-data-lake.html (accessed 2022-10-08).

[26] *Pricing | Methanex Corporation*. https://www.methanex.com/our-business/pricing (accessed 2022-10-08).

# 7 Municipal Water Reuse for Non-potable and Potable Purposes

*Singfoong "Cindy" Cheah*
Tourmaline Forest LLC, Lakewood, Colorado, USA

**CONTENTS**



## 7.1 INTRODUCTION

In the water cycle, also known as the hydrologic cycle or the hydrological cycle, water moves continuously above and below the surface of the earth. Water evaporates from the ocean or other water bodies, condenses to form clouds, precipitates as rain, snow, or hail, and then infiltrate soils, flows in groundwater aquifer, flows downstream in rivers, or temporarily resides in lakes or other water bodies.

With human beings and other natural plants and animals that consume then release waste, the water is continuously used by one group of users and then reused by downstream users. Water reuse has, in fact, occurred for many generations. With the lower density of the population in

DOI: 10.1201/9781003167693-7

centuries past, the dilution process by the river and the natural solid removal through settling, adsorption, and absorption rendered the addition of waste imperceptible to human vision or taste.

The world's population has increased 40-fold from 2,000 years ago, while the amount of freshwater available has not. In fact, due to pollution, the total available freshwater has probably decreased. Meanwhile, the available freshwater per capita has definitely decreased because of the population increase.

In many regions that lack proper sanitation or have insufficient wastewater treatment, multiple sources of untreated wastewater (unplanned wastewater discharge) are being released into drains and waterways. Such water eventually makes its way into the surface or groundwater system, and the large cumulative volume of untreated water can no longer be effectively diluted or absorbed by natural means. When the downstream population uses the water, they could potentially be exposed to the high pathogen and contaminant levels, resulting in higher rates of illness and loss of economic productivity. Such de facto reuse is termed *unplanned water reuse* and has been documented in several regions [1,2].

Wastewater treatment has to be in place for an effective, economical integrated water management system to protect the health and safety of downstream users, preserve the ecosystem's health, and ensure clean water availability for future generations.

In addition to the need to protect public health and the environment, the degree of water shortage may be more severe in the future due to population increase and economic development. Furthermore, certain regions receive less precipitation, or the precipitation that does arrive does not follow historical patterns because of climate change. The altered precipitation pattern can cause annual or intra-annual water shortages. Therefore, preserving and efficiently using freshwater sources to ensure a secure future is beneficial. As the technologies to reuse water become more mature and the treatment cost decreases in general, water reuse can play an important role in alleviating water scarcity in the future [3].

In the prevalent water management practices worldwide, water reuse is defined as the use of *treated* wastewater for beneficial purposes, which increases a community's available water supply and makes it more reliable, especially in times of drought. The term "planned" is usually left out of the phrase, with "water reuse" implying the planned use of treated wastewater.

This paper describes municipal water reuse for non-potable and potable purposes in Sections 7.2 and 7.3, respectively. Section 7.4 presents a framework for estimating the amount of water that can be reused for these two end purposes. Finally, Sections 7.5 and 7.6 present summary and other considerations of water reuse in comparison to seawater and brackish water desalination, two other common water treatment methods in arid regions.

## 7.2 MUNICIPAL WATER REUSE FOR NON-POTABLE PURPOSES

Water reuse for non-potable purposes is widely practiced in many regions that experience water scarcity, including the western United States, the European Union, and MENA (Middle East and North Africa).

To reuse the water for non-potable purposes, a system for wastewater collection is a prerequisite. After collection, the water is partially treated, i.e., not to human potable standard, but to a standard deemed adequate for the non-potable purpose chosen [4,5]. These chosen purposes include but are not limited to landscape irrigation, agriculture, wetland recharge, and industrial uses.

### 7.2.1 TECHNOLOGIES

The basic technologies used for treating municipal water for non-potable purposes can be the same as those used in conventional water treatment and sanitation practices. In the primary treatment step, a sedimentation process is used to remove solids and sludge. Secondary treatment is then employed to degrade the biological content substantially. This can be achieved using an aerobic biological process, trickling filter, activated sludge, etc. The treatment steps that follow will depend on the water quality standards to which the non-potable use is subject to.

### 7.2.2 Economy

The total cost for municipal water reuse for non-potable purposes is the sum of the cost of water treatment, the cost of infrastructure (piping and pumps), and the cost of the electricity for pumping. Because the water is not treated to potable standards, a separate piping and storage system for the treated water is necessary for the typical centralized water and wastewater treatment systems that are common in developed countries. In such cases, the cost for the separate piping and storage infrastructure must be accounted for when estimating the treatment cost.

A study on several European cities estimated that using treated wastewater for landscaping cost approximately between \$0.14 and 0.62/m$^3$ (in U.S. dollars (USD) estimated using an exchange rate of 1 Euro to 1.1 USD). Another report by the European Union estimates that treatment cost ranges from \$0.09 to \$0.23/m$^3$ [6]. An in-depth study conducted in the United States estimated that municipal water reuse for non-potable purposes costs approximately \$0.46/m$^3$ [7].

One way to lower the treatment cost is to take advantage of the economy of water treatment, which applies to even some of the most basic steps, such as coagulation, commonly used to assist the sedimentation process in water treatment. Figure 7.1 below shows the economy of scale of the coagulation process.

With the economy of scale issue, one might conclude that a small number of large treatment plants would provide a favorable cost. However, when the agricultural irrigation is outside of the

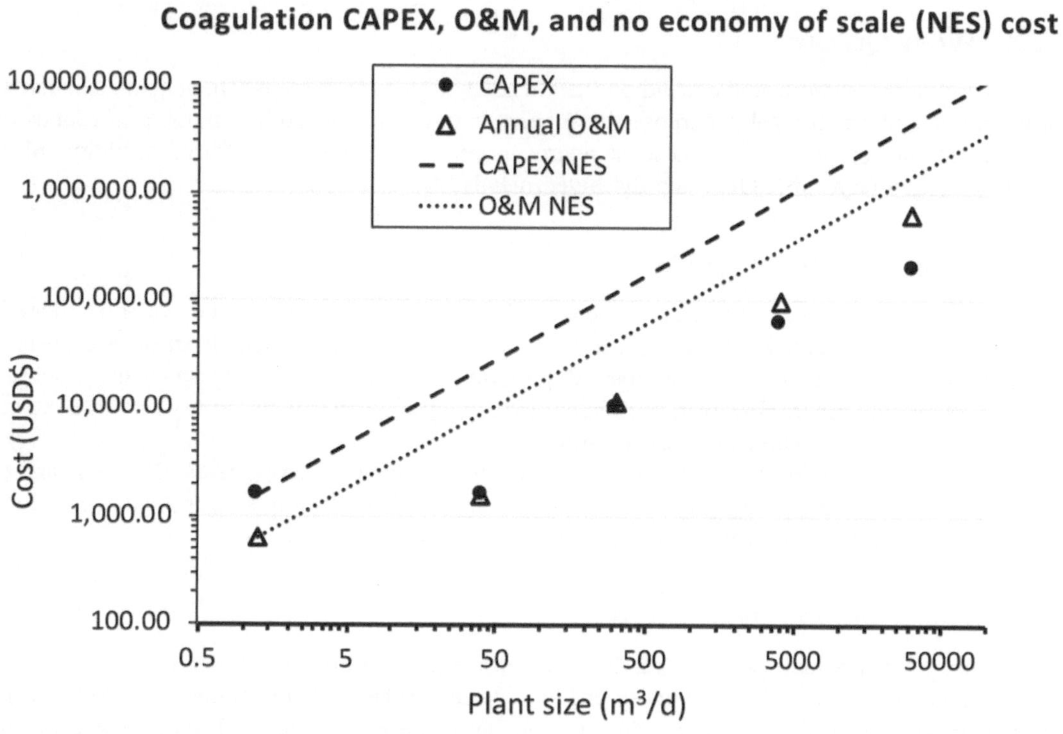


**FIGURE 7.1** Capital cost (solid black circles) and annual O&M (empty black triangles) for the coagulation process compared to capital cost and O&M without economy of scale (black dashed and dotted line, respectively). The data for the coagulation process is used in this plot, but the general economy of scale principle applies to several other water treatment methods.

municipal area, besides the cost of separate piping for non-potable water described above, there is an additional cost for conveyance and energy to pump the treated water to agricultural land.

The energy for conveyance could be 0.5 to 1.0 kWh/m$^3$ or even higher [6]. The combined treatment and electricity cost are \$0.33/m$^3$, assuming an electricity price of \$0.1 per kWh. The infrastructure cost is highly dependent on terrain and distance from the municipal area to the agricultural area, which is typically between \$0.1 and \$0.5/m$^3$, assuming it is not a highly hilly terrain.

Overall the cost for the reuse of municipal wastewater (including conveyance) for agricultural purposes in European countries can range from \$0.24/m$^3$ (taking the low of the treatment, electricity, and infrastructure) to a \$0.83/m$^3$ in cases where long conveyance and high cost are involved [8]. Because the conveyance cost depends on the terrain and other relevant local conditions and regulations, the generalization of the overall cost of municipal water reuse for non-potable purpose can be complicated. A publication by the International Water Association put the high estimate at \$1.6/m$^3$. In other words, in non-optimal conditions, delivering 1 m$^3$ of reused water for non-potable purposes could cost as much as delivering 1 m$^3$ of reused water for potable purposes described in the next section (Section 7.3) and summarized in Table 7.1.

In summary, it will be necessary to properly balance treatment and conveyance costs. In regions where conveyance cost is high, there could be advantages to forego economy of scale and focus on smaller treatment plants for local or regional water reuse to lower the requirements for separate conveyance (pipes, small canals).

For municipal areas where there are large areas of green space, e.g., parks, farms within the city, reuse of treated wastewater for landscaping, agriculture, etc., can be an excellent option with the relatively low treatment cost and minimal conveyance required.

### 7.2.3 Water Quality

The water quality needed would depend on the intended purpose of the reuse. In general, standards for water treated for non-potable purposes have been set or recommended by a number of countries and international organizations. These standards primarily pertain to coliform, turbidity, biochemical oxygen demand (BOD), and dissolved metals [4].

### 7.2.4 Environmental

Water for non-potable purposes can be beneficial for the environment if the water is treated properly to an adequate quality to preserve a healthy, sustainable environment. In most cases, using treated wastewater for non-potable purposes is preferable to using untreated wastewater, which is de facto, but unsafe reuse. However, prolonged use of recycled water with too high a salt content would contribute to the salinization of the soil.

Reusing wastewater for agriculture has been reported to increase crop yields [18]. An earlier study showed that with properly treated wastewater the benefit-cost ratio is on the order of 1.2 to 2.2, depending on the type of crops and treatments [19].

### 7.2.5 Legal and Institutional Constraints

Water reuse for non-potable purposes can involve considerable coordination. Often the standards for drinking water, agricultural drainage water, and wastewater discharge (both volume and the chemical and biological content be within a certain limit) are set at the national level. Therefore, successful water reuse involves the coordination of the private sectors and the departments of agriculture, health, water, natural resources, environment, infrastructure, and finance. Each of the government ministries can have jurisdiction over parts of the project; therefore, a national plan and an integrated water resource management approach will be critical for its success. For large projects that require

**TABLE 7.1**
**Performance Metrics of the Different Water Reuse and Desalination Method**

| Water reuse & desalination method | Water production cost ($/m$^3$) | Capital cost ($/m$^3$/day) | Operation and maintenance ($/m$^3$) | Energy usage (kWh/m$^3$) | Legal/institutional constraint | Beneficial for high population and GDP | References |
|---|---|---|---|---|---|---|---|
| Municipal non-potable | 0.3–1.6 | 9–400 (does not include conveyance) Data in Egypt 100–150 | 0.05–0.6 (European data) Most 0.3–0.6 | 0.3–1.2 | Need extensive coordination | Yes | [6,9–11] Low end is for favorable concentrate disposal; high end is with long conveyance |
| Municipal potable | 0.7–1.6 | 700–2,000 | 0.4–1.3 | 0.8–3.0 | Need extensive coordination. Strategic campaign needed for acceptance. | Yes | [7,9,12] Include Africa data; low end is for favorable concentrate disposal |
| Brackish water desalination | 0.3–1.1 | 300–1,200 | 0.2–0.9 | 0.8–1.6 | Need oversight so that do not over withdraw groundwater. Need transboundary coordination if aquifer goes over boundaries. | | [13,14] High end estimated with zero waste method |
| Seawater RO desalination | 0.5–2.3 | 800–2,300 | 0.3–1.0 | 2.4–4.5 (Includes pretreatment & intake) | | | [15–17] |

long-term planning of the country's monetary budget and potentially foreign aid and/or loan, agreement and cooperation between the Ministry of Finance and Foreign Affairs would ensure the country's long-term interests are being prioritized and articulated in a united fashion.

## 7.3 MUNICIPAL WATER REUSE FOR POTABLE PURPOSES

Many scientists and policymakers believe municipal water reuse for potable purposes is necessary in many regions of the world in the future. Singapore and Namibia are some of the most successful countries in municipal water reuse. In Singapore, currently, water reuse supplies 35% of the water need and is projected to supply up to 55% of the water need in the future [20]. Direct potable reuse has been practiced in Namibia for 50 years [12].

### 7.3.1 Technologies and Trends

To produce water suitable for potable purposes, after the primary and secondary treatment processes similar to those described in Section 7.2 of this chapter, the water needs to go through further tertiary and even fourth steps. In most cases, the final step is disinfection. Afterward, the high-quality multiple-treated water can be stored in a reservoir piped into the potable water system and blended with incoming treated "freshwater" for potable reuse [21,22]. Several studies have shown municipal water reuse for potable purposes to be economically efficient since it eliminates the need for much additional piping, as discussed in Section 7.2.

### 7.3.2 Economy

Estimated levelized cost of producing water for potable purposes from wastewater cost approximately \$0.7 to \$1.6/m$^3$ [9]. As described in Section 7.3.4 (Environmental), a treatment method that uses membrane will generate a concentrated waste stream which incurs disposal cost. The low-end levelized cost above is from the water reuse facilities in Namibia, which use conventional methods that do not require extensive concentrated waste stream disposal, and Singapore, which uses membrane techniques but has favorable concentrated waste stream disposal [12].

With the general improvement in membrane technologies such as reverse osmosis, which is used in desalination and can also be used in water reuse, the cost of water reuse for potable purposes is expected to drop further with time.

### 7.3.3 Water Quality

The treated wastewater can have as good a quality as any freshwater [22,23], provided proper treatment has been conducted and the conveyor system is of sufficient quality, e.g., not containing multiple broken junctures where contaminants can seep in.

### 7.3.4 Environmental

The reuse of municipal water for potable purposes has a positive environmental effect overall by minimizing the release of large volumes of untreated wastewater to nature. However, if a water recycling method that uses membrane technology is chosen, it will typically produce a concentrated waste stream that has a volume of 10 to 15% of the original volume of the water to be treated. To dispose of the concentrated waste stream, there are typically two options, which can also be combined. Option one is to discharge the concentrated waste stream into the ocean. Option two is to construct evaporation ponds to produce salt.

Typically, option one is the lowest cost. Though even in coastal cities, proper consideration and minimization of concentrated waste stream are necessary. The disposal of the concentrated waste stream in an inland area is even more challenging and expensive. For inland cities such as Windhoek, Namibia, conventional water treatment would be more environmentally feasible.

Recently, several companies developed different versions of zero or minimal discharge technologies. For example, the Bureau of Reclamation funded a pilot study on Zero Discharge Desalination, which is estimated to produce fresh water at the cost of \$0.6 to \$0.9/$m^3$ [13]. However, large-scale demonstration of this technology has not been undertaken by private companies.

### 7.3.5 Legal and Institutional Constraints

Municipal water reuse for potable purposes has obstacles, the primary two being public perception and coordination among different agencies. Even though highly treated wastewater is, in fact, biologically and chemically purer than what society generally associates with "original" or "pure fresh" water (e.g., water coming from a stream) [23,24], much of the population worldwide still generally considers treated wastewater to be inferior [25,26].

The second impediment to municipal water reuse, in many aspects, bears a resemblance to the institutional setup required for municipal water reuse for non-potable purposes. In this case, successful water reuse involves the coordination of the departments of health, water, natural resources, environment, infrastructure, and finance.

Such coordination is not limited to the national, large project level. At the local scale, there needs to be agreement among water utilities, water wholesalers and retailers (in a more decentralized system), wastewater management companies, public work officials, and the communities involved. If the waste collection and water treatment are owned and operated by different companies, then the two entities would need to reach a commercial agreement on pricing and funding the conveyance network. The network includes infrastructure for the conveyance of wastewater to the treatment site, further conveyance of partially treated water to tertiary and higher-level treatments, and conveyance to intermediate "storage" sites such as groundwater recharge or reservoir.

In general, the more the number of agencies and departments are involved, the more complex the negotiations could be. There could also be the potential for disagreement over financing and responsibilities, i.e., which agency controls the management of funds for infrastructure build and which entities are responsible for the final delivery of services and the quality of the services rendered. These issues need to be clarified clearly and at the early stage of the planning process.

In the implementation of water reuse in places such as Singapore, where the population density is high and decisions tend to be centralized within the government, the incentives to cooperate can chiefly come from government studies. Along the same vein, the decisions on how to implement the overall complex infrastructure could strictly be based on technology and economic data. In places such as California, USA, the pressure for the multitude of local organizations and water providers to cooperate came chiefly from water scarcity (e.g., drought that extended from 2011 to 2016), the need to improve the reliability of water supply in addition to existing sources (e.g., from the Colorado River and snow melt in the Sierra Nevada), and environmental impact of wastewater discharge [23].

### 7.3.6 Acceptance

This section lists some of the key lessons from countries that have successfully overcome the potential negative perception of the quality of the water treated for reuse.

A strategy of effective education and public awareness needs to be initiated and carried out broadly, ideally at least several years before the implementation of the water reuse project. In a survey conducted in California, USA, residents educated about the advanced wastewater treatment process are much more likely to favorably view water reuse for potable purposes [25].

Such educational and public awareness campaigns can include advertisements, social network campaigns, community outreach and engagement, and tours of facilities to experience water reuse firsthand [27]. In the case of water reuse in Singapore, the prime minister himself kicked off the campaign and created a photo op by drinking the recycled water, called NEWater by the Singapore government, in a national parade in 2002 [28].

The term "NEWater" also illustrates the tactic certain water utilities use to emphasize that it is water that has been purified and to simultaneously de-emphasize the treated water connection with wastewater.

Before the public awareness and educational campaign, it is relevant and valuable to conduct surveys to understand the local conception of water reuse. A report published by the WateReuse Foundation contains the results of an earlier study and provides a list of example questions that can be adapted for future site-specific surveys [29].

In a separate study by the University of Queensland, the researchers found that communication campaigns emphasizing the low risk of drinking recycled water can be the key to increasing public support [30]. However, the study was relatively small. More research on successfully integrating compelling messages and implementing new technology in different regions of the world would be helpful.

To have the substance that support or "back up" any successful public education and awareness campaign, a prerequisite is the successful implementation of the water reuse technology and a proven positive track record of the local utilities or authorities. In the case of Windhoek, Namibia, the wastewater to potable water treatment facility has been successfully operated for 50 years [31]. The successful operation instilled in the citizens of that city confidence in their local water utilities.

### 7.3.7 Key Lessons Learned

Municipal water reuse can be particularly successful and even be the best choice for a densely populated areas with limited water availability. There are several reasons that would be the case. From the perspective of wastewater being a resource, a densely populated area tends to produce a large volume of resource, the wastewater. Recycling the water within the municipal area would minimize the cost of a separate set of piping for non-potable water and building a conveyor system to transport the water to agricultural areas that may be far from the municipality. Lastly, because of the economy of scale of wastewater treatment and recycling technologies, treating a large volume of wastewater within the municipality is often cost competitive.

A second option for water reuse to potable water is to have the recycled water stored in an intermediate facility such as a surface reservoir or groundwater [23]. This option has several advantages. One is that though there is additional piping to convey the water to a reservoir or a groundwater percolation site, overall, there is no necessity for a separate set of piping to convey intermediately treated water to many separate locations.

The other advantage of this configuration is that by having a step that separates the treated wastewater from the drinking water, the public tends to receive the new setup more readily. Lastly, the storage site can also store for drought or low rainfall seasons and help balance out the production and consumption variability in a given year.

## 7.4 AMOUNT OF WATER THAT CAN BE PROVIDED THROUGH MUNICIPAL WATER REUSE—EFFECT OF NON-REVENUE WATER (NRW) AND OTHER CONSUMPTIVE, TREATMENT, CONVEYANCE, AND STORAGE LOSSES

### 7.4.1 Mathematical Framework for Municipal Water Reuse for Non-potable Purposes

To estimate the amount of recycled or reclaimed non-potable water that can be delivered to end-users, such as agricultural land, we need to follow the flow of the water and account for its losses in

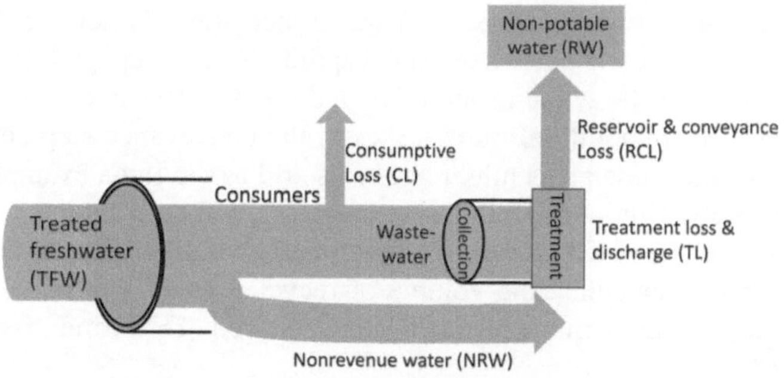


**FIGURE 7.2** Flow of treated freshwater to consumers, followed by collection and treatment of wastewater, which in turn is followed by storage and conveyance of reclaimed non-potable water. Losses in the different steps are labeled.

the intermediate steps (Figure 7.2). It can be illustrative to express the water delivered as a percent of the initial clean water produced by the water utilities.

However, it is important to note that this exercise is a simplified version of water balance because it only accounts for the losses at a single point in time. In reality, accurate water accounting needs to account for integrated flow through time, e.g., different seasons and through space, which can include seepage and recharge to the same or even different water basins.

Consider the case where we start with a certain volume of treated freshwater termed TFW (Figure 7.2). The water that is lost even before the clean water reaches the customers is termed non-revenue water (NRW) (Figure 7.2). Non-revenue water includes actual physical losses, e.g., leaks in disconnected pipes, and unaccounted-for water through theft or metering problems. Non-revenue water can be expressed in units of volume of water lost per km of distribution network per day or as a fraction of the water produced. In this chapter, we will use the second definition.

$$\text{Water delivered to consumers} = \text{TFW} * (1 - \text{NRW}) \tag{7.1}$$

Out of the volume of water delivered to the consumers, the percentage that is consumed and thus not collected by the wastewater collection system is termed "consumptive losses (CL)" (Figure 7.2). Consumptive loss, as an example, can be water used in cooking. After the food is consumed, the water is partially lost through perspiration and respiration of the consumers. Another example of consumptive loss could be water spilled on the soil during gardening and car washing, which then slowly percolates to groundwater or is lost through evaporation. Though the water seeping into the ground can contribute to long-term freshwater supply, it is not collected by the wastewater collection system for treatment. In most well-developed municipal water systems, the % that is returned to the wastewater collection system is in the range of 80 to 90%. For example, an EU study assumed a consumptive loss of 10% [6]. In this modeling exercise, a consumptive loss of 12% is used.

$$\text{Volume collected by wastewater collection system} = \text{TFW} * (1 - \text{NRW}) * (1 - \text{CL}) \tag{7.2}$$

After the wastewater is collected and during wastewater treatment, a certain percentage of water is lost (TL, shown in Figure 7.2). A study of traditional water treatment in the Damietta governorate in Egypt showed that the percent of water lost in treatment plants ranges from 6 to 24% [37]. Other treatment methods, e.g., reverse osmosis (RO), typically have 15 to 25% of the original volume rejected as concentrated waste stream. It is possible to use a combination of treatment technologies with lower losses, but the cost per volume of water produced might be higher. In this modeling exercise, an average of 25% loss in the water treatment plant is assumed.

Following treatment of the water for reuse, if the treated water is not used immediately, then often the water is stored in a reservoir, where the evaporative and seepage loss can be substantial. For example, studies in India show that evaporative and seepage loss in reservoirs alone can be as high as 33% per year [38]. In addition, more losses in the conveyance step could occur. For our calculation, we will assume some areas might not be as arid as the India example, and some water might be used immediately without intermediate storage. Therefore, a value of 33% for the sum of reservoir and conveyance loss (RCL, shown in Figure 7.2) would be reasonable.

A simplified equation to calculate the volume of recycled water that is delivered to its destination (RW) as a function of the volume of freshwater produced (TFW) and losses along the way is shown in Equation (7.3).

$$RW = TFW * (1 - NRW) * (1 - CL) * (1 - TL) * (1 - RCL) \qquad (7.3)$$

Using NRW = 5%, consumptive loss = 12%, treatment plant loss = 25%, reservoir and conveyance loss = 33%, the net recycled water that can be delivered to an irrigation scheme is 42% of the clean water that was initially produced by the utilities.

Once the recycled water arrives at its intended destination, it enters into a separate water cycle where the crop consumes most of the water, which is then lost through evapotranspiration. In other words, the recycled water for non-potable purposes can only be reused once. The actual fraction that is returned to the groundwater and surface water can only be accounted for in a hydrology model.

It should be noted that out of the four losses described, NRW is one of the parameters that water utilities can have a large potential to minimize. NRW worldwide ranges from a low of 5% to a high of more than 50% [32,33]. Reducing NRW is critical as it affects the actual volume that can be delivered to customers. Consequently, NRW reduction will also increase the revenue stream of the utilities and the volume of wastewater that can be collected. In the 2017 Singapore International Water Week, utilities were urged to have a "laser focus" on water loss [34]. Reducing NRW is of such importance that NRW should be given thorough attention *before* investments in water reuse is committed.

### 7.4.2 Mathematical Framework for Municipal Water for Potable Purposes

In water reuse for potable purposes, there is no evapotranspiration loss in the end-use step as in the case of reuse for non-potable purposes. Therefore, the water that is not lost as NRW, or in the consumption, treatment, storage, and conveyance steps, is theoretically cycled *indefinitely* within the municipal water cycle. The infinite cycle can be modeled as a new mathematical series expansion of Equation (7.3).

With the mathematical expansion and using the same estimated parameters as in Section 7.4.1 (NRW = 5%, CL = 12%. TL = 25%, RCL = 33%) we can derive the scope of municipal water reuse for potable purposes to be greater than 60% in an efficient system. In other words, approximately 60% of the demand can be met through water reuse for potable purposes. Note that this strictly engineering model is a simplistic accounting of water and is not a full hydrological water balance model.

Nevertheless, the estimate of a system that can deliver approximately 60% of its water through reuse is in line with the achievable target of highly successful water reuse programs. One such example is the stated goal of satisfying 55% of its water needs in Singapore by 2060 [35].

In summary, though the water can be reused multiple times, municipal water reuse for potable purposes will still require additional freshwater input from a river, groundwater, or desalinated seawater. To further minimize freshwater withdrawal in a circular economy, concurrent water demand management would likely be needed. Utilities and governments worldwide could optimize their decision making through technology research and case-specific detailed modeling.

## 7.5 SUMMARY OF RELATIVE COST AND ELECTRICITY USAGE

In many water-scarce areas, the citizens and governments may need to decide whether to pursue water reuse, desalination, or both. Therefore, a summary table (Table 7.1) is provided to compare the water production cost (expressed in dollars/$m^3$), capital cost, operation and maintenance (O&M) cost, energy usage (expressed in kWh/$m^3$), qualitative indications of potential institutional complexities and whether the method is suitable for high-density cities.

## 7.6 CONCLUSIONS

Water reuse can be very important to supplement and increase water supply in a region. In dense municipality, water reuse can potentially provide more than half of the water demand, freeing up water for other purposes such as agriculture in the country. In addition, water reuse can, in some cases, help a region deal with periodic drought if the system is designed in an integrated manner.

Water reuse, however, incurs additional financial and electricity costs. There are several decision steps in deciding whether they are a good option for a region.

1. Determine whether demand exceeds supply for at least part of the year.
2. Determine whether the excess demand can be met by efficiency improvement.
3. Determine whether the resource, i.e., wastewater, is being collected. If the wastewater is not being collected, the first step in the development is a proper wastewater collection and treatment system.
4. If the resource is available and the efficiency improvements cannot meet the demand, the decision moves into determining which water reuse or desalination method is the economically viable option.

The selection is highly dependent on local conditions in determining the economically viable option.

1. As shown in Figure 7.1, the economy of scale applies to many processes. Operating a water treatment above a certain size would provide water at a more favorable water production cost per $m^3$.
2. However, piping and conveyance are very high costs of the water infrastructure. Local, small-scale batch systems that provide safe drinking water and sanitation systems may be the favored option in the near to medium term.
3. In the case of water reuse, additional conveyance may be required if the intended destination is far from the source.
4. ***Optimization of treatment and distribution/conveyance costs is necessary.*** With data for a particular location, modeling and optimization can be performed.
5. The necessity to optimize the sum of treatment and conveyance costs points to dense municipalities with insufficient freshwater supply as areas likely to benefit from water reuse. A few other pointers in technology and implementation considerations are:
   a. The distance between resource (wastewater) and use is ideally short;
   b. A dense municipal area is also where wastewater, if untreated, can be seen and smelled easily, thus exerting pressure for actions to be taken. However, if the sanitation system is still insufficiently developed, wastewater reuse for potable purposes, in particular, is unlikely to gain the trust of the public in the near future;
   c. For development near the coast where the water supply is insufficient, seawater desalination can also supplement water reuse. However, water reuse is generally still more economical than seawater desalination.

For other locations and purposes, e.g., reuse of municipal wastewater for agricultural land outside of the municipal area or reuse of agricultural drainage water that needs substantial treatment, much more information is needed for cost-benefit analysis.

As the calculation framework in Section 7.4 shows, the percentage of the local water demand that can be met through water reuse is intimately tied to the efficiency and loss minimization measures taken in the water's passage. For municipalities, ***actively minimizing NRW*** is critically important to provide a higher percentage of the water through reuse and to decrease system cost, electricity requirement, and resource use.

As Whittington et al. [36] and others have pointed out, as communities experience economic growth, more households and firms will prefer the advantages of large-scale, reliable, piped water and sanitation services. As reliable water, energy, and communication infrastructure tend to increase business confidence and further investment, the benefit of water reuse, together with adequate wastewater infrastructure, would eventually outweigh the cost.

## NOTE

Singfoong "Cindy" Cheah obtained her Ph.D. in Geological and Environmental Sciences from Stanford University. She can be contacted at singfoong@alumni.stanford.edu. She has conducted interdisciplinary research in renewable energy and water resources in public and private sectors. At ITN Energy Systems, she managed projects funded by the U.S. Bureau of Reclamation and other private entities, developing new solar powered water desalination technology. As a senior scientist at the U.S. National Renewable Energy Laboratory, she managed energy science discoveries and analytical method development. More recently, she consulted on energy and water resources issues for the United Nations and other international organizations. She is currently a general engineer at the U.S. Energy Information Administration.

The statements and opinions in this article do not represent the viewpoints of DOE or the Federal government.

## REFERENCES


[1] J. Lautze, E. Stander, P. Drechsel, A. K. da Silva, and B. Keraita, "Global Experiences in Water Reuse," *CGIAR Research Program on Water, Land and Ecosystems*, 2014.

[2] A. Bahri, P. Drechsel, and F. Brissaud, "Water Reuse in Africa: challenges and opportunities".

[3] J. Freedman and C. Enssle, "Addressing water scarcity through recycling and reuse: a menu for policymaker s," *White Paper*, 2015. [Online]. Available: https://www.ge.com/sites/default/files/Addressing_Water_Scarcity_Recycle_Reuse_White_Paper.pdf

[4] H. Jeong, H. Kim, and T. Jang, "Irrigation Water Quality Standards for Indirect Wastewater Reuse in Agriculture: A Contribution toward Sustainable Wastewater Reuse in South Korea," *Water*, vol. 8, p. 169, Apr. 2016, doi: 10.3390/w8040169

[5] M. Gabr, "Wastewater reuse standards for agriculture irrigation in Egypt," presented at the Twenty-first International Water Technology Conference, IWTC21, Port Said, Egypt, Jun. 2019, pp. 234–246. Accessed: Apr. 02, 2020. [Online]. Available: https://www.researchgate.net/publication/333676905_WASTEWATER_REUSE_STANDARDS_FOR_AGRICULTURE_IRRIGATION_IN_EGYPT

[6] A. Pistocchi *et al.*, *The Potential of Water Reuse for Agricultural Irrigation in the EU: A Hydro-economic Analysis*. 2017. doi: 10.2760/263713

[7] A. Estevez-Olea, "Life Cycle Assessment of Reclaimed Water for Potable and Nonpotable Reuse in California," Master, University of San Francisco, San Francisco, 2015. Accessed: May 16, 2019. [Online]. Available: https://repository.usfca.edu/capstone/143/

[8] D. Hardy, F. Cubillo, M. Han, and H. Li, "Alternative water resources: a review of concepts, solutions and experiences," *International Water Association*, Feb. 2015. Accessed: Jun. 25, 2019. [Online]. Available: https://iwa-network.org/publications/alternative-water-resources-a-review-of-concepts-solutions-and-experiences/

[9] D. Hardy, F. Cubillo, M. Han, and H. Li, "Alternative water resources: a review of concepts, solutions and experiences," *International Water Association*, Feb. 2015. Accessed: Jun. 25, 2019. [Online]. Available: https://iwa-network.org/publications/alternative-water-resources-a-review-of-concepts-solutions-and-experiences/
[10] GWI, "Egypt's new $739m treatment plant to boost farming," *Global Water Intelligence*, May 02, 2019. Accessed: May 16, 2019. [Online]. Available: https://www.globalwaterintel.com/news/2019/18/egypt-s-new-739m-treatment-plant-to-boost-farming
[11] G. De Paoli and V. Mattheiss, "Cost, pricing and financing of water reuse against natural water resources," *Innovation Demonstration for a Competitive and Innovative European Water Reuse Sector (DEMOWARE), DEMOWARE GA No. 619040*, 2016. Accessed: May 16, 2019. [Online]. Available: http://demoware.eu/en/results/deliverables/deliverable-d4-7-cost-pricing-and-financing-of-water-reuse-against-natural-water-resources.pdf/view
[12] T. Guo, J. Englehardt, and T. Wu, "Review of cost versus scale: water and wastewater treatment and reuse processes," *Water Sci. Technol.*, vol. 69, no. 2, pp. 223–234, Nov. 2013, doi: 10.2166/wst.2013.734
[13] M. Cappelle, T. Davis, and E. Gilbert, "Demonstration of Zero Discharge Desalination (ZDD)," *U.S. Bureau of Reclamation*, 165, Jul. 2014.
[14] R. Raucher and G. Tchobanoglous, "The opportunities and economics of direct potable reuse," *WateReuse Foundation.*
[15] ALMAR Water Solutions, "Desalination Technologies and Economics: CAPEX, OPEX & Technological Game Changers to Come," *presented at the Mediterranean Regional Technical Meeting, Marseille CMI*, Dec. 2016.
[16] WateReuse Association, "Seawater Desalination Power Consumption," 2011.
[17] The WateReuse Desalination Committee, "Seawater Desalination Costs," *WaterReuse Association, White Paper*, Jan. 2012. Accessed: Apr. 26, 2019. [Online]. Available: https://watereuse.org/wp-content/uploads/2015/10/WateReuse_Desal_Cost_White_Paper.pdf
[18] C. Aoki, M. Memon, and H. Mabuchi, "Water and wastewater reuse: an environmentally sound approach for sustainable urban water management," *United Nations Environment Programme (UNEP), Global Environment Centre Foundation (GEC)*, 2004. Accessed: May 14, 2019. [Online]. Available: http://hdl.handle.net/20.500.11822/8390
[19] H. Yamagata, M. Ogoshi, Y. Suzuki, M. Ozaki, and T. Asano, "On-site water recycling systems in Japan," *Water Sci. Technol. Water Supply*, vol. 3, pp. 149–154, 2003, doi: 10.2166/ws.2003.0020
[20] PUB, "Singapore Water Story," *Public Utilities Board of Singapore.* Accessed: Jul. 02, 2019. [Online]. Available: https://www.pub.gov.sg/watersupply/singaporewaterstory
[21] H. L. Leverenz, G. Tchobanoglous, and T. Asano, "Direct potable reuse: a future imperative," *J. Water Reuse Desalination*, vol. 1, no. 1, pp. 2–10, Mar. 2011, doi: 10.2166/wrd.2011.000
[22] J. Lahnsteiner, P. van Rensburg, and J. Esterhuizen, "Direct potable reuse – a feasible water management option," *J. Water Reuse Desalination*, vol. 8, no. 1, pp. 14–28, Feb. 2017, doi: 10.2166/wrd.2017.172
[23] B. Sheikh, "Social-economic aspects of wastewater treatment and water reuse," in *Efficient Management of Wastewater*, Amman, Jordan: Springer, Berlin, Heidelberg, 2008, pp. 249–257. Accessed: Mar. 22, 2019. [Online]. Available: https://link.springer.com/chapter/10.1007/978-3-540-74492-4_21
[24] T. Asano, F. Burton, H. Leverenz, R. Tsuchihashi, and G. Tchobanoglous, *Water Reuse: Issues, Technologies, and Applications*, 1st ed. McGraw Hill Professional, 2007.
[25] S. Leung, "Overcoming the 'yuck' factor," *International Water Association*, Oct. 2016. Accessed: May 09, 2019. [Online]. Available: https://iwa-network.org/overcoming-the-yuck-factor/
[26] K. J. Ormerod, "Illuminating elimination: public perception and the production of potable water reuse," *Wiley Interdiscip. Rev. Water*, vol. 3, no. 4, pp. 537–547, 2016, doi: 10.1002/wat2.1149
[27] K. Lim and H. Safford, "Improving public perception of water reuse," *UCDavis Policy Institute for Energy, Environment, and the Economy*, Jan. 2019. [Online]. Available: https://policyinstitute.ucdavis.edu/improving-public-perception-of-water-reuse/
[28] "Treated sewage as water? They'll drink to that," *The Sydney Morning Herald.* Accessed: May 09, 2019. [Online]. Available: https://www.smh.com.au/national/treated-sewage-as-water-theyll-drink-to-that-20080627-gdsjrt.html
[29] B. M. Haddad, P. Rozin, C. Nemeroff, and P. Slovic, "The psychology of water reclamation and reuse," *WateReuse Foundation*, 2009. Accessed: May 10, 2019. [Online]. Available: https://www.waterboards.ca.gov/water_issues/programs/grants_loans/water_recycling/research/psychology_water_reclamation.pdf

[30] J. Price, K. S. Fielding, J. Gardner, Z. Leviston, and M. Green, "Developing effective messages about potable recycled water: The importance of message structure and content," *Water Resour. Res.*, vol. 51, no. 4, pp. 2174–2187, 2015, doi: 10.1002/2014WR016514

[31] C. Leong, "The role of emotions in drinking recycled water," *Water*, vol. 8, no. 11, p. 18, Nov. 2016, doi: 10.3390/w8110548

[32] Wikipedia, "Non-revenue water." Accessed: Oct. 12, 2022. [Online]. Available: https://en.wikipedia.org/wiki/Non-revenue_water

[33] (AIKP) African Infrastructure Knowledge Program, "Non-revenue Water Model (WSS)," *African Development Bank Group*, Database, 2016. Accessed: Sep. 22, 2019. [Online]. Available: http://infrastructureafrica.opendataforafrica.org/NRWM2016/non-revenue-water-model-wss-2016?

[34] T. Freyberg, "Utilities should have a 'laser focus' on reducing water system losses," *WaterWorld*, Singapore, Nov. 14, 2017. Accessed: Oct. 10, 2022. [Online]. Available: https://www.waterworld.com/drinking-water/distribution/article/16203061/utilities-should-have-a-laser-focus-on-reducing-water-system-losses

[35] E. Shin, S. H. Choi, A. K. Makarigakis, O. Sohn, C. Clench, and M. Trudeau, Eds., *Water Reuse within a Circular Economy Context*. France and Republic of Korea: United Nations Educational, Scientific and Cultural Organization (UNESCO) and the International Center for Water Security and Sustainable Management, 2020. Accessed: Oct. 11, 2022. [Online]. Available: https://unesdoc.unesco.org/ark:/48223/pf0000374715.locale=en

[36] D. Whittington, W. M. Hanemann, C. Sadoff, and M. Jeuland, "The Challenge of Improving Water and Sanitation Services in Less Developed Countries," *Found. Trends® Microecon.*, vol. 4, no. 6–7, pp. 469–609, 2009, doi: 10.1561/0700000030

[37] V. Khater, H. Fouad, A. El-Magd, and A. Hassanain, "Effect of Pipe Material and Size on Water Losses at Different Networks in Egypt," *Archives of Current Research International*, vol. 5, pp. 1–12, 2016, doi: 10.9734/acri/2016/29574.

[38] C. Sivapragasam, G. Vasudevan, J. Maran, C. Bose, S. Kaza, and N. Ganesh, "Modeling Evaporation-Seepage Losses for Reservoir Water Balance in Semi-arid Regions," *Water Resources Management*, vol. 23, pp. 853–867, 2008, doi: 10.1007/s11269-008-9303-3.

# 8 An Ethical Reflection on Water Management at the Community Level as a Contribution to Peace

*Tebaldo Vinciguerra*
Libera Università Maria SS. Assunta (LUMSA), Rome, Italy

## CONTENTS



## 8.1 INTRODUCTION

This chapter addresses water management from a peculiar perspective: a background in political science and international studies, a two-year water-related field experience in Peru, more than a decade working on natural resources and environmental issues in the light of the principles of the Catholic Church, the involvement in several water processes at the international level, and a particular interest in water economics as well as water as a commons.

The starting point of the reflection is a story taken from the Bible. Some hydric challenges are really ancient, indeed. The interconnectedness dimensions of sustainability (access to water, governance, peace) are well depicted in that story, which could help readers understand the need for an ethical approach to water. Hopefully, it could also help skilled inventors and experts in technologies or processes understand that improvements and innovations need to be carefully and patiently implemented within and by a given community. They should be implemented without neglecting the community's specificities, without neglecting its religious values and teachings since they can offer strong motivations for solidarity and for caring for water, without forgetting that water is a very special resource!

## 8.2 WATER ACCESS, ENGINEERING, AND CONFLICTS

Water access, engineering, and conflicts are ancient and still very tangible issues. The COVID-19 pandemic has radically and brutally changed the way we see the world. Perhaps never before had the interconnectedness between countries and the various challenges come so clearly to light. The exhortations to wash your hands—a planetary campaign—also highlighted how problematic access to water was for many populations. According to the United Nations (UN), in 2020, 2 billion people were without safely managed drinking water services, "including 1.2 billion people lacking

DOI: 10.1201/9781003167693-8

even a basic level of service. Eight out of 10 people who lack even basic drinking water service live in rural areas (…). At the current rate of progress, the world will reach 81 percent coverage by 2030, missing the target and leaving 1.6 billion people without safely managed drinking water supplies. (…) Universal access to drinking water, sanitation and hygiene is critical to global health. To reach universal coverage by 2030, current rates of progress would need to increase four-fold" [1].

Access to drinking water, a fundamental resource for human life, has not been prioritized enough. Yet, its access and management were a priority for all civilizations on all continents. Magnificent and capillary ancient engineering works provide a clear testimony of this. In the past, civilizations developed on the riverbanks and wells were a fundamental element of a city or village. This reflection on water management for peace takes its cue from the Bible in an episode that mentions wells specifically.

> At that time, Abimelek and Phicol the commander of his forces said to Abraham, "God is with you in everything you do. Now swear to me here before God that you will not deal falsely with me or my children or my descendants. Show to me and the country where you now reside as a foreigner the same kindness, I have shown to you." Abraham said, "I swear it." Then Abraham complained to Abimelek about a well of water that Abimelek's servants had seized. But Abimelek said, "I don't know who has done this. You did not tell me, and I heard about it only today." So Abraham brought sheep and cattle and gave them to Abimelek, and the two men made a treaty. Abraham set apart seven ewe lambs from the flock, and Abimelek asked Abraham, "What is the meaning of these seven ewe lambs you have set apart by themselves?" He replied, "Accept these seven lambs from my hand as a witness that I dug this well." So that place was called Beersheba, because the two men swore an oath there. After the treaty had been made at Beersheba, Abimelek and Phicol the commander of his forces returned to the land of the Philistines. Abraham planted a tamarisk tree in Beersheba, and there he called on the name of the Lord, the Eternal God and Abraham stayed in the land of the Philistines for a long time [2].

Abraham, the migrant (foreigner in the land of the Philistines), and his tribe faced quite an uncomfortable situation. King Abimelek welcomed them into his territory, but now, blatantly, he tolerated them grudgingly. On the one hand, he knew that Abraham is a prophet protected by the Lord; on the other hand, he was subject to a brawl with Abraham's people about wells. The ruler then spoke with Abraham accompanied by none other than Phicol, who was (we would say today) his Chief of Staff of the Army. A clear intimidating move. We can also doubt that the king was telling the truth when he explained, "I heard about it only today": he was probably aware (and that is precisely why he was escorted by Phicol). Abraham, for his part, feels the need to have it officially acknowledged that "I dug this well."

The scene narrates a dynamic of conflict over the control of a water source, specifically a well, i.e., a work of human ingenuity that allows one to draw water for quenching one's thirst, hygiene, and the flock and for crops—the most important work in an arid area. Daring to read between the lines, we can assume that it is also a matter of engineering! Abraham was perhaps better at finding water and building wells than Abimelek's people (otherwise what reason would they have for usurping a well? If Abraham's people were just taking too much water endangering the wells of the natives, the latter would have simply closed the well again). One could sense a certain jealousy, perhaps reinforced by demographic considerations.

Let us set aside, for now, the book of *Genesis* and consider some relatively recent news. The long Nile dam talks between Egypt, Ethiopia, and Sudan, failed again in 2020. These neighboring countries have once again failed to agree on a new negotiating approach to resolve their years-long dispute over the dam that Ethiopia started building in 2011 on the Blue Nile River [3].

In the Himalayan region (an area with rivers, lakes, and snowcaps), there have been decades of boundary disputes between India and China. During a skirmish in June 2020, soldiers from both armies died in the Galwan Valley [4].

After the occupation of Crimea by Russian troops in 2014, Ukraine responded by damming the Northern Crimean Canal (channeling water from Ukraine's Dnipro River) to prevent the Russian-occupied peninsula from receiving fresh water. Crimea's residents began to suffer chronic water shortages and occasional shut-offs at the tap. Russian troops blew open the dam in February 2022, days after Russian forces invaded Ukraine and took control of the area around Kherson [5]. Moreover, the Kakhovka dam, also in the region of Kherson, has been frequently mentioned during the military operation which took place in the Fall of 2022 since its destruction would cause massive floodings [6].

In the first case, another act in a prolonged series of tensions and mediations around dam sites on the Nile. In the other two cases, conflicts over control of water or taking place in areas crucial to the water supply. Water undoubtedly is a relevant factor in geopolitical analyses and strategies [7]. Indian activist Vandana Shiva, in her essay *Water Wars*, describes numerous conflicts related to the use of water, its supply, and its hoarding [8]. Cases of the use of water as a weapon or of armed violence targeting water infrastructure are well known [9] (despite prohibitions in international law). "In recent armed conflicts, state and non-state armed groups have destroyed and captured water installations. Water supply systems fail: supply lines were deliberately sabotaged, or water resources were poisoned to intimidate civilians. Non-state armed groups capture dams and barrages and use them to flood or starve downstream populations to defeat them" [10].

Migration dynamics can also trigger tensions. A sudden influx of migrants in chaotic conditions can create tensions over access to water: tensions with local residents (who have to share water or who see local streams sullied by the exponential increase in open defecation) and/or with authorities (it may not be easy to supply a new refugee camp with water). Various cases could be cited. Conversely, "the depletion of the basic natural resources that the earth provides, and water, in particular, can cause temporary or permanent displacement of families and communities" [11]; floods can destroy crops and assets; prolonged drought can endanger crops, flocks and the entire economy of a region [12].

A landmark study published by the United Nations further explains that "153 countries share rivers, lakes and aquifers. Transboundary basins cover more than half of the Earth's land surface, account for an estimated 60 percent of global freshwater flow and are home to more than 40 percent of the world's population" [13]. Of course, it is not only surface water that is transboundary but also groundwater reserves. Thus, there are many potential causes of conflict where two or more countries find themselves disposing of the same water. These countries have to deal not only with the rate of regeneration of the resource but also with the intentions of the other country. Water is rarely the single—and seldom the principal—trigger of controversy and violence, but it does have the potential to exacerbate existing tensions [14]. There are still cases that this text does not elaborate on, namely, tensions over the control of certain coastal areas and shipping lanes, as well as acts of piracy [15].

## 8.3 AN ESSENTIAL AND ENDANGERED RESOURCE

For various reasons related to our development model, water is taking on all the characteristics of a global commons that is scarce and rival [16]. The quantity and the quality of water that each person can access mark, on a daily basis, two frontiers; the first is between death and bare survival. The second frontier is between basic survival and various levels of well-being. The situation of hundreds of millions of people, poised on the first or second frontier, is a great disappointment and failure to humanity. Sanitation is closely related to water, and in a pandemic situation, these problems become even more dramatic.

Millions of people have unhealthy water available to them. This was the situation at the end of the UN Millennium Development Goals (MDGs) set for the period 2000–2015: 45 countries were not on track to meet the MDG drinking water target. The statistics relied on the so-called 'improved drinking water source' (assuming that the water supplied is drinkable), but many improved facilities were microbiologically contaminated [17]. We moved on to the Sustainable Development Goals (SDGs). In 2018, in SDG6 on clean water and sanitation, the United Nations warned that "only one in five countries below 95 percent coverage is on track to achieve universal basic water services by 2030" [18]! This is a great disgrace to humanity in the 21st century.

Today, in many cases, even where there is no source of drinking water, you can quench your thirst very well (such as an aircraft carrier at sea). The technologies exist, the physics knowledge is there, and people know how to build infrastructure. The problems are the lack of investment where it does not seem profitable to make it, lack of political will, inadequate awareness of this problem by some elites, and consideration of water as any commodity.

Water can tell us a lot about our societies, economics, institutions, and care for the environment. It often offers a reflection of humanity's behavior. Some pollutions are particularly dangerous, such as chemical and sewage wastes, mercury pollution in mineral extraction areas, or other products in fracking operations. Endocrine disruptors are also a topic of concern [19]. Wealthier countries often have plenty of water: they use it even for relatively unnecessary purposes (golfing in arid areas, ornamental pools), sometimes depriving developing neighboring countries of it. They also have sophisticated technologies and facilities for water purification. This abundance of water is linked to considerable energy use and, often, to a lifestyle characterized by heavy consumption of goods of various kinds, a certain culture of waste.

On the other hand, the inhabitants of many countries and areas that are struggling on the road to development are sometimes kept away from supply areas because of ongoing violence or the conflict that has rendered some water equipment unusable. They are often forced to operate or transport water using inefficient and time-consuming systems. They often have little drinkable water because there is little water due to unfavorable meteorology, withdrawals that do not respect the time needed to regenerate the resource, and competing and sometimes even conflicting uses [20] due to large projects (e.g., construction of dams, neighborhoods, mines) that have not duly carried out an adequate context analysis and an adequate impact study, or even due to the loss of certain wetlands ("Wetlands ensure fresh water, help replenish ground aquifers, and purify and filter harmful waste from water – such as fertilizers and pesticides, as well heavy metals and toxins from industry" [21]). Additionally, widespread pollution ends up making unhealthy the formerly potable water sources that rural populations traditionally draw from, and there is a lack of facilities to distribute water or ensure its adequate quality. Sometimes these pollutions come from intensive monocultures or mining projects for hydrocarbons or minerals destined largely for consumption by wealthier countries.

The inhabitants of privileged districts neighboring poor areas often represent a middle ground: they have, yes, relatively abundant water for their own consumption, even ornamental or recreational, but they usually use it with many waste and inefficient systems. In addition, their privileged standard of living is in sharp contrast to that of surrounding neighborhoods or rural areas, where water for sanitation or agriculture is scarce. Added to this is the fact that poor people living in slums or districts not connected to municipal pipelines usually pay significantly higher prices for drinking water than public ones [22] (even five or ten times more expensive while resorting to informal water vendors), they thus spend a high percentage of their income, without even the guarantee of having 'really potable' water (since they have no way to test its quality).

## 8.4 MANAGED WATER AS A COMMON GOOD

A famous article authored by Garrett Hardin (*The Tragedy of the Commons*) describes the destruction of "a pasture open to all" caused by "the inherent logic of the commons" in connection with demographic concerns and with the individualistic and (allegedly) rational thinking of each herdsman ("What is the utility to me of adding one more animal to my herd?" while all herdsmen share the effects of overgrazing). That is an open-access tragedy, occurring in a situation without rules to protect the common good or authorities capable of enforcing them, or a situation in which a particular person (legal or physical) is able to act undisturbed despite existing rules. Hardin, precisely, considers that "the laws of our society follow ancient ethics, and therefore are poorly suited to governing a complex, crowded, changeable world." He points out that it is possible, indeed, to take legal steps to prohibit certain behaviors (although it is not necessarily easy to enforce such prohibition). At the same time, it is difficult to promote temperance. So the remedy to commons "is the institution of private property" [23], which is precisely a kind of end of the commons.

Yet the so-called 'blue gold' is so essential to life (any life on our planet, not just human life) that it urges collaboration. According to a study commissioned by UNESCO, "Water has demonstrated its ability to build confidence and its utility as a convenor of parties even in situations where they otherwise are not talking" [24]. It can be a meeting point—a point from which sharing and collaboration flow. The importance of water and its necessity make it a central element of community. Everyone is and should be involved in water management. Let us go back to *Genesis*. Abraham followed the path of sincerity, of dialogue, of seeking compromise. And this precisely starts with water from the well, that is, water for drinking, for hygiene, for animals, for handcraft and for irrigation (at that time, given the very low rate of pollution in rivers and groundwater, we can assume that the same source of water was good for all these purposes). A population can also be united over water and through water if it is convinced that water is too valuable to be contended for, hoarded, severely polluted, or wasted. Pollution of a given water source, ultimately, "is fundamentally a water governance problem related to the institutional arrangements that are in place and their effectiveness in steering the behavior of people, firms and other organizations" [25]. This is because the good management of water for drinking, irrigation, or sanitation contributes to the pursuit of the common good of society as a whole. Ten years ago, Italian parliamentarian and jurist Stefano Rodotà insisted that common goods are functional to the exercise of fundamental rights, and hence must be safeguarded by removing them from the destructive logic of the short term, projecting their protection into the future world inhabited by next generations [26].

Therefore, one way forward is to take seriously the words of the Pope: "the protection of water as a common good, the use of which must respect its universal destination" [27]. He defined water as a common good par excellence [28]. The universal destination of goods is a challenging principle [29]. First and foremost, one must believe in the universal destination of water, that is, be convinced that water is meant for the whole generation and all generations; moreover, one must strive for it: vision and commitment. The challenge is to take care of a common good so that all may use it given the common vocation for integral development, protecting the sustainability of our common home where our heirs will dwell [30].

In addition to promoting 'water as a common good,' it is appropriate to reflect on what implies the common management of a given water source (i.e., spring, well, lake, sea, etc.). A water source can represent a different good depending on a given situation, such as more or less scarce, more or less rival, more or less excludable, and easy or difficult to defend. These reflections apply at various levels.

- The local level comes to regulating the use of a well or lake. One can also cite as examples the tribunals responsible for settling irrigation-related disputes, such as the Spanish ones inscribed as intangible cultural heritage by the United Nations Educational, Scientific and Cultural Organization (UNESCO): the Water Tribunal of Valencia and the Council of Wise Men of Murcia [31]. These legal institutions of water management introduced centuries ago are still in operation. In the past, water management at the community level (or, in any case, at the local level) succeeded mainly because the water was less scarce (lower per capita needs, smaller populations) and also because there was less pollution. Things have changed, and in many cases, water management is now more complex; (in certain urbanized areas, community management makes little sense). Nevertheless, this community management remains important in addressing water challenges in many rural areas lagging behind in universal access to safe drinking water. Elinor Ostrom, the first woman to receive the Nobel Prize in economics, elaborated on the importance of managing natural resources such as forests and irrigation water in common, thus enabling (in cases where management is effective) their perpetuity [32] and avoiding a tragedy (represented by resource destruction).

- The national level, for example, manages water transfers between regions of the same country.
- The bilateral or multilateral level manages rivers, dams, and groundwater reserves (cases of the Indus, Danube, and Senegal rivers discussed below, and also the Columbia River Treaty between the USA and Canada).
- Finally, at the international level: agreements on certain chemicals (which could pollute), shipping (such as the International Maritime Organization's rules limiting the sulfur in the fuel oil used on board ships), and protection of the seas. The United Nations Convention on the Law of the Sea affirms that a given Area of the high sea and its resources "are the common heritage of mankind," consequently the activities in the Area shall "be carried out for the benefit of mankind as a whole" [33]. Moreover, negotiations are ongoing for an international legally binding instrument under the United Nations Convention on the Law of the Sea on the conservation and sustainable use of marine biological diversity in areas beyond national jurisdiction.

Water management in many rural areas implies a high involvement for local communities. For example, when it comes to the decision to dig a well or the choice about how to use or share a source of water, the possibility to access a specific fishing zone, to tap a given quantity of water (generally, establishing a certain water tenure regime). Community management and the frequent presence of various committees in charge of settling disputes (if not a real court) is an integral part of social life and contribute to empowering the population [34]. Over time, the community becomes a ground for trust and social capital; thus, interdependence becomes a lived experience. The experience of shared living and peaceful relations is frequent in these situations [35]. Collective action enables the usual dichotomy drawn between market and state to be overcome [36].

In India, "collective water rights and management were the key for water conservation and harvesting. By creating rules and limits on water use, collective water management ensured sustainability and equity" [37]. This is precisely why experts who promote techniques for water recovery in arid areas suggest—as much as possible—to avoid managing these situations essentially through nationally established policies, but rather to encourage participatory dynamics for "planning of communal or common property, which is particularly important in many communities where communal land and water resources are seriously degraded" [38]. The collection of such information can also be done at the community level [39]. This is the principle of subsidiarity [40]. It is no coincidence that Elinor Ostrom lists as a key factor for good management of a commons the fact that the local community has the right to self-organize, recognized by the state [41]. The Food and Agriculture Organization of the United Nations (FAO) has urged governments to harmonize their laws and procedures by considering and respecting local communities that have their own customary tenure system, an important factor since water management often depends on these tenure systems [42]. In addition, FAO studied collective rights concerning water management [43]. The UN Special Rapporteur on the right to safe drinking water and sanitation advocates that people and communities be able to understand their rights to participate "both in the drafting of laws and regulations and in the management of drinking water and sanitation services, providing them with the means for such participation to be effective" [44].

Even the financing of access to water entails a community dimension. Initiatives such as AZURE and Water.org give out (through financial intermediaries) loans for access to clean water.

AZURE is a blended finance facility established in 2018; it mobilizes funding from a variety of institutional and private sources and promotes loans to small-size water utilities (1,000 houses or even less) in Central America. By doing so, it's filling a gap in the local financial market (chiefly through its smallest loans ranging from 50,000 to 100,000 USD). The water utilities are also supported through technical assistance (TA) covering both operational issues (such as engineering guidance for water management and maintenance) and administrative ones (business and finance). This mix of TA and financial capital is well suited for WASH investments, and it enables water service providers to upgrade their services and protect water sources while de-risking loans. It is worth highlighting that communities created and manage the majority of the water utilities (for example, neighboring creates their own cooperative entrusted, among other issues, of granting access to water, system maintenance, and collecting payments). As of February 2023, 26 loans have been disbursed supporting 19 water service providers [45]. This is an inspiring process. It should be noted that "there are many innovative technologies and business models to provide safe water or sanitation in remote areas on a paying basis. However, all these operations struggle with barriers such as willingness to pay and overall profitability. This limits their ability to scale up and replicate" [46] and to access credit.

Water.org, with its network of partners, promotes small loans that bring access to safe water and sanitation to people living in poverty. This initiative grew exponentially in a few years only. In total, more than 12,3 million loans have been disbursed so far (as of 2 June 2023) through local microfinance institutions (MFIs), which established water and/or sanitation loans in their portfolio of offerings in 13 countries. Many of the borrowers are Joint liability Groups or Self-help Groups [47]. This is not surprising since, in order to operate in poor areas and reduce the risk of default, microcredit lending organizations lend to groups: they expect borrowers to know each other and come together to weekly meetings to make their payments. In a group, borrowers may become more willing to help out a member who faces financial difficulty. At the same time, the fear of shame stimulates the diligence of each member [48].

That said, three critical aspects should be emphasized.

1. First, community management of a common good does not always succeed (Elinor Ostrom's essay also reviews failures), and it is often not easy. For instance, it can be complicated to organize everyone's participation, and it happens that participation slows down decision making [49].
2. Second, the water rights of a given community, regardless of how ancient they may be and how effectively they may have protected a given water source, should not violate solidarity and human dignity [50].
3. Finally (related to the previous aspect), even where a single community effectively and independently manages locally available water, it is often not possible to consider the community as an isolated microcosm. On the contrary, it is necessary to understand how it interacts and how other neighboring communities interact with communities in the area, with those farther downstream or upstream, and with higher levels of coordination (for policies, administration, infrastructure) and with the natural environment, since everything is connected.

## 8.5 EFFECTIVE INSTITUTIONS

Water resources have always been a concern of institutions and legal systems. A group of judges, meeting in 2018 in Brasilia, defined water as a 'Public Interest Good' [51]. Therefore, in light of the principle of subsidiarity, we will always need institutions and customs that are proportionate to the size and complexity of the challenges posed by water management as a common good. The reflection of engineers, economists, and administrators must question these issues! In Asia, Europe, Oceania, the Caribbean, etc., what mechanisms will foster good water management at the various levels, realizing the human rights to drinking water and sanitation for all? What technologies and policies can encourage and facilitate everyone's participation (contributing to the project, monitoring water quality) as well as traceability, transparency, sustainability, and efficiency? What software, what consultations, what education, what lifestyles, what costs, what health impacts?

The role of institutions is crucial, as recognized by the 16th SDG. Commons often have some sort of administration, customs, and rules. Consider a rural community [52] undertaking a borehole. There are those who dig, those who provide the pipes, those who set the rates, those who educate through school or family groups on water users, those who do the maintenance, and those who periodically analyze water samples (in order to assess water quality). A fertile harmony is (hopefully) created, and trust increases between those involved (trust obviously does not exclude institutions). In a long-enduring institution governing a commons, very likely, individuals affected by the rules can participate in adopting or modifying the rules [53]. One gets to know each other better, and therefore each person is better able to behave in a certain way within the community, knowing that one can expect others to behave in a certain way as well, thus contributing to the management of water really as a common good. Each person does his or her part, in a certain manner and time, for the benefit of the group as a whole [54]. One can expect an exceptionally high level of quasi-voluntary compliance with the decisions made by the local institution that arbitrates or otherwise manages the difficulties. All this is without omitting the fact that the various users know each other (expecting that almost every other member of that community will almost always choose to conform to the agreed decisions) and frequently communicate (so free-riders, those who 'cheat' are soon detected, and if the misbehavior is repeated, they are sanctioned). *Acequias*, for example, are centuries-old mutually managed irrigation channels in the Southwest of the United States of America. Each *acequia* is a customary democracy, with a governing commission and a manager elected by those who own water rights. "Things may seem easier when there's ample water; it's the dry years that prove the system works. [*Acequias*' members] may get angry, but they always recognize that a sustained drought means less for each individual and more intensive sharing" [55]. This contradicts the belief that "Common-pool resources correspond to open access regimes where there are no formal and informal laws that govern these resources. Anyone can use them without any restriction" [56].

Indeed, where there are no rules and no institutions, as Hardin pointed out in *The Tragedy of the Commons*, that absence leads to the destruction of the commons (think here of some multinational corporations and commodity traders who are able to act with opacity where rules and administrations are less effective, where the protection of human rights and natural resources are weaker, without taking on so-called negative externalities). Managing the many water-related issues has always been a field of work for the administration. Already the Code of Hammurabi, one of the oldest collections of written laws that have come down to us, contemplated water issues. The water resource, precisely, offers an excellent example of how governance should reflect the various levels of management (as explained above).

Since access to drinking water is a human right [57], states have a duty to contribute to its realization by consistently developing their infrastructure, legislation, and economy by controlling the actions of all actors involved in water operations. And also by ensuring that lenders do not irresponsibly promote over-indebtedness and by facilitating access to credit so that people are not tempted to turn to usurers. Ultimately, states have an obligation to ensure access to justice [58] since this right "has to be protected and promoted with a special legal framework and adequate institutions" [59]. The public authority maintains the regulatory and control function. It must therefore ensure that water maintains its universal destination. In line with what was said earlier, the needs of the poorest (those who suffer most from thirst in the most abandoned and neglected contexts) must be met urgently as a priority [60]. The ecosystems they rely on must be defended as a priority.

The administration must watch over the priorities to which water is allocated, according to a hierarchy that is, of course, inspired by respect for human dignity. It is not inherently wrong to water a golf course or consume water in the production of computer equipment. Still, it is a problem if doing so deprives a community of water for cooking and personal hygiene.

This reflection leads us to revisit the principle of justice articulated in its various forms building upon a book published by the Pontifical Council for Justice and Peace.

- Distributive justice implies that everyone can access the necessary resources and opportunities (including access to drinking water and sanitation).
- Contributive justice implies that everyone contributes to the development and well-being of a given group with professionalism, according to their own economic and technical possibilities (including financing for water infrastructure and training in areas in need). Contributive justice enlightens the Common but Differentiated Responsibilities concept often used during international negotiations about climate change and similar topics).
- Restorative justice means that those who have irresponsibly and excessively exploited or polluted water resources (or destroyed useful water infrastructure) work to fix the situation and mitigate the negative impact of their actions on people and nature. This aspect is crucial if wounds are to be healed and intergenerational justice fulfilled if relations of trust and peace are to be renewed [61]. Restorative justice can enlighten the so-called 'Polluter-Pays' principle.

That said, institutions are not the silver bullet. Technological and organizational innovations in this field (monitoring of indicators over time, data collection and analysis, information access and sharing, decision-making aid, and funding tracking) are also invaluable. Education, culture, and spirituality should also be considered [62] if we aim at inclusive and sustainable water management. Indeed, as Pope Francis taught, "Sustainability (…) is a multidimensional word. Aside from the environmental, there are also the social, relational, and spiritual dimensions" [63].

## 8.6 SHARING WATER FOR PEACE

Over time, taking common care of water creates customs, and you can experience "the challenge and the gift present in an encounter with those outside one's own circle" [64]. You get used to "identifying with others" [65] who also thirst, who also have to irrigate. The encyclical *Fratelli tutti* insists a great deal on this theme, explaining that human beings cannot fully know themselves apart from an encounter with other persons: "I communicate effectively with myself only insofar as I communicate with others." No one can experience the true beauty of life without relating to others, without having real faces to love. This is part of the mystery of authentic human existence. "Life exists where there is bonding, communion, fraternity; and life is stronger than death when it is built on true relationships and bonds of fidelity. On the contrary, there is no life when we claim to be self-sufficient and live as islands (…)" [66].

Encounters, and especially a relationship, always generate potential vulnerability, no question. Weaving bonds involves dismantling our fears, accepting our limitations, and even gaining a better knowledge of ourselves. But, as the exhortation *Christus vivit* makes clear, if we do it for the sake of helping others, we can have the magnificent experience of setting our differences aside and working together for something greater. If, as a result of our own simple and, at times, costly efforts, we can find points of agreement amid conflict, build bridges and make peace for the benefit of all, then we will experience the miracle of the culture of encounter [67].

Furthermore, it was previously said that 153 countries share rivers, lakes, and aquifers across the world. But "only 84 of these basins have joint water management bodies, and many of these are not considered effective. The number of shared aquifers without joint management bodies—more than 400—is significantly higher" [68]. Yet we should consider that, according to *Aqua fons vitae* (AFV):

> Well-proven transboundary water cooperation mechanisms are an important feature of peace and the prevention of armed conflicts. Joint river mechanisms and commissions established by water

> agreements may foster communication and dialogue, thus improving relationships among groups beyond their usual role of facilitating joint management of the water resources. A body of norms enshrined in treaties and customs applies to internationally shared rivers, lakes, seas, basins, and groundwater resources. These approaches – oriented towards mutual understanding, sharing information, searching, and implementing solutions together – constitute the so-called 'water diplomacy' [69].

There are widely studied cases. For example, cooperation between countries around the Senegal River or the Danube River is considered by various experts to be a valuable peace mechanism. Indeed, an "active water cooperation does not mean the mere signing of a treaty for allocation of water or data exchange or for establishing a river basin organization unless there is verifiable joint management of water resources" [70] highlights an initiative that has developed a method to measure the rate of cross-border cooperation over water and argues that countries with a high rate of cooperation will most likely not confront each other militarily [71]. The Indus Water Treaty is also considered an effective treaty as various disputes have been resolved over the years within the Permanent Indus Commission. During several decades, the treaty's provisions for the settlement of differences and disputes were tested and proved effective [72]. India and Pakistan, despite several political crises and climate change related challenges, haven't started a water war.

Unsurprisingly, the interest in the issue of reciprocity is re-emerging today. Some scholars promote multidimensional reciprocity that goes from gift exchange to contracts and rules, which are one and many simultaneously, without discarding either behavior of pioneers inspired by gratuity (otherwise, no sincere trust can be generated) or individualistic act [73]. A logic is created that some scholars have called "we-rationality"[74], especially if there is a large number of pioneers that are motivated people in the community willing to commit themselves to the common management of water. They obviously hope for reciprocity without making it a precondition to their commitment. And so, between those who trigger virtuous behavior and those who join at a later stage, the disappearance or mismanagement of water is avoided. Water is managed in a sustainable way.

The encyclical *Laudato si'* on the care of our common home, acknowledging the lack of awareness of our common origin, of our mutual belonging, and of a future to be shared by all, announces three major challenges for the human family: education, culture, and spirituality [75]. These challenges also apply to water. Education can help us appreciate the value of water and become familiar with itand learn about water challenges related to sustainability and development (this is consistent with SDG 4.7). A water culture can foster respect, sobriety, and solidarity in daily life. Spirituality—genuine religious values, rituals, and teaching—can offer great motivation to change, act, and engage. Therefore, educational, cultural, and spiritual activities, along with policies, institutions, and innovations, can facilitate sound water management as a common good, significantly contributing to peace.

In the text frequently considered, the first message entirely and extensively dedicated by a Pope to ecology—the *Message for the World Day of Peace 1990*—Pope John Paul II insisted precisely on collaboration, solidarity, and responsibility in the ecological sphere in relation to peace.

> The ecological crisis has assumed such proportions as to be the responsibility of everyone. As I have pointed out, its various aspects demonstrate the need for concerted efforts aimed at establishing the duties and obligations that belong to individuals, peoples, States and the international community. This not only goes hand in hand with efforts to build true peace, but also confirms and reinforces those efforts in a concrete way [76].

Peace is a gift from God but an ongoing responsibility of each person who must strive to establish and preserve it with the community and personal commitment [77]; it can also be sought around and from the well, river, transnational basin, or aquifer and sea.

## REFERENCES


[1] UN. (2022, July). *The Sustainable Development Goals Report 2022*: 38. https://unstats.un.org/sdgs/report/2022/The-Sustainable-Development-Goals-Report-2022.pdf (accessed 7 November 2022).

[2] *Genesis* 21, 22–33. The author's reflections on peace and collaboration proposed in this chapter and inspired by this passage from Genesis have in part already been published, in Italian, by the Dicastery for Promoting Integral Human Development in the 2023 book titled Aqua fons vitae. Valuing and caring for a common good: Water. Acta post webinar, March 22-26 2021, Libreria Editrice Vaticana, Vatican City: 137-147 (chapter "Il valore dell'acqua per la pace"). The present text, however, is not a translation but a reworking that contains different ideas and references from the Italian publication.

[3] (2020, November 5). "Nile Dam Talks Between Egypt, Ethiopia and Sudan Fail Again", *Aljazeera*. https://www.aljazeera.com/news/2020/11/5/egypt-ethiopia-sudan-fail-to-succeed-in-disputed-dam-talks (accessed 7 November 2022).

[4] (2021, February 19). "Ladakh: China Reveals Soldier Deaths in India Border Clash", *BBC*. https://www.bbc.com/news/world-asia-56121781 (accessed 7 November 2022).

[5] Anton Troianovski and Malachy Browne. (2022, June, 8). "Satellite Imagery Shows Ukrainian Water Flowing Again to Crimea, as Russia Nears Big Objective", *The New York Times*. https://www.nytimes.com/2022/06/08/world/europe/crimea-water-canal-russia.html (accessed 7 November 2022).

[6] (2022, November 7). "Kakhovka Dam in Moscow-Occupied Ukraine 'Damaged' by Kyiv Strike – Russian Agencies", *The Moscow Times*. https://www.themoscowtimes.com/2022/11/06/kakhovka-dam-in-moscow-occupied-ukraine-damaged-by-kyiv-strike-russian-agencies-a79300 (accessed 7 November 2022).

[7] Aymeric Chauprade. (2007). *Géopolitique. Constantes et changements dans l'histoire*, 3rd édition, Ellipses, Paris: 643–696.

[8] Vandana Shiva. (2002). *Water Wars. Privatization, pollution and profit*, India Research Press, New Delhi: ix–xiv.

[9] Geneva Water Hub. (2019). *The Geneva list of Principles on the protection of water infrastructure*, Geneva: 19–29.

[10] Global High-Level Panel on Water and Peace. (2017). Report *A Matter of Survival*, Geneva: 21.

[11] Migrants and Refugees Section and Integral Ecology Sector of the Dicastery for Promoting Integral Human Development. (2021). *Pastoral Orientations on Climate Displaced People*, Vatican City: 14.

[12] FAO. (2018). *Water stress and human migration: a global, georeferenced review of empirical research*, Land and Water discussion paper 11, Rome: 9.

[13] UN and UNESCO. (2018). *Progress on Transboundary Water Cooperation. Global baseline for SDG Indicator* 6.5.2, Paris: 9.

[14] Catholic Relief Services. (2019). *Water and Conflict. Incorporating Peacebuilding into Water Development*, Baltimore: viii.

[15] Francesco Fornari. (2009, April). "Upheaval in the Horn of Africa", *Freedom From Fear* Magazine, (3): 9–13. UNCTAD. (2014). *Maritime Piracy. Part I: an overview of trends, costs and trade-related implications*, New York and Geneva: 3-10.

[16] Luigino Bruni. (2011). "Il significato del limite nell'economia dei beni comuni", *Sophia* (III/2): 217.

[17] UN. (2014). *The Millenium Development Goals Report 2014*, p. 44.

[18] UN. (2018, May). *Sustainable Development Goal 6. Synthesis Report 2018 on Water and Sanitation*: 11.

[19] World Health Organization and UN Environment Programme. (2013). *State of the Science of Endocrine Disrupting Chemicals – 2012*. European Food Safety Authority. (2013). *Scientific Opinion on the Hazard Assessment of Endocrine Disruptors: Scientific Criteria for Identification of Endocrine Disruptors and Appropriateness of Existing Test Methods for Assessing Effects Mediated by These Substances on Human Health and the Environment.*

[20] Competition for water uses is a cross-cutting theme of the following report: FAO. (2020). *The State of Food And Agriculture. Overcoming Water Challenges in Agriculture*, Rome. Moreover, on the occasion of the recent UN Water Conference, experts have suggested to address the fragmentation in water governance. For example, Ines Dombrowsky, Annabelle Houdret and Olcay Ünver. (2023, March 20). "The UN Water conference – time to govern water as a global commons!", The Current Column. https://www.idos-research.de/en/the-current-column/article/the-un-water-conference-time-to-govern-water-as-a-global-commons/ (accessed 2 June 2023). Other experts, such the economist

Mariana Mazzucato in her numerous contributions easily retrievable online, have reflected on water as a (global) "common good".
[21] Convention on Biological Diversity. (2015). *Wetlands and Ecosystem Services*. CBD Press Brief. https://www.cbd.int/waters/doc/wwd2015/wwd-2015-press-briefs-en.pdf (accessed 20 November 2022).
[22] Cf. UN Development Programme. (2006). *Human Development Report 2006*, New York: 51–53.
[23] Garrett Hardin. (1968, December 3). The Tragedy of the Commons, *Science*, New Series (Vol. 162, No. 3859): 1243–1248. https://pages.mtu.edu/~asmayer/rural_sustain/governance/Hardin%201968.pdf (accessed 11 October 2022).
[24] UNESCO. (2004). *Ethics and Water Resources Conflicts*. Series on Water and Ethics, Essay 12, France: 6.
[25] Nigel Watson. (2021, January). *Written Evidence for the House of Commons Committee report on "Water quality in rivers"*, United Kingdom. https://committees.parliament.uk/writtenevidence/21746/pdf/ (accessed 7 November 2022).
[26] Stefano Rodotà. (2012, January 5). "Il valore dei beni comuni", *La Repubblica*.
[27] Francis. (2020, November 8). *Angelus*.
[28] Francis. (2015, March 22). *Angelus* (refer to the original text, pronounced in Italian).
[29] Pontifical Council for Justice and Peace (2004). *Compendium of the Social Doctrine of the Church*, Vatican City: § 171–184.
[30] Tebaldo Vinciguerra. (2020, November 14). "Il diritto all'acqua al tempo del covid", *L'Osservatore Romano*. https://www.osservatoreromano.va/it/news/2020-11/quo-264/il-diritto-all-acqua-al-tempo-del-covid.html (accessed 19 October 2022).
[31] UNESCO. (undated). *Irrigators' tribunals of the Spanish Mediterranean coast: the Council of Wise Men of the plain of Murcia and the Water Tribunal of the plain of Valencia*. https://ich.unesco.org/en/RL/irrigators-tribunals-of-the-spanish-mediterranean-coast-the-council-of-wise-men-of-the-plain-of-murcia-and-the-water-tribunal-of-the-plain-of-valencia-00171 (accessed 2 June 2023).
[32] Elinor Ostrom. (2015). *Governing the Commons*, Cambridge University Press, Reissue edition, United Kingdom: 83.
[33] UN. (1982). *Convention on the Law of the Sea*: articles 136 and 140.
[34] Dicastery for Promoting Integral Human Development. (2020). *Aqua fons vitae*, Vatican City: § 26
[35] Atul K. Shah. (2022, October). *Inclusive and Sustainable Finance. Leadership, Ethics and Culture*, Routledge, United Kingdom: 72 and 76.
[36] Stéphanie Leyronas & al. (2016, May). *Toward an Analytical Framework for the Governance of Natural Resources: The Case of Groundwater*, Papiers de Recherche AFD, n° 2016-24: 10-12.
[37] *Water Wars. Privatization, Pollution and Profit*: 12.
[38] IFAD & al. (2013). *Water Harvesting. Guidelines to Good Practice*, Bern: 20.
[39] Caterina de Albuquerque. (2019). *Droit au but. Bonnes pratiques de réalisation des droits l'eau et à l'assainissement*, Lisbon: 191.
[40] Benedict XVI. (2009, June 29). *Caritas in veritate*: § 47 and 57. *Compendium of the Social Doctrine of the Church*, § 185-191.
[41] Governing the Commons: 101.
[42] FAO. (2012). *Voluntary Guidelines on the Responsible Governance of Tenure of Land, Fisheries and Forests in the Context of National Food Security*, Rome: Iv and 8.
[43] FAO. (2020). *Unpacking Water Tenure for Improved Food Security and Sustainable Development*, Land and Water discussion paper 15, Rome: 10.
[44] Pedro Arrojo Agudo. (2021, July). *Report of the Special Rapporteur on the Human Rights to Safe Drinking Water and Sanitation "Plan and Vision for the Mandate from 2020 to 2023"*, A/HRC/48/50: § 15.
[45] https://www.azurewater.org/#/ (accessed 22 November 2022). Oksana Tkachenko and Goufrane Mansour. (2021, June 30). *Case study assessment: Blended finance in Water, Sanitation and Hygiene (WASH) – Lessons for Development Partners*. Australia: 34-38.
[46] Waterpreneurs. (2018, March). *White Paper Impact Investing for Water. Innovative Finance for Scaling-up. "Water, Sanitation and Hygiene" (Wash) Market-based Solutions. A Review of the Needs and Opportunities*, Geneva: 32.
[47] https://water.org/solutions/watercredit/ (accessed 22 November 2022).
[48] Abhijit V. Banerjee and Esther Duflo (2011). *Poor Economics. A radical rethinking of the way to fight global poverty*, 1st Edition, Public Affairs, United States of America: 166–167.

[49] Hydroaid, University of Turin & al. (2021, June). *Les processus participatifs de gouvernance environnementale en Afrique: expériences locales pour des perspectives globales*, Turin: 12.
[50] AFV: § 75.
[51] (2018. March 21). *Brasília Declaration of Judges on Water Justice*: 1st principle. http://www2.ecolex.org/server2neu.php/libcat/docs/LI/MON-093474.pdf (accessed 20 October 2022).
[52] Why this focus on the rural world? Usually water management in urban areas is the responsibility of the municipality, which traditionally receives financial and technical help from the government (and may contract private companies through procurement), while in rural areas the situation is often less clear. Michel Camdessus & al. (2004). *Eau*, Robert Laffont, France: 92.
[53] *Governing the Commons*: 90.
[54] Robert Sugden. (2018). *The Community of Advantage*, Oxford University Press, United Kingdom: 236–240; 256-281.
[55] Robert Neuwirth. (2019, May 17). "Centuries-old irrigation system shows how to manage scarce water", NationalGeographic.com. https://www.nationalgeographic.com/environment/article/acequias (accessed 2 June 2023).
[56] UNESCO Commission on the Ethics of Scientific Knowledge and Technology. (2018, September 14). *Water ethics: ocean, freshwater, coastal areas*. Report SHS/COMEST-10EXT/18/2 REV.2, Paris: § 156.
[57] United Nations General Assembly. (2010, July 28). *The human right to water and sanitation*, Resolution 64/292. United Nations. (1977). *Report of the United Nations Water Conference. Mar del Plata, 14-25 March 1977* E/CONF.70/29, New York: 66.
[58] Léo Heller (2015, July). *Report of the Special Rapporteur on the human right to safe drinking water and sanitation to the General Assembly*, A/70/203: § 35.
[59] Pontifical Council for Justice and Peace. (2012, March). *Water, an Essential Element for Life. Designing Sustainable Solutions. An Update*, Vatican City. https://www.humandevelopment.va/content/dam/sviluppoumano/pubblicazioni-documenti/archivio/ecologia-e-ambiente/2012NOTACQUAA_ENG_DEF.pdf (accessed 14 November 2022).
[60] *Report of the Special Rapporteur on the human right to safe drinking water and sanitation to the General Assembly*: § 88.
[61] Pontifical Council for Justice and Peace. (2015). *Land and Food*, Libreria Editrice Vaticana, Vatican City: § 76.
[62] Francis. (2015, May 15). *Laudato si'*: § 202.
[63] Francis. (2022, September 24). *Address in Assisi During the Event "Economy of Francesco"*.
[64] Francis. (2020, October 3). *Fratelli tutti*: § 90.
[65] *Fratelli tutti*: § 84.
[66] *Fratelli tutti*: § 87.
[67] Francis. (2019, March 25). *Christus vivit*: § 169.
[68] Report *A Matter of Survival*: 14.
[69] AFV: § 27 (Read also § 105).
[70] Strategic Foresight Group. (2013). *Water Cooperation for a Secure World*, India: 4.
[71] *Water cooperation for a secure World*: 2-23.
[72] Salman M.A. Salman. (2008). "The Baglihar Difference and Its Resolution Process - A Triumph for the Indus Waters Treaty?", *Water Policy* (10): 105–117.
[73] Luigino Bruni and Alessandra Smerilli. (2015). *The Economics of Values-Based Organisations. An introduction*, Routledge, Oxon and New York: 17, 18, 35 and 84.
[74] Alessandra Smerilli (2007). "We-rationality: per una teoria non individualistica della cooperazione", *Economia Politica*, (24/3): 407–425.
[75] *Laudato si'*: § 202.
[76] John Paul II, *Message for the World Day of Peace 1990*: § 15.
[77] John Paul II, *Message for the World Day of Peace 1982*: § 5.

# 9 Human Behavior Dynamics in Sustainability

*Eric C.D. Tan*[1]
National Renewable Energy Laboratory, Golden, Colorado, USA

**CONTENTS**



## 9.1 INTRODUCTION

Human behavior and actions play a direct and significant role in shaping pathways toward sustainability, resulting in many sustainability challenges, such as climate change, resource depletion, and biodiversity loss (Schill et al., 2019). These pressing problems represent social dilemmas involving a conflict between immediate self-interest and longer-term collective interests (Van Lange et al., 2013). To illustrate the human behavior dynamics in sustainability, this chapter examines the social dilemma associated with the disposal of unwanted medication. A social dilemma arose when a hypothetical Colorado rural town closed down most of its medicine take-back stations as a result of reduced tax revenue, creating a social situation where short-term self-interest was at odds with the long-term interests of the town community.

The United States is the largest market for both prescription and over-the-counter medications in the world, exceeding $200 U.S. billion in 2007, and the usage of these pharmaceuticals will continue to increase (Glassmeyer et al., 2009). With the increased consumption of medicines, the amount of unused medications, for reasons such as a change of medication by the doctor (49%) or self-discontinuation (26%), will also increase (Kümmerer, 2009). Thus, the challenge of unwanted medicine disposal is an important issue for humans and the environment. For one, improper handling of unwanted medications can have direct safety ramifications. One important reason to safely get rid of unused or expired medicines is to avoid accidental exposure. No longer needed medicines can create an unnecessary health risk in the home, especially if there are young children present. For example, over 450,000 children under six years old accidentally ingest medicine each year in the United States (McFee & Caraccio, 2006).

Furthermore, the improper disposal of unwanted medications allows active pharmaceutical ingredients to get into the environment, causing adverse effects on the environment. The two most common methods of disposal were to throw unwanted medicines in the trash (76.5%) or flush them down the drain (11.2%) (Kümmerer, 2009). Unfortunately, flushing the unused or expired medications into sewer systems or landfills causes certain active ingredients to leach into the water table (Vollmer, 2010). Pharmaceuticals have been detected in surface waters, groundwater, and treated

DOI: 10.1201/9781003167693-9

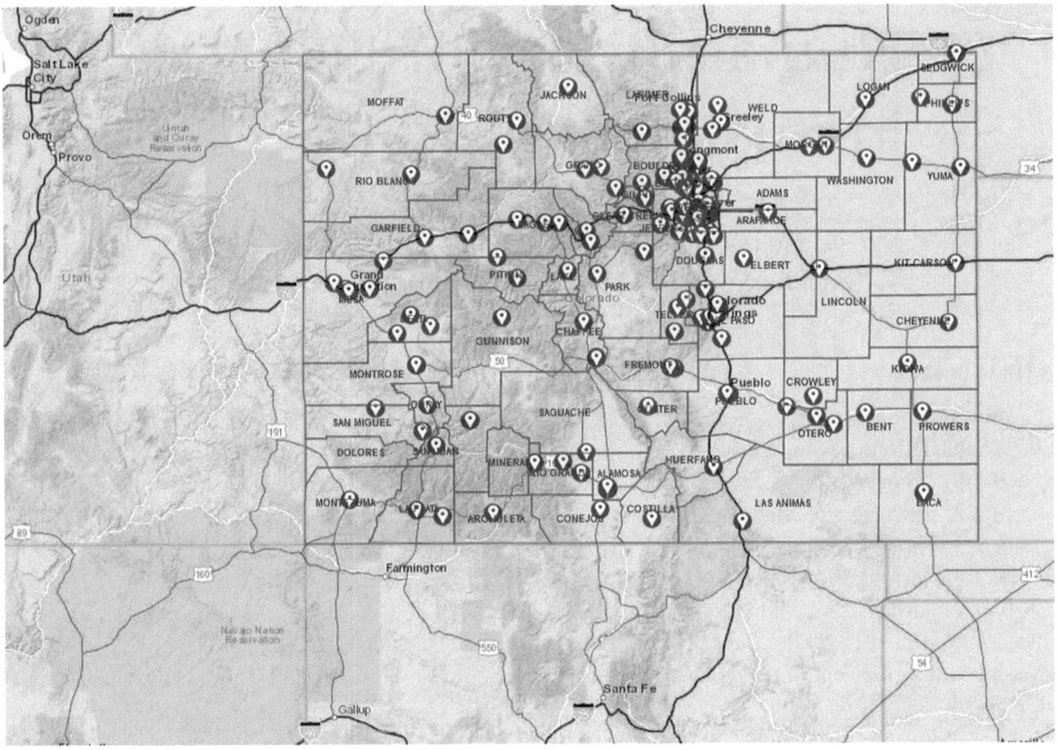


FIGURE 9.1 Interactive map of take-back locations in the Colorado state program.

*Source:* (CDPHE, n.d.)

drinking water. Some reported environmental effects include vitellogenin induction in male fish, gender and genital abnormalities in fish, and even population collapse (Glassmeyer et al., 2009). Thus, it is crucial to practice the safe disposal of unused or unwanted medications.

To reduce the amount of unwanted medicines entering the environment (namely, the negative environmental impact of improper medicine disposal) and causing harm to humans (including accidental poisoning and a public safety hazard), many state and local governments in the United States have established unwanted medicine collection programs, including setting up take-back stations. For example, the Colorado Department of Public Health and Environment (CDPHE) has established household medication take-back programs that accept and destroy unused and expired over-the-counter and prescription medications (CDPHE, n.d.). Figure 9.1 shows the locations in Colorado that accept a wide variety of unwanted medications, including anti-depressants as well as pain and cholesterol medications.

In this hypothetical scenario, a rural town in Colorado has to close down most of its medicine take-back stations due to reduced tax revenue. Because the closest take-back station is now located 25 miles away, which will cost \$10 in gas and take approximately 20 minutes to drive to, it has become a challenge for some residents to properly dispose of their unwanted medications. Consequently, the lack of easy access to take-back stations leads to a social dilemma. As human behavior is rich and heterogeneous (de Vries, 2012), some residents will continue to dispose of their unused medications at take-back stations; however, other residents will flush them down the sink or throw them in the trash, which ends up in the landfill. This chapter discusses the present resource-related social dilemma around the situation in which an individual "plays" against everyone else with respect to whether to drive to the take-back station to dispose of unused medication or to flush it or throw it in the landfill. The social dilemma discussion makes use of the

**TABLE 9.1**
**Pay-off Matrix in the Current Medication Disposal Social Dilemma**

| | | Everyone Else | |
|---|---|---|---|
| | | **Cooperate** | **Defect** |
| | | Everyone else drives to the take back station to dispose of medication | Everyone else flushes or throws medication in the landfill |
| Individual | **Cooperate** Individual who drives to the take-back station to dispose of medication | **Win-Win** | **You're the Sucker!** |
| | **Defect** Individual who flushes or throws medication in the landfill | **Free Rider** | **Tragedy of the Commons** |

structure of the pay-off matrix (as illustrated in Table 9.1), which encompasses four scenarios—*win-win*, *you're the sucker*, *free rider*, and *tragedy of the commons*—and is the most elementary device in the field of game theory (de Vries, 2012).

## 9.2 SOCIAL DILEMMA ASSESSMENT

### 9.2.1 Win-Win

The *win-win* scenario is when everyone is cooperative, and the outcome is where all of the participants involved are successful in achieving their objectives that benefit all. In this scenario, both the individual and everyone else choose to dispose of their unwanted medication by taking them to the take-back stations. Since nobody flushes or throws away the unwanted drugs, the whole community wins and enjoys the safe (i.e., non-pharmaceutically contaminated) drinking water.

The community members are dropping off their medication at the take-back stations because the people are generally educated on the importance of doing so. They are aware that the municipal wastewater treatment plant discharges into a stream that is for livestock irrigation, fishing, and a reservoir supplying the town's drinking water. As depicted in Figure 9.2, the rural town residents understand that their personal health depends on their water is free from contamination and food safety. Pharmaceutical residuals in water not only directly impact the residents' drinking water but also cause adverse impacts on the environment and food sources such as livestock and fish.

Moreover, one of the defining features of human beings is their capacity to exhibit high levels of cooperative behavior; human cooperation is much greater than any other species (Melis & Semmann, 2010). The success of the sustainability pursuit is largely context-specific and a collaborative affair (Matson, Clark, & Andersson, 2016). Additionally, various studies reveal that people tend to voluntarily organize themselves for a collective purpose (de Vries, 2012). Therefore, the majority of community members want to be an active contributor to preserving the quality of water and, thus, the well-being of all the town residents.

What motivates an individual to do the same (being cooperative) can be further attributed to the fact that they share the same value as the rest of the residents, namely, protecting the environment by not contaminating the water sources. The other possible reason is that the individual values social conformity and does not want to stand out, even though it takes resources (e.g., driving time and the cost of gas) to be cooperative. Individual who values social conformity will try to fit in a group or community and cooperate by changing their behavioral preferences when confronted with others' conflicting opinions (Huang, Kendrick, Zheng, & Yu, 2015). If everyone else in the town

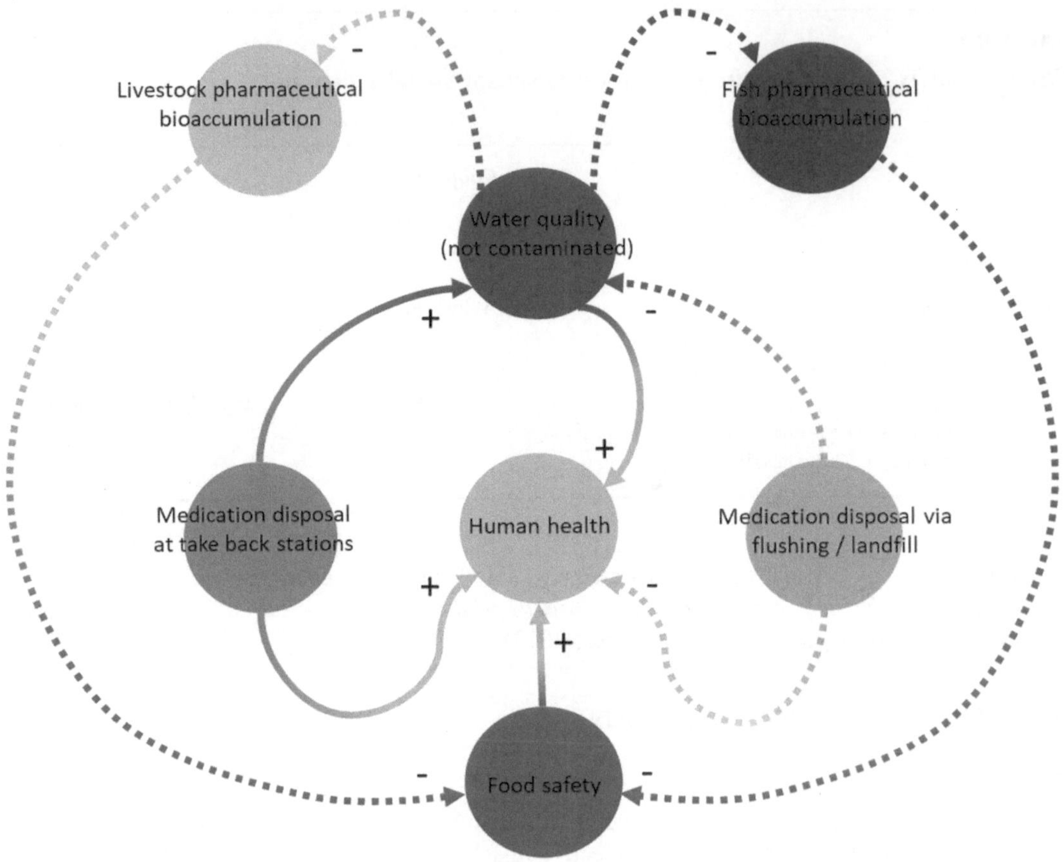


**FIGURE 9.2** Schematic illustration of the interrelationship between medication disposal and environment and human health.

drives to take back stations to dispose of their medication, it is deemed to be the right thing to do, and the desire to be liked and accepted prompts the individual to do the same. Group influence also plays a role here. The existence of group influence invokes people's herd psychology, which suggests that people are more willing to believe and follow the behavior of the majority (Yan & Jiang, 2018). Last but not least, individuals who think others will cooperate in social dilemmas are also more likely to cooperate themselves (Nowak & Highfield, 2011).

### 9.2.2 You're the Sucker

The *you're the sucker* scenario is a situation when only the individual is cooperative while everyone else defects, either flushing the medications or throwing them in the trash, which ends up in the landfill. Since it costs the individual time and money to drop off his or her unwanted drugs at the take-back station, yet everyone else (defectors) obtains instant collective benefits (safe and uncontaminated drinking water), the situation is also termed an investment (Melis & Semmann, 2010).

Knowing when to cooperate is an important skill to master. Most people tend to conform to group norms when the situation is unclear and turn to others for guidance. Unfortunately, due to the strong compelling force to conform, people frequently blindly follow others without identifying whether it is right or wrong, even if they disagree with the group's actions. There are many reasons for everyone else to defect and not practice proper medication disposal. For example, people might have the "ignorance is bliss" mentality. Choosing not to be aware of the collective damage the pharmaceutical ingredients can have on the communal water supply gives people little to no incentive to make an effort to dispose of the unwanted medications at the take-back station. The defectors regard this behavior as justified, as the costs outweigh the long-term benefits.

Doing the right thing here means safely and properly disposing of unwanted medication, which results in non-contaminated water for the livestock, fish, and town residents. It requires individual sacrifice (driving 50 miles per round trip and paying $10 for gas), which benefits everyone, but there is no guarantee that others will also be cooperative. It is intriguing why an individual would exhibit altruistic, cooperative behavior and want to be the "sucker" while knowing that everyone else is a defector.

The individual simply feels proud to cooperate and do the right thing. Social or moral emotions can play a role in cooperation in social dilemmas. Pride is a social emotion related to socially valued behaviors. This type of emotion arises when an individual feels responsible for a socially valued outcome and is likely to motivate pro-social behavior (e.g., cooperation) (Dorfman, Eyal, & Bereby-Meyer, 2014). It is noteworthy that pride is different from negative emotions such as guilt and shame or fear. People will also cooperate more when they experience these negative emotions (Dorfman, Eyal, & Bereby-Meyer, 2014). However, pride is a positive adaptive social emotion that benefits society. Suppose flushing or throwing medications in the landfill is morally wrong to the individual. In that case, the individual takes great pride in doing the right thing by driving to the take-back station to dispose of their medication. In this case, the individual is willing to be cooperative even when cooperating is costly.

The individual (the *sucker* for this scenario) can also be an uncalculating cooperator. Unlike conditional cooperators, who cooperate when the benefits outweigh the costs (i.e., making a decision based on a cost-benefit analysis), uncalculating cooperators are willing to pay the maximum possible cost of cooperation and are primarily motivated by reputation concerns (Jordan, Hoffman, Nowak, & Rand, 2016). In this case, the individual wants to be seen as an ethical person and not to be morally reprehensible as everyone else.

### 9.2.3 Free Rider

The *free-rider* scenario occurs when people take advantage of public goods without paying for them. In this case, the individual benefits from having safe and uncontaminated drinking water, although they are not cooperative in disposing of their unwanted medication at the drug take-back station. The free rider can get away with not being cooperative since unpolluted water is non-excludable and is available to everyone.

The individual is aware that flushing medication or throwing it into the trash (landfill) will eventually contaminate the groundwater. However, they think it makes little difference if one person, out of thousands (everyone else), in the town does not dispose of his or her unwanted medication at the take-back station. Therefore, there is an incentive to free-ride on the efforts of other people who safely and properly dispose of medication and not put in the effort themselves.

Some people may defect because they cannot get to the take-back stations. For example, the person is elderly or disabled, has no car, or cannot make it to the take-back station during open hours. Nevertheless, in the *free-rider* scenario, the individual is selfish and is insufficiently motivated by altruism. They are a *homo economicus* who always uses rational judgment and seeks to maximize profits and minimizes costs (Gintis, 2000). In the short term, the individual is best off acting according to their own individual selfish interest. It is undoubtedly easier (i.e., more comfortable, accessible, and manageable) and more cost advantageous (time and money) to flush the mediation down the sink rather than taking it to the take-back station. However, the individual fails to realize that in the long term, he or she is better off by acting for the collective interest to ensure safe drinking water for the community by helping keep the water free from pharmaceutical microcontaminants. Moreover, the individual exploits the free-riding opportunity since they know they will not get caught in flushing or dumping unwanted medication. In fact, research has shown that free riding can be pervasive under conditions of anonymous interactions (Fischbacher, Gächter, & Fehr, 2001).

One solution to the free-rider problem is to appeal to the free rider's altruism. It is essential to convince the individual that doing things for the benefit of others can also benefit him or herself. For example, high levels of antibiotics in drinking water can promote the evolution of highly antibiotic-resistant microorganisms that cause disease in humans (Fick et al., 2009).

### 9.2.4 Tragedy of the Commons

The current social dilemma is a situation where each person benefits by enjoying the safe and uncontaminated drinking water (the public good) as a result of others' contribution (disposing of unused drugs at take-back stations) and contributes himself or herself as little as possible (flush the unused drugs in the sink). While it is narrowly rational to free ride, if everyone behaves this way (inappropriate disposal of unused medications), safe drinking water will eventually be not available, and all are worse off. The scenario is known as the *tragedy of commons*, which illustrates one of the most interesting parts of the social dilemma.

The *tragedy of the commons* occurs when everyone in the community pursues their immediate self-interest and does not promote the collective good. Everyone does not cooperate as they feel that if "everyone else is not doing it" (i.e., not disposing of unused medications at the take-back stations), it is permissible for them to defect as well. Additionally, nobody wants to be a sucker since they would be the only ones spending the time and money to go to the take-back stations. Besides, there is no accountability (e.g., no punishment) associated with flushing unused medication down the sink. This type of dilemma arises when not taking unused drugs to the take-back stations initially produces rewards (saving time and money); however, continuing the same selfish behavior leads to consequences such as contaminated drinking water. Not immediately, but the whole point of the *tragedy of the commons* scenario is that eventually, the 'common pool resource' in this case, clean water, is used up and gone for good.

The *tragedy of the commons* scenario is a social trap. The town residents are aware that their selfish actions will lead to a disastrous long-term outcome. Unfortunately, they are motivated by short-term gratification; therefore, they simply cannot avoid acting in their self-interest, consequently creating a trap or tragedy.

When individuals flush unused drugs down the sink and cause water pollution, the costs of the water pollution for the rest of the residents in the town are not compensated for by these individuals and are externalized to the whole community. They do not pay for polluting the water and thus create a negative externality. One way to solve the central issue of the tragedy of commons is to connect the costs to the individuals' actions. This can be accomplished by regulating the system to ensure that those who create negative externalities will have to pay for them.

For illustration purposes, suppose the sole pharmacy store in the rural Colorado town has started an unwanted medication take-back program. The store will ask customers to pay a reasonable amount of deposit for the medication at the time of purchase. The customers can get their deposits back when they return their unused drugs or empty medicine bottle to the store. If they do not return their unused medication or empty medicine bottle, then as a "punishment," the customer will lose their entire deposits (which can then be used for supporting the take-back program). This take-back program at the local pharmacy not only provides people with a more convenient method to safely dispose of their unwanted medications, but it also creates an incentive for those who would otherwise inappropriately dispose of their medications (either flushing their medications or throwing them in the trash) to bring it to the pharmacy. The defectors would have to pay for their action that causes the negative externality. The lower the financial incentive for cooperating behavior, the higher the defecting propensity. Studies have identified punishment in social behavior as a mechanism that can effectively maintain cooperation in public goods experiments (Semmann, Krambeck, & Milinski, 2003). By including punishment (e.g., discipline and penalty), people tend to defect less, and highly cooperative outcomes emerge (Brandt, Hauert, & Sigmund, 2003). Additionally, allowing town residents to make small incremental commitments towards contributing to the communal good might also facilitate cooperation as it helps to prevent them from being free riders (Kurzban, McCabe, Smith, & Wilson, 2001).

## 9.3 CONCLUSIONS

Human behavior dynamics dictate sustainability, as demonstrated through the discourse of a situation in which an individual "plays" against everyone else concerning whether to drive to the take-back station to dispose of unused medication (i.e., cooperating) or to flush or throw it in the landfill (i.e., defecting). The discussion uses the structure of the pay-off matrix that encompasses four scenarios, *win-win, you're the sucker, free rider, and the tragedy of the commons.* Due to the lack of easy access to medicine take-back stations, individuals can choose to act in their own best interest (saving the trouble, time, and gas money), or act towards a collective better interest (keep water source from being contaminated with pharmaceutical ingredients from unused medicine by dropping off unused medication at the take-back station). The individual's cooperation has to do with the competing motives for personal gain and the desire to create a positive outcome (water free of pharmaceutical contamination) for the town residents.

The *win-win* scenario is where everyone is cooperative, disposing of their unwanted medication by taking them to take-back stations. What motivates the individual to cooperate can be attributed to the fact that they share the same value as the rest of the residents. The individual also values social conformity and does not want to stand out, even though it takes resources (e.g., driving time and the cost of gas) to be cooperative.

In the *you're the sucker* scenario, only the individual is cooperative, whereas everyone else defects and either flushes the medication or throws it in the trash. The principal motivation for the "sucker" to cooperate is that they take great pride in doing the right thing, even when cooperating is costly.

The *free-rider* scenario occurs when the individual benefits from having safe and uncontaminated drinking water. However, they are not cooperative in disposing unwanted medication at the drug take-back station. The free rider can get away with not being cooperative since unpolluted water is non-excludable and is available to everyone. The individual is defecting as they think that their actions will likely make little difference. Additionally, the individual is selfish or insufficiently motivated by altruism and feels that the bad action will not get caught. One solution to the free-rider problem is to appeal to the free rider's altruism. It is important to convince the individual that doing things for the benefit of others can also benefit him or herself.

The *tragedy of the commons* occurs when everyone in the community blindly pursues their immediate self-interest and does not promote the collective good, which later proves to have negative consequences. Everyone does not cooperate (i.e., not disposing of unused medications at the take-back stations) as they feel that "everyone else is not doing it." Additionally, nobody wants to be a sucker, and no accountability (e.g., no punishment) is associated with bad behavior.

A potential solution to the social dilemma is to make the defectors pay for their actions that cause negative externalities. The lower the financial incentive for cooperative behavior, the higher the defecting propensity. By including punishment, people tend to defect less, and highly cooperative outcomes emerge. When the individual always pays the full cost for not cooperating, the social dilemma will eventually cease to exist, and the rural town should have self-sustaining pharmaceutical contaminant-free water. The townspeople's water safety and public health of the depend on collective action, and that is what win-win cooperation is all about.

## NOTE

1 Contributions to the resulting publication were all on the author's own time.

## REFERENCES


Brandt, H., Hauert, C., & Sigmund, K. (2003). Punishment and reputation in spatial public goods games. *Proceedings of the Royal Society of London. Series B: Biological Sciences*, *270*(1519), 1099–1104. 10.1098/rspb.2003.2336

CDPHE. (n.d.). *Colorado Household Medication Take-Back Program*. Colorado Department of Public Health and Environment. https://www.colorado.gov/pacific/cdphe/colorado-medication-take-back-program

de Vries, B. J. M. (2012). *Sustainability Science*. Cambridge University Press.

Dorfman, A., Eyal, T., & Bereby-Meyer, Y. (2014). Proud to cooperate: The consideration of pride promotes cooperation in a social dilemma. *Journal of Experimental Social Psychology*, *55*, 105–109. 10.1016/j.jesp.2014.06.003

Fick, J., Söderström, H., Lindberg, R. H., Phan, C., Tysklind, M., & Larsson, D. G. J. (2009). Contamination of surface, ground, and drinking water from pharmaceutical production. *Environmental Toxicology and Chemistry*, *28*(12), 2522–2527. 10.1897/09-073.1

Fischbacher, U., Gächter, S., & Fehr, E. (2001). Are people conditionally cooperative? Evidence from a public goods experiment. *Economics Letters*, *71*(3), 397–404. 10.1016/S0165-1765(01)00394-9

Gintis, H. (2000). Beyond Homo economicus: Evidence from experimental economics. *Ecological Economics*, *35*(3), 311–322. 10.1016/S0921-8009(00)00216-0

Glassmeyer, S. T., Hinchey, E. K., Boehme, S. E., Daughton, C. G., Ruhoy, I. S., Conerly, O., Daniels, R. L., Lauer, L., McCarthy, M., Nettesheim, T. G., Sykes, K., & Thompson, V. G. (2009). Disposal practices for unwanted residential medications in the United States. *Environment International*, *35*(3), 566–572. 10.1016/j.envint.2008.10.007

Huang, Y., Kendrick, K. M., Zheng, H., & Yu, R. (2015). Oxytocin enhances implicit social conformity to both in-group and out-group opinions. *Psychoneuroendocrinology*, *60*, 114–119. 10.1016/j.psyneuen.2015.06.003

Jordan, J. J., Hoffman, M., Nowak, M. A., & Rand, D. G. (2016). Uncalculating cooperation is used to signal trustworthiness. *Proceedings of the National Academy of Sciences*, *113*(31), 8658–8663. https://www.pnas.org/content/113/31/8658.short

Kümmerer, K. (2009). The presence of pharmaceuticals in the environment due to human use – present knowledge and future challenges. *Journal of Environmental Management*, *90*(8), 2354–2366. 10.1016/j.jenvman.2009.01.023

Kurzban, R., McCabe, K., Smith, V. L., & Wilson, B. J. (2001). Incremental commitment and reciprocity in a real-time public goods game. *Personality and Social Psychology Bulletin*, *27*(12), 1662–1673. 10.1177/01461672012712009

Matson, P., Clark, W. C., & Andersson, K. (2016). *Pursuing Sustainability: A Guide to the Science and Practice*. Princeton University Press.

McFee, R. B., & Caraccio, T. R. (2006). "Hang Up Your Pocketbook"—An easy intervention for the granny syndrome: Grandparents as a risk factor in unintentional pediatric exposures to pharmaceuticals. *The Journal of the American Osteopathic Association*, *106*, 405–411.

Melis, A. P., & Semmann, D. (2010). How is human cooperation different? *Philosophical Transactions of the Royal Society B: Biological Sciences*, *365*(1553), 2663–2674. 10.1098/rstb.2010.0157

Nowak, M., & Highfield, R. (2011). *SuperCooperators: Altruism, Evolution, and Why We Need Each Other to Succeed*. Simon and Schuster.

Schill, C., Anderies, J. M., Lindahl, T., Folke, C., Polasky, S., Cárdenas, J. C., Crépin, A.-S., Janssen, M. A., Norberg, J., & Schlüter, M. (2019). A more dynamic understanding of human behaviour for the Anthropocene. *Nature Sustainability*, *2*(12), 1075–1082. 10.1038/s41893-019-0419-7

Semmann, D., Krambeck, H.-J., & Milinski, M. (2003). Volunteering leads to rock–paper–scissors dynamics in a public goods game. *Nature*, *425*(6956), 390–393. 10.1038/nature01986

Van Lange, P. A. M., Joireman, J., Parks, C. D., & Van Dijk, E. (2013). The psychology of social dilemmas: A review. *Organizational Behavior and Human Decision Processes*, *120*(2), 125–141. 10.1016/j.obhdp.2012.11.003

Vollmer, G. (2010). Disposal of pharmaceutical waste in households – A European survey. In K. Kümmerer & M. Hempel (Eds.), *Green and Sustainable Pharmacy* (pp. 165–178). Springer. 10.1007/978-3-642-05199-9_11

Yan, X., & Jiang, P. (2018). Effect of the dynamics of human behavior on the competitive spreading of information. *Computers in Human Behavior*, *89*, 1–7. 10.1016/j.chb.2018.07.014

# 10 Regional Sustainable Technology Systems

*Michael Narodoslawsky*
Technical University of Graz, Graz, Austria

**CONTENTS**



## 10.1 INTRODUCTION

There are many reasons for a fundamental change of the resource basis of human society. The most pressing is certainly climate change, which requires the de-carbonization of the human economy within a time span of a few decades. Other factors, however, intensify the pressure to transition away from our current mainly fossil resource basis. Examples are the increasing pollution, loss of biodiversity, and political as well as economic risks of dependency on unstable or outright hostile states. There is broad consensus that in the future, the human society shall predominantly utilize "renewable" resources, minimizing its consumption of fossil and mineral raw materials.

In order to understand the implications of such a transition of the resource basis for engineering, it is worthwhile to step back and look at the "bigger picture" of such a change. In particular, we may ask a) what resources we mean and b) what their characteristics are and how they differ from those currently used.

These differences in the characteristics of resources will have profound impact on technologies and the structure of economy. Engineers will have to rethink their role in societal development as well, away from just expert technology designers towards partners of other actors in the quest to achieve sustainable development. For this new role, they will need new guidelines for their decisions and innovative tools for their professional work.

## 10.2 CHARACTERISTICS OF RENEWABLE RESOURCES

The term "renewable" resources is actually an amalgam of different kinds of resources. They are lumped together in this term, although they differ in quality as well as in the reason why they shall be part of the future resource basis. The first group of resources subsumed under this heading is solar-based resources like solar heat, solar power, wind, hydropower, and biomass as well. The

DOI: 10.1201/9781003167693-10

reason to switch to these resources is clearly to de-carbonize the economy to avoid fatal climate change. The second group is secondary raw materials and products that originate from the recycling of used products. The reason for this group to become more prominent in the future is more complex: Recycling reduces or even avoids waste flows to the environment, thus reducing pollution and degradation of ecosystems and reducing the loss of biodiversity. Moreover, it reduces demand for primary resources and thus avoid their depletion and the impact on ecosystems linked to their provision. A third group consists of miscellaneous terrestrial resources, such as geothermal heat or tidal energy. This chapter will concentrate on the first and second group of renewable resources as the backbone of a sustainable human economy.

It is clear that there is exchange and synergy between these types of resources. Solar-based resources like used timber or biopolymers may become secondary raw materials and will be recycled. Electricity generated by photovoltaics or wind turbines may help utilize geothermal heat by powering heat pumps. These links between resource flows will add to the complexity of future technology systems. The two groups mentioned above will, however, shape the structure of technology systems and are therefore the focus of the argument in this chapter.

Let us first look at where we start: the current fossil and mineral resource basis and the mostly linear resource utilization from mining to processing to manufacturing to product use and finally disposal. The sources of our raw materials are point sources like mines, oil, or gas fields. Point source means that there is a small geographic region that is characterized by high productivity per unit area of the material in question, surrounded by large areas of negligible productivity of the material.

These point sources are usually far from the places where products from the material in question are used. This means that raw materials have to be transported, in many cases on a global scale, to places where they are processed and products are manufactured. As long as transport costs are reasonably low, substantial logistic advantages do not distinguish between different sites. The main comparative advantages between competitors for processing and manufacturing based on a resource then come from the efficiency to generate benefit. This, in turn, depends on technological advantage and economy of scale. This favours large-scale industries, often sited in industrial countries. From these industrial sites, precursors and pre-products will be further distributed to smaller scale industries, final products to retailers, and finally to consumers. After use, products will be disposed of. This typical flow of material is characterized by decreasing the size of transport means along the value chain, from cheap large-scale global transport means like oceangoing ships to smaller scale vehicles like railways and, finally, more expensive trucks and vans. This mode is typically for distribution logistics. Only after product use, there is a collection transport to disposal or recycling sites.

Besides their origin from point sources, fossil and mineral resources share other characteristics that influence the structure and shape of the economy. They tend to have good logistic properties like high transport densities and low humidity. This helps to make transport cheap and level out the disadvantages of processing and/or manufacturing sites far from the source of raw materials. As their sources are mines or oil and gas fields that operate year-round, they are continuously available. Storage, therefore, is not a necessity dictated by properties of resources but by considerations about the stability and economy of supply chains. Finally, the quality of these resources does not change widely, neither from source to source nor over time. This results in the utilization of technologies that are quite similar on a global scale.

Summing up, the characteristics of mineral and fossil resources have allowed for globally quite similar technologies, depending on cheap transport of materials with good logistic properties. In this fossil- and mineral-based economy, technological advantage and economy of scale are the most important factor for success.

The properties of most renewable resources differ markedly from those ascribed above to fossil and mineral raw materials. The first distinction is that solar-based and secondary resources are de-central resources. Solar-based resources depend on solar irradiation as the fundamental source.

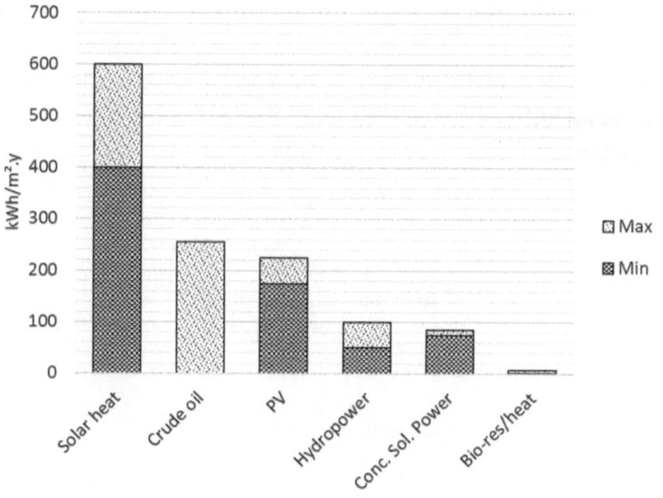


**FIGURE 10.1** Yield in kWh/m²y of different resources (adapted from Narodoslawsky, 2016a).[1]

This is an inherently ubiquitous and de-central source, and its utilization depends on the availability of terrestrial area.

Although secondary resources are not area bound like solar-based resources, their sources are consumers and businesses as well. These sources again are de-central; however, contrary to solar-based resources, they are of higher concentration in urban areas. For all renewable resources, it is therefore necessary to change the current shape of the supply chain. The first step now consists of a collection transport.

A second distinction is the kind of limitation of these resources. Fossil and mineral resources have finite sources that will run out over time. The yield of the sources in terms of flow per unit of time can, however, be increased by increasing investment and labour in their prospection and exploitation. Renewable resources, in contrast, have limited yields per unit of time and unit area for solar-based resources or per consumer or business unit for secondary recycling-based resources. These yields, however, may be sustained infinitely when sources are properly managed. Renewable resources are therefore limited in flow but infinite in time. Figure 10.1 compares some solar-based energy resources with crude oil according to their energy yield per unit area and unit time.

The third and most important distinction concerns the logistic characteristics of resources. Most renewable resources have poor properties for transportation over long distances. They are "short-legged" resources that do not easily run far or fast. Some solar-based resources like electricity from wind power, hydropower, or photovoltaic (PV) installations and heat from solar thermal installations as well are not material resources and require infrastructure like transmission lines or heat pipelines for their collection and distribution as well. Bio-resources, particularly lower-grade bio-resources such as silage and wood chips as well as agricultural waste and residue flows such as manure and straw, are often humid and/or bulky. This increases the transport effort as transport vehicles are not running at their capacity because of volume limitation or much dead weight (especially water) is transported. Table 10.1 compares transport distances achievable with transport means appropriate for the good in question, using 1%of the transported energy.

Table 10.1 highlights the inherent logistic difference of renewable, in particular solar-based resources and fossil raw materials (shaded gray). There are bio-resources that may be easily transported over longer distances, such as cereals or wood pellets. Most other solar-based resources, however, require considerable transport effort. This means that the site of their utilization, or at least their processing for further transport, should lay within a few kilometres from the place of their origin.

**TABLE 10.1**
**Transport Distance for Different Resources with Appropriate Means of Transportation, Using 1% of the Transported Energy (Using Data From Narodoslawsky, 2016a and Krozer & Narodoslawsky, 2019)**

| Resource | Transport Means | Transport Distance [km] |
|---|---|---|
| Heat | District heating pipe | <0.5 |
| Electricity | 10 kV transmission line | 3.3 |
| 380 kV transmission line | 100 | |
| Manure | Tractor | 5.3 |
| Corn silage | | 18 |
| | Loose straw | 23 |
| Wood chips | Truck | 40 |
| Split logs | | 100 |
| Wood pellets | Train (electric) | 475 |
| Corn | | 525 |
| Hard coal | Train (electric) | 1450 |
| Crude oil | Trunk pipeline | 2800 |
| | Ship | 7800 |

Availability of many solar-based resources is either cyclic or intermittent, requiring storage to align provision of resources and demand of products and services. Loss and cost always accompany storage. Table 10.2 shows average efficiencies for energy storage systems. Besides considerable loss of stored energy, all these storage systems show substantial investment cost. This is particularly true for electricity storage. Investment costs of heat storage tend to be at least one order of magnitude lower.

Cost and loss, however, is not restricted to the storage of energy. It also applies to bio-resources, which decay over time. As bio-resources usually are bound to cycles of planting, growth and harvest, they must be stored for months to ensure year-round supply. This requires either storehouses and/or conditioning of bio-resources, such as drying for many crops or silage for green biomass. Otherwise, technologies with a wide range in complexity are necessary to keep bio-resources from decaying. This ranges from adding conservation agents to cooling to keeping goods under protective atmosphere. Besides additional cost and in many cases additional energy, all these storage methods require space and infrastructure, which adds to the logistic disadvantage of renewable resources compared to fossil and mineral raw materials.

**TABLE 10.2**
**Efficiencies for Some Energy Storage Systems**

| | Storage System | Efficiency [%] |
|---|---|---|
| Power-to-gas[2] | Hydrogen | 57–73 |
| | Methane | 50–64 |
| Power-to-gas-to-elctricity[2] | Hydrogen | 34–44 |
| | Methane | 30–38 |
| Electricity[3] | Pumped Hydro | 70–85 |
| | Compressed Air | 40–95 |
| | Li-Ion Battery | 85–95 |

Summing up, renewable resources will fundamentally change the structure of future industry and business. Their sources are inherently de-central, either area based or dependent on consumers. They always require a collection transport first before they can be conditioned or processed. Many of these resources, in particular those that are currently underused such as waste heat, low grade bio-resources, or outright waste flows, have adverse logistic properties. This restricts their feasible distance of transport, in many cases, by orders of magnitude compared to fossil and mineral resources. Their cyclic or intermittent supply requires additional storage to align supply and demand, putting them at an additional economic disadvantage. These characteristics of renewable resources will become formative to future engineering solutions.

## 10.3 BASIC ENGINEERING GUIDELINES FOR RENEWABLE RESOURCE UTILIZATION

The characteristics of renewable resources require a re-thinking of tried and tested engineering concepts. Innovative guidelines for engineering solutions will be necessary to cope with restructuring industry, infrastructure, and supply chains that are sustainable and conform to the challenges of limiting climate change. Based on more than three decades of research covering sustainable development and renewable resource utilization, I propose four basic engineering guidelines for technologies based on renewable resources:

- Respect ecosystems and strive to enhance or at least preserve their quality
- Take responsibility for the whole value chain
- Adapt technical solutions to their local/regional context
- Increase resource utilization efficiency by integrating technologies

These guidelines will be briefly elaborated on in this section of the chapter.

### 10.3.1 Respect Ecosystems and Strive to Enhance or at Least Preserve Their Quality

Sustainable development can only be successful when ecosystems, both on a global and local/regional level, continue to function. Renewable resources, in particular bio-resources as the basis for food and many industries, are critically dependent on ecosystem functions. So is nature's ability to reintegrate residues from human consumption. Increasing the quality of ecosystems both globally and locally/regionally or at least preserving them and ensuring the co-evolution of humanity and nature is, therefore, not only an ethical imperative but also a necessity for human survival.

For decisions about utilizing particularly solar-based resources, it is therefore important to keep in mind that ecosystem functions are not restricted to cater to human consumption. Haberl et al., 2007, estimate that human consumption already appropriates almost a quarter of net primary production, which is the total carbon fixed by photosynthesis globally (as defined, e.g., by Beer et al., 2010). This means that humanity has become a serious contender for natural income from solar irradiation. Narodoslawsky (2014) argues that the utilization of bio-resources is arguably the most encompassing impact on ecosystems. It changes landscapes and puts enormous flows of nutrients like nitrogen and phosphorus in motion, and impacts local and global water cycles by changing the quality of soil and land cover. Loss of organic matter in soils due to tillage can contribute to climate change and severely reduce soil's capacity to retain water, disrupting regional water cycles and increasing the threat of flooding, erosion, and landslides. Finally, agriculture and forestry are responsible for a large part of biodiversity loss.

Ecosystems, however, offer a wide range of services encompassing the fundament of both human life and natural evolution. Narodoslawsky (2014) lists a number of services of ecosystems.

Social services for humankind include the provision of food, the basis for jobs, and enabling recreation, to name just a few. Environmental services include regulation of climate and sequestering of carbon, regulation of water cycles and water supply, reduction of risk of natural catastrophes like the impact of storms, landslides, and floods and, of course, support of biodiversity and natural evolution.

Ecosystems are competing for solar irradiation, area, fertile soil, and clean water on land and in oceans as well, with society's hunger for solar-based resources. For engineers, it is imperative to minimize the overall impact of value chains their technologies are part of on ecosystems. This means that they minimize resource use as well as waste generation across the whole value chain. Moreover, this imperative call for avoidance of infringement on pristine ecosystems and careful evaluation and preservation of ecosystem functions of those areas that are already under human management.

### 10.3.2 Take Responsibility for the Whole Value Chain

Technologies are never stand-alone entities but are always part of supply and value chains, which, in turn, are embedded in social systems that are themselves embedded in the ecosphere. Any decision taken on the level of a certain technological step, therefore, has implications on the social and ecological performance of the whole value chain. Utilizing a certain raw material causes social and ecological impacts upstream along the supply chain. The design of conditioning, manufacturing processes, and their efficiency define the "ecological Rucksack" (Lettenmeier, 2009) that the material, product, or service carries all the way down the supply chain. Finally, product design will define the social impact and ecological performance of the product during its lifetime and the pressure it exerts on the environment when it is finally disposed of. For engineers who strive for sustainable solutions, taking responsibility for all impacts up and down the value chain of their decisions is a professional imperative.

In the light of using renewable resources, this imperative even gets more important. The previous section on the properties of renewable resources already mentioned the profound difference in the logistic between conventional fossil and mineral resources compared to renewable resources. Temporal variability of the availability of renewable resources and their generally adverse logistic parameters requires a greater technological effort for transport and storage along the supply chain, increasing the complexity of engineering decisions. Moreover, both used products as the basic resource for secondary materials and bio-resources tend to have higher variability in composition and quality as well, resulting in more diverse and complex supply chains with less standardized intermediate products compared to conventional fossil-based value chains. This makes decisions about sourcing raw materials and energy at the same time non-trivial and decisive for the sustainability of the whole value chain. "Life cycle thinking and responsibility" thus becomes a fulcrum for successful engineering when renewable resources are concerned.

### 10.3.3 Adapt Technical Solutions to Their Local/Regional Context

The previous section on the properties of renewable resources already highlighted the fact that supply chains will become spatially closer because of the poor logistic properties of many renewable resources. This, in turn, means that the sources of raw materials and energy will become more local/regional.

Reorienting supply chains away from global, "long-legged" fossil and mineral resources towards "short-legged" local/regional resources has many implications for engineering. It is obvious that regions differ according to their natural endowment and hence according to the resources they may provide. Therefore, technologies utilizing these varied resources will be more diverse than current technologies based on fossil resources and, in particular, highly standardized intermediates that are currently globally traded. Combined with the generally higher complexity of supply chains as discussed above, this results in the need for technology solutions that are custom-made for the resources offered and the existing social, cultural, and economic structure in the region.

A second obvious aspect of sourcing regionally is the smaller size of industrial processes. Restriction in transport distance due to adverse logistic parameters, combined with limited yields of renewable resources either with respect to an area or to a consumer, result in limited resource capacity within a certain region. This means that the current tried and tested strategy to increase the efficiency of industrial processes by making them bigger does not apply anymore. Engineers must thus become even more innovative, supplanting the restricted dimension of increasing the size of processes by more integration along the value chain and/or creative utilization of by-products and services as well as with technological innovations in order to stay competitive.

Finally, as required above, taking responsibility along the whole value chain, now takes the form of taking concrete local/regional responsibility in cooperation with concrete local/regional actors. Those actors include regional farmers and logistic companies to provide resources, utilities to provide energy or distribute excess energy from the process, other companies who provide intermediates or utilize by-products, and consumers who use products or services and/or provide disposed of products and materials. Engineering decisions thus have to be embedded in a broader socio-economic planning process to utilize regional natural as well as secondary resources in order to supply regional citizens with necessary goods and services while optimizing regional benefit on the resources provided. These planning processes have to take the carrying capacity of local/regional ecosystems and the need for their improvement into account.

A particular aspect of this regional responsibility is the management of by-products and services. Many industrial processes use high quality energy and degrade it within their processes to lower-grade energy, such as off-heat. In order to stay competitive, this "by-service" has to be put to best use, either supplying residential areas with heat or using it for other processes with lower temperature requirements, such as drying of bio-resources (Stoeglehner et al., 2016). Material by-products like ashes or solid and liquid organic residues that cannot be used in other processes should be reintegrated into ecosystems in order to return nutrients and improve the quality of soils. The quality and quantity of these residual flows have to be carefully controlled, and they have to be conditioned if necessary to ensure their benefit to maintaining the productivity of local/regional ecosystems.

The adaptation of technologies to local/regional contexts requires from engineers' innovation and willingness to cooperate with regional actors in various sectors and with diverse backgrounds. Traditional engineering guiding principles, such as focussing on optimizing single technologies and applying economy of scale for efficiency improvement, are no longer useful in this context. As diverse as local and regional settings are the challenges to engineers. Although solutions that are successful in other regional settings may be inspirational in the search for technologies adapted to a certain region, they can almost never be simply copied, given the differences in natural settings and actor configurations between regions.

### 10.3.4 Increase Resource Utilization Efficiency by Integrating Technologies

Renewable resources have limited yields per area unit or consumer. Following the argument of previous sections, their logistic parameters limit their feasible transport distance. Thus, the resource capacity available at a certain industrial site is limited and cannot easily be increased. Increasing the capacity of an industrial process is, therefore, no promising strategy to increase efficiency and stay competitive. This leaves increasing the output from a unit of a resource by utilizing all by-products and residues as sellable products as a way to improve competitiveness, in particular for material resources such as bio-resources and used products. These resources are complex with regard to their composition and thus offer a broad portfolio of possible product lines.

This approach becomes even more interesting (and arguably more confusing) by the fact that on the regional level there are usually more than one renewable resource. The value chains of these resources may be combined and integrated in order to utilize synergies and/or generate more products. In the field of bio-resources, this approach is known as the "bio-refinery"[4] approach, using fossil oil refineries that generate a broad portfolio of products and services from a single resource as a model.

There are many examples of this approach already realized. They range from the complex (e.g., the industrial ecosystem around an enzyme factory in Kalundborg, Denmark[5]) to the down-to-earth bio-energy system in Mureck, Austria.[6] In the latter, a small town of roughly 3,500 inhabitants is supplied with heat, electricity, and (partly) bio-diesel by a combination of technologies, including a bio-diesel plant based on used vegetable oil, a bio-gas plant using the glycerol phase of the bio-diesel plant, together with manure and corn silage. This plant supplies the town with heat in summer and a baseload of heat in winter while producing 8,500 MWh of electricity per year. Heat in winter is supplied by a biomass heating plant with two 2 MW biomass boilers. Demand peaks can be met by an additional 2 MW biogas boiler. In total, these plants provide about 6,000 MWh of heat. In addition to these bio-resource-based plants, the system also includes a PV park with almost 3 $MW_p$, partly mounted on industrial buildings and greenhouses that supply the town with vegetables. Farmers in the immediate surroundings use the biogas manure as fertilizer.

Using renewable resources within a limited regional setting to address regional demand in goods and services and utilizing available resources thoroughly not only calls for efficient single technologies but also optimized resource-technology-supply networks. These networks include a multitude of actors. Planning such networks is a socio-economic-political process in which engineers provide crucial input. The operation of such networks, in turn, alters regional societies and economies, as is shown by Tomescu (2005) for the example of Mureck.

## 10.4 TOOLS TO HELP ENGINEERS ESTABLISH REGIONAL SUSTAINABLE TECHNOLOGY SYSTEMS

The previous sections have explained that a change in the resource basis away from fossil towards renewable resources will have a profound impact on the structure and logic of industrial technology. Sustainable technological solutions will have to take into account the spatial context; they will become smaller, more varied, and integrated. Engineering will become an integral part of regional development planning, requiring engineers to open up to other actors and disciplines. For these new tasks, engineers will need innovative tools. In this last section, we discuss requirements for as well as characteristics of some tools in two crucial areas: evaluation of ecological performance and optimization of resource-technology-supply networks. These two fields are particularly important for the contribution of engineers to regional development planning.

There is no comprehensive evaluation tool for sustainability that measures the overall performance of a technological solution. This requires using separate measures for ecological, economic, and social aspects. For engineers, ecological performance is of particular interest as it is directly linked to design decisions. The offer of evaluation methods in this area is overwhelming and confusing. Therefore, it is necessary to provide some methodological help to enable engineers to pick the right tool for their needs.

As the very term "evaluation" implies, sustainability measures are always reflecting a certain value system and a certain assumption of what sustainable technologies should achieve (Narodoslawsky & Shazahd, 2015). Many measures evaluate various forms of efficiencies of technologies. This applies to Material Intensity (Lettenmeier et al., 2009) which measures efficiency in terms of material flows triggered by a certain technology. More advanced measures, such as EMERGY (Odum, 1996), measure the efficiency of utilizing low-entropy input, usually with reference to solar irradiation. This amounts to evaluating the efficiency of the use of natural income. All these measures have in common that they consider sustainable engineering as increasing efficiency. They do not have a direct link to the limitations of ecosystems regarding either their productivity or their ability to absorb residues from human activity.

Another class of measures touted as sustainability evaluation methods may be called problem-oriented. The most prominent of them is the carbon footprint (ISO, 2013), measuring the emission of greenhouse gases of a technology or a value chain. There exist other comparative measures, such as the water footprint (Hoekstra & Chapagain, 2008), that evaluate the water consumption of technologies. Guinée et al. 2001 combined many such measures into a common framework, the CML method, that is currently the basis of many life-cycle analysis (LCA) studies.

All problem-oriented methods have in common that they focus on a certain environmental problem (climate change for the carbon footprint, water scarcity for the water footprint, etc.) and then measure the impact of a technology on this particular ecological aspect. The measures again do not have any regard for the limitations of ecosystems. They regard sustainable engineering as providing solutions that have a low impact on particular environmental problems. The choice of the problems to focus on, as well as the way of weighing those indicators if more than one problem is evaluated, reflect the value set of the user.

Finally, there are measures that base evaluation explicitly on sustainability. These are, in particular, evaluation methods in the ecological footprint (EF) family, notably the EF (Rees, 1992) itself and the sustainable process index (SPI) (Narodoslawsky & Krotscheck, 1995). These evaluation methods measure the area necessary to embed technologies or whole value chains sustainably into the biosphere. The larger the area is, the less environmentally competitive the evaluated technology. As the sum of all available area is finite on our planet, these measures also take the natural limitations of ecosystems into account. The major difference between the two methods is that the EF does not evaluate the impact of residues from human activities, whereas the SPI does. This makes the latter more comprehensive. Moreover, the EF does not effectively discern different forms of energy provision, making it hard to distinguish between processes based on renewable resources and those that utilize fossil materials. For the SPI, there exists an open software to evaluate technologies[7] that contains data from almost 2,000 processes already evaluated. With this database, most value chains can easily be put together from existing data. The software is also capable of evaluating value chains containing internal loops, e.g., if a bio-diesel production shall be evaluated where agriculture and transport are themselves based on the bio-diesel produced.

In general, engineers who want to employ ecologic evaluation have to choose their measuring tool carefully. The following guideline may help them come up with a feasible choice (see Narodoslawsky & Shazahd 2015):

- Define the normative framework to which decisions are oriented. Establish an explicit, clearly stated and well-argued goal for the activities that are subjected to evaluation, including what is within and outside the responsibility of the actors.
- Deduce from this normative framework the evaluation method whose normative basis conforms best to the defined framework.
- Explain the normative framework and the reasoning why you choose this particular evaluation method whenever relaying evaluation results to other actors in the development process.
- When comparing different evaluation results on the same subject, make sure that the normative frameworks of the evaluations are commensurable. If they are not, comparisons make no sense!

A second field where engineers have to choose the right tool is the optimization of regional technology networks. As already emphasized earlier, utilizing renewable resources within the regional context that is often defined by adverse logistic properties of many resources leads to closely interlinked technology networks rather than coexisting stand-alone technologies.

Optimizing these networks that operate within the ecological, societal, and economic framework and limitations of a concrete region is a joint multidisciplinary effort of many actors. The optimization is not a conventional optimization of parameters within a given structure. It should identify the optimal structure of the technology network linked by the most advantageous flows of energy, material, and benefit between its nodes.

A method that has proven its merits in optimizing regional technology networks is the process network synthesis (PNS) method (Friedler et al., 1995). Application of this method to regional sustainable technology networks has been reported by Narodoslawsky et al., 2016b and Lam et al. 2010, among many others. This method is based on the bi-partite graph (P-graph) theory, using combinatorial rules to define pathways from resources to products. A thorough explanation and helpful software for optimization can be found on the webpage of the p-graph organization.[8] The advantage of this method is that it is quick, does not fall for local optima, and provides not only the most optimal solution but also any number of the runner-ups. The latter property is particularly important for optimization within the framework of regional development planning, as it allows actors to see trade-offs as well as solutions that are not the overall optimum but nevertheless preferable because of ecological, social, or cultural considerations. Optimization in the framework of regional development planning is never a simple calculation of a technology structure that is then subsequently realized. It is much more of a learning process. Within the discourse, actors will want to change boundary conditions such as expectations of future prices for goods or changes in productivity due to climate change and other ecological restrictions. This makes optimization an iterative process, and the optimization and evaluation methods are ingredients in the encompassing development discourse on how to utilize regional resources best.

## 10.5 CONCLUSIONS

The utilization of renewable resources, regardless if they are secondary materials collected after product use or solar-based resources, will alter the future structure of the economy in general and industry in particular. The sources of these resources are inherently de-central, either consumers or planetary surface area, to capture solar irradiation. The current structure of fossil and mineral resource-based economy, where point sources generate resources usually far away from the point where products from these resources are demanded, has led to a globalized economic system. Raw materials and processed intermediates do not differ widely in quality; resources are constantly available but ultimately limited by finite sources. Good logistic properties and long distances make these resources "long-legged" with far and fast global transport. As transport from source to processing and/or manufacturing is necessary in any case, no site is distinguished by geographical advantage. Together with highly standardized raw materials and intermediates, this leaves the economy of scale as the most promising strategy for high efficiency and economic success.

Renewable resources, in contrast, exhibit infinite availability in terms of time but are restricted by productivity either of ecosystems or recycling. Solar-based resources often have cyclic or intermittent sources such as cyclic solar irradiation, intermittent wind power, or seasonal growing cycles for bio-resources. This requires the integration of storage in value chains, making them more costly. Moreover, many promising resources such as solar heat, off-heat from industry, or lower grade bio-resources such as straw and manure have adverse logistic parameters. This means that they must be utilized close to the place where they are generated. This makes many renewable resources "short-legged," restricting transport distances to local and regional dimensions. Together with the necessity to manage the sources of renewables within the natural, social, and economic limitations of the spatial context of their generation, this makes renewable resources inherently local/regional.

Utilizing these resources within a specific regional context requires new approaches to engineering. Regional ecosystems have to be managed, and their quality improved. The complex value chains require the engineer to take responsibility for the whole life cycle. Technologies have to be adapted to regional natural, social, cultural, and economic frameworks. Finally, as regional technologies are limited in scale by restricted transport distances and yields per area, efficiency must be obtained by technology integration instead of the economy of scale. Regional resource utilization will thus lead to technology networks, linking resources and demand while optimizing regional benefits. The shape of such integrated networks will be strongly influenced by the resource with the strongest limitation on transport, which is often heat. Engineering will become an integral part of regional sustainable development planning, requiring engineers' cooperation with many actors across diverse sectors. For this, new tools such as efficient ecological evaluation and optimization of the structure of regional technology networks are necessary. Reshaping the economy and industry to conform to the requirements of renewable resources is a joint enterprise by many actors, with engineers playing a crucial role in the transformation process.

## NOTES

1 Yields are related to overall area, e.g., the area of the whole oil field or a solar energy park.
2 Sterner et al. 2015.
3 EESI 2019.
4 See https://task42.ieabioenergy.com/ for more information on bio-refinery systems.
5 See more detail on http://www.symbiosis.dk/en/; last retrieved October 2022.
6 For a list of involved technologies and information on the plant see http://energieschaustrasse.at/index.php/en/energy-resting-area/mureck; last retrieved October 2022.
7 See https://spionweb.tugraz.at/; last retrieved October 2022. This webpage also provides more information on the method, literature concerned with the SPI, and examples of the application of the evaluation method.
8 See http://p-graph.org/ last retrieved October 2022.

## REFERENCES


Beer, C., Reichstein, M., Tomelleri, E., Ciais, P., Jung, M., Carvalhais, N., Rödenbeck, C., Arain, M.A., Baldocchi, D., Bonan, G.B., Bondeau, A., Cescatti, A., Lasslop, G., Lindroth, A., Lomas, M., Luyssaert, S., Margolis, H., Oleson, K.W., Roupsard, O., Veenendaal, E., Viovy, N., Williams, C., Woodward, F.I. and Papale, D., *Terrestrial Gross Carbon Dioxide Uptake: Global Distribution and Covariation with Climate*. Science, 2010, 329 (5993), 834–838.

EESI: *Fact Sheet Energy Storage, 2019*; https://www.eesi.org/papers/view/energy-storage-2019. Last retrived October 2022.

Friedler, F., Varga, J.B., and Fan L.T., *Decision-mapping: a tool for consistent and complete decisions in process synthesis*. Chemical Engineering Science, 1995, 50, 1755–1768.

Guinée, J. B., Huppes, G., and Heijungs, R., *Developing an LCA guide for decision support*. Env. Mgnt. & Health 2001, 12, 301–311.

Haberl, H., Erb, K.-H., Krausmann, F., Gaube, V., Bondeau, A., Plutzar, Ch., Gingrich, S., Lucht, W., and Fischer-Kowalski, M., *Quantifying and mapping the human appropriation of net primary production in earth's terrestrial ecosystems*. Proc. Natl. Acad. Sci. USA, 2007, 104 (31), 12942–12947.

Hoekstra, A.Y., and Chapagain, A.K., *Globalization of water: Sharing the planet's freshwater resources*. Oxford, UK: Blackwell Publishing. 2008.

ISO, ISO/TS 14067. *Greenhouse gases — Carbon footprint of products — Requirements and guidelines for quantification and communication*. London/GB: BSI Standards Limited, 2013.

Krozer, Y., Narodoslawsky, M. (Eds.), *Economics of Bioresources –Concepts, Tools, Experiences*. Cham (Switzerland): Springer Nature, 2019.

Lam, H.L., Varbanov, P.S., and Klemeš, J.J., *Optimisation of regional energy supply chains utilising renewables: P-graph approach*. Computer Aided Chemical Engineering, Elsevier, 2010, 34(Issue 5), 782–792.

Lettenmeier, M., Rohn, H., Liedtke, C., and Schmidt-Bleek, F., *Resource Productivity in 7 steps: How to develop eco-innovative products and services and improve their material footprint*. Wuppertal Institute for Climate, Environment and Energy, 2009, Wuppertal/Germany.

Narodoslawsky, M., and Krotscheck, C., *The sustainable process index (SPI): Evaluating processes according to environmental compatibility*. J. of Hazardous Materials, 1995, 41 (2+3), 383–397.

Narodoslawsky, M., *Utilising Bio-resources: Rational Strategies for a Sustainable Bio-economy*. Vienna/Austria: Manu:scripts, ITA, 2014. http://epub.oeaw.ac.at/ita/ita-manuscript/ita_14_02.pdf Last retrieved October 2022.

Narodoslawsky, M., and Shahzad, K., *What Ecological Indicators Really Measure – The Normative Background of Environmental Evaluation*. Chem. Engn. Transact., 2015, 45, 1807–1811.

Narodoslawsky, M., *Towards a Sustainable Balance of Bio-resources Use between Energy, Food and Chemical Feed-stocks*. Foundations and Trends® in Renewable Energy, 2016a, 1(2), 3–68.

Narodoslawsky, M., Cabezas, H., Maier, St., and Heckl, I., *Using Regional Resources Sustainably and Efficiently*. Chem. Engng. Progr., 2016b, Oct., 48–54.

Odum, H.T., *Environmental Accounting. EMERGY and Environmental Decision Making*. Toronto: John Wiley & Sons, 1996.

Rees, W. E., *Ecological footprints and appropriated carrying capacity: what urban economics leaves out*. Environment and Urbanisation, 1992, 4(2), 121–130.

Sterner, M., Thema, M., Eckert, F., Lenck, T., and Götz, P, *Bedeutung und Notwendigkeit von Windgas für die Energiewende in Deutschland, Forschungsstelle Energienetze und Energiespeicher (FENES) OTH Regensburg*. Energy Brainpool, Studie im Auftrag von Greenpeace Energy, Regensburg/Hamburg/Berlin, 2015; http://www.greenpeace-energy.de/fileadmin/docs/sonstiges/Greenpeace_Energy_Gutachten_Windgas_Fraunhofer_Sterner.pdf. Last retrived October 2022.

Stoeglehner, G, Neugebauer, G., Erker, S., Narodoslawsky, M., *Integrated Spatial and Energy Planning-Supporting Climate Protection and the Energy Turn with Means of Spatial Planning*. Springer Cham (Switzereland), 2016.

Tomescu, M., *Innovative Bioenergy Systems in Action: The Mureck bio-Energy Cycle: Synergistic Effects and Socio-economic, Political and Sociocultural Aspects of Rural Bioenergy Systems,* in IIIEE Reports 2005:6, University of Lund/S, 2005, https://lup.lub.lu.se/search/publication?q=%22tomescu%22; last retrieved October 2022.

# 11 Renewable Microgrids as a Foundation of the Future Sustainable Electrical Energy System

*Anna Trendewicz*
Future Ventures Management, Berlin, Germany

*Eric C.D. Tan and Fei Ding*
National Renewable Energy Laboratory, Golden, Colorado, USA

## CONTENTS



## 11.1 INTRODUCTION

As illustrated by P. Hafner [1], sustainability is the dynamic balance between efficiency and resilience. Too much focus on efficiency leads to brittleness and too much emphasis on resilience results in stagnation. Our electricity system has historically been dominated by the large-scale fossil fuel based generation plants and long-distance transmission and distribution lines. The power generation patterns were stable and predictable. However, there are currently several disruptions to that picture at play.

Firstly, as the capacity of intermittent renewables (wind and solar PV) in the system increases, the electricity generation patterns exhibit more fluctuations that require enhanced balancing capabilities. According to the International Energy Agency (IEA), an increase in renewable energy generation capacity of 12% per year over the next decade is required to reach the decarbonization goals for the electricity sector in net-zero energy (NZE) scenario [2]. This trend is expected to be accompanied by a strong increase in energy storage capacity and demand-side flexibility services.

Secondly, the increase in the exposure to extreme weather events driven by climate change greatly challenges the ability of the centralized power system to restore from shocks quickly. This can be illustrated in the example of the power outage in Texas in the winter of 2021, where more than 4.5 million (mln) households and businesses were without power for several days, with the overall damages estimated of over $195 bn [2]. The analysis of the situation concluded that the underlying cause was the lack of interconnection and the lack of contingency investment driven by price competition in the wholesale electricity market. According to the Center for Climate and Energy Solutions, the top 10 most costly disasters since the year 2000 were all due to hurricanes

DOI: 10.1201/9781003167693-11

and jointly accounted for $772 bn [3]. In addition, the Carbon Brief analysts found that 70% out of 405 mapped extreme weather events are amplified by human-caused climate change [4]. Adapting to progressing climate change is becoming essential, and microgrids equipped with battery storage capable of operating as energy islands could play a key role as solutions to natural disaster response.

Lastly, the role of the power system is becoming increasingly important for the global economy. Electricity is steadily replacing oil in many sectors, including transport, heating, and industry (e.g., electric vehicles, heat pumps, air-conditioning, etc.). As a result, the global electricity demand is expected to rise by over 40% from today's 23,300 TWh to 33,200 TWh in 2030 in the IEA NZE scenario [5]. In addition, the demand patterns are likely to change (e.g., increased peak demand due to EV charging and air conditioning), potentially introducing additional stress on the power grid and further amplifying system balancing challenges.

The convergence of these three trends leads to a crisis in our electrical energy system. This crisis is both a danger and an excellent opportunity to tap into the potential of decentralized renewable energy sources and energy storage by creating adaptive, quick-responding, and resilient microgrids. These microgrids also have the unique opportunity to leverage the potential of coupling electricity, mobility, and heating and cooling sectors and apply circular economy principles. The circular economy strives to slow down and close material loops, use renewable energy as a foundation for all activities, and deploy non-toxic materials [6]. Although over 100 various definitions of circular economy exist, the most commonly known was developed by the Ellen MacArthur Foundation and defines circular economy as a framework for a restorative and generative economy based on three main principles: 1) design out waste and pollution, 2) keep products and materials in use, and 3) regenerate natural systems. An overview of additional approaches has been summarized by E. Tan and P. Lamers [6]. Microgrids greatly support the circular economy mission by supplying the demand for electricity, mobility and thermal comfort with renewable energy and thus eliminating the use of limited natural resources (e.g., oil, gas, coal) for these purposes. Coupled with innovative, local business models, they could also contribute to thriving local economies and thus increase the social well-being.

## 11.2 MICROGRIDS' MARKET POTENTIAL

Microgrids are defined as local power systems that are equipped with their own interconnected generation sources. They can work together with and independent of the medium- and high-voltage grid. Microgrids can be very diverse in size and form as they deploy combinations of local generation sources to supply specific energy needs. According to an National Renewable Energy Laboratory (NREL) study [7], they could be classified based on the market segment as follows: campus, community, commercial projects, utility microgrids, and remote microgrids. Based on the case studies analyzed in the report, distributed energy sources (DERs) include diesel, natural gas, combined heat and power (CHP), biofuel, solar photovoltaic (PV), wind, and fuel cell and energy storage. Microgrid sizes can be in the range of 100 kW to multiple MW scale [8].

The benefits of microgrids include the following:

- Integration of renewable energy sources (e.g., rooftop solar PV, wind power, local bioenergy sources, etc.),
- Integration of grid flexibility solutions (e.g., battery storage, flywheels, EV charging, and demand-side flexibility),
- High efficiency due to local energy generation and consumption, thus avoiding transmission and distribution losses,
- Deferred investment in power transmission and distribution infrastructure due to improved local energy management and balancing,

- Increased grid resilience to disruptions (e.g., extreme weather events) by providing local power systems with the capability to operate in an island mode,
- Reduced electricity prices in locations depending on the use of diesel generators (e.g., islands, remote communities).

The challenges of microgrids include the following:

- Uniqueness of each project and the need for a customized approach introduce challenges in financial assessment and might increase the risk perception of some investors,
- Financial incentives (e.g., tax credits) differ by region, technology, and ability of a developer to access them,
- Regulatory and legal challenges differ by region (e.g., lack of harmonized definition of a microgrid) and could lead to increased costs, higher risk due to regulatory uncertainty, or restrictions potentially hindering project development,
- High complexity of the system and specialized skills required for operation and maintenance.

The overview of existing microgrids in the United States performed by NREL[4] provides detailed information about the type of generation sources and cost breakdown classified per market segment based on 80 case study projects. The general conclusions indicate that the majority of microgrids (80% of installed capacity) still leverage fossil fuel generation, namely, combined heat and power (CHP), natural gas, and diesel generators. These are being increasingly complemented by solar PV and energy storage. The mean normalized microgrid costs (in $/MW) are lowest for the community segment ($2.1 mln/MW) followed by utility scale ($2.5 mln/MW), campus ($3.3 mln/MW), and commercial segment being the highest ($4.1 mln/MW). The project specific cost can vary widely within each segment, driven by unique, local conditions. Community, utility, and campus cases account for 90% of the capacity covered in the database, while the campus segment accounts for more than 50% of the covered power capacity alone. While the database might not be exhaustive, it sheds some light on the status quo of microgrid development today, which is the very onset of the journey of the energy system transformation.

The critical enabler for the successful integration of microgrids into the existing power grid and harmonious, stable operation is the forming inverter-based resource control [9]. The term relates to non-traditional generation sources (e.g., solar PV and wind) and energy storage technologies (e.g., batteries), which are connected to the power grid through power electronic inverters. Grid-forming controls provide the much-needed capability to restore the system after a blackout and enhance grid resilience.

As the penetration of renewable generation sources and microgrids increases, the power grid will need to synchronize and orchestrate the traditional control responses achieved with rotating turbine generators with the inverter-based resources in the emerging hybrid power system. Integration of inverter-based resource controls is expected to progress in stages, starting from local microgrids, through islands, smaller and weaker grid systems, all the way to the national grid system, according to a joint research project led by NREL [8].

Microgrids are currently in the early stage of development marked by propagation of small-scale demonstration projects e.g., Ta'u Island (American Samoa), King Island (Australia), El Hierro (Canary Islands, Spain), which provide practical knowledge for further development and scale up. The mid-term phase (3–15 years) is expected to test inverter-based control strategies integration on larger islands (e.g., Oahu, Hawaii) and address sub-transmission security, inter-operability, and system stability issues. The final long-term phase (7–20 years) is expected to be performed on mainland grids to address issues associated with the transmission security, inter-operability, and stability of the entire system. The full integration of inverter-based controls in the power system is expected to take up to 30 years.

The practical experience should be complemented with feasibility studies and roadmaps to assess technical aspects (e.g., communication protocols, security), future system architecture, and control strategies. Finally, successful integration of microgrids requires adaptation of the regulatory frameworks as well as development of technical standards for the new system architecture. The regulatory aspects are discussed in detail in Section 11.4.

According to the Center for Climate Solutions, microgrids currently supply less than 0.2% of the electricity in the United States [8]. However, the microgrids capacity is expected to grow because of its ability to incorporate renewables and respond to natural disaster events. At the time of this writing, the United States expects microgrids to grow by 3.5 times to reach 32.5 GW capacity by 2030 [10]. Additionally, the forecast spending on renewable microgrids assets is expected to generate $72.3 bn in GDP and create 500,000 jobs [11].

According to the European Commission Report [10], energy communities in the EU could own 17% of wind capacity and 21% of solar capacity by 2030, and nearly 50% of EU households are expected to generate renewable energy by 2050. Collectively local energy communities formed by households, businesses, and public buildings could potentially own up to 45% of Europe's renewable energy generation by 2050 [8]. Most of the community-owned microgrids are expected to be grid-connected, with some stand-alone systems on islands and in remote areas. The estimated number of renewable energy cooperatives in Europe is currently around 3,500 [8], found mainly in Western Europe, with Germany and Denmark jointly accounting for 70% of the total number of this type of energy organization, contributing 1,750 and 700, respectively.

## 11.3 MICROGRIDS OF THE FUTURE

The future renewable microgrids will incorporate advanced information and communication technology, including smart meters, artificial intelligence (AI)-driven control algorithms, and blockchain, to create a transparent, intelligently balanced, and economically optimized system. When integrated with the utility grid, they will provide cost savings for users and flexibility for the grid operators. The main building blocks and technologies that comprise the emerging microgrids of the future are rooftop PV, battery storage, EV charging, heat pumps, and intelligent meter-based infrastructure. The meter-based IT infrastructure enables real-time data analysis for system balancing and dynamic price structure. Although additional renewable sources might be deployed (e.g., biogas, hydrogen, hydropower, etc.), these are expected to play a rather minor, local role. This section provides an overview of leading case studies and start-ups, summarized in Table 11.1 and Table 11.2, respectively, that push the frontier of possibilities and make that vision a reality.

Quartierstrom project in Switzerland [14] focuses on the development of a local energy market for direct peer-to-peer (P2P) trading of excess renewable electricity. The project used blockchain technology to facilitate the transactions for 37 households and a retirement home in Walenstadt. The participating prosumers and consumers were able to actively shape energy prices. The local utility played a balancing role, providing additional capacity and buying excess power at fixed tariffs. Over the pilot phase of 12 months, the rooftop solar panels produced 250 MWh, which accounted for 53% of the total energy consumption of participating households. Peer-to-peer trading volume was 70 MWh (28% of the total renewable power generation). The project brought new insights into developing innovative energy tariffs and applying blockchain technology to local energy markets [15]. The efforts are continued by a start-up Exnaton, which further develops the IT infrastructure and business models.

The Kit Carson project in Taos, NM enables 100% daytime solar power by integrating a software platform for increased transparency, understanding of power flows, and improved grid management delivered by Camus Energy start-up [16]. The project has been a great success by helping to reduce energy costs from 9.5 c/kWh to 4.5 c/kWh and generating $10 mln in annual savings while also contributing to achieving renewable energy goals and providing economic benefits to the local community.

**TABLE 11.1**
**Summary of Microgrid Case Studies**

| Name | Country | Key features |
|---|---|---|
| Quartierstrom | Switzerland | Peer-to-peer (P2P) local energy market, innovative energy tariffs, and blockchain implemented. |
| Kit Carson | USA | 100% daytime solar and reduced energy price achieved with innovative grid management software. |
| Easton Energy | UK | Community-owned private microgrid implemented. |
| Simris | Sweden | Several days on 100% renewables achieved, demand side flexibility services possible for participants. |
| Kodiak | USA | Near 100% renewables achieved, reduced energy cost, flywheel energy storage implemented. |
| Island of El Hierro | Spain | 100% renewable energy achieved, pump hydro storage implemented in an extinct volcano. |
| Isle of Eigg | Scotland | Community designed, owned and maintained grid system, peak load restrictions introduced. |
| Basalt Vista | USA | Net-zero energy homes with fully electrified energy demand (solar PV, batteries, EVs and heat pumps). |

**TABLE 11.2**
**Summary of Start-ups Actively Shaping Renewable, Decentralized Energy Systems**

| Name | Country | Description |
|---|---|---|
| Exnaton | Switzerland | Start-up founded in 2020, offers flexible software as a service (SaaS) platform for decarbonized, decentralized and digital energy communities. |
| Camus | USA | Start-up founded in 2019, offers an open-source grid management SaaS platform for utilities to enable community-led decentral renewable energy. |
| Lumenaza | Germany | Start-up founded in 2013 offers an energy-as-a-service software platform to accelerate green, distributed electricity. |
| Energy Web Foundation | Germany, Switzerland, USA | Non-profit founded in 2017 by the Rocky Mountain Institute and Grid Singularity focused on accelerating blockchain in energy sector by building open-source, decentralized operating systems. |
| Kiwigrid | Germany | Start-up founded in 2008, offers a modular energy- IoT platform to implement new products and business models along the entire value chain. |
| Solshare | Bangladesh | Start-up founded in 2015, develops ICT solutions to enable peer-to-peer (P2P) electricity markets and provide affordable access to clean electricity and micro-mobility. |
| Enphase | USA | Energy technology company, one of the market leaders in advanced microinverters for solar PV applications, additional solutions include software for home energy management, including solar generation, energy storage, and cloud-based monitoring and control systems [12]. |
| SolarEdge | Israel | Power technology company, one of the market leaders in power optimizers, solar inverters, and monitoring systems for photovoltaics [13]. |

The project launched by Easton Energy Group aimed to install 120 kW of rooftop solar PV with 60 participating households in Easton, UK. The individual generators were connected with a community-owned private grid, using the exemption from license requirement for suppliers smaller than 2.5 MW [17].

A local renewable energy system pilot project in Simris, Sweden, run by a utility company E.ON aims to test a 100% renewable energy system with the capacity to operate in islanding mode. The system is comprised of a 500 kW wind turbine, 400 kW solar PV, battery storage (800 kW power and 330 kWh storage capacity), and a diesel backup. It supplies 150 households, which can also provide grid flexibility services from their water heaters and heat pumps [18].

The island of Kodiak in Alaska, with a population of 13,000 people and a total installed capacity of 75 MW, managed to reach an average renewable energy share of 99.7%. It also reduced and stabilized electricity rates to generate savings of $4 mln per year [19]. The renewable energy sources include wind and hydropower supported by a backup diesel capacity and an energy storage system comprised of a 3 MW (2 MWh) battery storage and a 1 MW flywheel.

The island of El Hierro, Canary Islands (Spain), with a population of 11,000 and an installed capacity of 35 MW, managed to reach 100% renewable energy generation. The energy demand was met with wind power supported by pumped hydro energy storage situated in an extinct volcano with 700 meters in elevation [12].

The Isle of Eigg, Scotland, with a population of 100 people and installed capacity of 250 kW, managed to reach a share of renewable energy of 87%. The microgrid is developed, actively owned, and maintained by the community. The system is comprised of wind, solar, hydro, diesel generators, and battery energy storage. The community also decided to set a limit for peak power at 5 kW for residential and 10 kW for commercial users [12].

Led by Habitat for Humanity of the Roaring Fork Valley, 27 all-electric, energy-efficient homes are constructed in Basalt Vista, Colorado. In the first four homes, distributed real-time control algorithm developed by the National Renewable Energy Laboratory was adopted by a microgrid controller manufactured by Heila Technologies to control rooftop solar PVs, home battery energy storage systems, HVAC loads, electric water heater loads, and electric vehicles. Smart controls and on-site solar and energy storage systems can operate the four homes as microgrids that do not need to extract electricity from the grid.

A promising start-up, Lumenaza provides software-as-a-service (SaaS) solutions to enable local renewable energy markets by introducing new energy tariff designs, facilitating the billing and provider switching processes for utilities. It also provides the guarantee of the origin of renewable energy. The flagship projects include Regionah Energie in Reutlingen and Ulm area, Stadtwerke Wunsiedel, Jurenergie in Bavaria, Stadtwerke Karlsruhe, and more [20].

In addition to start-ups, the European Commission launched collaborative publicly funded research project entitled ''Empowering local renewable energy communities for the decarbonization of the energy systems'' with a total funding of €7.2 mln [21]. The project's main objectives are to deploy and demonstrate community-driven, local, renewable energy systems, which incorporate sector coupling (electricity, heating, mobility, etc.) for optimizing renewable energy share, flexibility, and security of supply. The expected completion date is the second quarter of 2025.

Renewable microgrids are building blocks of the grid of the future. With the increasing number of distributed assets integrated into the grid, a hierarchical, decentralized architecture, shown in Figure 11.1, offers a more effective and scalable option for the grid to operate. At the lowest level are individual renewables and conventional generation resources, energy storage systems, building loads, electric vehicles, and other types of loads. Generation sources, storage, and loads co-located in a section of a distribution feeder are collectively managed by the next level of control, forming a "cell." Each cell is equipped with advanced controls that can manage its intrinsic assets and enable the multi-cell coordination and operation to provide grid services. These cells form the building blocks of the future grid. Grid-forming DERs, including inverter-based renewable resources and

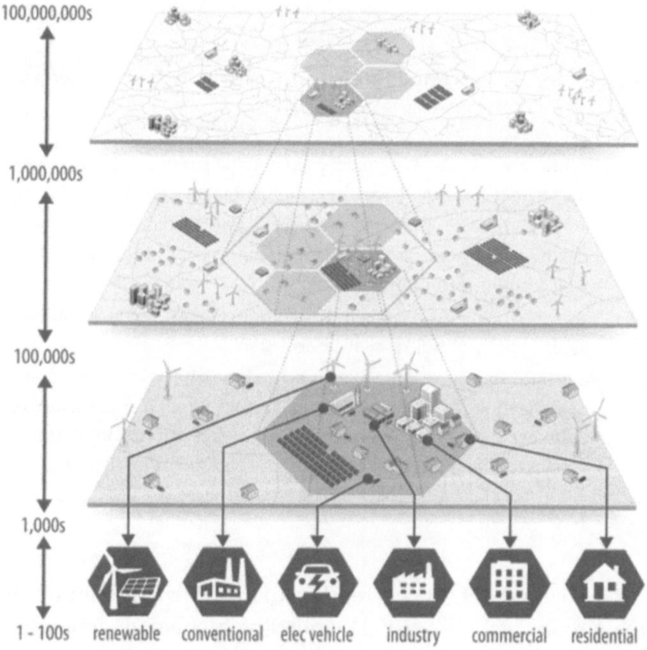


**FIGURE 11.1** A hierarchical, decentralized structure for the grid of the future with the large number of distributed assets. Source: National Renewable Energy Laboratory, published in Pong et al. (2021) [22], *Advances in Applied Energy*. Used with permission according to the Creative Commons license.

energy storage, are increasingly adopted by customers and utilities. The presence of grid-forming DER(s) in a cell makes the cell essentially a microgrid. During major disruptions, the hierarchical cell structure provides a starting point for adaptive microgrid formation and operation to respond to different system conditions. Such autonomous microgrids significantly increase the flexibility and resilience of the energy system.

## 11.4 REGULATORY ASPECTS OF MICROGRIDS

Local energy systems, microgrids, and energy communities are emerging concepts in the power sector. The lack of clear definitions, market operating principles, and roles and responsibilities of all participants are among the major challenges to address for successful deployment and scale-up around the world. The status quo of the regulatory environment in the EU is summarized in a recent Council of European Energy Regulators (CEER) report [23] and a European Commission research report [8]. An overview of the regulatory environment in the United States is provided in an NREL research report [24].

The regulation in the European Union distinguishes Renewable Energy Communities (RECs) and Citizen Energy Communities (CECs) as market actors in the attempt to provide the first set of high-level guidelines for a new market design. Both types of communities are set up as legal persons with the primary non-commercial objective to provide environmental, economic, and social benefits. CECs are limited to activities in the field of electricity with no geographical or technology constraints. At the same time, RECs can be active in all energy sectors and are limited to members' geographical proximity to projects and to renewable energy technologies only. Both types of communities can actively engage in market activities, including generation, distribution, storage, and provision of other energy-related services. The communities must comply with market rules and regulations applicable to respective market participants in a non-discriminatory way.

**TABLE 11.3**
**Summary of Regulatory Needs for Energy Communities in Focus Areas**

| Focus Area | Regulatory Needs |
|---|---|
| Customer Protection | Ensure free choice of an energy supplier, security of supply, access to high-quality service, transparent terms and conditions including energy tariffs and grid fees |
| System Balancing | Define market structure and price incentive mechanisms for communities to provide flexibility services that support regional system balancing and avoid self-optimization, which could lead to grid constraints |
| Market Design | Ensure effective management of multiple energy suppliers reflected in clear contractual agreements and transparent information |
| Power Grid Stewardship | Adhere to the same DSO standards and principles in terms of service quality (e.g., digitalization, advanced data management through smart meters), unbundling, handling of customer data, equal treatment of all market actors, long-term financial viability through system maintenance, define classification rules of a microgrid as a DSO |

According to CEER research [21], the currently existing types of communities can generally be categorized into three groups as follows: 1) collectively owned generation assets; 2) virtual communities, where members are not constraint by physical proximity; and 3) local microgrids, e.g., islands and remote areas.

The goal of the regulatory framework is to create a level playing field for energy communities while adhering to existing market principles and without introducing any distortions. Although the European Clean Energy Package provides high-level guidelines, it remains relatively open and requires transposition into national laws to develop specific instructions in the areas of customer protection, system balancing, market design, and power grid stewardship, as summarized in Table 11.3.

Although the aforementioned regulatory adaptations are currently not yet reflected in the electricity market regulation, there are several initiatives on a national level in the EU striving to encourage local energy systems. The United Kingdom and the Netherlands introduced regulatory sandboxes to allow and encourage experimentation with energy communities [8]. These include exemption from the supplier license, innovative grid tariff structures, and peer-to-peer trading. The governments in Scotland and Denmark actively promote community ownership of renewable generation assets. Scotland targets 2 GW of locally owned renewables by 2030, while Denmark requires project developers to offer 20% of wind projects ownership to local communities [8]. In addition, Belgium, France, and Spain introduced a clause allowing collective self-consumption of energy.

The existing legal forms of energy communities include energy cooperatives, limited partnerships, community trusts and foundations, housing associations, non-profit organizations, public-private partnerships, and public utility companies. Cooperatives are common in Germany and Sweden. Limited partnerships are suitable for larger scale projects, such as collective ownership of wind parks. Trusts are popular in Scotland, with the flagship project of the Isle of Eigg. Housing associations exist in the United Kingdom, Denmark, and Sweden, and non-profit organizations are preferred in Denmark.

Regulatory incentives are the primary driver for fostering local energy systems. The growth of renewable energy prosumers and energy communities is correlated to feed-in-tariffs (FiTs), tax incentives, and grants, as shown by the example of Germany, Denmark, and the United Kingdom [11]. According to the research, additional drivers include investment in sustainable infrastructure, production of green electricity and heat, social and environmental sustainability, self-sufficiency, energy efficiency, and security of supply [11].

In the United States, there are also no microgrid specific regulations in most states. New York, New Jersey, Connecticut, and California started developing the regulatory frameworks [24]. Under the current framework, utilities operate as regulated monopolies in electricity sales and distribution markets. They are also responsible for ensuring compliance with all safety and reliability standards. Microgrid owners are required to become a regulated utility, except for small-size microgrids (e.g., maximum five customers in Iowa and maximum 25 customers in Minnesota) [24]. The current regulatory environment also prohibits physical and virtual direct peer-to-peer energy trading. Interconnection standards (e.g., IEEE 1547), need to be updated to allow connection of a microgrid as a whole instead of individual generation assets.

The efforts focused on regulatory reforms in Connecticut are focused on providing a definition of microgrids and enabling microgrid owners to cross public rights-of-way [24]. Several states, including New Jersey, Illinois, and Hawaii, have launched feasibility studies and demonstration projects.

In addition, scientific efforts at NREL propose a hypothetical future market structure for microgrids integration called "Networked Microgrids-friendly, NMG-friendly." In this setup, there are no barriers to exchanging energy and services during both normal operations and outages for microgrids. Peer-to-peer markets are enabled, and utilities allow microgrid owners to use the grid infrastructure under negotiated agreements.

## 11.5 SUMMARY AND CONCLUSIONS

Renewable microgrids are an integral part of the future sustainable energy system. They provide resilience and flexibility required for adaptation to climate change and changing supply and demand patterns. In addition to environmental benefits, they also offer social and economic benefits by enabling residents, communities, and businesses to become active participants in the energy market while reducing energy costs. Microgrids could account for up to 45% of renewable energy generation in Europe by 2050 and over 30 GW of renewable generation capacity in the United States by 2030. Therefore, there is vast potential and enormous business opportunities.

Renewable microgrids are currently at the very early stage of development, driven mainly by funded research projects, pilot projects, local government initiatives, and start-ups. The early results show that nearly 100% renewable energy can be reached while reducing energy costs. To successfully scale up and commercialize this concept, further development of information and communication (ICT) technologies (e.g., control algorithms, data management, and blockchain), market design, business models, and regulatory frameworks that clearly specify the rights and responsibilities of microgrid owners in relation to customers and the large utilities are required.

## 11.6 DISCLAIMER

The work performed in part by the employees of the National Renewable Energy Laboratory, operated by Alliance for Sustainable Energy, LLC, was supported by the U.S. Department of Energy (DOE) under Contract No. DE-AC36-08GO28308. The views expressed in the article do not necessarily represent the views of the DOE or the U.S. Government. The U.S. Government retains and the publisher, by accepting the article for publication, acknowledges that the U.S. Government retains a nonexclusive, paid-up, irrevocable, worldwide license to publish or reproduce the published form of this work, or allow others to do so, for the U.S. Government purposes.

## REFERENCES

[1] P. Harper, (2016). Alternative Technology and Social Organization in an Institutional Setting. Science as Culture. 25. 415–431. 10.1080/09505431.2016.1164406

[2] https://en.wikipedia.org/wiki/2021_Texas_power_crisis#:~:text=In%20February%202021%2C%20the%20state,17%2C%20and%2015%E2%80%9320.&text=More%20than%204.5%20million%20homes,power%2C%20some%20for%20several%20days
[3] https://www.c2es.org/content/extreme-weather-and-climate-change/
[4] https://www.carbonbrief.org/mapped-how-climate-change-affects-extreme-weather-around-the-world
[5] IEA, World Energy Outlook 2021.
[6] E. Tan, P. Lamers, (2021). Circular Bioeconomy Concepts – A Perspective, frontiers in Sustainability, doi: 10.3389/frsus.2021.701509
[7] J. Giraldez, F. Flores-Espino, S. MacAlpine, P. Asmus, (October 2018). Phase I Microgrid Cost Study: Data Collection and Analysis of Microgrid Costs in the United States, Technical Report, NREL/TP-5D00-67821
[8] https://www.c2es.org/content/microgrids/
[9] Y. Lin, J.H. Eto, B.B. Johnson, J.D. Flicker, R.H. Lasseter, H.N. Villegas Pico, G.S. Seo, B.J. Pierre, A. Ellis, (2020). Research Roadmap on Grid-Forming Inverters. Golden, CO: National Renewable Energy Laboratory. NREL/TP-5D00-73476. https://www.nrel.gov/docs/fy21osti/73476.pdf
[10] A. Caramizaru, A. Uihlein, (2020). Energy Communities: an overview of energy and social innovation, JRC Science for Policy Report.
[11] Guidehouse, The Renewable Energy Economic Benefits of Microgrids, November 2021.
[12] https://enphase.com/
[13] https://www.solaredge.com/
[14] Swiss Federal Office of Energy, Community energy network with prosumer focus, Quatierstrom, 2020
[15] https://quartier-strom.ch/index.php/en/homepage/
[16] https://www.camus.energy/blog/kit-carson-electric-cooperative
[17] http://www.eastonenergygroup.org/
[18] https://www.eon.se/en_US/samhaelle—utveckling/local-energy-systems/we-are-renewing-simris.html
[19] Rocky Mountain Institute, Carbon War Room, Renewable Microgrids, Profiles from Islands and remote communities across the globe, 2015
[20] https://www.lumenaza.de/en/references/
[21] https://cordis.europa.eu/project/id/957819
[22] P.W. T. Pong, A.M. Annaswamy, B. Kroposki, Y. Zhang, R. Rajagopal, G. Zussman, H. Vincent Poor, (2021). Cyber-Enabled Grids: Shaping Future Energy Systems. *Advances in Applied Energy* 1 (February): 100003. 10.1016/j.adapen.2020.100003
[23] CEER, Regulatory Aspects of Self-Consumption and Energy Communities, Report number: C18-CRM9_DS7-05-03, 2019.
[24] F. Flores-Espino, J. Giraldez, A. Pratt, (2020). Networked Microgrid Optimal Design and Operations Tool: Regulatory and Business Environment Study. Golden, CO: National Renewable Energy Laboratory. NREL/TP-5D00-70944. https://www.nrel.gov/docs/fy20osti/70944.pdf

# 12 Applications of Electrochemical Separation Technologies for Sustainability

## *Case Studies in Integrated Processes, Material Innovations, and Risk Assessments*

*Yupo J. Lin, Matthew L. Jordan, and Thomas Lippert*
Argonne National Laboratory, Lemont, Illinois, USA

*Tse-Lun Chen*
ETH Zurich, Zurich, Switzerland

*Li-Heng Chen*
Industrial Technology Research Institute, Chutung, Hsinchu, Taiwan

**CONTENTS**



DOI: 10.1201/9781003167693-12

## 12.1 INTRODUCTION

Electrochemical processes offer research and development (R&D) opportunities toward decarbonization and resource recovery in circular economies, while innovative materials and advanced manufacturing techniques further widen their applications. In this chapter, we will discuss electrochemical process designs and material innovations that can address the technical and economic challenges of separations in biochemical/biofuel production, resource recovery, $CO_2$ utilization, and impaired water treatment. We will also discuss the assessment and reduction of risk related to the scale-up of electrochemical separation technologies (ESTs) in industrial environments to help develop a path forward to commercialization.

Electrochemical separations allow for selective capture of charged species and/or in-situ pH manipulation. By applying an electric current, ions can be separated from a liquid, while the number of ions removed is proportional to the invested electrical energy. Such selective separation of ions against other non-charged species enables a highly efficient "fit-for-purpose" operation, which has the potential to significantly reduce the costs for various desalination applications, such as (i) cooling water supply for power plants, (ii) nutrient removal/recovery, or (iii) inorganic salts ratio control in irrigation discharge water. Common ESTs, such as electrodialysis (ED), are used in various industrial applications; electrodeionization (EDI) is used in ultrapure water production. The rest of the ESTs, like capacitive deionization (CDI), cation intercalation desalination (CID), and ion concentration polarization (ICP), are used in more specific applications.

Compared to other separation technologies (such as pressure-driven, thermal, or biological separations), electrochemical separations are cost-effective and have a small physical footprint. While pressure- and temperature-driven membrane and distillation processes are effective in removing water from high-titer streams, they have much lower energy efficiency in dilute aqueous streams. Biological separations, on the other hand, utilize microbial bio-electrochemical reactions to drive the removal of ions from the solution. However, their ability to produce "fit-for-purpose" water has not been explored yet.

Interest in "fit-for-purpose" separations has increased in recent years, and the application of selective separation technologies will become more important in the future to address the challenges of climate change and necessary technology adaptations. For example, the production of biofuel and bio-products to reduce greenhouse gas emissions from fossil fuels or the exploitation of non-conventional water supplies in the context of the water-energy-nexus will require energy-efficient and cost-effective separation technologies. Innovative electrochemical separations can provide transformational impacts in advancing selective separations for highly energy-efficient, small-footprint, and low-cost operations and allow for a paradigm shift to use alternative energy and water supplies in industrial applications.

In the following sections, separation performances of ESTs for various process streams are discussed, including 1) applications of EST in biorefinery, 2) selective removal and recovery of ammonia from biological wastewater via in-situ pH control, 3) R&D of innovative material synthesis and their impacts on the separation performance, and 4) risk-reduction for commercial applications.

## 12.2 APPLICATIONS OF EST IN BIOREFINERY

This section includes discussion and comments based on summarized results taken from the previously published book chapter *Bioprocessing of Cost-competitive Biobased Organic Acids* in *Commercializing Biobased Products* (see Lin, Y. J.; Hestekin, J. A.; Henry, M. P.; Sather, N. Bioprocessing of Cost-Competitive Biobased Organic Acids. In Commercializing Biobased Products: Opportunities, Challenges, Benefits, and Risks; Green Chemistry Series; Royal Society of Chemistry, 2016. Reference [1]). Together with new figures, some figures modified or revised from Reference [1] are used to illustrate the applications of EST for sustainability.

Biorefineries play a vital role in sustainable chemical and fuel production. However, conventional bioprocesses are energy-intensive and characterized by high costs due to the purification challenges of low product titer in bioconversion processes. This limitation prohibits the production of cost-competitive, bio-based chemicals from replacing oil-based chemicals in the commodity chemicals market. The high processing energy demand further reduces the sustainability of biorefineries when fossil fuels are used to provide the required thermal energy (e.g., for distillation).

Compared to competing separation technologies such as membrane-based and solvent extraction, ESTs enable an integral bioprocessing platform to reduce energy demand, processing cost, and carbon footprint, thus increasing the sustainability of a biorefinery. For instance, the integration of a bioreactor with an EST for in-situ product capture provides a new approach in bioprocessing to produce cost-competitive bio-based chemicals. So-called *separative bioreactors* (SBs) integrate upstream bioconversion and downstream product separation into a continuous process. It reduces product inhibition due to the simultaneous production and in-situ removal of organic acids, requires fewer unit operations, and eliminates the need for neutralizing chemical additives. SBs significantly increase the organic acid product titers (>10X) as well as bioconversion rates and yields. Detailed information and technical results on several platform technologies developed to apply the new bioprocessing technique can be found in the literature [1], [2].

In the following case studies, resin-wafer electrodeionization (RW-EDI) was applied to enable the in-situ capture of organic acids from a continuous bioreactor. EDI is a membrane-based, electrically driven separation technology to extract charged species, such as organic acids, from liquids. The description of the RW-EDI technology is discussed in the following sections. The different SB-based bioprocesses that will be discussed include succinic acid production by anaerobic fermentation and gluconic acid production by aerobic fermentation from glucose. Pilot-scale operation of SBs was also conducted to investigate long-term performance.

### 12.2.1 Current Technologies for Organic Acid Production with Bioconversion Processes

Biocatalysis provides sustainable chemical production processes that utilize renewable feedstocks, and organic acids made from biomass feedstock are potential products to replace petroleum feedstocks for the commodity chemicals market [3]–[5]. Citric acid produced by fermentation is a prime example of such a production pathway and has an annual global market value of over 2 billion dollars [6]. Various examples of organic acid production, including feedstock preparation, bioconversion, and organic acid purification, can be found in the literature [7]–[11].

The replacement of petrochemicals with bio-based chemicals requires very efficient and cost-effective bioprocesses. Due to the nature of bioconversion reactions with product inhibition and acidification, buffering or neutralization is often necessary, which can cause the produced acids to be in their salt rather than their pure acid form. Accordingly, conventional bioconversion processes often require further purification steps, such as the conversion of the organic salt into its pure acid form and, subsequently, the increase of the acid product concentration, for instance, by dewatering. These product refinery steps add significant energy demands and processing costs. Some examples of unit operations for product purification are multi-stage filtration or chromatography, ED, liquid extraction, salting out, distillation, and crystallization [12]–[17]. Other non-conventional unit separations for the chemical industry were also investigated for bio-based organic acid productions, including reverse osmosis [18], nanofiltration [3], [19], ion exchange [20], ion chromatography [21], Donnan dialysis [22], and supercritical fluid extraction [20]. However, the dilute nature of biologically produced organic acids typically renders these separation approaches economically impractical [3].

Therefore, more cost-effective processes are needed in product recovery and energy consumption. According to the literature, ED separations require approximately 0.2–3.5 kWh per kg of product and cost about $0.1–0.4 per kg of acid recovered [23] [24]. For example, the separation costs for bio-converted succinic and lactic acid have been reported at $0.85–2.20/kg and $0.55/kg, respectively

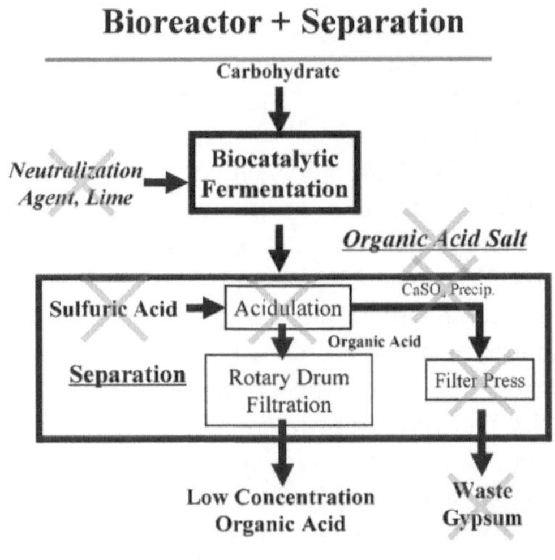


**FIGURE 12.1** Process block diagram of conventional bioprocessing to produce organic acids and possible elimination of unit operation blocks by integrating bioconversion and separation. Figure modified from reference[1].

242 [4], [25]–[27]. Reliable cost information for organic acid purification by other separations techniques is scarcely available, but costs are deemed to be high. Conventional chemical purifications such as crystallization include multiple unit operations, e.g., refrigeration, filtration, centrifugation, drying of crystals, and/or solvent extraction to remove impurities, which are all troublesome. Therefore, the use of these unit operations is expected to be prohibitive for application to biochemicals.

### 12.2.2 Integrated Bioprocess Design

As mentioned in the last section, acidification and product inhibition are the two self-limiting constraints to most biological reactions. Adding neutralizing chemicals to maintain optimal pH is essential to overcome the self-limitation. However, this leads to the formation of organic salts instead of organic acids that create complexity for downstream bioprocessing challenges, i.e., extraction, acidification/purification. Furthermore, by-product inhibition results in low titers of organic salts. Figure 12.1 displays a schematic of the conventional fermentation processes for the production and recovery of organic acids. As shown in the conventional bioprocess, equal moles of waste gypsum are produced from every mole of organic acid produced in low product titers. The biocatalytic reaction could be sustained without neutralization by acid; the organic acid products need to be removed in situ to low levels from the bioreactor when the product was prevented from inhibiting and maintaining optimum pH. The in-situ product separation during bioconversion is the principle of SBs, as illustrated in the process boxes in Figure 12.1. Actual applications of SB on various organic acid production by bioreactors are discussed in the next section.

### 12.2.3 Separative Bioreactor (SB)

The separative bioreactor (SB) process design offers a pathway for producing cost-competitive, bio-based organic acids. It addresses the barriers of high energy consumption with low acid titers in the traditional bioproduction of organic acids. As illustrated in the process flow diagram of Figure 12.1, several unit operations can be skipped by avoiding the neutralization step. Figure 12.2 shows a process block flow diagram (subfigure a) and the actual system integration of fermentation and in-situ product capture (separation) device (subfigure b). In this case, succinic acid production using an SB can provide a substantial environment favoring biological activities without the limitations of pH shifts and product inhibition. With a continuous supply of substrate, integrated SB provides a continuous operation of fermentation compared to conventional batch operation. In such a process integration arrangement, a highly concentrated ("pure") organic acid product is continuously extracted from the broth and captured in a separate reservoir. It, thus, can greatly

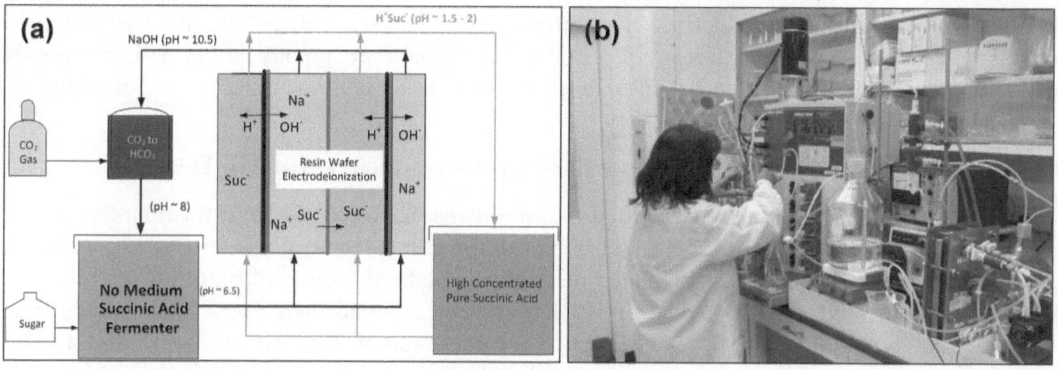


**FIGURE 12.2** (a) Schematic of an integrated fermentation and electrodeionization separative bioreactor (IF-EDI-SB) for succinic acid production and (b) photograph of the laboratory setup.

reduce the need for energy-intensive dewatering to achieve a concentrated final product. The performance of succinic acid production using SB is discussed in the following sections.

### 12.2.4 Innovative Electrodeionization Technology to Capture Organic Acids

Selective extraction of only pure organic acid is needed for an effective SB system. ED and EDI resemble such selective extraction technologies, while EDI is even more energy efficient compared to ED when weak organic acids need to be extracted. In EDI, organic acids are ionized simultaneously by the ion-exchange resin bed inside the reaction chamber and selectively extracted from the fermentation broth across the ion-exchange membranes by electric field forces. EDI is commercially applied for producing ultrapure water in the semiconductor and pharmaceutical industries, which was re-engineered to porous matrices containing immobilized ion-exchange resin beads called ion-exchange resin wafers (RW) [28]. The use of wafer material extends the potential applicability of EDI-based separation processes beyond ultra-pure water production. It is generally designed to reduce the consumption of electric energy while maintaining high rates of charged species separation in low-conductivity aqueous solutions [28]–[30]. The separation productivity, i.e., the acid capture rate and energy consumption, are the main determinants of process performance and economics.

The acid capture rate determines the size (and cost) of the RW-EDI capital equipment and the total membrane area needed, while the energy consumption determines the electrical energy cost. These three factors determine the economics of separation cost, which is roughly around 50–70% of the final organic production cost.

Examples of operations and their process performance of different separative bioreactor configurations for producing organic acids are discussed. In the system setups of these two fermentation applications of separative bioreactors, a filtered slipstream of fermentation broth is pumped through the RW-EDI stack for organic acids extraction from the broth stream. The effluents (broth), along with the bacteria, which were recaptured from the retentate of filtration step, were then returned to the fermenter.

### 12.2.5 Demonstration of Separative Bioreactor Performance for Organic Acid Production

SB platform processing improves both operation and control compared to conventional bioconversion processes to produce organic acids. First, it enables direct conversion of the batch into continuous fermentation with significant improvement of throughput and process control. Second, the in-situ extraction of products (i.e., the organic acids) significantly eliminates product inhibition and helps biological acid productivity with elevated acid or salt concentrations. Therefore, the overall unit operations of bioprocess using the SB system are greatly simplified. Successful examples of up-scaling from the bench-scale to the pilot-scale are discussed in the following.

### 12.2.6 Integrated Fermentation and EDI Separative Bioreactor (IF-EDI-SB)

RW-EDI processes can be easily retrofitted into any existing fermentation to form an SB system and simplify the operation and control of a bioreactor compared to conventional fermentation. In addition, SB also provides self-regulated pH control via the recycling of buffering chemicals in the fermentation (see example in Figure 12.2). These factors and their impacts on process performance and economics are discussed next, using the examples of the anaerobic and aerobic fermentation of glucose to produce succinic acid and gluconic acid, respectively.

### 12.2.7 Anaerobic Fermentation to Produce Succinic Acid

Succinic acid is an important building-block chemical feedstock for polymers with many large market uses, including acyl halides, anhydrides, esters, amides, and nitriles applied in drugs, agriculture, and food products, and other industrial uses. It is a sustainable source to replace maleic anhydride or butanediol. Current bioprocesses of generating succinic acid produce succinate in the form of a salt and require many post-fermentation unit operations, as discussed in previous sections. Since it takes one mole of sugar and $CO_2$ to produce succinic acid, the recycled hydroxide ion, shown Figure 12.2 (a), is not just helping to maintain pH but also improves the $CO_2$ gas uptake and utilization in the fermentation. Therefore, compared to other organic acids production, SB offers even more profound impacts on the operation and economical way for direct production of pure succinic acid that can be sold as a commodity chemical in the marketplace.

The anaerobic fermentation was conducted in a temperature-controlled, stirred bioreactor. As discussed above, fermentation combines one mole of $CO_2$ with one mole of glucose to produce one mole of succinic acid and other side products, such as ethanol and acetic acid, etc. Currently, $CO_2$ is provided from a $CO_2$ capture reactor in the form of carbonate, which is formed by bubbling $CO_2$ gas into a solution at basic pH. It provides neutralization of succinic acid in the broth while also supplying the $CO_2$ needed for the fermentation pathway. In the IF-EDI-SB system, the fermentation broth is sent to the RW-EDI device for the extraction of succinic acid in pure acid form and recycles the resultant basic solution to the $CO_2$ capture reactor (see Figure 12.2 (a)).

Because of the in-situ extraction of succinic acid from the broth, IF-EDI-SB provides effective biological activities in water without the use of complex fermentation broth composition such as conventional Luria broth (LB) (see Figure 12.3). This minimizes the introduction of salts and other by-products that are components of the LB that could interfere with the recovery of a purified succinic acid product. The non-LB fermentation produces succinic acid with much less co-product contamination, such as acetic acid.

A "bipolar" EDI membrane configuration was used to separate and convert succinates and other organic salts to their organic acid form. The succinate concentration could be maintained at a very low level in the fermenter using the IF-EDI-SB system. Most of the succinate salts were captured as pure acid from the EDI device. Figure 12.4 (a) shows succinic acid titers in the fermentation broth and in the extracted succinic acid solution tanks. Depending on the anion-exchange membrane used for succinate transport in EDI, as much as 48 wt.% succinic acid can be extracted directly from the broth into the capture streams. The product concentration is limited by the water that is transported along with the acid through the membranes. The RW-EDI also provided selective separation among the co-products in the fermenter, as depicted in Figure 12.4 (b). Succinic acid was preferentially recovered over the acetic acid.

The separation selectivity of RW-EDI is dependent on the relative concentrations and mobility of the differently charged species in the fermentation solution. The preferred extraction of succinic

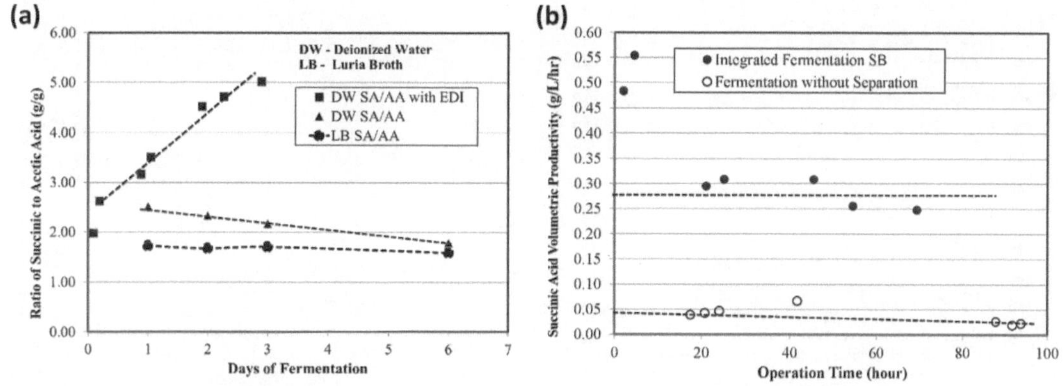


**FIGURE 12.3** (a) Enhanced production ratio of succinic acid (SA) over co-product (acetic acid, AA); (b) profound increase of succinic acid biological conversion activity in an integrated fermentation and electrodeionization separative bioreactor (IF-EDI-SB). Figure modified from reference [1].

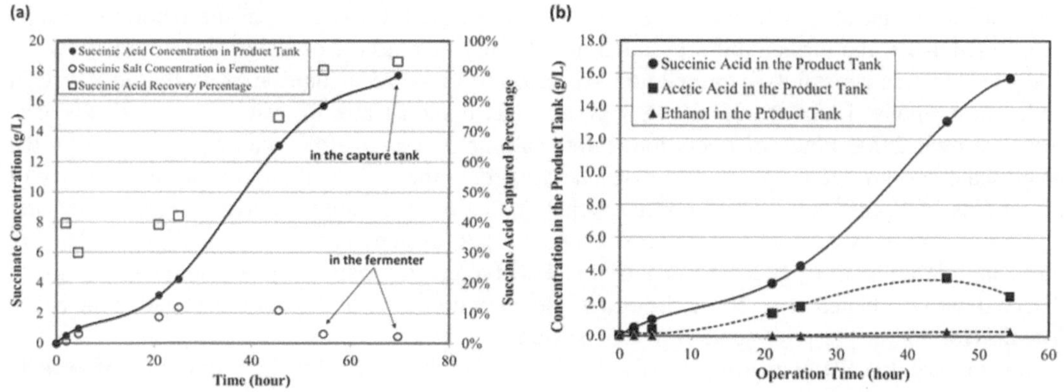


**FIGURE 12.4** (a) Concentrations and capture percentage of succinic acid in the fermenter and acid product capture tank and (b) comparison of recoveries among succinic acids and co-products distribution influenced by the integrated fermentation and electrodeionization separative bioreactor system (IF-EDI-SB). Figure modified from reference [1].

acid in-situ compared to other co-products from the fermentation broth promotes favorable biological activity to succinic acid production since it was removed faster than the competing species, as shown in Figure 12.3 (right) where it is compared with the product distribution from fermentation without separation. The coupled of RW-EDI to fermentation improved the productivity more than fivefold compared to fermentation without integrated product separation. For succinic acid produced from bioconversion to be cost-effective in the commodity market (e.g., to replace maleic anhydride), the fermentation bioreactor must achieve a minimum productivity goal of 0.25 g/L/h [31]. The IF-EDI-SB process has demonstrated enhanced productivity that can meet or exceed this goal, suggesting that it has the potential to improve the economic viability of fermentation organic acids production more generally.

## 12.2.8 Aerobic Fermentation to Produce Gluconic Acid

The production of gluconic acid from aerobic fermentation using IF-EDI-SB was demonstrated in both bench- and pilot-scale systems in extended operation to assess process robustness and its resistance to process upset. In this case, there was no addition of pH buffer needed because of the

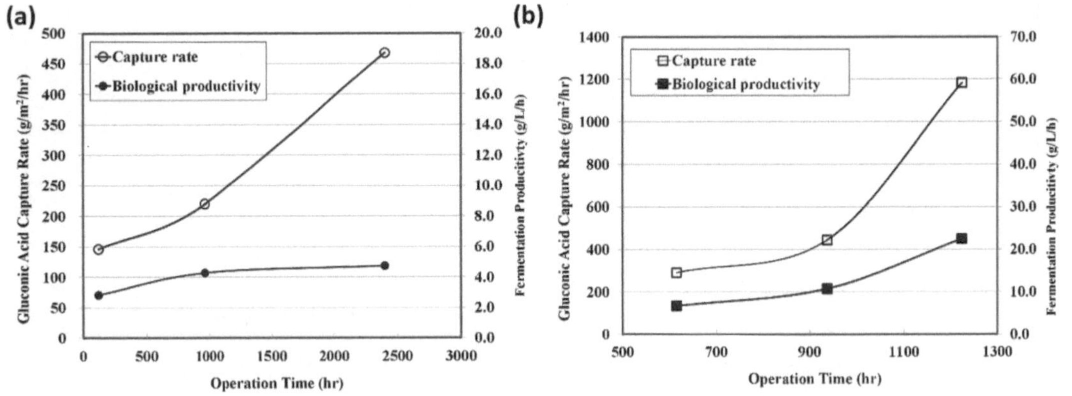


**FIGURE 12.5** Biological activities and capture productivities of gluconic acid (GA) production from integrated fermentation separative bioreactor (SB). (a) Laboratory-scale performance; (b) pilot-scale performance.

use of acid-tolerant strains in the fermenter. It was operated at pH 2–3. In the laboratory scale, using IF-EDI-SB, the fermentation productivity steadily increased from 3 g/L/h to 5 g/L/h, while gluconic acid was constantly captured and separated from the fermentation broth by the RW-EDI device (see Figure 12.5 (a)). IF-EDI-SB system has demonstrated robustness with steady performance over 2,000 hours of operation. Pilot-scale units, i.e., 30-fold scale-up in size from the bench-scale system, were operated to assess and validate the technical and economic viability of the process and potential process upsets. Acid production and capture performance of the integrated pilot-scale system compared to the bench-scale system are shown in Figure 12.5 (b). Fermentation productivity was significantly increased to greater than 20 g/L/h compared to 5 g/L/h observed with the bench-scale SB at the end of the operation. One of the reasons is due to the better flow distribution in the pilot-scale RW-EDI system.

In summary, the SB platform provides the following advantages for the production of organic acids in bioconversion processes:

- Eliminates or greatly reduces product inhibitions in fermentation
- Eliminates or significantly reduces the need for neutralization chemicals
- Directly produces organic in pure acids form, it eliminates acidification in conventional bio-organic acid production
- Significantly reduces both energy consumption and processing costs to produce organic acids
- Can potentially reduce fermenter size by several folds (owing to increased fermentation productivity)
- RW-EDI can be retrofitted into the fermenter and allows for a continuous fermentation

### 12.2.9 Conclusion

The use of sustainable feedstocks for fuel and chemical production is a critical strategy to address climate change. Bio-based chemical production is one way to reduce $CO_2$ emissions and is gradually penetrating the commodity market. This section has described the technological viability of the SB platform that has been demonstrated to produce organic acids from biomass feedstocks. The SB overcomes the challenges of high energy and processing costs for the relatively low product titers of organic acids. The RW-EDI unit operation fits well into an SB system to overcome these hurdles. In-situ capture of acid products significantly stabilizes and increases the bioconversion productivity. It enables continuous product formation and recovery of organic acids while avoiding product inhibition. In addition, the SB can produce organic acids in the acid form, which avoids the need and cost for neutralization and subsequent acid regeneration. At both the bench scale and pilot scale, the long-term operation of SB for organic acid production has demonstrated process performance advantages and the robustness to resist membrane fouling and contaminations in fermentation.

## 12.3 AMMONIA REMOVAL AND RECOVERY FROM NUTRIENT-RICH WASTEWATER

Ammonium ($NH_4^+$) recovery from wastewater is a crucial pathway for dealing with resource shortages in human society. Wastewater contains multiple nutrients, which have attracted the attention of scientists and engineers [32]. Discharge of nutrient-rich wastewater has resulted in environmental pollution and deterioration of water quality and ecosystem structure [33]. In general, the degree of eutrophication of lakes and reservoirs in Asia and the Pacific, Europe, North America, South America, and Africa were 54%, 53%, 48%, 41%, and 28%, respectively. In the United States, potential economic damages due to eutrophic water bodies are estimated to be $2.2 billion annually [34]. The excessive ammonium discharge from wastewater causes severe environmental problems, and the production of ammonia as fertilizer accounts for 1.5–2.5% of annual energy consumption and for 1.6% of global $CO_2$ emissions [35]. Ammonia production has been growing and reached 140 million tons in 2018, for which China, Russia, the United States, and India accounted for 31.4%, 10.0%, 8.9%, and 7.8%, respectively [36]. The simultaneous removal and recovery of ammonium from wastewater allows not only remediating aquatic environments, but also helps reducing greenhouse gas emissions that originate from traditional ammonia production [37].

Ammonia-nitrogen (ammonium and ammonia, $NH_3$-N) is the most important and widely produced inorganic chemical. The applications of ammonia-nitrogen are (i) production of fertilizers such as ammonium nitrate, ammonium phosphate, and urea; (ii) absorbance of waste acid gas; (iii) use as a refrigerant in buildings and industry; (iv) use as raw material for plastic, polymer, and acid production; and (v) use as fuel for power generation [38]. The largest contributors to ammonia-nitrogen emissions to the aquatic environment are the food, fertilizer, and chemical industries. Traditional biological treatment schemes, including nitrification and denitrification, have been used to treat high-strength ammonia-nitrogen wastewater by reducing ammonia to nitrogen into the atmosphere [39] and still remain the most common approach for ammonia wastewater treatment [40]. However, large amounts of sewage sludge are generated as secondary pollution, which is one of the major challenges of wastewater treatment plants (WWTPs). It also requires long hydraulic retention times and large basins, thus increasing the difficulty of implementation. Excess $NH_3$-N in raw water may increase oxygen demand and interfere with the chlorination and manganese filtration processes, thus reducing the treatment performance in WWTPs. Due to the toxic effects of $NH_3$-N in water bodies, the U.S. Environmental Protection Agency regulates the limitation for $NH_3$-N discharge [41]. Therefore, both removal and recovery of ammonium from wastewater have attracted considerable attention in recent years due to stricter discharge regulations for $NH_3$-N in WWTPs [42].

Traditionally, in WWTPs, the wide use of the anaerobic digestion (AD) process aims to stabilize organic wastes, produce biogas, and reduce sludge generation. The residuals of the AD process contain high total ammonia nitrogen (TAN) concentrations, which have a negative effect on the performance of AD treatment as the C/N ratio decreases [43]. On the other hand, large amounts of ammonia production consume the coal-derived hydrogen, which results in high energy consumption and a large carbon footprint. To improve the efficiency of ammonia production associated with $CO_2$ abatement, a chemical looping process designed for $H_2$ and $NH_3$ co-generation using coke-oven gas has been investigated [44]. This is a future potential application by integrating the hydrogen production and biomass gasification for ammonia production, namely biomass-to-ammonia.

Thus, simultaneous removal and recovery of $NH_3$-N is an emerging approach in the development of a circular resource economy. Implementation of electrochemical separation technologies is one of the most attractive approaches for $NH_3$-N removal and recovery. However, limitations in scale-up, cost effectiveness, and market acceptance require further attention in the future [45]. The electrochemical separation/recovery process has been proven as an energy-efficient pathway for a sustainable nitrogen cycle. For example, electrodialysis (ED) and electrodeionization (EDI) can achieve $NH_3$-N removal ratios of 90% at an energy consumption of around 5 kWh per kg nitrogen. In addition, ED or EDI could reduce chemical consumption for pH adjustments using electrical energy during $NH_3$-N removal/recovery. Electric water splitting can produce protons and hydroxyl ions that achieve in-situ pH control under an appropriate electric field. Bipolar membranes used in the ED or EDI cells could separate the inflow stream into acid and base streams, which is favorable for generating multiple by-products, including pure water, cations, and anions [46]. Another electro-adsorption process called capacitive deionization (CDI), which has been investigated for $NH_3$-N removal and recovery, shows high energy efficiency and cost-benefits for ammonium recovery. A bench-scale CDI has been set up to remove $NH_4^+$ from wastewater, demonstrating a high removal ratio of 95% during continuous charging/discharging cycles [47]. Furthermore, the integration of biological and electrokinetic processes (i.e., bio-electrochemical systems, BES) is an alternative to recover $NH_4^+$ at low cost and energy consumption. In BES, bio-degradable organic matter providing the electricity conversion could drive $NH_4^+$ ions across the cation exchange membrane to separate it from wastewater. BES is similar to electrokinetic processes, which offers the chance to concentrate $NH_4^+$ with the high purity via the process integration of air stripping, transmembrane chemisorption (TMCS), and struvite precipitation [48]. In general, the $NH_4^+$ removal/recovery ratios using electrokinetic processes from real urine were reported to range from 38% to 92%, with the energy consumption ranging from 2.9 kWh/kg-N to 46.3 kWh/kg-N. BES has a lower energy consumption of 1.1 kWh/kg-N to 20.5 kWh/kg-N [49]. In conclusion, the demonstration of BES and electrokinetic process for $NH_4^+$ removal/recovery should be implemented at a commercialized scale under the available economic benefits.

## 12.4 MATERIAL INNOVATIONS IN ELECTROCHEMICAL SEPARATIONS

Material innovations have driven the development of electrochemical separations and broadened applications since the technology's inception. The creation of charged ion-exchange polymers during the 1940s that only allowed cation or anion transport opened the door toward a continuous ion-exchange process known as electrodialysis (ED) [50], [51]. Eventually this invention of cation and anion selective membranes led to the first commercialization of ED for water treatment by *Ionics, Inc* [52], [53]. Since these early days of the electrochemical separation industry (notably less than 100 years ago), further material innovation in ion-exchange materials, spacer channel conductors, electrochemical cell design, and electrode design have broadened the applications beyond water treatment.

At the heart of ED and electrosorption technologies are ion-exchange materials composed of positively or negatively charged functional groups or surfaces. An external potential field attracts charged species toward these charged functional surfaces and removes them from the process stream. The earliest applications of electrochemical separation technologies were toward single domain separations such as desalination. As ion-exchange materials advanced, more complex separations such as organic acid recovery in bioprocessing, lithium refining from brine sources, and selective water purification were empowered through selective ion-exchange materials. The development of more energy efficient processes through ionic conductors, such as the invention of electrodeionization (EDI) [54], has enabled more energy-efficient separation technology that is competitive with reverse osmosis for brackish water treatment [55], [56]. The field of electrochemical separations evolved alongside material innovations.

This section discusses case studies of material advancements that have expanded the competitiveness and application of electrochemical separations. The use of tailored functionalities of ion-exchange membranes to affect selectivity and transport flux will be discussed for targeting lactic acid from impure fermentation streams. Organic acid recovery is a critical technology to enable direct extraction of the chemical intermediates necessary for renewable, bio-based chemicals. In addition, recent developments in ionic conductor technologies (i.e., resin wafers) will be discussed, and their implications toward separation rate and energy efficiency. EDI combines the advantages of ion-exchange with electrodialysis to enable an energy-efficient separation platform for ultrapure or dilute separations. Moreover, electrosorption technologies (i.e., capacitive deionization, CDI) have made significant developments recently through material developments in selective electrode materials. Lastly, the final section describes the recent development of a hybrid material known as membrane wafer assembly (MWA) that was created to solve transport barriers between the ion-exchange materials used in EDI (resin column and membranes) to increase the recovery of bulky ionic species such as phenolic acids. These material solutions to separation challenges increased the overall competitiveness of electrochemical separations and created separation solutions for applications not practical with any other unit operation technology.

### 12.4.1 Case Study in Ion-exchange Membrane Development

Ion-exchange membranes form the core of ED devices. The original ion-exchange membranes were designed to indiscriminately transfer ions from the feed to the concentrated process streams and, by doing so, complete an electrical circuit between the anode and cathode electrodes. Traditionally, commercial ion-exchange membranes fail to discriminate against ions of similar charge and valence. Targeting one type of ion from a mixture of ions with the same charge and different valences poses significant separation challenges. The nonselective nature of these early ion-exchange membranes limited the applications toward strictly deionization applications such as demineralization of whey protein or simple desalination of water. As the ion-exchange membrane technology advanced, various polymer design strategies have emerged to enhance the selectivity of these materials. In turn, these selective materials opened an opportunity for more advanced separation applications to selectively screen one ion for another.

In this example, a separation process to capture lactic acid from a bioreactor fermentation stream is considered [57]. The process is conveyed in Figure 12.6 (a) and depicts the removal and concentration of organic acid anions along with competing inorganic salt ions such as phosphate and sulfate that are present in fermentation streams. These inorganic salts, such as phosphate and sulfate anions, can poison the catalysts used in downstream processes converting lactic acid into biochemical products. Selective separation processes are therefore essential that can remove organic acid anions over inorganic anions. The target metrics for this study were to design anion exchange chemistries that can provide high transport rates of the lactate anion and minimize the crossover of competing inorganic anions.

The separation targeted lactic acid anions with a dopant of contaminant sulfate and phosphate ions which are commonly found in bioreactor fermentation streams. To target the carboxyl group within the lactic acid molecule, tethered imidazolium functional groups were investigated. Literature on $CO_2$ electrolysis and ionic liquid separations reported high affinity for $CO_2$ molecules to imidazolium functional groups [58]–[61]. The chemical similarity between a $CO_2$ molecule and the carboxylate moiety in organic acid anions was explored as a strategy for promoting the transport and selectivity of carboxylic acids. An imidazolium functionalized poly(arylene ether sulfone) (QIPSf, structure shown in Figure 12.6 (b)) anion exchange membrane (AEM) was benchmarked against the performance of a quaternary ammonium functionalized poly(arylene ether sulfone) (QAPSf, structure shown in Figure 12.6 (c)). The quaternary ammonium AEM served as a control in this study due to the prevention of the formation of a stable carbene which is reported to be the primary active site on imidazolium groups for $CO_2$ absorption. Further, the ion-

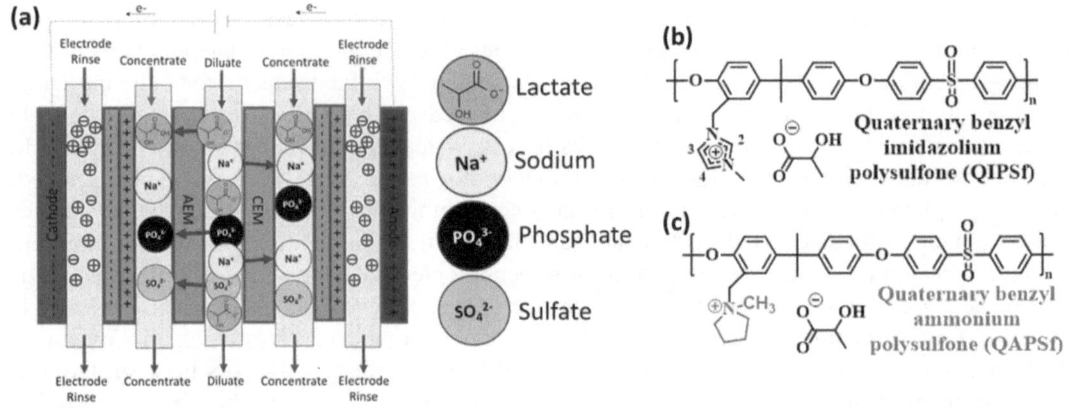


**FIGURE 12.6** Illustration of the single cell pair electrodialysis stack arrangement (subfigure a) for benchmarking the performance of the two anion exchange membrane (AEM) head group functionalizations (imidazolium vs. quaternary ammonium), with CEM = Cation exchange membrane and BPM = bipolar membrane. Chemical structure comparison between the imidazolium functionalization (QIPSf) (subfigure b) and quaternary ammonium functionalized (QAPSf) polysulfone anion exchange membranes (subfigure c). Figure reproduced from reference [57].

exchange capacity and polymer backbone chemistry were controlled to directly compare any differences in transport or selectivity properties to be attributable only toward the cation head group chemistry.

A simple one cell-pair ED unit that is illustrated in Figure 12.6 (a) was used to measure the rate of lactic acid capture and specific energy consumption of the two different polysulfone AEMs. A synthetic fermentation broth containing 33 g/kg$^{-1}$ lactic acid, 1.25 g/kg$^{-1}$ sodium sulfate, and 0.72 g/kg$^{-1}$ sodium phosphate was used as the feed solution for the diluate chamber and a 10 g/kg$^{-1}$ sodium chloride solution in the concentrate chamber. The ionic flux rates of each ionic species and specific energy consumption values were computed using Equations 12.1 and 12.2.

$$J_i = \frac{m_{conc}}{A}\frac{dC_i}{dt} \tag{12.1}$$

$$E = \frac{V\int I dt}{m} \tag{12.2}$$

The imidazolium tethered groups demonstrated a 99% greater lactic acid transport rate compared to the quaternary ammonium tethered groups (Figure 12.7 (a)). However, the imidazolium functionalization also promoted greater sulfate and phosphate fluxes by 62% and 54%, respectively, when compared to the quaternary ammonium functional groups. Due to the substantially improved lactic acid transport rates, the relative amount of lactate captured over sulfate and phosphate (i.e., selectivity) was still improved by 27% and 23%, respectively, using the imidazolium functionalization despite also having greater contaminant transport. The increased flux of lactic acid also contributed to improved specific energy consumption metrics by 62% (0.704 kWh kg$^{-1}$ versus 1.83 kWh kg$^{-1}$ for QIPSf and QAPSf, respectively, Figure 12.7 (b)). Electrochemical organic acid extractions are typically conducted at constant voltage to maintain control of the pH of the process streams. Due to the constant voltage operation, membrane design strategies to increase the rate of organic acid transport are the primary contributor toward improving the overall specific energy consumption metrics for electrochemical organic acid extraction technologies.

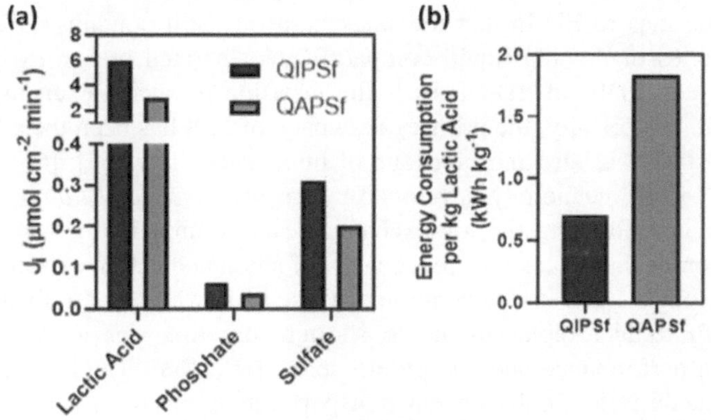


**FIGURE 12.7** Ionic flux results of lactic acid, phosphate, and sulfate across imidazolium functionalized polysulfone (QIPSf) and quaternary ammonium functionalized polysulfone (QAPSf) anion exchange membranes with a single-cell pair electrodialysis experiment (subfigure a) and specific energy consumption for lactic acid capture for the two membrane head group chemistries (subfigure b). Figure reproduced from reference [57].

Device-level ED studies demonstrated that imidazolium tethered groups promote lactate transport rates over quaternary ammonium groups in polysulfone-based AEMs. The greater lactic acid transport rates translated to greater selectivity when utilizing a QIPSf AEM over a QAPSf AEM in a single-cell ED unit fed with a model fermentation broth. The greater flux of lactic acid anions through imidazolium-type of AEMs also reduced the energy consumption for the organic acid recovery in ED. In terms of overall process economics, an almost doubling of the lactic acid transport rate implies an ED stack with half the required membrane area could be designed with equivalent performance if an imidazolium functionalized membrane was utilized. Ion-exchange membranes are commonly cited as the dominant cost in ED separations, and the reduction in the required membrane area directly leads to reductions in the capital costs of the separation platform.

Broadly speaking, ED applications have used historically applied a standard ion-exchange membrane (i.e., *Neosepta* AMX & CMX or *Ionics* AR/CR membranes) for all separation applications. However, as recent membrane advancements have shown, the tailoring of membrane chemistry for targeted separation applications has become advantageous to improve process economics and product purity. Strategies to discriminate the difference in solvation energies, valences, and size all can be taken advantage of with rationalc polymer design. In turn, these tool sets in membrane design will further expand the possible applications of electrochemical separations for targeted downstream processes.

### 12.4.2 Innovations in Spacer Channel Conductors

EDI is a commercial separation technology primarily deployed for ultrapure water production and remediation of industrial process waste streams [62], [63]. EDI employs the use of an ion-exchange resin column in the diluate liquid chambers [64]. The resin column augments the ionic conductivity of the diluate chamber when the solution conductivity becomes depleted below the ion exchange resin's material conductivity. By enhancing the solution conductivity, the EDI stack is more energy efficient for removing ions in the challenging dilute concentrate regime by lowering ohmic resistances within the diluate compartment. The modular nature of EDI makes the technology easily expandable between lab-scale to commercial scale based on the number of cell pairs (i.e., the size of each stack) and the number of stacks. Moreover, versatile operating conditions such as process flow rates, operating voltages, or stack design create a range of applications that require dilute operating conditions or enhancing the ionization of weak acids.

EDI carries similarities to ED in that the stack hardware still contains an anode and cathode electrodes with a series of flowing liquid compartments separated by ion-exchange membranes. The difference between EDI and ED stacks is the insertion of an ion-exchange resin bed in the diluate compartment. Historically, the primary drawback of EDI has been the added complexity of stack construction, stack leakage, and blockage of bulk liquid flow [65], [66]. Resin wafer electrodeionization (RW-EDI) technology developed by *Argonne National Laboratory* has addressed these challenges by immobilizing the ion-exchange resin column into a porous sheet [67]. The second-generation ion-exchange resin wafers consist of anion and cation exchange resins bound by a polyethylene (PE) binder. The first generation of these resin wafer materials incorporated a latex binder which the PE binder replaced due to shorter cure time for wafer manufacturing and improved separation performance and energy efficiency [30], [68]. The ion-exchange resins supplement ionic conductivity and enhance ion transport across the resin bed while the PE binder maintains the resin beads stationary. During material manufacturing, a sacrificial porosigen is added to generate a porous flow matrix and allow the process stream to flow through the ion-exchange resin wafer. The second-generation RW-EDI technology has been applied toward organic acid recovery [67], [69]–[71], brackish water desalination [55], [66], pH modulation of process streams [72], $CO_2$ direct air capture [73], methane upgrading [74], and selective metal purification [75]. Further, the second-generation RW-EDI technology has been demonstrated at pilot-scale by Argonne National Laboratory for industrial bioprocessing applications.

Although the second-generation RW-EDI technology has been successfully applied to a range of applications, the presence of the non-conductive PE binder was suggested to impede ion transport and contribute toward ohmic resistances in EDI [76]. This section describes the third generation of RW-EDI technology, where a conductive ionomer supplanted the PE binder (illustrated in Figure 12.8). The ionomer binder-based resin wafer can be further formulated with either cation exchange or anion exchange tethered functional groups to aid in the ion transport of either cations or anions. By using an ionomer-based binder, the binder adds another layer of material design that can be accomplished with the ion-exchange resin bed (along with the desired anion-to-cation resin ratio, types of resins used and porosity of the matrix). This section explores

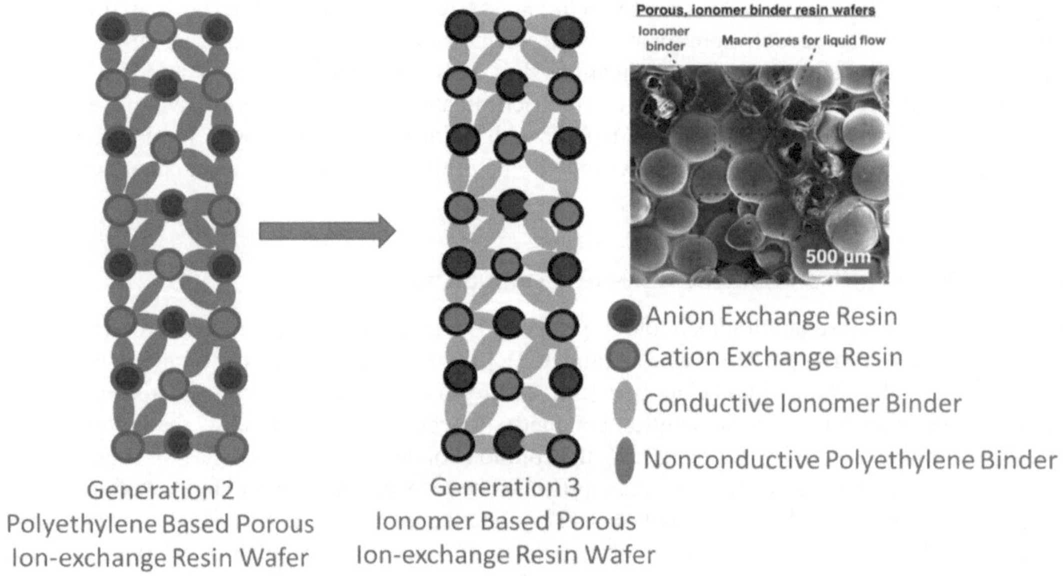


**FIGURE 12.8** Transition of ion-exchange resin wafer technology from "Generation 2" containing a non-conductive thermoplastic to "Generation 3" containing a conductive ionomer binder.

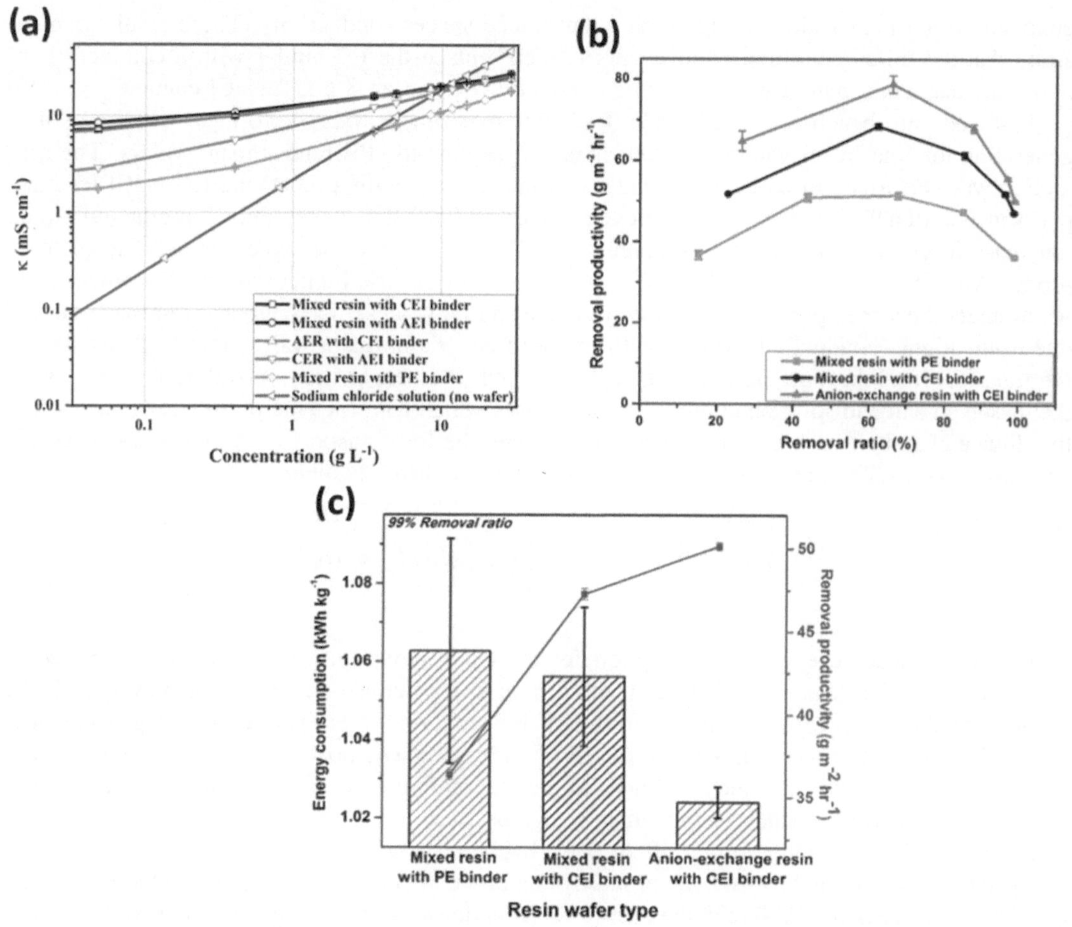


**FIGURE 12.9** Ionic conductivity (κ) of third-generation ionomer resin wafers compared to the second-generation polyethylene binder resin wafers at different NaCl concentrations (subfigure a). Removal productivity results (i.e., ion removal flux) vs. removal ratio for a 4-cell pair resin wafer electrodeionization benchmark study for NaCl desalination under batch mode at 99% removal (subfigure b). Energy consumption for Cl- removal (subfigure c). Two electrodeionization runs were performed with each material. The average result is given in each plot, and the error bars represent the absolute difference between trial runs. CEI = Cation exchange ionomer, AEI = Anion exchange ionomer, AER = Anion exchange resin, CER = Cation exchange resin, PE = Polyethylene. Figure reproduced from reference [76].

the performance of two cation exchange ionomer binder (CEI) resin wafers formulated with either a mixed cation and anion resin bed or an anion exchange resin (AER) only bed. The performance of the third-generation ionomer binder resin wafers is compared with the performance of a second-generation PE binder resin wafer with a mixed resin bed.

Two-point ionic conductivity measurements were conducted on the third-generation ionomer resin wafers at different salt concentrations (Figure 12.9 (a)). For the entire sodium chloride concentration regime, the ionic conductivity for each third-generation ionomer binder RW was higher compared to the benchmark PE binder RW. Further, the mixed bed ionomer binder resin wafers displayed the highest ionic conductivities. The conductivity measurements also reveal the maximum total dissolved salt content that the ion-exchange resin wafers still augment the solution conductivity. From Figure 12.9 (a), the ionomer binder resin wafers augment the spacer compartment conductivity up to 8 g $L^{-1}$ NaCl while the PE binder resin wafer up to 3.5 g $L^{-1}$ NaCl. Once the solution conductivity exceeds the resin wafer material conductivity, then material conductivity no longer aids in augmenting the diluate spacer conductivity. These results demonstrate that the third-generation resin wafers (which replace the PE binder with a conductive ionomer binder) contribute toward the ionic conductivity up to an 8 g $L^{-1}$ NaCl content.

A 4 cell-pair bench-scale (14 $cm^2$) RW-EDI experiment was performed with the third-generation ionomer resin wafers and benchmarked against the PE binder resin wafers. The separation was performed in batch mode under constant voltage with a 5,000 mg $L^{-1}$ NaCl feed and performed until a 99% removal was achieved. Figure 12.9 (b) shows the removal productivity (i.e., ion mass flux) vs. removal ratio (Equation 12.3) results from these experiments. The removal productivity for the ionomer binder resin wafers was almost 30% higher than the PE binder resin wafer under the same operating conditions (voltage and flow rates). These greater removal rates are due to the higher conductivity values with the third-generation resin wafers that further augment the transport rates through the ion-exchange resin bed. Further, as the removal ratio is increased, the dissolved salt and solution conductivity decreases which increases the ohmic resistance within the diluate chamber. At a constant operating voltage, the ion transport (and total stack electrical current) is reduced as the resistance is increased in the diluate chamber.

$$Removal\ \ ratio(\%) = \left(1 - \frac{C_{diluate}}{C_{feed}}\right) \times 100 \tag{12.3}$$

The effect on specific energy consumption for the $Cl^-$ removal at 99% removal is explored in Figure 12.9 (c). Up to a 4.3% reduction in specific energy consumption was achieved with the ionomer binder resin wafers compared to the PE binder resin wafer. The modest improvement is due to the greater transport rates and reduced ohmic resistances provided by the ionomer binder resin wafers. Although this study did not optimize operating conditions to minimize the energy consumption metrics, further improvements can be accomplished by adjusting the operating conditions, such as the electric field, flow rates, back pressure, etc.

Operating expenses in RW-EDI are primarily composed of energy consumption and ion-exchange membrane replacements. Although the energy consumption was only marginally improved, the 30% greater removal productivity metrics with the ionomer binder resin wafers have greater implications. An RW-EDI stack that can achieve faster rates of ion transport requires lower capital expenses (due to a smaller overall stack array size) and lower operational expenses (due to less required area of the ion-exchange membrane and resin wafers). In short, the third generation of resin wafer technology which employs a conductive ionomer binder to immobilize the ion-exchange resin bed generates faster ionic separations with less energy and lower overall capital and operating expenses.

### 12.4.3 Innovations in Hybrid Materials

ED and EDI have been demonstrated as an effective separation technology for capturing small organic acid molecules such as lactic, acetic, or glycolic acid [77]–[79]. More notably, these electrochemical separation technologies can be operated in-operando with a bioreactor to simultaneously produce and purify organic acids as they are made. So-called separative bioreactors have the advantages of improving the bioreactor performance by preventing product inhibition, maintaining the bioreactor pH by removal of the acid molecules, producing the acid in its protonated form without additive chemicals, and achieving greater titers of product than otherwise possible due to the simultaneous removal of acid from the reactor [67].

Applications of electrochemical separations toward larger, bulkier organic acids, such as those containing aromatic rings like phenolic acids and linear $C_5$ to $C_{10}$ organic acids, have been less effective. Limited permeability and transport resistances of the bulkier organic acid molecules through commercial anion exchange membranes have prevented applications of electrochemical separations in this area. The literature that has explored phenolic acid capture has encountered

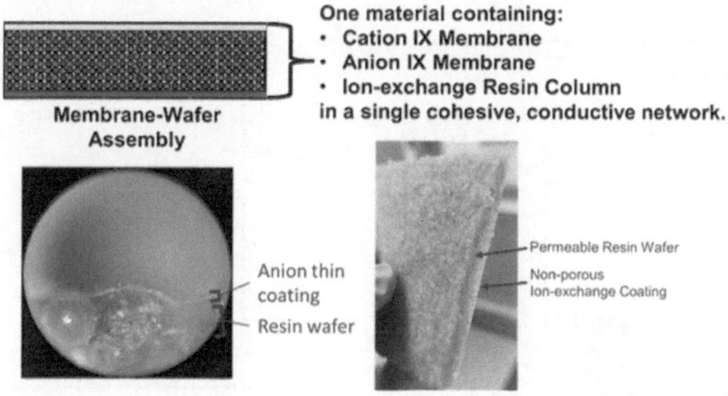


**FIGURE 12.10** Illustration of the integration of the ion-exchange membranes with an immobilized resin bed and a conductive ionomer binder. The integrated material eliminates resistances between the ion-exchange membrane and resin bed. Optical microscopy picture (100x) (bottom left) and a cutout of the material (bottom right) showing the membrane wafer assembly material containing a porous, permeable resin wafer with a nonporous membrane coating on the surface. IX = ion exchange.

either high energy usage or low recovery rates of these species. For example, an ED study removing phenoxide ions required 10–100 kWh of power per kg phenoxide recovered [80]. This value represents 10–100x greater energy consumption than is typical of the ~1 kWh $kg^{-1}$ required to capture lactic acid. Another study reported diminished ED efficiency due to the irreversible binding of gallic acid to the anion exchange membrane [79]. Clearly, there is a need for material solutions that promote greater transport rates of bulky organic acid molecules, such as phenolic acids, in order for electrochemical separations to branch into this area.

This case study describes an MWA material that was developed to increase transport rates of bulky phenolic acids by eliminating transport resistances between the ion-exchange membranes and an ion-exchange resin bed in electrodeionization. MWAs combine 60 μm thick AEMs and CEMs that are thermally laminated to an ionomer binder resin wafer which creates a single, cohesive conductive material. Figure 12.10 illustrates the new MWA, which is a single material against a conventional resin wafer electrodeionization configuration that consists of three separate materials (i.e., CEM, AEM, and RW). The ohmic interfacial resistances between the membranes and resin wafer are also illustrated along with a photo of an actual material with the anion exchange thin coating applied to the resin wafer. The interior of the MWA remains porous to allow liquid flow to permeate the resin bed, while the AEM and CEM thin coatings are impermeable to retain the liquid flow within the material. The anion exchange membrane and ionomer binder also employed the quaternary benzyl 1-methyl imidazolium groups described in the earlier section to further improve the transport rates of the organic acids.

A three-compartment RW-EDI cell was used to benchmark the performance of the MWA material against the second-generation PE binder RW-EDI setup. p-coumaric acid was applied as a model compound to benchmark the capture rate and energy consumption metrics. The capture productivity and specific energy consumption values are compared in Figure 12.11 (a). Benchmarking the MWA against the second-generation PE binder RW-EDI setup, a 7-fold improvement in the p-coumaric acid flux rate was observed while also consuming 70% less energy. The substantially improved capture rate is partially attributed to the tethered imidazolium groups on the AEM backbone. The previous section demonstrated a doubling of the ionic transport rates of lactic acid with this type of AEM functionalization.

The elimination of contact resistances between the membrane and resin bed was also thought to aid in the improved transport properties. To explore this idea, a two-point conductivity measurement in a 2.5 g/$kg^{-1}$ p-coumaric acid supporting electrolyte was used to characterize the interfacial contact resistance before and after mechanical-thermal lamination of the membrane-resin wafer assemblies

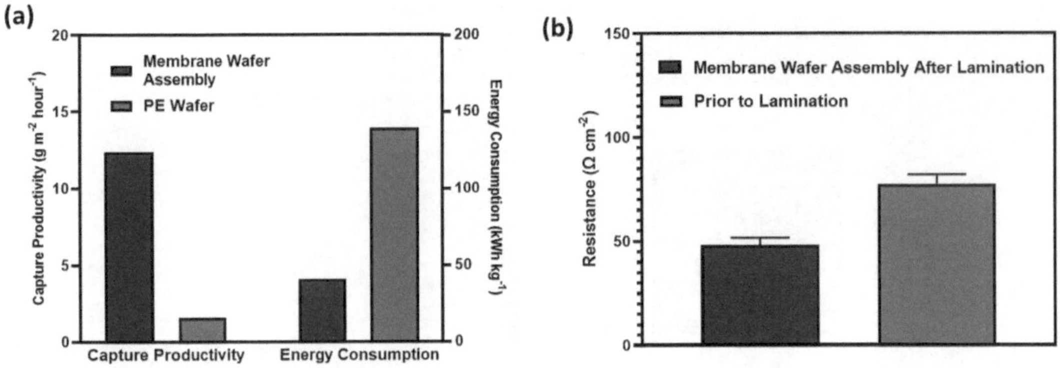


**FIGURE 12.11** Results from a three-cell pair benchmark separation of p-coumaric acid between the membrane wafer assembly (MWA, red) and second-generation polyethylene (PE) binder resin wafer electrodeionization (blue). Capture productivity is normalized to active area and energy consumption to kg of product produced (subfigure a). Material (ohmic) resistance measurements of the MWA before and after thermal-mechanical lamination in 2.5 g/kg-1 p-coumaric acid (subfigure b). The difference of these values was used to characterize the interfacial transport resistance between the ion-exchange membrane and resin wafer materials.

(Figure 12.11 (b)). The difference between the total resistance of the MWA and the resistance of the individual component prior to thermal lamination is to determine the interfacial resistance eliminated by the integration of materials. The estimated interfacial resistance for the MWA with p-coumaric acid was reduced 38% with thermal-mechanical lamination (48 ± 4 Ω $cm^2$ without mechanical-thermal lamination versus 78 ± 5 Ω $cm^2$ with mechanical-thermal lamination). In total, the amount of eliminated interfacial contact resistances between the ion-exchange membrane and ionomer binder resin wafer amounted to 30 Ω $cm^2$, which contributes toward the improved transport rates and specific energy efficiency performance. The reduced interfacial resistance amounts to a 5.5% reduction in cell voltage based on the 3.7 mA $cm^{-2}$ operating current and 6 V cell voltage (for a total of three cell pairs).

In the above case studies, material advancements in both ion-exchange membrane and spacer channel conductors have been shown that provides a major leap forward in separation efficiencies. These downstream purification processes are commonly cited as consuming 50–70% of overall production costs for organic acids [77]. Therefore, understanding rationale material designs that either accelerate the transport rate of organic acid molecules or reduces material resistances are critical areas in order to improve ion-exchange materials for bioprocessing applications. These material advancements, in turn, lead to breakthrough technologies that have substantial impacts on the overall process economics.

## 12.5 REDUCING RISKS IN WATER-ENERGY INTERDEPENDENCY NETWORKS

Contemporary water and energy systems are highly interdependent. They are regarded as one water-energy interdependency network (WEIN) where the impacts on either one of the individual systems can have repercussions to its counterparts [81]. Optimal coordination of the WEIN is much needed to counter climate variability. Climate change may change the regional precipitation pattern, which directly and negatively impacts the water system. In the WEIN, repercussions pass on to the energy system with less water available for cooling purposes in energy generation, causing a network-wide impact and leading to uncertainty in the availability and quality of water and energy. A schematic of the interdependency of water treatment and power plants is presented in Figure 12.12.

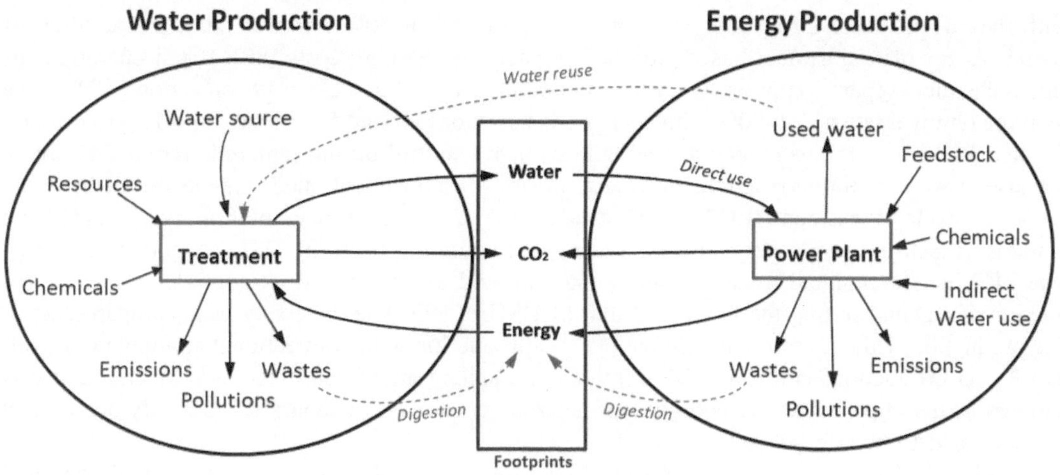


**FIGURE 12.12** Water-energy interdependency network (WEIN) for water treatment and power plants.

In the United States alone, 14 million metric tons of $NH_3$ are utilized directly and indirectly as fertilizer [82], [83]. The volume of N nutrient demand is substantial. The sources of nitrogen-rich anthropogenic sources are urban (20–40 mg/L), industrial (10–400 mg/L), and animal/dairy (500–3,500 mg/L) [84], of which are subject to government-mandated standards of discharge,. The ammonia nitrogen's discharge standard into surface water is 17 mg/L in the United States, 1.5~10 mg/L in the European Union, and 1.5~5 mg/L in China. Technologies that recover nutrient wastes from anthropogenic sources not only assist in complying with the regulations but also retain essential resources close to the production sources.

Industrial and agricultural anthropogenic sources are close to manufacturing and production processes, so it is logistically feasible and ideal for nutrient recycling, and they usually have a higher concentration. Different from domestic sources, industrial and agricultural sources are common nutrient-consuming hotspots. The strategic transformation of their roles from pure nutrient consumers to prosumers that participate both in the production and consumption of nutrients through in-situ nutrient recycling will improve the resilience of production.

As climate change amplifies the significance of alternative fuel options, fuel production accompanied by possible nutrient and high-value organic by-products recovery becomes very attractive. This section will focus on the urban mining potential of nitrogen nutrients from the wastewater of microalgal biofuel production.

The ammonia demand is 2.7 kg $NH_3$ per million British thermal units of renewable diesel (RD) for microalgae hydrothermal liquefaction (HTL), which implies that 3.3 million metric tons of ammonia are required to produce 10 billion gallons per year of RD by HTL. HTL of algae feedstock at 300°C led to the nitrogen distribution of roughly 30% of the feedstock nitrogen in the oil and solid phases and 70% in the HTL aqueous phase (HTL-AP) [85]. HTL-AP is generally characterized by high concentrations of ammonium, phosphates, and organic carbon [86]. Besides the prospective recovery of $NH_4$-N, the separation of other high-value organic compounds, i.e., acids, further improves the economic upside of algae biofuel production. HTL oil requires significant upgrading before it can be used as RD for transportation, because of high oxygen, nitrogen, and sulfur contents, together with acidity and viscosity issues [82]. Depending on the upgrading process of HTL oil, nitrogen incorporated into HTL oil, which roughly accounts for 6 wt.%, may be dissolved in the process wastewater and can be further recovered.

Controlling energy and operating costs are the key to the economic viability of ammonia recovery from wastewater. The unit energy consumption for conventional stripping and production of $(NH_4)_2SO_4$ from wastewater is approximately 9 kWh/kg-N, while ammonia fertilizer production with the Haber–Bosch process is 10 kWh/kg-N [87]. At an 80% removal ratio, the addition of chemicals for pH adjustment has a profound impact on operating costs [88]. Batch operations of electrochemical separation processes, e.g., electrodialysis (ED) and electrodeionization (EDI), have reported removal ratios over 90% with energy consumptions around 5 kWh/kg-N [46]. Furthermore, ED and EDI may eliminate chemical additives for pH control during ammonia removal/recovery because of water-splitting reactions that cause in-situ pH adjustment under the electric field.

A life cycle assessment (LCA) conducted by the U.S. Department of Energy evaluated the benefits of EDI-based recovery of mixed acids and ammonia from the HTL aqueous phase [89]. The LCA was completed assuming ammonia and acid are recovered as coproducts in the production of RD and setting the functional unit as 1 MJ of RD. The three scenarios compared are 1. RD production without ammonia removal; 2. RD production with conventional ammonia removal; and 3. RD production with RW-EDI ammonia and acids removal and recovery. Acetic acid was adopted as the coproduct. The energy and materials consumed to separate and purify acetic acid was assumed to be negligible.

Fossil fuel consumption is lower for the EDI case than for the conventional ammonia removal case. If there were no displacement credit for acetic acid, the EDI case would still outperform conventional $NH_3$ removal.

Water consumption of RW-EDI is drastically less than the conventional ammonia removal case. In the conventional case, quick lime (CaO) consumption mainly drives water consumption. Without the need of additional chemicals, the water consumption of RW-EDI treatment is less than 10% of the consumption of conventional ammonia removal and close to the no removal scenario.

The study has shown that EDI-based treatment has an indisputable advantage over the conventional ammonia removal method in wastewater nutrient recovery. Not only will there be a decent improvement in greenhouse gas emissions and fossil fuel consumption, but also be a significant difference in water consumption. In many cases, water consumption is the major concern of a region, and the application of EDI-based treatment in production processes would be beneficial to local WEINs.

## 12.6 SUMMARY

EST is an enabling technology with near zero carbon footprint to enhance sustainability because of its use of renewable energy. The integration of bioreactors and EST for in-situ product separation can significantly enhance biological reactivities and reduce the unit operations steps of "downstream" separation. It enables a cost-effective organic acid production to economically replace petroleum-based chemicals for the commodity chemical market from biomass feedstock. The results of energy and process cost savings of the integrated "separative bioreactor" (SB) concept was demonstrated by sugar fermentation of succinic and gluconic acid in separate case studies. In the case studies, technical and economic viabilities of biological productions of succinic and gluconic acid have been successfully demonstrated in laboratory-scale and pilot-scale operations using an integrated separative bioreactor. A several-fold increase in the selective production of succinic acid over the by-products of acetic acid and ethanol was demonstrated using the SB platform process. Succinic acid is a polymer building block that can replace the petroleum-based maleic anhydride. Gluconic acid is commonly used in the food industry. Over 2,500 hours of continuous operation in laboratory-scale SB and 1,200 hours of continuous gluconic acid fermentation enabled by the SB platform process at the pilot-scale demonstrate the technology's robustness and process performance consistency in a real-world environment. The low cost and energy consumption to produce organic acids using SB platform processes are due to multiple factors: Application of SBs 1) greatly reduces product inhibition in fermentation; 2) eliminates or significantly reduces the need for neutralization chemicals; and 3) enables a continuous fermentation. Ammonium recovery from bioprocessing waste was demonstrated using resin wafer electrodeionization (RW-EDI) technology that provided in-situ pH control capability without chemical additives. The energy consumption of ammonium recovery using RW-EDI technology was found to be only half of that of ammonium synthesis by the Haber-Bosch process. The results of improved bioprocessing and resource recovery from the waste show a significant reduction of carbon and water footprints, which both profoundly contribute to sustainability. Material innovations of the membranes and resin wafer adsorbent can profoundly enhance the separation performance by reducing energy consumption and increasing the separation rate. The improvement of material properties to address membrane fouling, lower energy consumption, and increased separation productivity were demonstrated in several case studies. Ionic-liquid-based additives to reduce ion transport resistances on ion-conductive membranes offers a simple and effective technique to improve separation performance. The innovative synthesis of complete ion-conductive resin wafer material and membrane-wafer assembly (MWA) technique not only improves the separation performance but also overcomes the challenges of high resistance transport of high carbon chain organic acids separation by EST. Its success enables the application of EST to produce sustainable aviation fuel production. A case study of risk assessment on RW-EDI applications for resources recovery from waste liquid also confirms sustainability improvement in carbon emission, fossil fuel consumption, and water footprint.

## REFERENCES


[1] Lin, Y. J.; Hestekin, J. A.; Henry, M. P.; Sather, N. Bioprocessing of Cost-Competitive Biobased Organic Acids. In *Commercializing Biobased Products: Opportunities, Challenges, Benefits, and Risks*; Green Chemistry Series; Royal Society of Chemistry, 2016.

[2] Datta, S.; Lin, Y. J.; Snyder, S. W. Current and Emerging Separations Technologies in Biorefining. In *Advances in Biorefineries*; Woodhead Publishing Limited, 2014.

[3] Blinco, J. A.; Doherty, W. O. In *Proceedings of the Conference of the Australian Society of Sugar Cane Technologists*; 2005.

[4] López-Garzón, C. S.; Straathof, A. J. J. Recovery of Carboxylic Acids Produced by Fermentation. *Biotechnol. Adv.* 2014, *32* (5), 873–904. 10.1016/j.biotechadv.2014.04.002

[5] Magarifuchi, T.; Goto, K.; Iimura, Y.; Tadenuma, M.; Tamura, G. Effect of Yeast Fumarase Gene (FUMl) Disruption on Production of Malic, Fumaric and Succinic-Acids in Sake Mash. *Journal of Fermentation and Bioengineering 8* (4), 355–361. 10.1016/0922-338X(95)94204-5

[6] Ling, L.-P.; Leow, H.-F.; Sarmidi, M. R. Citric Acid Concentration by Electrodialysis: Ion and Water Transport Modelling. *J. Membr. Sci.* 2002, *199* (1–2), 59–67. 10.1016/S0376-7388(01)00678-0

[7] García-Fraile, P.; Silva, L. R.; Sánchez-Márquez, S.; Velázquez, E.; Rivas, R. Plums (Prunus Domestica L.) Are a Good Source of Yeasts Producing Organic Acids of Industrial Interest from Glycerol. *Food Chem.* 2013, *139* (1–4), 31–34. 10.1016/j.foodchem.2012.12.043

[8] Yoshida, S.; Yokoyama, A. Identification and Characterization of Genes Related to the Production of Organic Acids in Yeast. *J. Biosci. Bioeng.* 2012, *113* (5), 556–561. 10.1016/j.jbiosc.2011.12.017

[9] Kutyła-Olesiuk, A.; Wawrzyniak, U. E.; Ciosek, P.; Wróblewski, W. Electrochemical Monitoring of Citric Acid Production by Aspergillus Niger. *Anal. Chim. Acta* 2014, *823*, 25–31. 10.1016/j.aca.2014.03.033

[10] Yu, S.; Huang, D.; Wen, J.; Li, S.; Chen, Y.; Jia, X. Metabolic Profiling of a Rhizopus Oryzae Fumaric Acid Production Mutant Generated by Femtosecond Laser Irradiation. *Bioresour. Technol.* 2012, *114*, 610–615. 10.1016/j.biortech.2012.03.087

[11] Zhang, Z. Y.; Jin, B.; Kelly, J. M. Production of Lactic Acid from Renewable Materials by Rhizopus Fungi. *Biochem. Eng. J.* 2007, *35* (3), 251–263. 10.1016/j.bej.2007.01.028

[12] Gu, B. H.; Zheng, P.; Yan, Q.; Liu, W. Aqueous Two-Phase System: An Alternative Process for Recovery of Succinic Acid from Fermentation Broth. *Sep. Purif. Technol.* 2014, *138*, 47–54. 10.1016/j.seppur.2014.09.034

[13] Matsumoto, M.; Mochiduki, K.; Fukunishi, K.; Kondo, K. Extraction of Organic Acids Using Imidazolium-Based Ionic Liquids and Their Toxicity to Lactobacillus Rhamnosus. *Sep. Purif. Technol.* 2004, *40* (1), 97–101. 10.1016/j.seppur.2004.01.009

[14] Oliveira, F. S.; Araújo, J. M. M.; Ferreira, R.; Rebelo, L. P. N.; Marrucho, I. M. Extraction of L-Lactic, l-Malic, and Succinic Acids Using Phosphonium-Based Ionic Liquids. *Sep. Purif. Technol.* 2012, *85*, 137–146. 10.1016/j.seppur.2011.10.002

[15] Ali, T.; Bylund, D.; Essén, S. A.; Lundström, U. S. Liquid Extraction of Low Molecular Mass Organic Acids and Hydroxamate Siderophores from Boreal Forest Soil. *Soil Biol. Biochem.* 2011, *43* (12), 2417–2422. 10.1016/j.soilbio.2011.08.015

[16] Cherkasov, D. G.; Il'in, K. K. Salting out of Butyric Acid from Aqueous Solutions with Potassium Chloride. *Russ. J. Appl. Chem.* 2009, *82* (5), 920–924. 10.1134/S1070427209050346

[17] Wu, D.; Chen, H.; Jiang, L.; Cai, J.; Xu, Z.; Cen, P. Efficient Separation of Butyric Acid by an Aqueous Two-Phase System with Calcium Chloride. *Chin. J. Chem. Eng.* 2010, *18* (4), 533–537. 10.1016/S1004-9541(10)60255-8

[18] Drewes, J. E.; Mitterwallner, J.; Gruenheid, S.; Bellona, C. A Novel Approach Using Reverse Osmosis/Electrodialysis (RO/ED) to Concentrate and Isolate Organic Carbon from Water Samples. In *Proceedings of the American Water Works Association* 2002; pp. 252–269.

[19] Timmer, J. M. K.; Kromkamp, J.; Robbertsen, T. Lactic Acid Separation from Fermentation Broths by Reverse Osmosis and Nanofiltration. *J. Membr. Sci.* 1994, *92* (2), 185–197. 10.1016/0376-7388(94)00061-1

[20] Lemba, J.; Kārkliņš, R.; Lapele, I. Purification and Isolation of Organic Acids by Ion Exchange and Electrodialysis (Organisko Skābju Attīrīšana Un Izdalīšana Ar Jonapmaiņu Un Elektrodialīzi). *Materialzinatne un Lietiska Kimija* 2000, *1*, 73–80.

[21] Ullah, S. M. R.; Takeuchi, M.; Dasgupta, P. K. Versatile Gas/Particle Ion Chromatograph. *Environ. Sci. Technol.* 2006, *40* (3), 962–968. 10.1021/es051722z

[22] Wiśniewski, J.; Różańska, A.; Winnicki, T. Removal of Troublesome Anions from Water by Means of Donnan Dialysis. *Desalination* 2005, *182* (1–3), 339–346. 10.1016/j.desal.2005.02.032

[23] Wang, Y.; Huang, C.; Xu, T. Which Is More Competitive for Production of Organic Acids, Ion-Exchange or Electrodialysis with Bipolar Membranes? *J. Membr. Sci.* 2011, *374* (1–2), 150–156. 10.1016/j.memsci.2011.03.026

[24] Vertova, A.; Aricci, G.; Rondinini, S.; Miglio, R.; Carnelli, L.; D'Olimpio, P. Electrodialytic Recovery of Light Carboxylic Acids from Industrial Aqueous Wastes. *J. Appl. Electrochem.* 2009, *39* (11), 2051–2059. 10.1007/s10800-009-9871-9

[25] Zeikus, J. G.; Jain, M. K.; Elankovan, P. Biotechnology of Succinic Acid Production and Markets for Derived Industrial Products. *Appl. Microbiol. Biotechnol.* 1999, *51* (5), 545–552. 10.1007/s002530051431

[26] Datta, R.; Henry, M. Lactic Acid: Recent Advances in Products, Processes and Technologies—a Review. *J. Chem. Technol. Biotechnol.* 2006, *81* (7), 1119–1129. 10.1002/jctb.1486

[27] John, R. P.; Nampoothiri, K. M.; Pandey, A. Fermentative Production of Lactic Acid from Biomass: An Overview on Process Developments and Future Perspectives. *Appl. Microbiol. Biotechnol.* 2007, *74* (3), 524–534. 10.1007/s00253-006-0779-6

[28] Datta, R.; Lin, Y. J.; Burke, D.; Tsai, S. P. Electrodeionization Substrate, And Device For Electrodeionization Treatment. *US Patent #6495014*, 2002.

[29] Lin, Y. J.; Hestekin, J. A.; Arora, M. B.; St. Martin, E. J. Electrodeionization Method. *US Patent #6797140*, 2004.

[30] Lin, Y. J.; Henry, M. P.; Snyder, S. W. Electronically And Ionically Conductive Porous Material And Method For Manufacture Of Resin Wafers Therefrom. *US patent #7452920*, 2008.

[31] Werpy, T.; Petersen, G. *Top Value Added Chemicals from Biomass: Volume I – Results of Screening for Potential Candidates from Sugars and Synthesis Gas*; Technical Report; 2004. https://www.nrel.gov/docs/fy04osti/35523.pdf.

[32] Shannon, M. A.; Bohn, P. W.; Elimelech, M.; Georgiadis, J. G.; Mariñas, B. J.; Mayes, A. M. Science and Technology for Water Purification in the Coming Decades. *Nature* 2008, *452* (7185), 301–310. 10.1038/nature06599

[33] Kavvada, O.; Tarpeh, W. A.; Horvath, A.; Nelson, K. L. Life-Cycle Cost and Environmental Assessment of Decentralized Nitrogen Recovery Using Ion Exchange from Source-Separated Urine through Spatial Modeling. *Environ. Sci. Technol.* 2017, *51* (21), 12061–12071. 10.1021/acs.est.7b02244

[34] Nikitin, O. V.; Stepanova, N. Yu.; Latypova, V. Z. Human Health Risk Assessment Related to Blue-Green Algae Mass Development in the Kuibyshev Reservoir. *Water Supply* 2015, *15* (4), 693–700. 10.2166/ws.2015.022

[35] Galloway, J. N.; Townsend, A. R.; Erisman, J. W.; Bekunda, M.; Cai, Z.; Freney, J. R.; Martinelli, L. A.; Seitzinger, S. P.; Sutton, M. A. Transformation of the Nitrogen Cycle: Recent Trends, Questions, and Potential Solutions. *Science* 2008, *320* (5878), 889–892. https://www.science.org/doi/10.1126/science.1136674

[36] USGS. Nitrogen Statistics and Information. 2019.
[37] Kim, T.; Gorski, C. A.; Logan, B. E. Ammonium Removal from Domestic Wastewater Using Selective Battery Electrodes. *Environ. Sci. Technol. Lett.* 2018, *5* (9), 578–583. 10.1021/acs.estlett.8b00334
[38] Zhang, H.; Wang, L.; Van Herle, J.; Maréchal, F.; Desideri, U. Techno-Economic Comparison of Green Ammonia Production Processes. *Appl. Energy* 2020, *259*, 114135. 10.1016/j.apenergy.2019.114135
[39] Kurniawan, T.; Lo, W.; Chan, G. Physico-Chemical Treatments for Removal of Recalcitrant Contaminants from Landfill Leachate. *J. Hazard. Mater.* 2006, *129* (1–3), 80–100. 10.1016/j.jhazmat.2005.08.010
[40] Ioannou, L. A.; Puma, G. L.; Fatta-Kassinos, D. Treatment of Winery Wastewater by Physicochemical, Biological and Advanced Processes: A Review. *J. Hazard. Mater.* 2015, *286*, 343–368. 10.1016/j.jhazmat.2014.12.043
[41] Yeung, A. T.; Gu, Y.-Y. A Review on Techniques to Enhance Electrochemical Remediation of Contaminated Soils. *J. Hazard. Mater.* 2011, *195*, 11–29. 10.1016/j.jhazmat.2011.08.047
[42] Wang, H.; Zhou, Q.; Zhang, G.; Yan, G.; Lu, H.; Sun, L. A Novel PSB-EDI System for High Ammonia Wastewater Treatment, Biomass Production and Nitrogen Resource Recovery: PSB System. *Water Sci. Technol.* 2016, *74* (3), 616–624. 10.2166/wst.2016.254
[43] Lee, G.; Kim, K.; Chung, J.; Han, J.-I. Electrochemical Ammonia Accumulation and Recovery from Ammonia-Rich Livestock Wastewater. *Chemosphere* 2021, *270*, 128631. 10.1016/j.chemosphere.2020.128631
[44] Xiang, D.; Zhou, Y. Concept Design and Techno-Economic Performance of Hydrogen and Ammonia Co-Generation by Coke-Oven Gas-Pressure Swing Adsorption Integrated with Chemical Looping Hydrogen Process. *Appl. Energy* 2018, *229*, 1024–1034. 10.1016/j.apenergy.2018.08.081
[45] Chen, T.-L.; Chen, L.-H.; Lin, Y. J.; Yu, C.-P.; Ma, H.; Chiang, P.-C. Advanced Ammonia Nitrogen Removal and Recovery Technology Using Electrokinetic and Stripping Process towards a Sustainable Nitrogen Cycle: A Review. *J. Clean. Prod.* 2021, *309*, 127369. 10.1016/j.jclepro.2021.127369
[46] van Linden, N.; Bandinu, G. L.; Vermaas, D. A.; Spanjers, H.; van Lier, J. B. Bipolar Membrane Electrodialysis for Energetically Competitive Ammonium Removal and Dissolved Ammonia Production. *J. Clean. Prod.* 2020, *259*, 120788. 10.1016/j.jclepro.2020.120788
[47] Zhang, C.; Ma, J.; Waite, T. D. Ammonia-Rich Solution Production from Wastewaters Using Chemical-Free Flow-Electrode Capacitive Deionization. *ACS Sustain. Chem. Eng.* 2019, *7* (7), 6480–6485. 10.1021/acssuschemeng.9b00314
[48] Georg, S.; Schott, C.; Courela Capitao, J. R.; Sleutels, T.; Kuntke, P.; Heijne, A. ter; Buisman, C. J. N. Bio-Electrochemical Degradability of Prospective Wastewaters to Determine Their Ammonium Recovery Potential. *Sustain. Energy Technol. Assess.* 2021, *47*, 101423. 10.1016/j.seta.2021.101423
[49] Kuntke, P.; Sleutels, T. H. J. A.; Rodríguez Arredondo, M.; Georg, S.; Barbosa, S. G.; ter Heijne, A.; Hamelers, H. V. M.; Buisman, C. J. N. (Bio)Electrochemical Ammonia Recovery: Progress and Perspectives. *Appl. Microbiol. Biotechnol.* 2018, *102* (9), 3865–3878. 10.1007/s00253-018-8888-6
[50] Shaposhnik, V. A.; Kesore, K. An Early History of Electrodialysis with Permselective Membranes. *J. Membr. Sci.* 1997, *136* (1–2), 35–39. 10.1016/S0376-7388(97)00149-X
[51] Meyer, K. H.; Straus, W. La Perméabilité Des Membranes VI. Sur Le Passage Du Courant Électrique à Travers Des Membranes Sélectives. *Helv. Chim. Acta* 1940, *23* (1), 795–800. 10.1002/hlca.19400230199
[52] Grebenyuk, V. D.; Grebenyuk, O. V. Electrodialysis: From an Idea to Realization. 2002, *38* (8), 4.
[53] Juda, W.; McRae, W. A. Coherent Ion Exchange Gels and Membranes. *J. Am. Chem. Soc.* 1950, *72* (2), 1044–1044. 10.1021/ja01158a528
[54] Walters, W. R.; Weiser, D. W.; Marek, L. J. Concentration of Radioactive Aqueous Wastes. Electromigration Through Ion-Exchange Membranes. *Ind. Eng. Chem.* 1955, *47* (1), 61–67. 10.1021/ie50541a027
[55] Pan, S.-Y.; Snyder, S. W.; Ma, H.-W.; Lin, Y. J.; Chiang, P.-C. Energy-Efficient Resin Wafer Electrodeionization for Impaired Water Reclamation. *J. Clean. Prod.* 2018, *174*, 1464–1474. 10.1016/j.jclepro.2017.11.068
[56] Alkhadra, M. A.; Su, X.; Suss, M. E.; Tian, H.; Guyes, E. N.; Shocron, A. N.; Conforti, K. M.; de Souza, J. P.; Kim, N.; Tedesco, M.; Khoiruddin, K.; Wenten, I. G.; Santiago, J. G.; Hatton, T. A.; Bazant, M. Z. Electrochemical Methods for Water Purification, Ion Separations, and Energy Conversion. *Chem. Rev.* 2022, *122* (16), 13547–13635. 10.1021/acs.chemrev.1c00396

[57] Jordan, M. L.; Kulkarni, T.; Senadheera, D. I.; Kumar, R.; Lin, Y. J.; Arges, C. G. Imidazolium-Type Anion Exchange Membranes for Improved Organic Acid Transport and Permselectivity in Electrodialysis. *J. Electrochem. Soc.* 2022, *169* (4), 043511. 10.1149/1945-7111/ac6448. Licensed under CC BY 4.0. To view a copy of this license, visit http://creativecommons.org/licenses/by/4.0/.

[58] Kutz, R. B.; Chen, Q.; Yang, H.; Sajjad, S. D.; Liu, Z.; Masel, I. R. Sustainion Imidazolium-Functionalized Polymers for Carbon Dioxide Electrolysis. *Energy Technol.* 2017, *5*, 929–936. 10.1002/ente.201600636

[59] Cadena, C.; Anthony, J. L.; Shah, J. K.; Morrow, T. I.; Brennecke, J. F.; Maginn, E. J. Why Is $CO_2$ So Soluble in Imidazolium-Based Ionic Liquids? *J. Am. Chem. Soc.* 2004, *126* (16), 5300–5308. 10.1021/ja039615x

[60] Carlisle, T. K.; Bara, J. E.; Lafrate, A. L.; Gin, D. L.; Noble, R. D. Main-Chain Imidazolium Polymer Membranes for CO2 Separations: An Initial Study of a New Ionic Liquid-Inspired Platform. *J. Membr. Sci.* 2010, *359* (1–2), 37–43. 10.1016/j.memsci.2009.10.022

[61] Bara, J. E.; Carlisle, T. K.; Gabriel, C. J.; Camper, D.; Finotello, A.; Gin, D. L.; Noble, R. D. Guide to $CO_2$ Separations in Imidazolium-Based Room-Temperature Ionic Liquids. *Ind. Eng. Chem. Res.* 2009, *48* (6), 2739–2751. 10.1021/ie8016237

[62] Wood, J.; Gifford, J.; Arba, J.; Shaw, M. Production of Ultrapure Water by Continuous Electrodeionization. *Desalination* 2010, *250* (3), 973–976. 10.1016/j.desal.2009.09.084

[63] Arar, Ö.; Yüksel, Ü.; Kabay, N.; Yüksel, M. Various Applications of Electrodeionization (EDI) Method for Water Treatment—A Short Review. *Desalination* 2014, *342*, 16–22. 10.1016/j.desal.2014.01.028

[64] Alvarado, L.; Chen, A. Electrodeionization: Principles, Strategies and Applications. *Electrochimica Acta* 2014, *132*, 583–597. 10.1016/j.electacta.2014.03.165

[65] Zheng, X.-Y.; Pan, S.-Y.; Tseng, P.-C.; Zheng, H.-L.; Chiang, P.-C. Optimization of Resin Wafer Electrodeionization for Brackish Water Desalination. *Sep. Purif. Technol.* 2018, *194*, 346–354. 10.1016/j.seppur.2017.11.061

[66] Pan, S.-Y.; Snyder, S. W.; Ma, H.-W.; Lin, Y. J.; Chiang, P.-C. Development of a Resin Wafer Electrodeionization Process for Impaired Water Desalination with High Energy Efficiency and Productivity. *ACS Sustain. Chem. Eng.* 2017, *5* (4), 2942–2948. 10.1021/acssuschemeng.6b02455

[67] Arora, M. B.; Hestekin, J. A.; Snyder, S. W.; St. Martin, E. J.; Lin, Y. J.; Donnelly, M. I.; Millard, C. S. The Separative Bioreactor: A Continuous Separation Process for the Simultaneous Production and Direct Capture of Organic Acids. *Sep. Sci. Technol.* 2007, *42* (11), 2519–2538. 10.1080/01496390701477238

[68] Arora, M. B.; Hestekin, J. A.; Lin, Y. J.; Martin, E. J.; Snyder, S. W. Porous Solid Ion Exchange Wafer for Immobilizing Biomolecules. *Patent number #7306934*, 2007. https://www.osti.gov/doepatents/biblio/921019.

[69] Gurram, R. N.; Datta, S.; Lin, Y. J.; Snyder, S. W.; Menkhaus, T. J. Removal of Enzymatic and Fermentation Inhibitory Compounds from Biomass Slurries for Enhanced Biorefinery Process Efficiencies. *Bioresour. Technol.* 2011, *102* (17), 7850–7859. 10.1016/j.biortech.2011.05.043

[70] Lopez, A. M.; Hestekin, J. A. Improved Organic Acid Purification through Wafer Enhanced Electrodeionization Utilizing Ionic Liquids. *J. Membr. Sci.* 2015, *493*, 200–205. 10.1016/j.memsci.2015.06.008

[71] Datta, S.; Lin, Y. J.; Schell, D. J.; Millard, C. S.; Ahmad, S. F.; Henry, M. P.; Gillenwater, P.; Fracaro, A. T.; Moradia, A.; Gwarnicki, Z. P.; Snyder, S. W. Removal of Acidic Impurities from Corn Stover Hydrolysate Liquor by Resin Wafer Based Electrodeionization. *Ind. Eng. Chem. Res.* 2013, *52* (38), 13777–13784. 10.1021/ie4017754

[72] Jordan, M. L.; Valentino, L.; Nazyrynbekova, N.; Palakkal, V. M.; Kole, S.; Bhattacharya, D.; Lin, Y. J.; Arges, C. G. Promoting Water-Splitting in Janus Bipolar Ion-Exchange Resin Wafers for Electrodeionization. *Mol. Syst. Des. Eng.* 2020, *5* (5), 922–935. 10.1039/C9ME00179D

[73] Datta, S.; Henry, M. P.; Lin, YuPo. J.; Fracaro, A. T.; Millard, C. S.; Snyder, S. W.; Stiles, R. L.; Shah, J.; Yuan, J.; Wesoloski, L.; Dorner, R. W.; Carlson, W. M. Electrochemical $CO_2$ Capture Using Resin-Wafer Electrodeionization. *Ind. Eng. Chem. Res.* 2013, *52* (43), 15177–15186. 10.1021/ie402538d

[74] Snyder, S. W.; Lin, Y. J.; Urgun-Demirtas, M. Methane Production Using Resin-Wafer Electrodeionization. *US Patent #8679314B1*, 2014. https://patents.google.com/patent/US8679314B1/en.

[75] Ulusoy Erol, H. B.; Hestekin, C. N.; Hestekin, J. A. Effects of Resin Chemistries on the Selective Removal of Industrially Relevant Metal Ions Using Wafer-Enhanced Electrodeionization. *Membranes* 2021, *11* (1), 45. 10.3390/membranes11010045

[76] Palakkal, V. M.; Valentino, L.; Lei, Q.; Kole, S.; Lin, Y. J.; Arges, C. G. Advancing Electrodeionization with Conductive Ionomer Binders That Immobilize Ion-Exchange Resin Particles into Porous Wafer Substrates. *Npj Clean Water* 2020, *3* (1), 5. 10.1038/s41545-020-0052-z. Licensed under CC BY 4.0. To view a copy of this license, visit http://creativecommons.org/licenses/by/4.0/.
[77] Kim, N.; Jeon, J.; Chen, R.; Su, X. Electrochemical Separation of Organic Acids and Proteins for Food and Biomanufacturing. *Chem. Eng. Res. Des.* 2022, *178*, 267–288. 10.1016/j.cherd.2021.12.009
[78] Handojo, L.; Wardani, A. K.; Regina, D.; Bella, C.; Kresnowati, M. T. A. P.; Wenten, I. G. Electro-Membrane Processes for Organic Acid Recovery. *RSC Adv.* 2019, *9* (14), 7854–7869. 10.1039/C8RA09227C
[79] Bober, M.; Crespo, J. G.; Velizarov, S. Electromembrane Processing for the Recovery of Low–Molecular Weight Bioactive Compounds from Model Solutions. *Procedia Eng.* 2012, *44*, 714–716. 10.1016/j.proeng.2012.08.542
[80] Wu, D.; Chen, G. Q.; Hu, B.; Deng, H. Feasibility and Energy Consumption Analysis of Phenol Removal from Salty Wastewater by Electro-Electrodialysis. *Sep. Purif. Technol.* 2019, *215*, 44–50. 10.1016/j.seppur.2019.01.001
[81] Chen, L.-H.; Li, P.-C.; Lin, Y.; Chen, I.-C.; Ma, H.; Yu, C.-P. Establishing a Quantification Process for Nexus Repercussions to Mitigate Environmental Impacts in a Water-Energy Interdependency Network. *Resour. Conserv. Recycl.* 2021, *171*, 105628. 10.1016/j.resconrec.2021.105628
[82] Frank, E. D.; Elgowainy, A.; Han, J.; Wang, Z. Life Cycle Comparison of Hydrothermal Liquefaction and Lipid Extraction Pathways to Renewable Diesel from Algae. *Mitig. Adapt. Strateg. Glob. Change* 2013, *18* (1), 137–158. 10.1007/s11027-012-9395-1
[83] Glauser, J.; Kumamoto, T. *Chemical Economics Handbook, Marketing Research Report: Ammonia*; SRI Consulting, 2010.
[84] Canter, C. E.; Blowers, P.; Handler, R. M.; Shonnard, D. R. Implications of Widespread Algal Biofuels Production on Macronutrient Fertilizer Supplies: Nutrient Demand and Evaluation of Potential Alternate Nutrient Sources. *Appl. Energy* 2015, *143*, 71–80. 10.1016/j.apenergy.2014.12.065
[85] Yu, G.; Zhang, Y.; Schideman, L.; Funk, T.; Wang, Z. Distributions of Carbon and Nitrogen in the Products from Hydrothermal Liquefaction of Low-Lipid Microalgae. *Energy Environ. Sci.* 2011, *4* (11), 4587–4595. 10.1039/C1EE01541A
[86] Barbera, E.; Bertucco, A.; Kumar, S. Nutrients Recovery and Recycling in Algae Processing for Biofuels Production. *Renew. Sustain. Energy Rev.* 2018, *90*, 28–42. 10.1016/j.rser.2018.03.004
[87] Maurer, M.; Schwegler, P.; Larsen, T. A. Nutrients in Urine: Energetic Aspects of Removal and Recovery. *Water Sci. Technol.* 2003, *48* (1), 37–46. 10.2166/wst.2003.0011
[88] Wang, L. K.; Hung, Y.-T.; Shammas, N. K. *Advanced Physicochemical Treatment Technologies*; Humana Press, 2007; Vol. 5.
[89] U.S. DOE. Bioprocessing Separations Consortium: Three-Year Overview. U.S. Department of Energy, Office of Energy Efficiency & Renewable Energy. 2020.

# 13 All-electric Vertical Take-off and Landing Aircraft (eVTOL) for Sustainable Urban Travel

*Raffaele Russo*
Joby Aviation, Santa Cruz, California, USA

*Eric C.D. Tan*
National Renewable Energy Laboratory, Golden, Colorado, USA

## CONTENTS



## 13.1 INTRODUCTION

Congestion in urban centers around the world has an adverse effect on the economy, health, and environment [1,2]. The dramatic costs to people and businesses of congestion in urban centers across the world, both time and fuel wasted, accounted for up to $88 billion in direct losses in the United States alone in 2019, while the movement of fuel and goods also suffers from congestion, estimated at $74 billion in indirect losses. In addition, the authors recognize that pattern shifts in how people live and work – especially due to the emergence of remote and hybrid work – may create latent demand for longer-distance transportation in sprawling urban centers. The efficiency of these trips is often defined by the existence of effective ground-based infrastructure, which is costly to build, expand, and maintain as populations continue to grow.

The emerging industry of electric vertical take-off and landing (eVTOL) aircraft can provide a critical time saving opportunity to travelers within urban areas. A wide range of companies is working to develop these aircraft around the United States and the world, with most concepts currently being explored focused on delivering aerial ridesharing services with a maximum occupancy of between three and eight passengers. Existing heliport and regional airport infrastructure are

DOI: 10.1201/9781003167693-13

expected to support the early adoption of eVTOL technology across markets, with companies looking to the construction of additional take-off and landing locations enabled by the low environmental and noise footprint of new aircraft technologies to enable future growth.

The present-day development of eVTOL aircraft is enabled by several long-term technological innovation trends:

- Lithium-ion cells have reached an energy density (expressed as kilowatt hours per kg of the cell) that enables their use in aviation: an application where weight sensitivity is very high. Prior to these advancements, in part pushed by the booming electric vehicle industry, for an eVTOL flight to be possible, an excessive amount of total aircraft weight would have had to be dedicated to batteries, in turn reducing the weight left for payload.
- The widespread diffusion of carbon fiber as an engineering material in the last ten years has enabled larger and larger parts within vehicles to achieve significant weight savings as they move away from metallic solutions and into lightweight composites. Aviation quickly embraced the opportunity. For example, in 2016, Boeing began preproduction of its wide haul 777 aircraft wings made of carbon fiber [3].
- While initial eVTOL aircraft will be piloted, future development towards autonomy – with a more predictable environment in the sky than on the ground – will also contribute to a gradual improvement in unit economics, removing the cost item of the pilot's salary and freeing up payload for an additional paying passenger.

The ultimate value proposition to customers is time savings versus an increased price over other forms of mobility, so companies are incentivized to operate routes that minimize operating costs and maximize improved trip time over existing transportation options. For that reason, many eVTOL under development today will operate flights between 5–50 miles across congested cities such as New York, Los Angeles, and London, and many eVTOL companies have identified routes within this window that offer promising passenger demand and time savings. Additionally, shorter-distance trips are ideal for eVTOL aircraft as achieving a high utilization rate is important to achieving positive unit economics, as well as minimal downtime to recharge and condition batteries in between flights.

The emergence of eVTOL aircraft, as well as all-electric conventional aircraft for short regional flights, is seen as one element of the broader solution to transitioning all forms of mobility to more environmentally sustainable modes.

With nearly all countries committing to the Paris Agreement to limit further global temperature increases to at most 1.5 degrees Celsius [4], all technologies that may aid in achieving said goal have received both financial support and environmental scrutiny. eVTOLs are one of the said technologies: their widespread adoption will help combat greenhouse gas emissions but will also require a strong case for their true impact on the planet. Hence, standardized, well-executed life-cycle assessments (LCAs) for eVTOL operations are the best tool to support such endeavors.

## 13.2 EVTOL LIFE-CYCLE ASSESSMENT: A CASE STUDY

Sustainability can be part of the very roots of eVTOL aircraft design, provided tools, methodology, and significant precedents exist to enable it. The nascent eVTOL industry offers a prime opportunity for life-cycle assessment (LCA) to support the development of the technology in a direction that benefits sustainable transportation. As the design space is vast and still being explored, LCAs can offer quantitative knowledge of the impact of eVTOL aircraft design choices on the environmental impacts, including its life-cycle greenhouse gas (GHG) emissions that contribute to global warming and climate change. LCA can assist in identifying opportunities to improve the environmental performance of eVTOL aircraft at various life cycle stages. LCA can also inform decision makers for strategic planning, priority setting, product or manufacturing process design or redesign. Unlike other industries already mature and fixed in their customs, LCA finds a fertile ground in eVTOL to set the status quo.

Beyond the broad societal benefit described above, there is strong self-interest for any company developing eVTOL aircraft to place sustainability and LCAs front and center in their strategy. First, eVTOL businesses prioritizing LCAs will set the standard for the rest of the industry since an agreed standard today does not exist. Second, LCAs provide concrete feedback on areas of focus and opportunities for climate footprint reduction across the businesses' products and services. Even if a company does claim to prioritize sustainability in its corporate goals, without an LCA, it risks devising roadmaps that indeed do not optimize for the life-cycle climate footprint. This is particularly critical for eVTOLs, touted as the substitutes of ground transportation and aviation: the two industries that have come under intense scrutiny in the past decades as governments pressure businesses to prioritize emissions reduction. Finally, the publication of rigorously construed LCAs provides businesses with tangible proof to back their sustainability claims and strengthen the company's brand image.

The LCA study that follows aims to provide an example of methodology, assumptions, and results for an eVTOL manufacturing and service business. The entire life of the aircraft is considered: raw materials, eVTOL aircraft manufacturing, in service operations, and (in part) end-of-life. All materials, processes, and activities that make a relevant contribution to the life-cycle climate footprint are included. The goal is to estimate and assess eVTOL GHG emissions (also referred to as carbon intensity) per passenger per kilometer traveled, expressed in $CO_2$e/pax.km and defined as follows:

- "CO2e" refers to carbon dioxide ($CO_2$) equivalent. While $CO_2$ emissions cause global warming, other GHG emissions (e.g., methane emissions) also contribute to global warming. The term "$CO_2$ equivalent" is a common unit for measuring the climate footprint of various greenhouse gases, and not just the carbon footprint. It is the equivalent mass of $CO_2$ that would warm the earth as much as a greenhouse gas calculated from its global warming potential. It is noteworthy that, in addition to GHGs, physical phenomena (e.g., aircraft contrails) *de facto* can also contribute to global warming [5,6]. However, aircraft contrails or any non-$CO_2$ emissions do not apply to eVTOLs, which are *no-emissions* aircraft with no combustion emissions.
- "Per passenger" indicates that the emissions generated by one aircraft are distributed across the passengers traveling in said aircraft. The purpose is to enable comparison among various aircraft classes and other forms of transport: a critical requirement for eVTOL businesses looking to establish themselves as a more sustainable option. For example, an internal combustion engine (ICE) public bus has a much greater footprint per vehicle than an ICE car, but since it carries, on average, a larger number of passengers, the ICE bus has much lower $CO_2$e per passenger per kilometer.
- "Per kilometer" indicates that the whole life-cycle emissions of the aircraft have amortized across the total number of kilometers the aircraft will travel in its lifetime. Several factors weigh into this calculation, discussed later.

The core benefit of using $CO_2$e/pax.km as the reference unit (also known as the functional unit in LCA) is to provide a simple way for the end user to estimate their emissions. By simply multiplying the trip distance by the reference unit, they can estimate their footprint for that trip, regardless of how many others traveled with them.

Beyond the life cycle GHG emissions for climate change, a complete LCA shall evaluate multiple additional impact categories, such as acidification, eutrophication, resource depletion, and water use. These environmental impact categories are critical for estimating the consequences of human activity on human health, biodiversity depletion, and more. However, for simplicity, the scope of this study focuses on quantifying and assessing the life-cycle carbon intensity. This case study claims in no way to be an exclusive methodology for LCA; other assumptions and methods present their benefits.

### 13.2.1 Methodology and Assumptions

The eVTOL company of this case study runs an electrical aerial rideshare service. Passengers book a trip and travel on a piloted flight between two (likely urban) destinations. The business is first an aircraft OEM (or original equipment manufacturer), purchasing raw materials and certain components from tier suppliers but manufacturing other parts in-house, and integrating them all into the complete aircraft. The same business then operates the airline transporting passengers to their destinations. Not all eVTOL companies are pursuing vertical integration of both manufacturing and operations; some manufacturers intend to sell aircraft to leasing companies or airlines. In these cases, it is necessary to reach out to both the OEM and the operator in order to properly conduct a full LCA.

The LCA methodology was used to quantify and assess the carbon intensity throughout all life-cycle stages, from raw material extraction and production of eVTOL vehicles. The carbon intensity measures the relative potential impact of GHG emissions with respect to a given timeframe. GHG emissions are represented in the mass of carbon dioxide equivalent (CO2e) using a 100-year GHG emission factor [7]. The key function of an eVTOL vehicle is to provide a relatively faster and more flexible means of transportation in large urban centers. Therefore, the functional unit for the current LCA is the transportation of passengers (pax) for a given distance (kilometer) (or pax.km$^{-1}$).

The LCA results on GHG emissions can be used for a corporate for its GHG accounting and reporting using the concept of "scope":

- Scope 1 emissions are related to direct emissions from the company facilities and vehicles. They have a minimal footprint relative to the aircraft operation.
- Scope 2 emissions are associated with electricity and power purchased for own use, which is non-negligible due to the high energy consumption in the manufacturing plants.
- Scope 3 emissions represent the majority of the footprint. They are attributed to raw materials and parts purchased, and for an eVTOL business, electricity used to recharge aircraft in the field dominates over other contributors.

The eVTOL aircraft production processes over the life cycle require consideration of inputs and outputs of materials and energy through upstream and downstream stages, encompassing raw material acquisition, manufacturing, transport, and vehicle utilization. The system boundary for the LCA study is depicted in Figure 13.1. System boundaries are established in LCA in order to

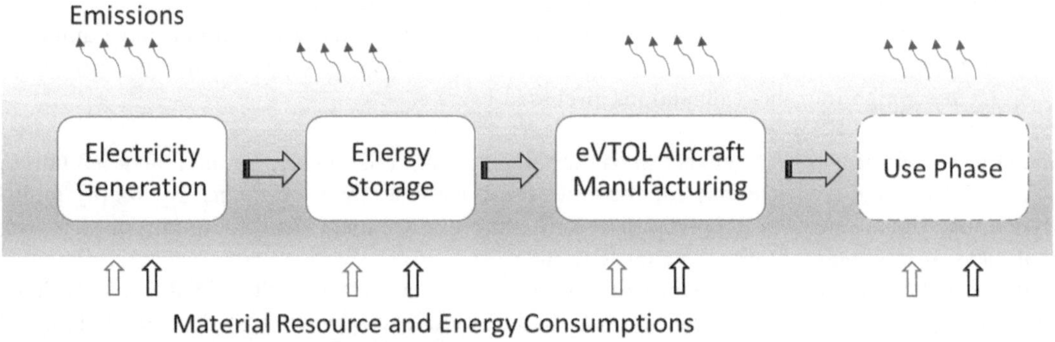


**FIGURE 13.1** System boundary and scope of the life-cycle assessment.

include the significant life-cycle stages and unit processes, as well as the associated environmental flows in the analysis. This lays the groundwork for a meaningful assessment where all important life-cycle stages, and the flows associated with each alternative, are considered.

For this analysis, we account for the stages in the life cycle of the manufacturing of the aircraft, including airframe materials used, manufacturing processes, energy storage (i.e., battery production), and vehicle operation (i.e., inflight energy usage and charging energy losses). As mentioned previously, this case study evaluates all the relevant contributors to climate footprint, focusing only on Scope 3 emissions and some Scope 2 emissions. The study also does not tackle end-of-life (EoL) treatment quantitively but does offer strategies for its climate footprint reduction.

The aircraft manufacturing stage is usually the central focus, which is known as a foreground process in LCA terminology. The foreground process is the place where decision makers can change the design and operating attributes to affect the manufacturing process and, indirectly, other parts of the life cycle (i.e., minimizing resource consumption and environmental releases) [8,9]. There are also background processes, like processes for the production of generic materials, energy, transport, and waste management. The reference flows and life-cycle inventories linked to the foreground process are based on material and energy inputs and outputs that capture the impacts of input raw materials and outputs, such as emissions and wastes.

### 13.2.2 Life-cycle Inventory

This section discusses the actual input variables that will be used to generate results. Compilation of the life-cycle inventory (LCI) of a product or service often is the most time-consuming part of the LCA, requiring broadly two sets of inputs: foreground data and background data.

#### 13.2.2.1 Inputs: Foreground Data

Foreground data is typically obtained from within the company or external tier suppliers. The specific data related to aircraft manufacturing, performance, and in-field operational utilization are internal data of the business. It is required to work with various company departments, both on the design and manufacturing sides, to collect and compile the necessary data. This section describes what information is needed for eVTOL LCI and the recommended methodology for information gathering across the business.

The step for gathering foreground is to characterize a nominal eVTOL aircraft mission: this is a critical step that will enable the engineers and LCA analysts to communicate to understand exactly what the request is. The table below (Table 13.1) lists the most critical values that are worth defining.

There are some caveats and comments:

- The example values provided in Table 13.1 are for demonstration purposes only and in no way represent any known aircraft operation.

**TABLE 13.1**
**Key Parameters Associated with a Nominal eVTOL Aircraft Mission**

| Parameter | Units | Example |
|---|---|---|
| Aircraft iteration | – | V5_3.1 |
| Aircraft year | – | 2030 |
| Trip year | – | 2032 |
| Trip distance | km | 30 |
| Block speed | km/hour | 200 |
| Payload (passenger count) | – | 3.5 |

- The aircraft iteration explicitly is critical in aviation as often various serial designs are made. Especially during the R&D phases, each iteration can present different features. For example, battery life alone can improve a lot across versions and has a strong impact on the total climate footprint. An established business may iterate less frequently but will likely contain different aircraft in its fleet.
- For the same reasons as above, quoting the year of aircraft production adds a definition to which production processes and parts were used. Sometimes the same aircraft iteration may undergo design or manufacturing improvements over the years, for example switching from hand layup to automated fiber placement (AFP) machines. If relevant, distinguishing aircraft production year and flight year provides an added level of clarity. For example, a business might move onto charging its aircraft through more sustainable energy by a certain year of operation, in turn reducing the contribution to the total climate footprint.
- For eVTOLs, trip distance has a significant impact on aircraft footprint: short trips still require a (vertical) take-off and landing, which consume substantial amounts of energy but are amortized over a smaller number of total kilometers. Similarly, varying aircraft speed can vary efficiency significantly (in terms of kWh per km). Block speed—the average speed given over the distance from the departure gate to the destination parking spot[10]—is also needed to convert any input provided as flight hours, typical in aviation, into a per-km figure.
- The payload or passenger count typically affects the answers engineering design and R&D teams will provide since the aircraft is designed for maximum occupancy. Passenger count may have a non-negligible effect on the aircraft's efficiency since flying a greater payload decreases efficiency. By far, the passenger count has the most significant direct impact on the life-cycle climate intensity as its considerable contribution to the final total climate footprint is divided by the payload, expressed in $CO_2e$ per passenger per kilometer (or $CO_2e$/pax.km). Though apparently nonphysical, passenger count can be a non-unit number, such as 3.5, if this better approximates a median value.

The choice of the actual values for the parameters in Table 13.1 to be used for LCA depends on several factors. Aircraft iteration should choose the most used variant first, conduct the LCA with such aircraft, and then move on to examine other variants. It is better to select production year and trip year at a high production and air traffic because this will be when more absolute emissions are produced anyway. Choosing to run the LCA based on launch year may account for operational inefficiencies (e.g., low passenger count since ride matching is not yet optimized) that increase the footprint estimate yet hold only for a year or two.

Trip distance and passenger count should be estimated by looking at median estimates of trips, whether in historical data or future projections. It is worth involving any operations or data analytics teams within the company that can provide the best accuracy on this since the choice has large repercussions on the results. The calculation for eVTOL aircraft used for cargo is identical, but the passenger count is replaced with kilograms transported.

When being asked to provide details about their designs and systems, it is likely engineers will ask for more information anyway, so preparing a defined list ahead of time will save time. Alternatively, and more concerningly, not providing enough definition may lead engineers to make assumptions different from those the LCA analyst is considering. For example, they might describe a manufacturing process as it happens today rather than when once scale production is achieved: the result is likely a more manual and labor-intensive process with higher scrap rates that will be achieved over the following years. The footprint estimate will then exceed the true estimate.

Once the LCA analyst communicates with the design and manufacturing engineers, the latter will provide the former with the type of information illustrated in Table 13.2. The requests are repeated for every single part of the system.

**TABLE 13.2**
**Foreground Data Needed for LCA for Each Part or Assemble Associated with eVTOL Aircraft Manufacturing**

| Foreground Data | Units |
|---|---|
| Bill of materials | kg |
| Manufacturing bill of materials | kg |
| Manufacturing process | – |
| Manufacturing electricity consumption | kWh |
| Life of part | Flight hours |

- **Bill of materials (BOM):** a list of materials and components for making an assembly or part. This is the starting point for any part-specific LCA analysis, as it is the first introduction to the materials and manufacturing processes involved. The BOM will contain weights for each part, so it suggests the likely carbon intensity (but not the total footprint) of the part and the complexity of the part (multiple materials made in multiple ways). The BOM is often not sufficient as it only provides information about the finished part and not what was used to make it. This is given in the manufacturing bill of materials.
- **Manufacturing bill of materials (MBOM):** all the parts, raw materials, and expendable materials used to manufacture each part in the BOM. Values in the MBOM can often be significantly different from those in the BOM. One example is machining, where as much as 95% of the starting billet material is wasted. Using the data from the BOM alone would grossly underestimate the footprint of machined parts. Large enough companies with established internal processes tend to keep accurate track of BOM and MBOM, so this information is typically readily available to the LCA analyst upon request.
- **Manufacturing process:** going in hand with the MBOM is a detailed explanation of the manufacturing process undergone by each part in the LCA. The LCA analyst is to gather a step-by-step account of the manufacturing procedures, how much energy, materials, and expendable materials (e.g., inert gases or fuels). Each step of the process will also have nonzero scrap rates, such that for, say, 100 final parts made, 110 undergo the first step, 105 the second, etc. Detail account of the location at which the operation is performed (internally, within the country, or overseas) will affect the carbon intensity of all elements involved. Any transportation step should be considered as well, especially those happening over long distances and for heavy parts (i.e., steel billets, but not printed circuit boards or displays). The manufacturing process should go as upstream as needed until the materials involved have a known carbon intensity. For example, the carbon intensity of rolled steel bars is well known and quantified, so the LCA analyst can just plug in such value without working back to iron ingots or further upstream.
- **Manufacturing electricity consumption:** if not already accounted for in the previous step, total electricity consumption will account for a variable size footprint depending on how said electricity is produced. Knowledge of the location of electricity use is critical in estimating its carbon intensity.
- **Life of part:** the life of a part affects the number of kilometers that the aircraft will travel before that part is replaced, directly impacting an eVTOL's life-cycle carbon intensity.

Table 13.2 above is required for each part in the eVTOL aircraft and recursively for each part or subassembly making up the main part. For example, in its simplest approximation, a resistance heater will be made of a metallic core that heats up, fins to release the heat, and several other smaller components, like a controller, mainly made of a printed circuit board. Typically, the BOM and MBOM of the whole heater will loosely specify a controller but not break down the parts within it. The LCA analyst should go deeper and repeat the process for the controller. It is up to the LCA analyst to determine how many layers deep into a part it is sensible to reach down. Unless subassemblies represent a significant percentage by weight of the total assembly or appear in multiple instances, it is generally not worth the effort to estimate footprints more than one layer deep, as they do not change the total footprint significantly.

Conversely, it is possible to aggregate some parts under the same request and the same Table 13.2. For example, all carbon fiber manufacturing will likely be centralized in one department, producing parts for design engineers across various teams (fuselage, wing, tail, etc.). Provided the BOM and MBOM account for all the parts accurately, it will be much simpler for the LCA analyst to obtain one set of information from such a team rather than going by the department to obtain information for each component. Moreover, manufacturing processes for different size composite parts will likely differ in that larger parts may be mechanically spun in an automatic fiber placement (AFP) machine and cured in an autoclave, whereas smaller parts are hand laid up and cured in a hot press.

For those products or services that are not produced internally (e.g., outsourcing of a pump), it is required to work with the tier supplier and obtain their carbon intensity (Scope 3 emissions), results of any LCA they have conducted on their products. Alternately, the tier supplier can also provide the associated inventory, i.e., a breakdown of raw materials and manufacturing processes.

#### 13.2.2.2 Inputs: Background Data

Background data are secondary data related to underlying processes, such as the production of raw materials and the generation of electricity. For example, while the foreground data are the amount of carbon fiber needed to manufacture an eVTOL aircraft frame, the associated background data are resources it takes to make the carbon fiber.

The carbon intensity of the background data looks at the footprint of all steps: extraction of ore from the mine (for metals), the processes used to extract the metal (smelting, refining), and all downstream processes (hardening, alloying), and surface treatments, as well as all the transportation steps. The calculation of the carbon intensity values for these underlying processes from scratch is not recommended as they are generally widely available. Table 13.3 shows the carbon intensity values for the materials most found in the aircraft and were obtained mainly from the Ecoinvent database [10], a common source of LCI data and one of the most consistent and transparent databases that support environmental assessments of products and processes. Other sources, such as U.S. LCI [11], open literature, or grounds-up estimation, can fill the data gap.

### 13.2.3 Calculation and Processes

Having defined a methodology and the categories of inputs needed, we turn to the process itself, which requires input from several parties.

Aircraft and ground vehicles have thousands of parts developed through complex manufacturing processes. Studying all of them to capture them accurately in an LCA can turn into an endeavor of prohibitive magnitude. It is paramount that the LCA analyst prioritizes the systems and processes with the largest footprint impact and moves into the more minor details only in a second moment. However, prior to completing the LCA, it is impossible to know which factors are more relevant. The list below is a good starting point.

- **Inflight energy:** the energy the batteries deplete during the flight. These are best estimated by looking at a full aircraft model simulation (using tools such as Simulink/Simscape) if the aircraft designers have developed these. If not available, the alternative is to estimate the efficiency losses at each step in the process: starting with the lift-to-drag

**TABLE 13.3**
**Greenhous Gas Emission (Carbon Intensity) Factors**

| Material | Units | Footprint |
|---|---|---|
| Carbon Fiber | kgCO2e/kg | 30.1 |
| Aluminum | kgCO2e/kg | 8.76 |
| Titanium | kgCO2e/kg | 31.5 |
| Copper | kgCO2e/kg | 2.13 |
| Stainless Steel | kgCO2e/kg | 1.9 |
| Polycarbonate | kgCO2e/kg | 8.23 |
| Epoxy/Polymers | kgCO2e/kg | 6.9 |
| Rubber (synthetic elastomers) | kgCO2e/kg | 2.94 |
| Tedlar | kgCO2e/kg | 4.6 |
| Foaming Epoxy | kgCO2e/kg | 6.9 |
| Syntactic Silicone | kgCO2e/kg | 5.18 |
| Foaming silicone | kgCO2e/kg | 4.56 |
| $TiO_2$ | kgCO2e/kg | 4.67 |
| 2-heptanone | kgCO2e/kg | 1.36 |
| Propanol | kgCO2e/kg | 4.56 |
| Toluene | kgCO2e/kg | 1.56 |
| Polyisocyanate | kgCO2e/kg | 6.59 |
| 3-butoxypropan-2-ol | kgCO2e/kg | 5.4 |
| 2-methoxy-1-methylethyl acetate | kgCO2e/kg | 5.4 |
| Acetone | kgCO2e/kg | 3.89 |

ratio, calculating the power needed at the propellers, and then working back the power exiting the batteries while adding up all the efficiencies along the way. The battery power multiplied with total time provides a good first-order estimate, provided take-off and landing times are minimal compared to the rest of the flight. If separate batteries are used for powering propulsion versus running other functions, the depletion of all should be considered, with attention given to double counting (e.g., in the case one battery is used to charge another). Special care should be taken to estimate the carbon intensity of the grid electricity. This will depend on the energy sources used to power the grid, which vary greatly by location; more later.

- **Charging energy losses:** the energy losses from the electrical grid (or any other power source used upon charging) into the battery often contribute substantially. Losses in the grid itself going into the charger, losses into the charger, and from the charger into the batteries can add up to 5–10% of the total energy entering the cell. Battery cells overheat during charging, the heat representing further losses. All of the above data are typically available from the charging equipment and cell manufacturers. Finally, it is not uncommon for charging processes to include an external cooling loop, which can greatly improve charging times. Energy consumption from such cooling – typically coming from a compressor and a pump – should be included.
- **Cell manufacturing:** represents the largest source of footprint alongside their charging. To a first approximation, it is sufficient to work with the cell manufacturer to obtain all the relevant information and estimate footprint contribution from cells alone, i.e., neglecting contributions from other components of the battery module (spacers, potting, structures). Accuracy and precision in this step of the LCA are paramount as cell manufacturing is

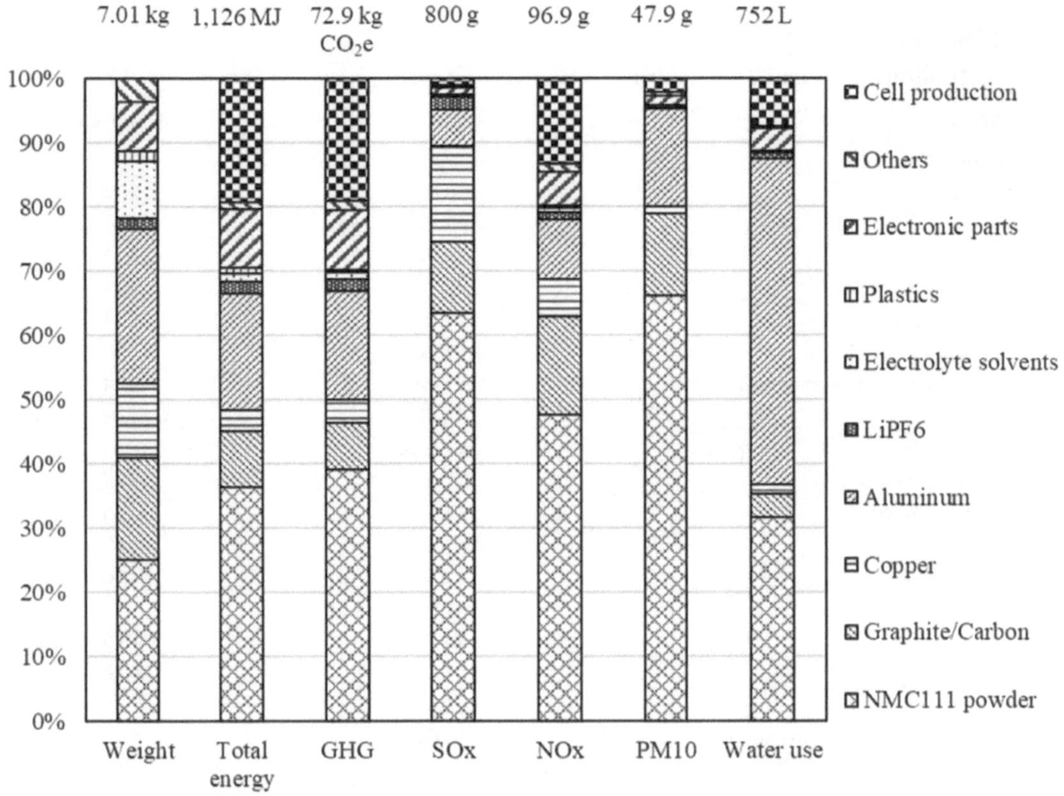


**FIGURE 13.2** Cradle-to-gate impact breakdowns and bill of materials (BOM) of 1 kWh NMC111 battery. Reprinted from Dai et al., [12] Copyright 2019, with permission from Elsevier.

often a point of criticism for the sustainability of electric vehicles (EVs) as it comprises a large portion of their overall emissions. An LCA that can back its claims with clear data and references is much more likely to endure public scrutiny and criticism. Figure 13.2 shows the breakdown of NMC111 cell LCA contributions, which yields 72.9 kg $CO_2e$ per/kWh (cell LCA literature is typically quoted for kWh than per kg for more effective comparison) [12]. Both raw materials and manufacturing processes are significant and should be accounted for. Lastly, perhaps most importantly, an accurate cell lifetime must be accounted for. Cell life is often much lower than any other component in the aircraft, such that battery-swapping maintenance cycles are planned at most eVTOL companies. Cell life is usually measured in full charge/discharge cycles, so the LCA analyst must first estimate how many flights can be achieved per cycle to work out the number of flight hours or kilometers in a battery lifetime. With eVTOL companies claiming anywhere from 1,000 to 15,000 lifetime cycles, accurate knowledge of one's own cells' life expectancy can affect the LCA result by orders of magnitude (more in sensitivity analysis later).

- **Machined parts:** machining of titanium, aluminum, and steel billets is frequent in aviation due to its tight geometrical tolerances, low wall thicknesses, and flexibility in design, unlike the cheaper and more scalable casting. Typical aviation production volumes at scale are often well suited for machining, such that a dozen or more automated CNC machines can suffice for full-scale production. When estimating the footprint of machined parts, one must look at the initial billet weight, not the weight of the final part.

The two can differ greatly: machined parts can weigh as little as 3–5% of the initial billets. If the BOM – which contains the finished parts – is used, the footprint calculated can be as low as 20 to 30% less than the actual footprint, as included in the MBOM. While the chips generated during machining are recycled and melted to form new billets, they are often not reused in aviation applications. Certified flight articles require aerospace-grade billets which do not come from recycled machined scraps.

Further, recycling is generally accounted for as a reduction in the footprint of the next part made rather than that from which the recycled material first came from. Depending on the mass of each, it is typical that titanium machined parts contribute more than those in steel or aluminum since the footprint of titanium can be 15 times that of steel and nearly four times that of aluminum (Table 13.3). It's expected that most eVTOL OEMs will have a centralized department or shop executing machining operations for all the components across systems. Therefore, it is easiest for the LCA analyst to work directly with such group to get a total MBOM for machined parts rather than obtain each MBOM from each system owner (e.g., communicating directly to the company's internal motor team, landing gear team, etc.).

- **Airframe carbon fiber:** Often a significant system by mass alongside the battery and machined parts, modern airframes are made of carbon fiber thanks to their high stiffness-to-density ratio. Carbon fiber has a high carbon intensity, on par with titanium (Table 13.3). The manufacturing climate footprint can also be substantial depending on how the autoclaves used for curing are powered (gas or electric). However, airframes often tend to have the largest life of any component of the aircraft, so their contribution is typically lower than the systems already discussed.
- **Additive manufacturing:** 3D printing offers even greater flexibility than machined parts, with printed parts making their way into more and more hardware engineering applications. In the near-term, it is expected that 3D-printed parts will make up a very small percentage of the overall weight of eVTOL aircraft, but as mass reduction is a key driver of aircraft performance and efficiency and therefore a primary objective for product improvement, the authors predict additive manufacturing will play a larger role in future aircraft designs. While 3D printing typically incurs far less material waste than machining processes, it nonetheless can amount to a sizeable footprint contribution and thus should be examined. As per machining, we look at the raw input material weight, since wasted material accounts for about 20% of the starting weight.

The following contributors are typically smaller than the ones listed above but may be non-negligible. The case study showed that the contributors above account for more than 90% of the total footprint, while the list below is the remainder. It is suggested to examine them only as a second pass.

- **Maintenance:** depending on the maintenance cycle frequency and the scope of work in each cycle, maintenance can account for a sizeable footprint contribution. Helicopters and aircraft have several moving parts under harsh temperature and vibration conditions, resulting in maintenance intervals as low as a hundred or tens of hours. eVTOLs, instead, with their lower temperature powertrain and less harsh vibration environment, can be expected to handle more extended periods between maintenance. Estimates for maintenance schedules are likely to become clearer as eVTOLs begin to enter operations and more empirical data on performance becomes available, but it is expected they will improve on maintenance schedules over the industry status quo due to reduced numbers of moving parts and the higher reliability of electric motors over combustion engines. Typical materials used for maintenance are oil and coolant refills (often outweigh the initial fluid fills), lubricants and grease, and paint touch-ups.

- **Radiators and heat exchangers:** aluminum-brazed radiators can add sizeable weights and thus the overall life cycle climate footprint. Radiators are typically outsourced and made by brazing together aluminum extrusions. They, therefore, do not appear in the machining and additive manufacturing MBOMs. Not too much waste is generated from machining, so the dry weight of the radiator is a good proxy for the total aluminum used.
- **Printed circuit boards:** eVTOL aircraft contain several dozens or hundreds of printed circuit boards (PCBs) used to run the inverters, batteries, flight computers, etc. Quantifying the carbon intensity of PCBs is nearly impossible since the circuitry and devices on a populated board vary from application to application. For the case of eVTOL, just estimating the total surface area of all boards, along with the number of layers on each, is a good starting point. For additional accuracy, the larger devices can be accounted for, namely, the large silicon carbides, processors, and capacitors.
- **Landing gear:** landing gear comprises various components and materials, though the focus should be put on the wheels and tires. As they are made of rubber and replaced more frequently than other parts, they can accrue to a sizeable footprint. Most eVTOL aircraft carry standard landing gear in case they are forced into a conventional landing (CTOL) for emergency reasons.
- **High voltage wiring:** large copper wires are used to propagate the electrical current from the batteries to the propeller motors, the total length of which can add up to several meters or tens of meters. The largest contributor is the copper core rather than the insulation layers.
- **Wiring:** low voltage cables and ethernet cables can run even longer but typically constitute only a small fraction of the footprint. Their main footprint contribution derives mostly from the difficulty in their recycling.
- **Interiors:** all the cosmetic components covering the interior cabin may well contribute to sizeable footprints. With weight savings being a central focus of designers, it is unlikely that interiors add up to a significant weight, but as aircraft get to market and begin to incorporate customer experience feedback, this may change. Most likely, all interior parts are only 10–20 kg per passenger, meaning their footprint is minimal.
- **Battery pack manufacturing:** as mentioned previously, the cells represent the greater contributor to pack footprint. The first column of Figure 13.3 shows the life-cycle GHG emissions (expressed in global warming potential or GWP) of pack and module packaging, which exhibit minimal contribution in a Li-S battery [11]. However, since pack design varies from one OEM to another and battery types, module, and pack materials and manufacturing can be significantly different.
- **Paint and adhesives:** Most of the eVTOL airframe, wings, and propellers are coated with several layers of paints and primer and recoated regularly. However, since paint is made of various complex compounds and materials (such as polyisocyanate or fluorinated polyurethane), it is often very difficult to track back their carbon intensity. The low volumes of paint used means that unless paint cycles are applied more frequently than the norm, it is generally not worth dedicating the time to working out the LCA for paint and adhesives.
- **Refrigerant:** Many eVTOL aircrafts plan to have an active refrigeration cycle utilizing a compressor to generate sub-ambient temperature airflow. Typical refrigeration systems contain only 1 or 2 kilograms of refrigerant mass per passenger, so their relevance in the footprint calculation depends on the carbon intensity. In the past, refrigerants were high GWP hydrofluorocarbons, with global warming potentials on a 100-year time scale (GWP100) in excess of 5000. R134a is a common refrigerant still in use and has a GWP100 of 1300 [13]. Further, the refrigerant may leak out and require constant top-ups. Today, however, high GWP refrigerants are nearly out of use, with common options like R1234YF having a GWP100 of just 4. More and more systems are running on $CO_2$ as a

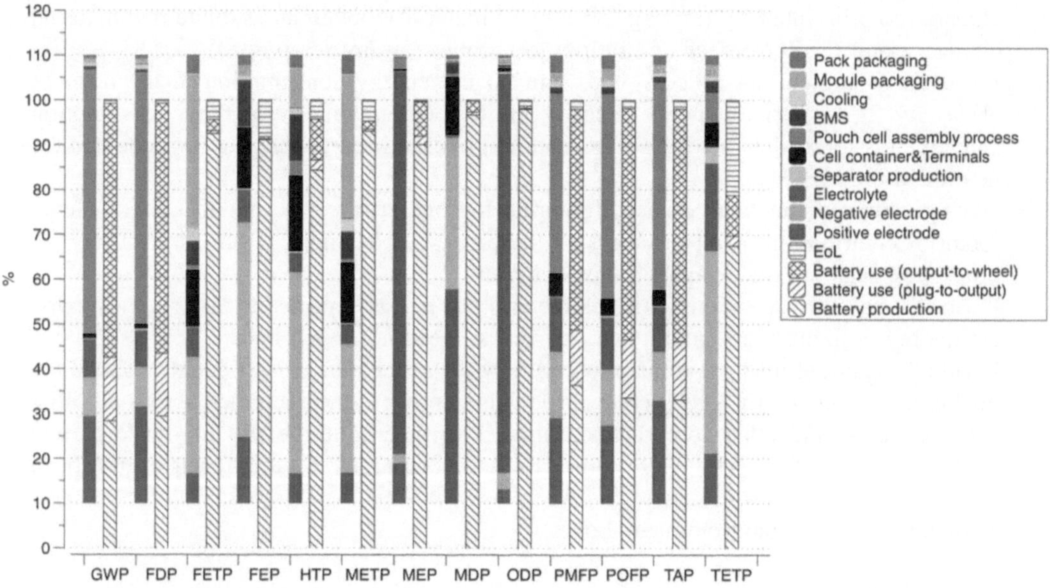


**FIGURE 13.3** Contributions of life-cycle environmental impacts of the Li-S battery. Reprinted from Deng et al., [11] Copyright 2017, with permission from Elsevier.

refrigerant, with a GWP of 1 by definition. The refrigeration may be used for sole cabin cooling or also for battery cooling. Depending on which, the volume of refrigerant used can vary greatly.

Having lined up all the raw data discussed so far, the LCA analyst can proceed to estimate the relevant footprint contributors. The calculation was described in the earlier sections and generally consisted of summing a flight-hour adjusted list of all the footprint contributions, see Figure 13.4. Steps are as follows:

1. Utilizing the MBOM obtained from the engineering part owner (see "Inputs: Foreground data" section), obtain an account of all the raw materials. By multiplying each with their

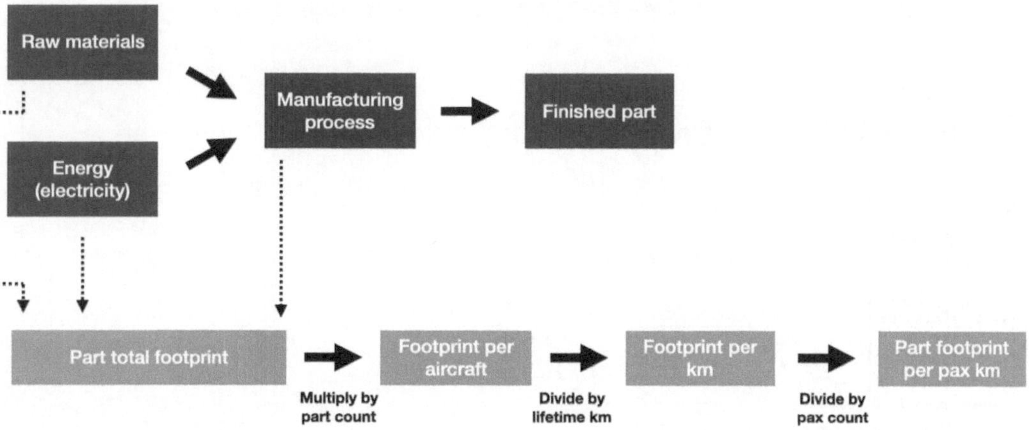


**FIGURE 13.4** Core schematic representation of climate footprint calculation.

climate footprint intensity (i.e., kg $CO_2$e/kg of material), obtain an absolute raw material climate footprint. Remember to multiply to account for any scrap rates.

2. For each manufacturing process step, sum up the energy consumption in the form of electricity. The electricity energy intensity will depend on the location of its consumption and how it is sourced. Other fuels (e.g., natural gas used in furnaces) can be accounted for here or in the MBOM.
3. Having obtained the total climate footprint for one part, work out the aircraft-level footprint contribution of all instances of the parts. For example, if one aircraft has ten propellers, scale up the per-part footprint by ten.
4. Divide the result by the lifetime (in kilometers) this part is expected to last. As most parts are quoted in flight hours, multiplying by block speed is needed first.
5. Divide the result of the total climate footprint by the average passenger count to obtain the final carbon intensity (in kg$CO_2$e/pax.km). This number is then summed across all other parts, systems, and other footprint contributors to determine the full life-cycle carbon intensity of an eVTOL. The figure can be used to compare with other forms of transport.

Two comments on the method outlined above:

- The calculation above is for physical parts. For the electricity consumption during the flight, the flow chart is still valid. However, raw materials and the manufacturing process do not apply to electricity generation. Therefore, the electricity's carbon intensity can be determined by dividing the total electricity usage by the trip length rather than by the lifetime distance traveled.
- Since the passenger count is the same for all parts, it is possible to sum up all contributions from parts and systems before the last step, when they are still expressed as contributions per aircraft per km, and then divided by passenger count. The intermediate result of climate footprint per aircraft per km is also often used for comparison purposes.

The calculation process described above is convenient as it circumvents the need to estimate how many replacement cycles each part experiences throughout the aircraft's lifetime. An alternative method of calculation is shown graphically in Figure 13.5 below. Starting from the airframe

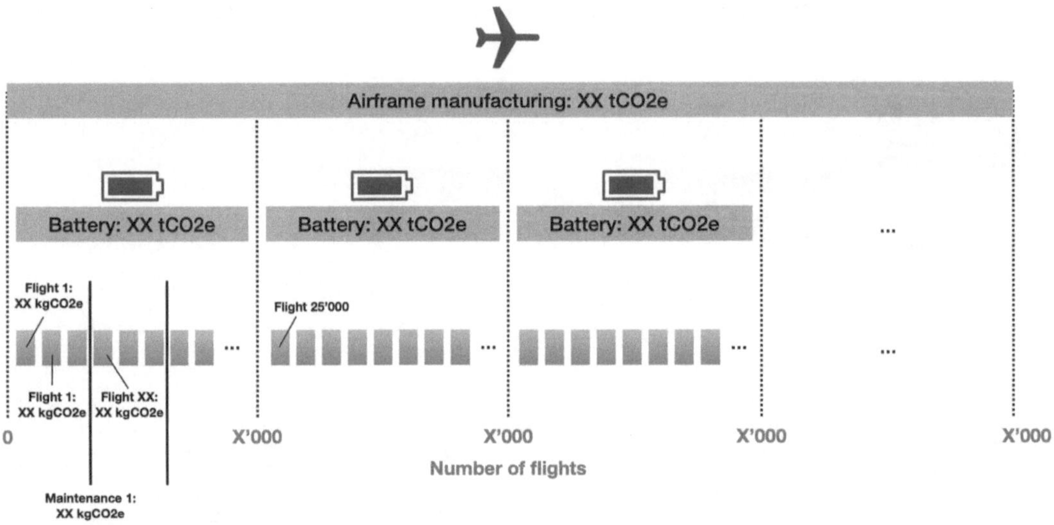


**FIGURE 13.5** Battery climate footprint contribution over flight count.

footprint – generally the longest life part of the aircraft – amortized over aircraft life, it breaks down into smaller and smaller parts by looking at the number of battery replacements, number of maintenance services, number of fights, etc. Each footprint is then multiplied by the number of times it is replaced in the aircraft's lifetime and summed up across parts. The total number is the lifetime emissions of the whole aircraft. This number is in itself of huge value as it quantifies the true impact on the aircraft in the most absolute sense – i.e., what ultimately affects climate change. Dividing the total climate footprint by lifetime distance traveled and passenger count will yield the same number as per the other calculation mentioned above, in $kgCO_2$/pax.km.

### 13.2.4 Results and Reporting

A robust LCA report shall include as much detail as possible, including the breakdown of the footprint contributions, sensitivity studies, and uncertainty characterization.

Figure 13.6 shows some suggestions for the presentation of the LCA results. The values shown are for demonstration purposes only, and do not represent exact results from the LCA this case study is based on:

a. A pie chart describing the relative contributions of each system and operation. This format allows the reader to understand immediately how the scale of the contributions compares to each other. It is suggested only the top 4–6 contributions are shown; all other contributors can be collapsed into an "other field" provided none of its contributors add up to a value in excess of 5%.
b. Equivalent to the above, a donut chart allows for the final climate footprint value to be shown in the middle, ultimately providing the user with all the information needed in one image.
c. Equivalent to a), the format in c) adds the breakdown of the "others." The value of this chart format over the previous versions is to convey thoroughness at the cost of simplicity: the smaller pie chart's purpose is to show the extent to which all potential contributions were considered, to confirm that they did not sum up to a significant amount.

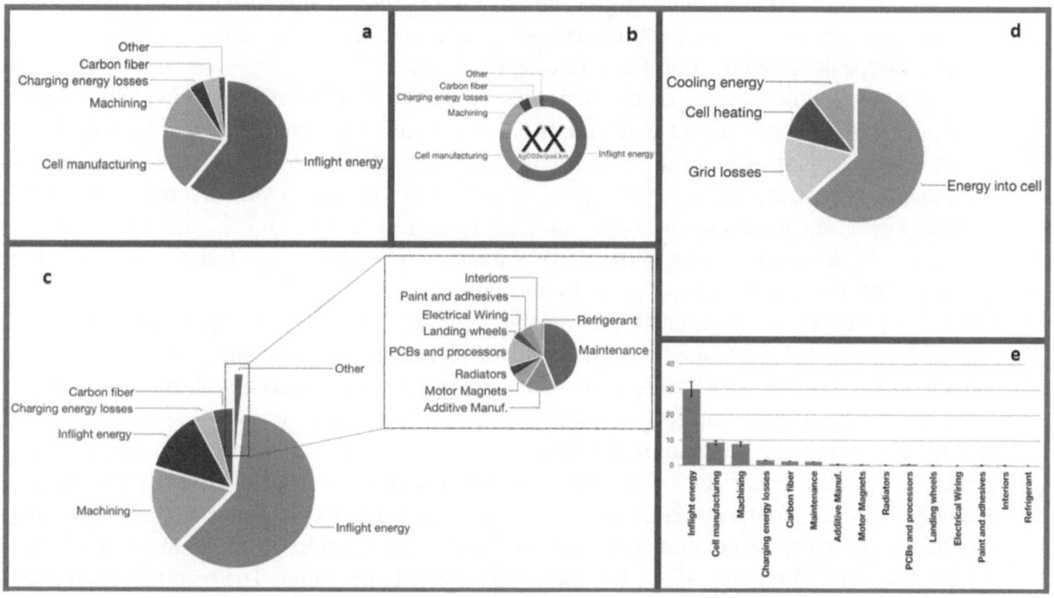


**FIGURE 13.6** Examples of the graphic display of LCA results.

d. Also, with the goal of providing additional detail, it is possible to show the further breakdown of any of the footprint contributors. In format d, the “inflight energy” and “charging losses” are further expanded, showing exactly how the losses add up. It is recommended to make use of detailed charts like this one for any sizeable wedge shown in formats a and b, especially in those where the LCA analyst may expect follow-up questions, for example, if the footprint was larger than expected.
e. Bar charts are better at representing total contributions rather than the relative ones of pie charts. An added benefit of bar charts is the ability to append error bars. Error analysis is an integral part of the LCA and will be discussed in the “Uncertainty characterization” section.

### 13.2.5 End-of-Life (EoL) Treatment

This case study does not address the end-of-life (EoL) life-cycle stage in much detail. A few qualitative notes are made on contributions that could have a meaningful impact on the climate footprint.

- **Batteries:** As the EV and eVTOL industries are booming, a greater number of batteries reach the end of life, at which point the battery is unable to hold the same level of charge during the early part of life. The useful life of an eVTOL battery will be defined by two primary factors: trade-offs taken by the company between necessary specific power, specific energy, and cell life; and FAA regulations governing the introduction of electric aircraft. The batteries can experience two EoL treatments:
  - Stationary storage: aircraft batteries can remain on the ground near charging stations, providing energy during periods of peak grid demand. This is referred to as “peak shaving,” whereby the peaks in power demand a grid sees daily are softened by the batteries, which act as buffers. The batteries then recharge during the period of low demand (e.g., at night). While the total energy going into the aircraft still all comes from the grid, peak shaving can have positive climate footprint effects. Most commonly, utility providers respond to peak demands by sequentially turning on generators and power plants with higher emissions relative to the usual energy sources. Therefore, the greater the peak in grid power demand, the higher the carbon intensity: bringing peak shaving to de facto reduce footprint.
  - Battery recycling: several of the materials contained in the cells are recyclable. The copper in the anode current collector and the cathode active materials can be recycled via a process that uses up the electrolyte itself. The aluminum in the cathode current collector is typically not economical to collect, but future improvements might change this. Recycling the cathode active material is critical as the nickel, manganese, and cobalt contained in it have high carbon intensity. The process generates some emissions, but they are far outweighed by the benefits.
- **Machining chips:** as mentioned previously, as much as 95% of the initial billet is removed during the machining process, both for aluminum and titanium. These chips are collected and recycled, turned into new ingots. As the chips are mixed with oil from CNC machines (for lubrication and cooling), the first step is the removal of such oil. Natural gas is then used for remelting the aluminum chips, followed by direct chill casting and rolling into sheets. The process generates a non-negligible footprint; however, when compared to using new aluminum, the footprint reduction is still significant. The steps of alumina refining and smelting, required for new aluminum and the biggest contributors to the climate footprint (as much as 80% [14]), are skipped when using recycled aluminum. The resulting recycled aluminum is not aerospace grade, so it cannot be used for eVTOL application. Even if it were, LCA standards dictate that the climate footprint reduction is not to be claimed by the

application generating the recycled aluminum but rather by using the recycled aluminum as input material. A similar process as above is also true for titanium scraps.

- **Carbon fiber:** recycling carbon fiber is still costly today, so it is typically not economically advantageous. Two techniques are common to remove the cured resin and extract the original fibers: pyrolysis, i.e., burning off the resin using high heat. The second is solvolysis, whereby a solvent dissolves the resin. The recycled fibers obtained are of lower quality than the starting ones, so they cannot be used in the same eVTOL application. As discussed for machining scraps, the climate footprint reduction is to be claimed by the next product, not the eVTOL LCA.

### 13.2.6 Uncertainty Characterization

The application of LCA as a decision support tool can be hampered by the numerous uncertainties embedded in the calculation [15]. It is fair to state that all data in life cycle models have some sort of uncertainties, namely, variation in the data, correctness or representativeness of the model, and incompleteness of the model. Data variation is the focus here. LCA uncertainty can be associated with input data that are uncertain (for instance, the origin of raw materials, such as compositions), to input data that are variable (for instance, the lifetime of aircraft and battery), and to choices that must be made by the analyst (for example, how many Monte Carlo runs in comparative probabilistic LCA [16]). The reliability and hence the applicability of the LCA results are dependent upon the quality of the original data provided for the LCA study, such as reference flows, life-cycle inventory, and characterization factors. Input data used in this study were obtained from the aircraft OEM, and the emission factors were obtained from various sources. Both could potentially influence the reliability and applicability of the results. Therefore, the treatment of uncertainty is necessary to enhance the reliability and credibility of the current LCA results. Uncertainty characterization is the quantification and propagation of input uncertainties to output uncertainties [17]. Although a great deal of research is undergoing in the field of LCA, there is yet no consensus on how to undertake quantifying uncertainty [18].

The uncertainty characterization can be performed by adopting the approach using a data quality pedigree matrix [19] and Monte Carlo analysis in the SimaPro software program [20]. This method proposes a matrix of data quality indicators (DQI) and corresponding coefficients of variation for determining the basic uncertainty of a parameter that can be adjusted to reflect other sources of variation (Table 13.4).

Although data quality indicators are semi-quantitative values associated with the data set, they still represent the quality of the data. The data quality indicators are reliability (i.e., relating to the sources, acquisition methods, and verification procedures used to obtain the data), completeness (i.e., relating to the statistical properties of the data: how representative is the sample), temporal correlation (i.e., representing the time correlation between the year of study and the year of the obtained data), geographical correlation (i.e., illustrating the geographical correlation between the defined area and the obtained data), and further technological correlation (i.e., concerning with all other aspects of correlation than the temporal and geographical considerations).

After assigning pedigree matrix values to the underlying unit process data used to model the LCA, uncertainty factors were calculated. The value of $Cv$ is the coefficient of variation, and $(1 + Cv)$ is equal to the square of the geometric standard deviation for the lognormal distribution. The lognormal distribution typically seems to be a more realistic approximation for the variability in fate and effect factors than the normal distribution [21]. Unlike normal distributions, lognormal distributions stop at 0, which is much more physical as a footprint cannot have a negative contribution. The lognormal distribution is assumed by default to all process steps in this study. If actual field measurements or reliable and verified data suggest that a parameter has indeed a normal distribution. In that case, the coefficient of variation is determined by dividing the sample standard deviation by the sample mean.

**TABLE 13.4**
**Pedigree Matrix with Five Data Quality Indicators**

| Indicator Score | 1 | 2 | 3 | 4 | 5 |
|---|---|---|---|---|---|
| Reliability | Verified data based on measurement | Verified data partly based on assumptions or non-verified data based on measurements | Non-verified data partly based on assumptions | Qualified estimate (e.g., by industrial expert) | Non-qualified estimate |
| Completeness | Representative data from a sufficient sample of sites over an adequate period to even out normal fluctuations | Representative data from a smaller number of sites but for adequate periods | Representative data from an adequate number of sites but from shorter periods | Representative data but from a smaller number of sites and shorter periods or incomplete data from an adequate number of sites and periods | Representativeness unknown or incomplete data from a smaller number of sites and/or from shorter periods |
| Temporal correlation | Less than three years of difference to year of study | Less than six years' difference | Less than 10 years' difference | Less than 15 years' difference | Age of data unknown or more than 15 years of difference |
| Geographical correlation | Data from area under study | Average data from larger area in which the area under study is included | Data from area with similar production conditions | Data from area with slightly similar production conditions | Data from unknown area or area with very different production conditions |
| Further technological correlation | Data from enterprises, processes and materials under study | Data from processes and materials under study but from different enterprises | Data from processes and materials under study but from different technology | Data on related processes or materials but same technology | Data on related processes or materials but different technology |



As the resulting uncertainty values are lognormal values, they must be interpreted through a lognormal process such as Monte Carlo analysis [22]. This was done prior to applying Monte Carlo analysis to the entire product system and was performed in SimaPro software. The carbon intensity uncertainty profiles for the individual areas of the system can be found, and for each a 95% confidence interval can be calculated. Higher confidence levels can be achieved at the expense of a wider range, and vice versa. The absolute uncertainty of a single system is usually much higher than the differential uncertainty due to the correlations. Combining all the footprints and systems together yields a unique number for total footprint, alongside a range of certainty.

### 13.2.7 Sensitivity Analysis

In LCA, a sensitivity analysis consists in varying within reasonable ranges certain inputs and monitoring the impact they have on the final result. The goal is to help understand the main sources

**TABLE 13.5**
**Suggested Sensitivity Parameters for eVTOL LCA**

| | |
|---|---|
| Electricity source | Renewable, state/U.S. average, generator |
| Electricity source | Variation over the years based on projections |
| Passenger count | Low vs full occupancy, deadheads per trip |
| Battery life cycles | 20%–200% of nominal value |
| Flight distance | <10 km vs max range |
| Airframe life | 20%–200% of nominal value |
| Maintenance intervals | 20–1,000 hours |
| Machining scraps | No recycling vs full recycling |

of uncertainty and their relevance. The outcomes can then be used to prioritize future research and development efforts, decreasing the uncertainty of the LCA model and its inputs by refining the most critical elements in an iterative approach [15]. Sensitivity analysis differs from uncertainty analysis, discussed previously. The former's focus is on the impact of choices, such as electricity options, data sources, treatment of data gaps, analysis boundaries, and process assumptions. The latter examines the implications associated with data quality using knowledge about the statistical distribution of input variables. Further, since analyzing the impact of every single choice made in this project is not practical (or not necessarily warranted), sensitivity analyses should focus only on those parameters truly at risk of variation over the years, based on design changes, or for other unknown factors.

The simplest analysis is a single point sensitivity analysis representing a deviation in a single project parameter with all other parameters remaining constant at the base case value. Moreover, each sensitivity scenario has an associated deviation value from the base case. A suggestion on what sensitivity analyses for eVTOL LCA should focus on is listed in Table 13.5.

Deviation from the base case electricity source can have significant impacts on the carbon intensity, since typical renewable electricity has a GHG emission factor of 28 $gCO_2e/kWh$, whereas in low renewable grids such as West Virginia GHG exceeds 900 $gCO_2e/kWh$ [23]. On the other hand, Vermont's electricity mix is the cleanest (100% renewable) and has the lowest climate footprint (less than 30 $gCO_2e/kWh$). The local electricity mix can potentially play an important role in influencing the life-cycle impacts of electric aircraft, which in turn can be influenced by the share of renewable energy.

Varying the battery life, cycle routine, and flight distance can have an impact on battery life, which will, in turn, impact the overall footprint. Sensitivity around changing battery cell selection does not provide significant changes to climate footprint, as most of the newer generation lithium-ion cells have low variance in footprint among them (see Figure 13.7).

## 13.3 LEVERAGING NEW TECHNOLOGIES IN LCA

New technologies such as blockchain can play a role in eVTOL LCA. Blockchain is a distributed ledger technology (DLT), a consecutive list of time-stamped recorded (usually digital transaction data) sequentially linked using cryptography [25]. Blockchain exhibits four key characteristics: decentralization, persistency, anonymity, and auditability [26]. Business operations using blockchain technology which could function as a distributed database without third parties by virtue of its distributed and decentralized characteristics, are essentially not prone to malicious attacks, malfunction and artificial alterations constantly faced by traditional business operations, which rely heavily on a centralized authority or third parties (such as a bank).

Recently, blockchain applications have dramatically increased, emerging and expanding to many different areas, such as blockchain-enabled physical distribution and logistics, business

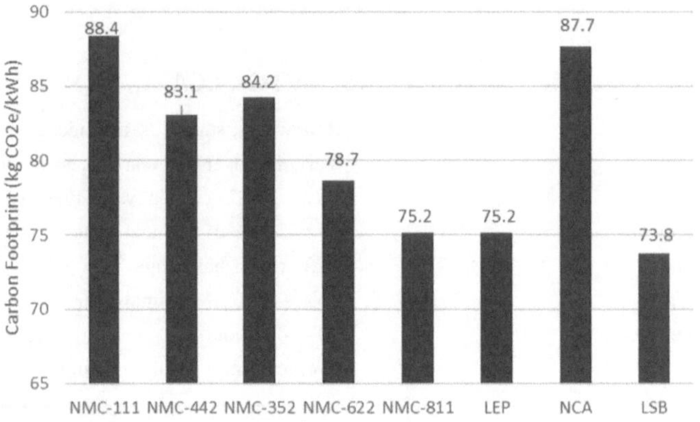


**FIGURE 13.7** Potential battery systems for eVTOL aircraft.

***Note: Figure created based on data obtained from Barke et al. (2021)*** **[24]*****. NMC – lithium-ion batteries (LIB) based on lithium manganese cobalt oxides; LEP–LIB based on lithium iron phosphate; NCA–LIB based on lithium nickel cobalt aluminum oxides; LSB – lithium-sulfur battery. Specific energy: 195–290 Wh/kg. Mass: 22–15 t. Number of cells: 13k–23 k.***

process management, information sharing, business operations, and risk analysis [25]. For example, transportation sector decarbonization and alternative energy that demands a more comprehensive life-cycle assessment (LCA) will need technology like blockchain (by its distributed architecture and immutability of records characteristics) to link multiple stakeholders and track data [27]. Real-world applications of blockchain technology in different sectors were also reported [28].

Blockchain technology can play an important role in LCA, including the life-cycle inventory and uncertainty. LCA is an effective tool for assessing the potential environmental impacts of a product, service, or system. A complete LCA or "cradle-to-grave" LCA encompasses the material and energy inputs and outputs of the entire supply chain. LCA studies can help improve environmental impacts, such as greenhouse gas emissions and resource consumption, and guide decisions and policymakers. The availability of reliable supply chain data holds the key to high-quality LCA results. However, due to globalization and market expansion, the increasingly complex manufacturing of goods and product portfolios create supply chain inventory challenges, both data availability and uncertainty associated with the product origins, processing, and transportation. The challenge becomes both quantitative and qualitative, and the supply chain's main challenge remains in the traceability and data management system [29]. Blockchain represents a new approach in the supply chain area, where the principal challenges are visibility and transparency of product flows [29]. The blockchain can be integrated into the supply chain architecture to create a reliable, transparent, authentic, and secure system.

Karaszewski et al. illustrated a blockchain-based LCA framework integrating blockchain technology to ensure traceability and transparency of the goal and scope definition, use the Internet of Things (IoT) concept to collect and integrate data collected in real time at the inventory analysis level, and create an analytic form at the level of impact assessment [30]. Integrating blockchain and other smart-enabling technologies (e.g., sensors and devices that generate a huge amount of data in real time) into multiple LCA stages can make the processes more efficient and effective [31].

## 13.4 SUMMARY AND FUTURE WORK

All-electric vertical take-off and landing (eVTOL) aircraft for sustainable aerial ridesharing can provide a critical time saving opportunity to travelers within and around urban areas, in turn reducing congestion for ground travel and greenhouse gas and other air emissions. This case study served as an example for conducting an LCA of an eVTOL aircraft, looking at its manufacturing processes, its in-life operation as a passenger transportation service between urban locations, and (in part) end-of-life treatment. The case study aimed to provide guidance on eVTOL LCA, particularly on how to define the problem, how to obtain necessary and relevant information from engineers inside and outside their own organizations, and what areas to focus on.

As the LCA study aims to support an eVTOL company's development of eco-conscious manufacturing practices and to better understand its operational climate footprint, LCA can assist in identifying opportunities to improve the environmental performance of the aircraft at various life-cycle stages. LCA can also inform decision makers for strategic planning, priority setting, and product or manufacturing process design or redesign. Last but not least, LCA can also enable marketing, for example, by making an environmental claim. To this end, this case study also provided examples of how to report and present the LCA results, optimizing for communicability among those not non-LCA experts.

Moreover, the results of any LCA are only as accurate as the assumptions that were made. A more rigorous LCA can help further understand the limitations of those assumptions. The case study also emphasized the importance of sensitivity analysis and uncertainty characterization. New technologies such as blockchain may also play an important role in LCA, including the life-cycle inventory and uncertainty.

Future work shall consider the climate footprint of the infrastructure required to support eVTOL operation. Much like cars traveling on roads, which have a (large) footprint that is to be accounted for in the car's LCA, installing eVTOL charging stations in existing landing sites, or building new landing sites, will carry an associated footprint. A future LCA shall account for this footprint by amortizing it onto the aircraft footprint. It is estimated, however, that the footprint of infrastructure will be low relative to the aircraft footprint, as it is spread over several aircraft (possibly 5–15 aircraft per site), flights (10–40 flights per day), and a long period of time (vertiport lifetime can be expected in the 10–20 year time frame).

Another factor that should be considered when comparing eVTOL emissions to ground transportation is route (in)directness. In other words, when traveling from A to B within an urban center, it is likely for a ground vehicle to follow set routes dictated by highways, bridges, tunnels, and congestion. Within urban trips, this lengthens the trip by an additional 40% of distance on average, relative to a straight line. The number decreases as the route length increases. For the same A to B trip, an eVTOL is more likely to follow a direct line path, with at most the occasional stadium or airport imposing a minimal detour. This difference should be accounted for when comparing cars and eVTOL on a direct $kgCO_2e$/pax.km basis, since ultimately, the "km" element is de-facto different on a per-trip basis. The simplest way is to use the aforementioned 40% to calculate that eVTOL footprint is 30% (1–1/(1+40%) = 30%) lower than calculated thus far, or alternatively to increase all ground vehicle footprints by 40%.

Similarly, we note that ground vehicles are typically traveling directly from door to door. The traveler will enter the rideshare car and departure location and get off at the destination. In ideal scenarios, eVTOL vertiports will similarly take passengers directly from their starting point to their final destination, such as from an airport to a hotel downtown. However, it is expected that many routes will require ground transportation for either the first or last leg of the trip, including walking, biking, public transportation, or ground vehicles. In these cases, this should be accounted for using a weighted average between the footprint of the two modes and the miles traveled via each.

As eVTOLs are a nascent industry, many LCAs will be conducted on a forward-looking plan based on manufacturing and financial projections rather than historical data. This is perfectly valid and indeed critical since it allows for any roadmap changes to be made before heaving investments are made into manufacturing at scale and certain operational profiles. It is strongly encouraged that LCA analysts treat their reports as a work in progress and revisit or entirely overhaul their LCA every 18–24 months as new information is made available, either regarding aircraft design or maturity of operations.

## 13.5 DISCLAIMER

The work performed in part by the employees of the National Renewable Energy Laboratory, operated by Alliance for Sustainable Energy, LLC, was supported by the U.S. Department of Energy (DOE) under Contract No. DE-AC36-08GO28308. The views expressed in the article do not necessarily represent the views of the DOE or the U.S. Government. The U.S. Government retains and the publisher, by accepting the article for publication, acknowledges that the U.S. Government retains a nonexclusive, paid-up, irrevocable, worldwide license to publish or reproduce the published form of this work, or allow others to do so, for the U.S. Government purposes.

## REFERENCES


[1] Reed, T. *INRIX Global Traffic Scorecard*. https://trid.trb.org/view/1456836 (accessed 2022-12-02).

[2] Samal, S. R.; Mohanty, M.; Santhakumar, S. M. Adverse Effect of Congestion on Economy, Health and Environment Under Mixed Traffic Scenario. *Transp. in Dev. Econ.* 2021, *7* (2), 15. 10.1007/s40890-021-00125-4.

[3] *Boeing: State of the art 777X Composite Wing Center Completes Parts*. Boeing 777X Reveal. https://www.boeing.com/777x/reveal/state-of-the-art-777x-composite-wing-center-completes-parts/ (accessed 2022-12-02).

[4] *Key aspects of the Paris Agreement | UNFCCC*. https://unfccc.int/most-requested/key-aspects-of-the-paris-agreement (accessed 2022-12-02).

[5] Brazzola, N.; Patt, A.; Wohland, J. Definitions and Implications of Climate-Neutral Aviation. *Nat. Clim. Chang.* 2022, *12* (8), 761–767. 10.1038/s41558-022-01404-7.

[6] Sanz-Morère, I.; Eastham, S. D.; Speth, R. L.; Barrett, S. R. H. Reducing Uncertainty in Contrail Radiative Forcing Resulting from Uncertainty in Ice Crystal Properties. *Environ. Sci. Technol. Lett.* 2020, *7* (6), 371–375. 10.1021/acs.estlett.0c00150.

[7] IPCC. *International Panel on Climate Change (IPCC) Fifth Assessment Report – Impacts, Adaptation and Vulnerability*. http://www.ipcc.ch/report/ar5/wg2/ (accessed 2017-04-16).

[8] Smith, R. L.; Tan, E. C. D.; Ruiz-Mercado, G. J. Applying Environmental Release Inventories and Indicators to the Evaluation of Chemical Manufacturing Processes in Early Stage Development. *ACS Sustainable Chem. Eng.* 2019, *7* (12), 10937–10950. 10.1021/acssuschemeng.9b01961.

[9] *Block Speed. Paramount Business Jets*. https://www.paramountbusinessjets.com/aviation-terminology/block-speed (accessed 2022-12-04).

[10] US LCI. *U.S. Life-Cycle Inventory, v.1.6.0, National Renewable Energy Laboratory, Golden, CO, 2008.*

[11] Deng, Y.; Li, J.; Li, T.; Gao, X.; Yuan, C. Life Cycle Assessment of Lithium Sulfur Battery for Electric Vehicles. *Journal of Power Sources* 2017, *343*, 284–295. 10.1016/j.jpowsour.2017.01.036.

[12] Dai, Q.; Kelly, J. C.; Gaines, L.; Wang, M. Life Cycle Analysis of Lithium-Ion Batteries for Automotive Applications. *Batteries* 2019, *5* (2), 48. 10.3390/batteries5020048.

[13] Xiang, B.; Patra, P. K.; Montzka, S. A.; Miller, S. M.; Elkins, J. W.; Moore, F. L.; Atlas, E. L.; Miller, B. R.; Weiss, R. F.; Prinn, R. G.; Wofsy, S. C. Global Emissions of Refrigerants HCFC-22 and HFC-134a: Unforeseen Seasonal Contributions. *Proceedings of the National Academy of Sciences* 2014, *111* (49), 17379–17384. 10.1073/pnas.1417372111.

[14] Gautam, M.; Pandey, B.; Agrawal, M. Chapter 8 - Carbon Footprint of Aluminum Production: Emissions and Mitigation. In *Environmental Carbon Footprints*; Muthu, S. S., Ed.; Butterworth-Heinemann, 2018; pp 197–228. 10.1016/B978-0-12-812849-7.00008-8.

[15] Igos, E.; Benetto, E.; Meyer, R.; Baustert, P.; Othoniel, B. How to Treat Uncertainties in Life Cycle Assessment Studies? *Int J Life Cycle Assess* 2019, *24* (4), 794–807. 10.1007/s11367-018-1477-1.

[16] Heijungs, R. On the Number of Monte Carlo Runs in Comparative Probabilistic LCA. *Int J Life Cycle Assess* 2020, *25* (2), 394–402. 10.1007/s11367-019-01698-4.

[17] Cucurachi, S.; Blanco, C. F.; Steubing, B.; Heijungs, R. Implementation of Uncertainty Analysis and Moment-Independent Global Sensitivity Analysis for Full-Scale Life Cycle Assessment Models. *Journal of Industrial Ecology n/a* (n/a). 10.1111/jiec.13194.

[18] *Environmental Life Cycle Assessment: Measuring the Environmental Performance of Products*; Scheneck, R., White, P., Eds.; American Center for Life Cycle Assessment, 2014.
[19] Weidema, B. P.; Wesnæs, M. S. Data Quality Management for Life Cycle Inventories—an Example of Using Data Quality Indicators. *Journal of Cleaner Production* 1996, *4* (3), 167–174. 10.1016/S0959-6526(96)00043-1.
[20] PRé Consultants. *SimaPro*; PRé Sustainability: Amersfoort, the Netherlands, 2019.
[21] Hofstetter, P. Perspective in Life Cycle Impact Assessment: A Structured Approach to Combine of the Technosphere, Ecosphere and Valuesphere. *Int. J. LCA* 2000, *5* (1), 58–58. 10.1007/BF02978561.
[22] Tan, E. C. D. Sustainability Benefits of Valorizing Associated Flare Gas for the Production of Transportation Fuels, Harvard University Division of Continuing Education, Cambridge, Massachusetts, 2022. https://dash.harvard.edu/handle/1/37371424 (accessed 2022-05-01).
[23] US EPA. *Emissions & Generation Resource Integrated Database (eGRID2019).* https://www.epa.gov/egrid (accessed 2021-05-06).
[24] Barke, A.; Thies, C.; Popien, J.-L.; Melo, S. P.; Cerdas, F.; Herrmann, C.; Spengler, T. S. Life Cycle Sustainability Assessment of Potential Battery Systems for Electric Aircraft. *Procedia CIRP* 2021, *98*, 660–665. 10.1016/j.procir.2021.01.171.
[25] Chang, S. E.; Chen, Y. When Blockchain Meets Supply Chain: A Systematic Literature Review on Current Development and Potential Applications. *IEEE Access* 2020, *8*, 62478–62494. 10.1109/ACCESS.2020.2983601.
[26] Zheng, Z.; Xie, S.; Dai, H.; Chen, X.; Wang, H. An Overview of Blockchain Technology: Architecture, Consensus, and Future Trends. In *2017 IEEE International Congress on Big Data (BigData Congress)*; 2017; pp 557–564. 10.1109/BigDataCongress.2017.85.
[27] Rolinck, M.; Gellrich, S.; Bode, C.; Mennenga, M.; Cerdas, F.; Friedrichs, J.; Herrmann, C. A Concept for Blockchain-Based LCA and Its Application in the Context of Aircraft MRO. *Procedia CIRP* 2021, *98*, 394–399. 10.1016/j.procir.2021.01.123.
[28] Dutta, P.; Choi, T.-M.; Somani, S.; Butala, R. Blockchain Technology in Supply Chain Operations: Applications, Challenges and Research Opportunities. *Transportation Research Part E: Logistics and Transportation Review* 2020, *142*, 102067. 10.1016/j.tre.2020.102067.
[29] Azzi, R.; Chamoun, R. K.; Sokhn, M. The Power of a Blockchain-Based Supply Chain. *Computers & Industrial Engineering* 2019, *135*, 582–592. 10.1016/j.cie.2019.06.042.
[30] Karaszewski, R.; Modrzyński, P.; Müldür, G. T.; Wójcik, J. Blockchain Technology in Life Cycle Assessment—New Research Trends. *Energies* 2021, *14* (24), 8292. 10.3390/en14248292.
[31] Zhang, A.; Zhong, R. Y.; Farooque, M.; Kang, K.; Venkatesh, V. G. Blockchain-Based Life Cycle Assessment: An Implementation Framework and System Architecture. *Resources, Conservation and Recycling* 2020, *152*, 104512. 10.1016/j.resconrec.2019.104512.

# 14 Current Progress in Sustainability Evaluation, Pollution Prevention, and Source Reduction Using GREENSCOPE

*Selorme Agbleze and Shuyun Li*
Department of Chemical and Biomedical Engineering, West Virginia University, Morgantown, WV, United States

*Erendira T. Quintanar-Orozco*
Centro de Investigaciones Químicas, Universidad Autónoma del Estado de Hidalgo, Mineral de la Reforma, Mexico

*Gerardo J. Ruiz-Mercado*
Office of Research and Development, U.S. Environmental Protection Agency, 26 W Martin L. King Dr., Cincinnati, OH, United States

Chemical Engineering Graduate Program, Universidad del Atlántico, Puerto Colombia, Colombia

*Fernando V. Lima*
Department of Chemical and Biomedical Engineering, West Virginia University, Morgantown, WV, United States

## CONTENTS



DOI: 10.1201/9781003167693-14

## 14.1 INTRODUCTION

Limited and decreasing availability of resources consumed by the current and projected population, as well as the environmental impact of resource utilization, have caused frequent discussions about sustainability, especially in sectors and industries immediately associated with the transformation of raw materials into valuable products and services. Sustainability is described as the efficient use of resources in a manner that does not expend resources depriving future generations while maintaining ecological balance (Ruiz-Mercado, Gonzalez, and Smith 2013). In the chemical industry, sustainability is of paramount importance as this industry is the channel through which many products used domestically or exported abroad are produced. In addition, this industry relies immensely on quantities of raw materials that are extracted from the environment. In certain instances, parts of the life cycle of chemical products and associated services impact local or global ecosystems. To this end, this industry sector impacts the economy, society, and ecology in general. To preserve the benefits from resources over an extended period, sustainability metrics must be considered to enable process improvements systematically and select sustainable process alternatives. One of the main challenges to performing sustainability analysis for a chemical process is the definition of quantifiable metrics, data needs, and measuring scale to complete the analysis. To address this challenge, Gauging Reaction Effectiveness for the ENvironmental Sustainability of Chemistries with a multi-Objective Process Evaluator (GREENSCOPE) was developed (Smith, Ruiz-Mercado, and Gonzalez 2015). The objective of this chapter is to introduce GREENSCOPE and show a new practical intuitive interface between GREENSCOPE and chemical process simulators (e.g., CHEMCAD). Then, examples of sustainability analysis with GREENSCOPE indicators for single or multiple processes are presented, including cases at steady-state and dynamic operations.

## 14.2 BACKGROUND

GREENSCOPE is a versatile tool that performs sustainability analysis of chemical processes during the design or operating phase, for a single equipment or process flowsheet in steady-state or dynamic operations. GREENSCOPE provides sustainability indicators for quantifying the process performance. In addition to indicators, the equations and types of input required to evaluate these indicators are provided. There are a total of approximately 140 indicators in four main categories; Material Efficiency (26), Energy (14), Economics (33), and Environmental (66) (Ruiz-Mercado, Gonzalez, and Smith 2014). Indicators under material efficiency relate material inputs to metrics that represent the amounts, usage of these materials, and their content in product streams. Indicators under the energy category link energy input and generation in the process to metrics that explain different aspects of energy quality and usage. The economic indicators provide metrics that show a variety of economic aspects of the process such as costs and revenues. These range from broad metrics describing the net return of an entire process to specific metrics such as material cost.

The environmental indicators relate the specific process to metrics that show the material releases, and the potential effect of the process on humans and the environment (air, soil, and water). These indicators are then transformed using a measuring scale that provides insight into the sustainable performance of the specific process use of inputs (i.e., mass, energy) to obtain specific output materials and conditions. Two versions of GREENSCOPE have been developed, a version in Microsoft Excel (Ruiz-Mercado, Carvalho, and Cabezas 2016) and another with a graphical user interface that is introduced in this work. These two versions share common features in data requirement and indicator calculations for sustainability quantification and analysis, which are discussed in the next section.

For conducting the analyses, users will perform three main steps in GREENSCOPE: data entry, indicator calculations, and data analysis. Data requirements for sustainability analysis of a process or equipment have been grouped in GREENSCOPE under the following sections: (i) Stream and Compound Data; (ii) Equipment and Cost Data; and (iii) Utility Data. The indicators available for determining the process sustainability performance can be grouped into four main categories as briefly mentioned above, which include Material Efficiency, Energy, Economics, and Environmental, for assessing the sustainability of a process using radar plots involving these indicators. The details associated with the data requirements, sustainability metrics, and interfaces developed for GREENSCOPE are provided in the next subsections.

### 14.2.1 Data Requirements

The Stream and Compound Data requirements for sustainability analysis of a process or equipment in GREENSCOPE are in the form of stream mass and energy flows, temperature, pressure conditions, and compositions at the boundaries of the process that characterize inputs and outputs. The other data requirement under Stream and Compound Data section in GREENSCOPE includes reaction stoichiometry of reacting stream species, specific properties of stream components (e.g., thermodynamic), and other properties relating the interaction and effects of the stream components on living and non-living organisms and the environment. The next data requirement section is the Equipment and Cost data, which allows for equipment cost, and data about the energy produced or consumed to be entered for the most common equipment found in the chemical industry. This section also allows for the evaluation of uncommon/new equipment introduced in the process. Next, data on the types, flow rates, and costs of utilities used in the process are required.

### 14.2.2 Sustainability Metrics

To calculate the sustainability performance of a process or equipment, the data described in the previous section need to be provided. From this data, the sustainability performance of the process or equipment is calculated in the four main categories: Material Efficiency, Energy, Economics, and Environmental. Selected indicators will be presented in the case studies below, but the definitions of this extensive list can be seen in previous work (Ruiz-Mercado, Smith, and Gonzalez 2012a; Ruiz-Mercado, Smith, and Gonzalez 2012b). To assess the sustainability performance of the specific process, each indicator value is transformed into a dimensionless form and scaled into a percentage, as defined in equation (14.1).

$$Indicator\ \ Score = (Actual - Worst) \times 100/(Best - Worst) \tag{14.1}$$

Here, the actual value of the indicator is transformed with its best- and worst-case values for the specific process. This scaling allows for the analysis and comparison of multiple processes considering a common basis.

### 14.2.3 GREENSCOPE Interfaces

As mentioned, GREENSCOPE is available in two versions, the original Microsoft Excel version, and a graphical user interface (GUI) version introduced in this work. The flow chart of the steps for the GREENSCOPE Excel Version can be seen in Figure 14.1.

Some of the main features in Figure 14.1 for the GREENSCOPE MS Excel version is shown in Figure 14.2–Figure 14.4. These show the different sections in Figure 14.1 and the relevant data fields required.

Next, a schematic of additional functionalities introduced and available in the GUI GREENSCOPE version are shown in Figure 14.5. These functionalities include new interfaces that allow data input into various fields, as well as a COM Server import interface to enable data transfer between the process simulation (e.g., in CHEMCAD) and GREENSCOPE (Chemstations 2017; Ruiz-Mercado, Carvalho, and Cabezas 2016).

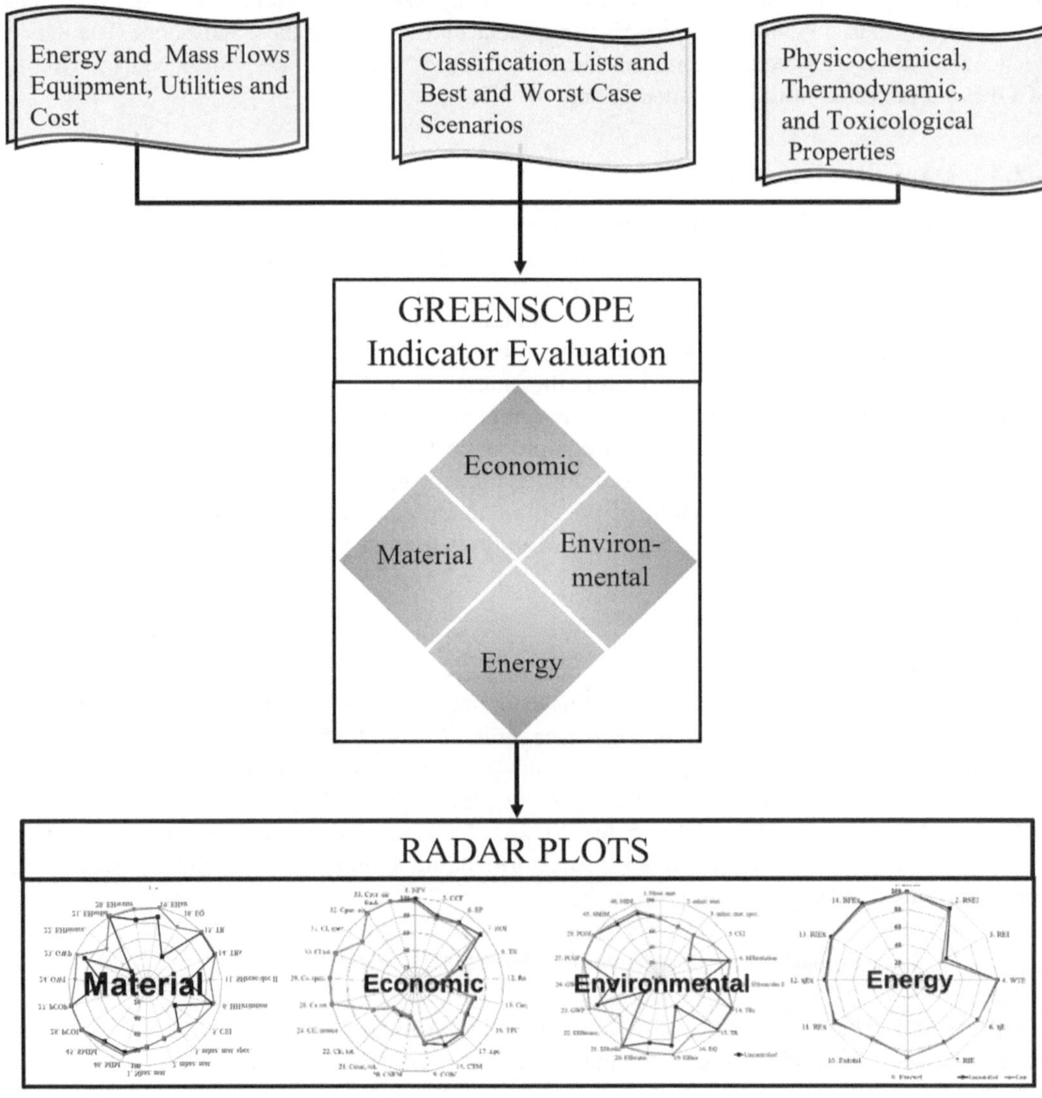


**FIGURE 14.1** Components in GREENSCOPE MS Excel version.

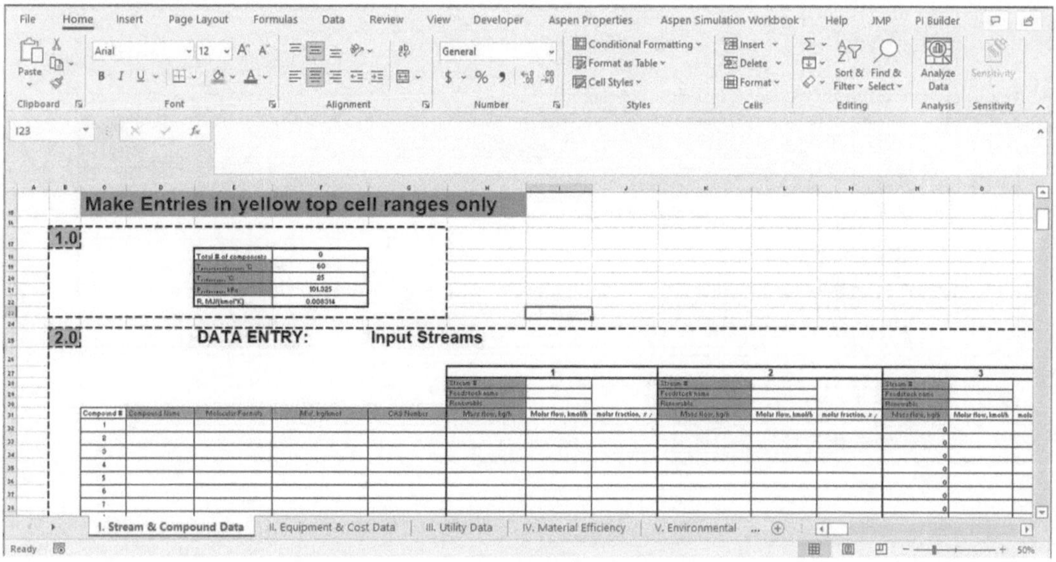

FIGURE 14.2 GREENSCOPE MS Excel version stream and compound data inventory.

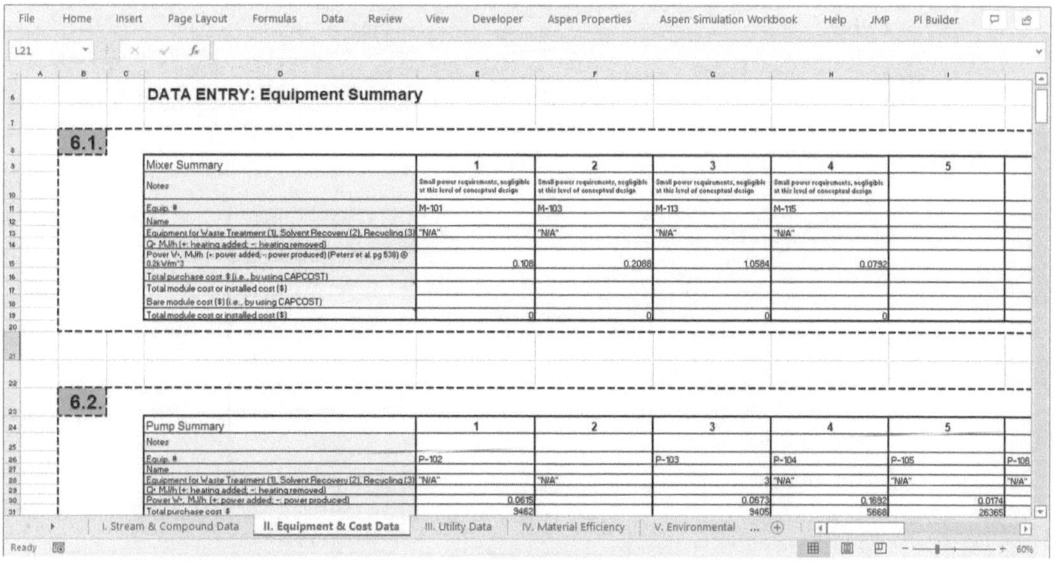

FIGURE 14.3 GREENSCOPE MS Excel version equipment and cost data inventory.

Some of the main interfaces referenced in Figure 14.5 are shown in Figure 14.6. The main menu interface shows three button options, which enable the selection of the appropriate analysis to be performed, i.e., GREENSCOPE itself, pollution control unit (PCU) calculations, and both GREENSCOPE with pollution control analysis respectively (Li et al. 2018). The PCU calculations enable the simulation of waste treatment processes to determine the appropriate technology, stream composition, and cost associated with the process.

Selecting the first icon in the general main menu opens the GREENSCOPE Main Menu, which allows for data import, analysis, and plotting of the results. In particular, the Import Simulation button in this menu allows the direct importing of data from a chemical process simulation. This significantly reduces the amount of time needed to provide the required stream data manually.

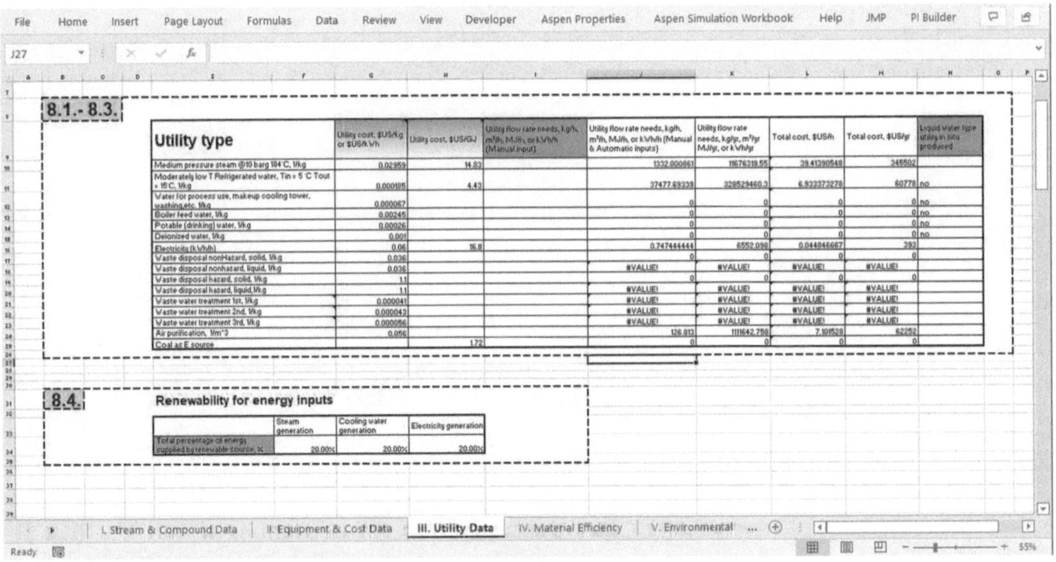


**FIGURE 14.4** GREENSCOPE MS Excel version process utility inventory.

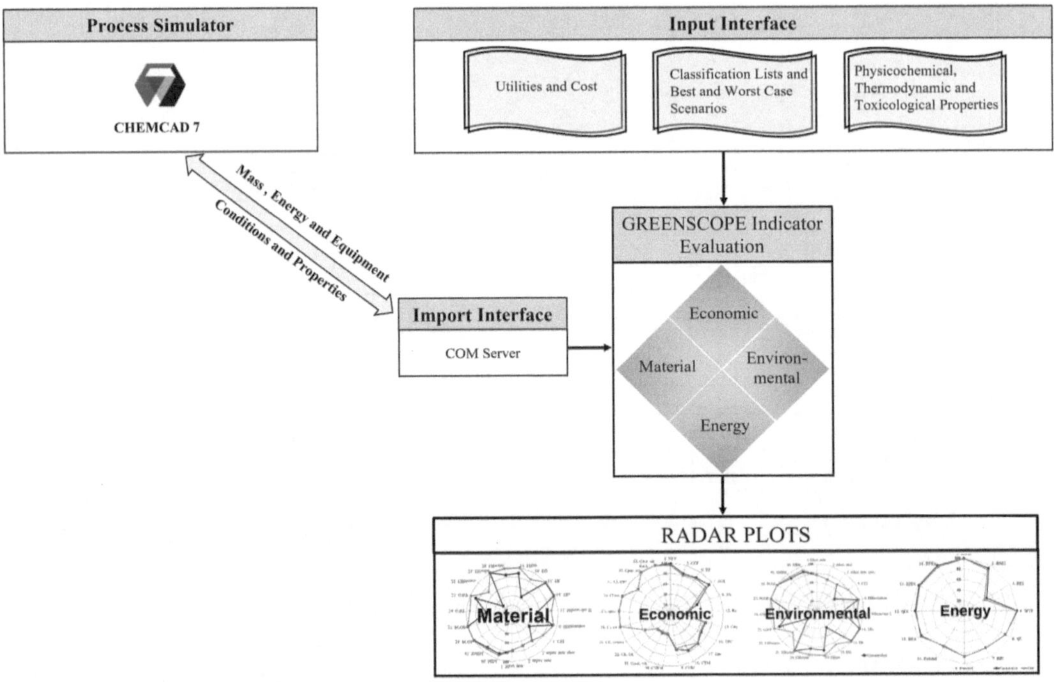


**FIGURE 14.5** Components in new GUI GREENSCOPE version.

In summary, GREENSCOPE enables sustainability analysis of the process during the design phase, which provides information that can guide decision-making in terms of sustainability performance for local equipment or entire flowsheets. In addition, the sustainability performance may be evaluated for a process in real-time using corresponding dynamic equations of the indicators of interest. To show some of the capabilities of GREENSCOPE, steady-state and dynamic cases are presented in the next sections.

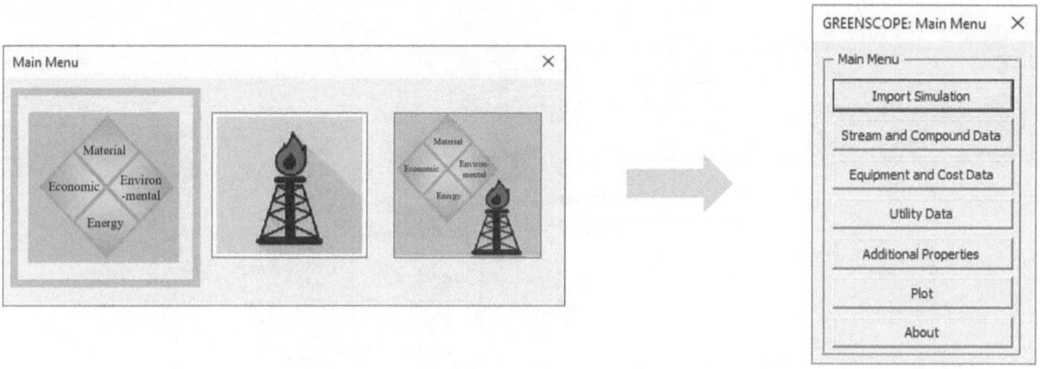


**FIGURE 14.6** General main menu (left), GREENSCOPE new main menu (right).

## 14.3 STEADY-STATE PROCESS CASE STUDIES

### 14.3.1 Acetic Acid Production Process Optimization

#### 14.3.1.1 Introduction and Process Description

Monsanto and Cativa$^{TM}$ processes are the two main methods of acetic acid manufacturing. Other approaches involve acetaldehyde and ethylene oxidation (Yoneda et al. 2001; Pal and Nayak 2017). In particular, the Monsanto process involves the carbonylation of methanol in the presence of a rhodium catalyst while the Cativa$^{TM}$ process relies on an iridium catalyst. Equation (14.2) shows the general carbonylation of methanol ($CH_3OH$) to acetic acid ($CH_3COOH$). Equations (14.3)–(14.6) are side reactions involving carbon monoxide (CO), carbon dioxide ($CO_2$), methane ($CH_4$), methyl acetate ($CH_3COOCH_3$), water ($H_2O$), and propionic acid ($CH_3CH_2COOH$). Equation (14.6) shows the reaction involving the promoter methyl iodide ($CH_3I$), carbon monoxide, and water to produce acetic acid and hydrogen iodide (HI).

$$CH_3OH + CO \rightarrow CH_3COOH \tag{14.2}$$

$$CO + CH_3OH \rightarrow CO_2 + CH_4 \tag{14.3}$$

$$CH_3OH + CH_3COOH \rightarrow CH_3COOCH_3 + H_2O \tag{14.4}$$

$$CH_3OH + CH_3COOH \rightarrow CH_3CH_2COOH + H_2O \tag{14.5}$$

$$CO + CH_3I + H_2O \rightarrow CH_3COOH + HI \tag{14.6}$$

The base case acetic acid process flowsheet including reactions is taken from a work on emission estimation and life-cycle inventory of the process (Smith et al. 2017). In this work, a sustainability analysis of process alternatives for an acetic acid process is presented. This includes improvement of the separation process after reaction, resulting in reduced waste gas components, recycling rate, and energy usage.

#### 14.3.1.2 Acetic Acid Process Base Case

The flowsheet of this process was modeled in CHEMCAD as in Figure 14.7, which shows the process flow diagram (PFD) of the base case process.

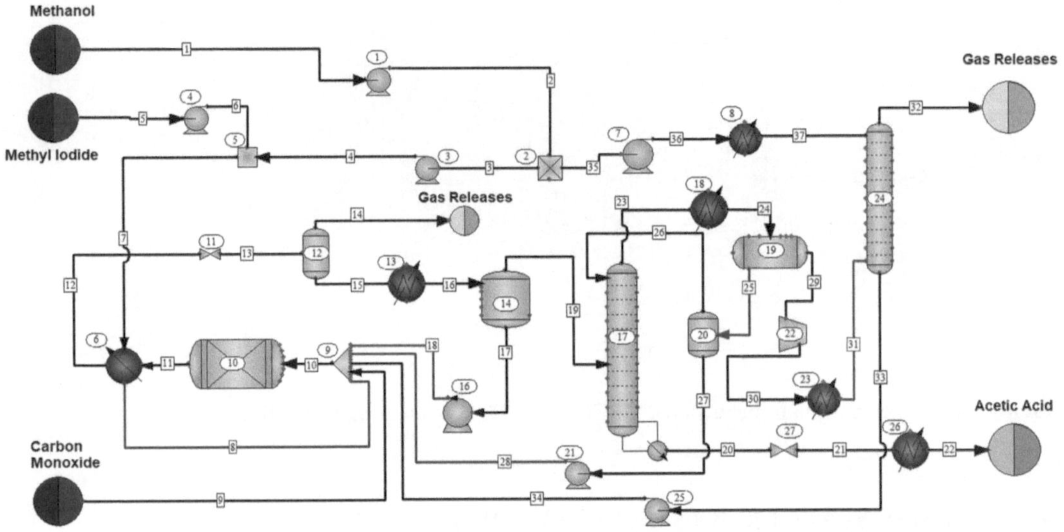


**FIGURE 14.7** Base case process flow diagram for acetic acid production.

In this process, methanol and methyl iodide both at 1 bar and 25°C enter the process. Methanol and methyl iodide are then combined in Stream 7 and heated in Stream 8. Part of the input methanol is used as a solvent for an absorber (Unit 24). The effluent from the reactor (Unit 10) is separated with a component separator (Unit 12) which removes half of all the non-condensable gases. A flash vessel (Unit 14) routes the liquid stream which contains some of the reactants back to the reactor and the gas stream to the distillation column (Unit 17) for the recovery of acetic acid in Stream 20. This stream is then cooled for storage. The production rate of acetic acid is 36,111 kg/h. The distillate (Stream 23) from the distillation column is sent to a flash vessel (Unit 19), and the liquid stream (Stream 25) which contains a small amount of vapor is partly separated in a component separator (Unit 20) that goes back to the distillation column. Stream 17 from the component separator and the methanol stream from the absorber are recycled. The combined flows at the boundaries of this process are summarized in Table 14.1.

**TABLE 14.1**
**Inventory of Base Case Acetic Acid Process**

| Compound name | Input (kg/h) | Output (kg/h) |
|---|---|---|
| Carbon Monoxide | 18695.00 | 1600.15 |
| Carbon Dioxide | 0.47 | 128.54 |
| Methane | 0.09 | 46.78 |
| Water | 10.06 | 108.18 |
| Methanol | 19800.00 | 70.15 |
| Acetic Acid | 0.00 | 36110.89 |
| Methyl Acetate | 0.00 | 426.19 |
| Hydrogen Iodine | 0.00 | 73.99 |
| Methyl Iodide | 450.00 | 374.40 |
| Propionic Acid | 0.00 | 23.45 |

#### 14.3.1.3 Acetic Acid Process Optimized Case

Figure 14.8 shows the changes made to the acetic acid process flowsheet to reduce the waste gases being generated while keeping a similar production rate of acetic acid. The optimized process involves a reconfiguration of the separation process after the reactor, resulting in a reduced number of separation equipment and waste stream contents.

In this new process topology, the component separator is replaced by a heat exchanger (Unit 13) and flash vessel (Unit 19) before the distillation column (Unit 16). Next, reactions (14.2)–(14.6) are analyzed and the required methyl iodide for acetic acid is specified. With this simplified flowsheet, the representation in Figure 14.8 was obtained with Stream 23 as the final recycle stream. The combined flows at the boundaries of this process are summarized in Table 14.2.

In the performed analysis, the process was optimized to improve raw material usage and reduce waste gas production which resulted in reduced utility usage. In the sustainability assessment in the next section, the specific areas of improvement become apparent between the base and optimized cases.

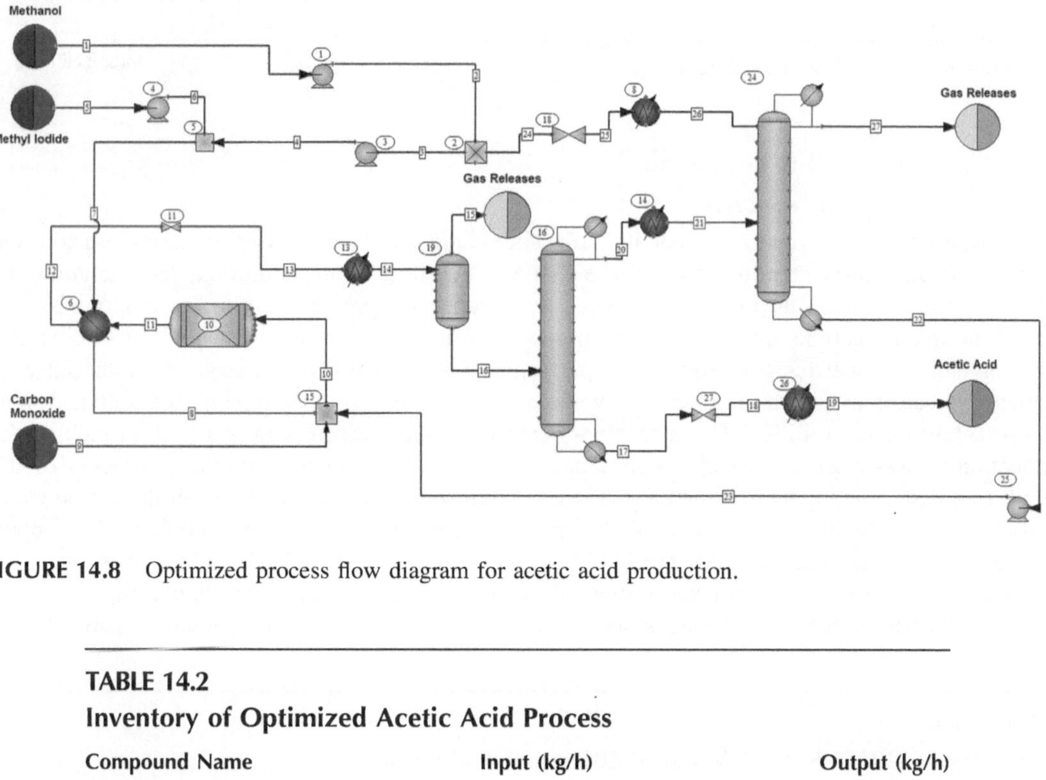


**FIGURE 14.8** Optimized process flow diagram for acetic acid production.

**TABLE 14.2**
**Inventory of Optimized Acetic Acid Process**

| Compound Name | Input (kg/h) | Output (kg/h) |
|---|---|---|
| Carbon Monoxide | 18694.88 | 1612.23 |
| Carbon Dioxide | 0.47 | 20.05 |
| Methane | 0.09 | 7.23 |
| Water | 10.06 | 151.59 |
| Methanol | 19800.00 | 6.54 |
| Acetic Acid | 0.00 | 36094.59 |
| Methyl Acetate | 0.00 | 588.96 |
| Hydrogen Iodine | 0.00 | 67.62 |
| Methyl Iodide | 75.00 | 0.00 |
| Propionic Acid | 0.00 | 32.69 |

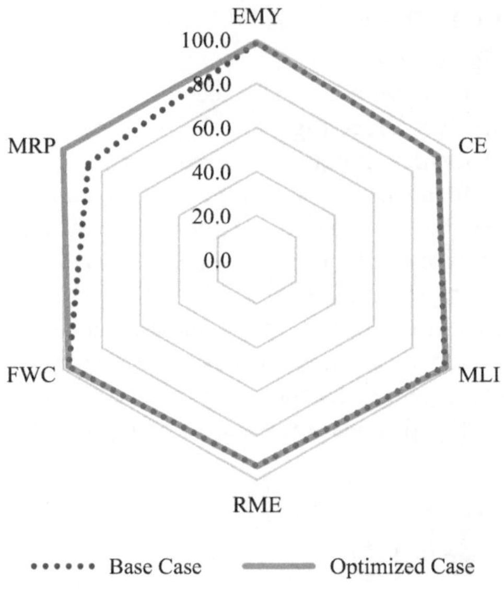


**FIGURE 14.9** Radar plot of selected material efficiency indicators for base and optimized cases.

#### 14.3.1.4 Sustainability Evaluation

##### *14.3.1.4.1 Material Efficiency*

Six indicators were chosen to evaluate the differences between the two cases of interest. In addition to the input and output stream data from the CHEMCAD simulations, additional data including the classification of streams as products or waste, the presence and properties of toxic compounds, data about the main reactions occurring in the process, thermodynamic data, as well as handling and environmental properties are required to perform the sustainability analysis. The material efficiency indicators assess the efficient conversion of raw materials to products, utilization of solvents, catalysts, and utilities. The main factors affecting the indicators in this section include the reduction in waste gas in the outlet streams and improvement in the separation process of the optimized case. The inputs into the base case and optimized case are relatively similar (same when possible) except for the promoter (methyl iodide). A sensitivity analysis was performed to obtain the required enough methyl iodide needed in the optimized case. The data similarity between both cases results in most of the selected indicators being constant, as shown in Figure 14.9.

The indicator names, actual values, and scores used in Figure 14.9 are shown in Table 14.3.

**TABLE 14.3**
**Summary of GREENSCOPE Material Efficiency Evaluation**

| Indicator | Symbol | Base Case Value | Optimized Case Value | Units | Base Case Score (%) | Optimized Case Score (%) |
|---|---|---|---|---|---|---|
| Effective Mass Yield | EMY | 0.551 | 0.539 | kg/kg | 98.60 | 98.70 |
| Carbon Efficiency | CE | 0.936 | 0.936 | kg/kg | 93.60 | 93.60 |
| Mass Loss Index | MLI | 0.060 | 0.049 | kg/kg | 97.00 | 97.60 |
| Reaction Mass Efficiency | RME | 0.938 | 0.937 | kg/kg | 93.80 | 93.70 |
| Fractional Water Consumption | FWC | 2.741E-07 | 2.738E-07 | $m^3$/kg | 97.30 | 97.30 |
| Material Recovery Parameter | MRP | 0.435 | 1.00 | $m^3$/$ | 87.00 | 100.00 |

The effective mass yield is highly dependent on the amount of potentially hazardous materials entering the process; in this process, this includes methanol and methyl iodide. Excellent scores of approximately 98.6% and 98.7% were obtained for both base and optimized cases. Although a reduced amount of methyl iodide exists in the optimized case, the similarity in score is due to methanol initially present 44 times more by mass compared to methyl iodide. This results in the mass ratio of hazardous compounds methanol and methyl iodide calculated as similar quantities by mass in both cases. The next set of indicators shows similar results because the input reagents and main product rates were kept constant. A carbon efficiency score of approximately 93.6% was obtained. This indicator tracks the carbon transfer from the reagents to the main product acetic acid. This excellent score implies that a significant amount of the initial carbon in the reagents ends up forming the main product. A mass loss index score of 97.0% for the base case and 97.6% for the optimized case is based on the ratio of the sum of undesired and unreacted compounds to acetic acid. The difference in scores results from excess methyl iodide present in the base case but reduced in the optimized case. Scores of approximately 93.8% and 93.7% for both cases show good performance for reaction mass efficiency. This indicator provides a different approach to evaluate the amount of input reagents that gets converted to desired products instead of undesired products. The fractional water consumption score of 97.3% for both cases indicates good performance for this indicator, which evaluates the amount of water used as a reagent in the process. This score results from relatively low water utilization as a reagent in both cases. Finally, the material recovery parameter indicator values of approximately 87% and 100% for base and optimized cases respectively result from reducing methyl iodide from 450 kg/h to 75 kg/h amount required to achieve the same production of the main product. An alternative approach could be the use of additional units to separate the original 450 kg/h to improve this indicator.

### *14.3.1.4.2 Environmental Efficiency*

The environmental efficiency indicators assess the potential impact of the process on the environment and ecosystem services. In addition, this assessment area allows for analysis of the impact on the health of living organisms including humans and animals. In this process, the compounds methanol and methyl iodide were considered hazardous. Methanol was considered hazardous due to its flammability and effect on various organs and nervous system. In addition, methyl iodide was considered hazardous due to known acute neurological illness and various toxicity symptoms that it may cause. Several eco-toxicity properties are needed to assess the effects of these elements on the environment. A few of these properties include potency factors that relate the known effect of certain chemicals to the potential effect of different chemicals, the classification lists such as Risk Phrase, which provides potential chemical hazards along with severity. The results of the environmental efficiency analysis for both base and optimized cases are shown in Figure 14.10.

The indicator names, actual values and scores used in Figure 14.10 are shown in Table 14.4.

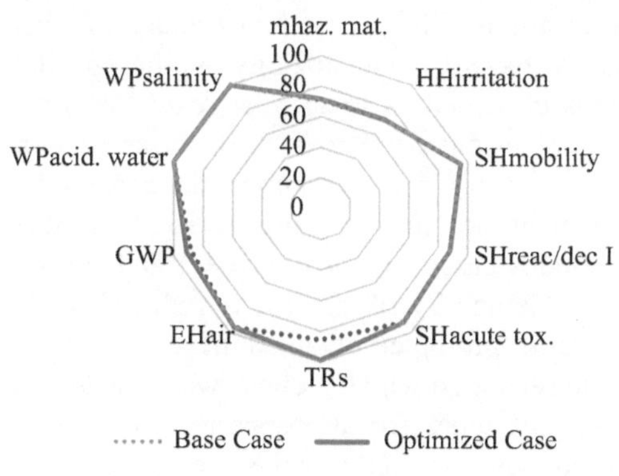


**FIGURE 14.10** Radar plot of environmental efficiency indicators for base and optimized cases.

**TABLE 14.4**
**Summary of GREENSCOPE Environmental Efficiency Evaluation for Acetic Acid Example**

| Indicator | Symbol | Base Case Value | Optimized Case Value | Units | Base Case Score (%) | Optimized Case Score (%) |
|---|---|---|---|---|---|---|
| Mass of Hazardous Materials Input | $m_{haz.\ mat.}$ | 20250 | 19875 | kg/h | 71.07 | 71.61 |
| Health Hazard, Irritation Factor | $HH_{irritation}$ | 312944 | 313114 | $m^3$/kg | 71.55 | 71.54 |
| Safety Hazard, Mobility | $SH_{mobility}$ | 0.495 | 0.486 | kg/kg prod | 95.05 | 95.14 |
| Safety Hazard, Reaction / Decomposition | $SH_{reac/dec}$ | 0.120 | 0.120 | 1 | 88.00 | 88.00 |
| Safety Hazard, Acute Toxicity | $SH_{acute\ tox.}$ | 8392 | 8345 | $m^3$/kg | 91.61 | 91.65 |
| Specific Toxic Release | $TR_s$ | 8.82E-03 | 6.19E-04 | kg/kg prod | 85.54 | 98.99 |
| Environmental Hazard, Air Hazard | $EH_{air}$ | 43268 | 43024 | $m^3$/kg | 95.67 | 95.70 |
| Global Warming Potential | GWP | 0.118 | 0.089 | kg $CO_2$ equivalent/kg prod. | 88.19 | 91.12 |
| Aquatic Acidification Potential | $WP_{acid.\ water}$ | 4.49E-13 | 9.79E-06 | kg H+ equiv./ time | 100 | 99.98 |
| Aquatic Salinization Potential | $WP_{salinity}$ | 0 | 0 | kg Salt-forming ion equiv./time | 100 | 100 |

The scores of approximately 71.1% and 71.6% for mass of hazardous materials are due to the presence of methanol and methyl iodide in the base and optimized case processes. The score of 71.5% for health hazard irritation factor for both cases is due to the classification of certain components to be irritants (in both cases, these compounds are present in the same quantity). The safety mobility score assesses the quantity of volatile compounds above their boiling points present in the process with scores of 95.05% and 95.14% obtained for both cases due to the similar amounts present. The score of 88.0% for the safety hazard for reaction/decomposition is due to methyl iodide and hydrogen iodide being above their auto-ignition temperatures. The score of 91.6% for safety hazard, acute toxicity is due to the presence of compounds classified as toxic. The indicator scores are highly dependent on the toxicity classification and availability and accuracy of toxicity properties including immediately dangerous to life or health (IDLH) concentrations. The score shows equal weights of effect in the base and optimized cases. Unlike safety hazard, acute toxicity of the specific toxic release is calculated based on mass of compounds classified as toxic relative to the mass of product. This does not consider concentrations required to be considered toxic or the classification category. The scores of 85.5% and 99% represent relatively fewer compounds considered toxic by mass in the optimized case. The environmental hazard for air score of 95.7% is due to the compounds categorized as air pollutants, this includes carbon monoxide (CO) and carbon dioxide ($CO_2$). The global warming potential scores of 88.1% and 91.1% are due to reduced waste because of using the appropriate amount of methyl iodide promoter obtained from sensitivity analysis. This reduces the amount of excess methyl iodide separated which reduces waste materials generated from the base and optimized case. The excellent score for atmospheric acidification potential for both cases is due to a small number of compounds classified as acids remaining in the waste stream. The 100% aquatic salination potential is due to no compounds forming high concentrations of salts in water.

### *14.3.1.4.3 Energy Efficiency*

The energy efficiency indicators consist of 14 indicators that assess sustainable energy usage for specific applications such as waste treatment, solvent recovery, and recycling, in addition to overall energy requirement. Some of these indicators use readily available process data including flowrates, enthalpies, entropies, work, and heat available for the process to provide several energy-related properties and variables. Given process conditions, these properties and variables can be obtained from thermodynamic references/databases or commercial process simulation software including Aspen Plus, HYSYS, and CHEMCAD. These variables are then combined using efficiency or equivalency factors to provide sustainability scores. In addition to providing insight into sustainability, these indicators provide a general overview of energy distribution in the process of focus. Energy efficiency results for the acetic acid process are available in Figure 14.11.

Next, a summary of indicator names, actual values, and scores from Figure 14.11 are shown in Table 14.5.

The specific energy intensity score assesses the overall energy usage through utilities such as combustion fuels, steam, and electricity. Scores of approximately 94.0% and 96.9% for the base case and optimized case, respectively, were obtained. The main source of difference between these scores is the additional steam demand in the base case process for the separation section. Similarly, the energy intensity indicator assesses the overall energy usage through utilities normalized by the sales revenue

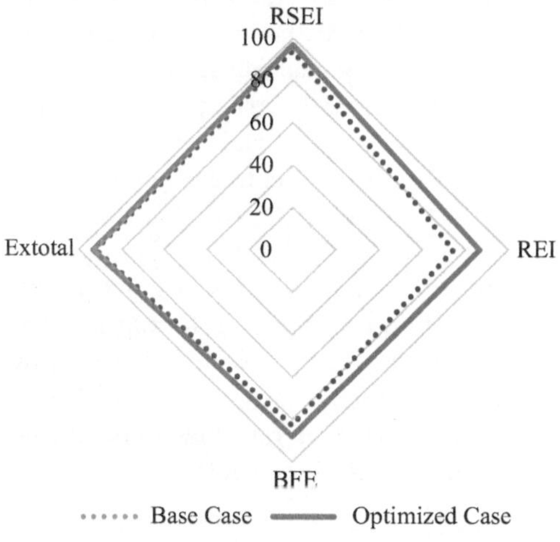


**FIGURE 14.11** Radar plot of selected energy efficiency indicators for acetic acid process.

**TABLE 14.5**
**Summary of GREENSCOPE Energy Efficiency Evaluation for Acetic Acid Process**

| Indicator | Symbol | Base Case Value | Optimized Case Value | Units | Base Case Score (%) | Optimized Case Score (%) |
|---|---|---|---|---|---|---|
| Specific Energy Intensity | $R_{SEI}$ | 5.97 | 3.10 | MJ/kg | 94.03 | 96.90 |
| Energy Intensity | $R_{EI}$ | 9.55 | 4.96 | MJ/$ | 74.40 | 86.70 |
| Breeding-energy Factor | $BF_E$ | 1.24 | 1.32 | – | 82.72 | 87.93 |
| Exergy Intensity | $Ex_{total}$ | 19.91 | 17.71 | MJ/kg | 92.36 | 93.48 |

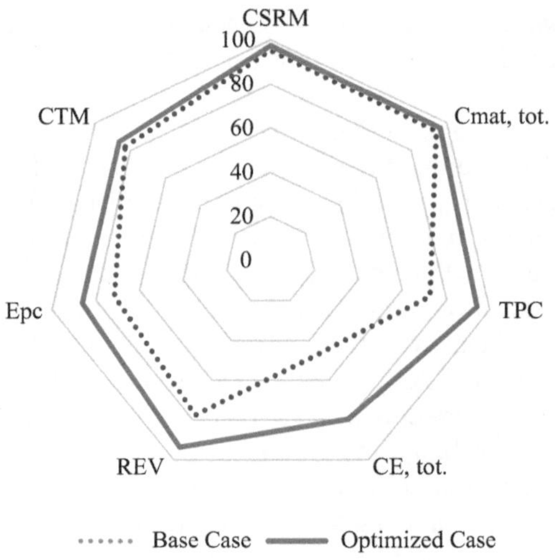


**FIGURE 14.12** Radar plot of selected economic efficiency indicators for acetic acid process.

from products. As a result of the normalization of this indicator, it may be used for comparing and evaluating multiple processes. Scores of 74.4% and 86.7%, respectively, were obtained for the base and optimized cases. The breading-energy factor score relates total energy in output streams to the non-renewable material energy. This indicator rewards the process that uses less non-renewable materials, with that process being considered more sustainable. For similar acetic acid rates in both base and optimized cases, scores of 82.7% and 87.9% are results of slight differences in input and the narrow range of the best- and worst-case values used for this indicator. The exergy intensity score assesses the remaining resources in terms of exergy after estimated losses used to deliver the main product. For this indicator, a smaller number of resources in terms of exergy units is rewarded and considered more sustainable. Scores of 92.4% and 93.5% were obtained for the base case and optimized case.

##### *14.3.1.4.4 Economic Efficiency*

The economic efficiency indicators provide insight into economic trends in terms of costs, profit, and savings. These indicators provide comprehensive insights into economic performance, considering capital costs of one-time expenses, operating costs associated with utilities, raw materials, and revenue from product sales in addition to other economic factors associated with operating the process for the entire project life. Generally, the price of raw materials can be estimated using data available from several sources including VWR, Independent Commodity

**TABLE 14.6**
**Summary of GREENSCOPE Environmental Efficiency Evaluation for Acetic Acid Process**

| Indicator | Symbol | Base Case Value | Optimized Case Value | Units | Base Case Score (%) | Optimized Case Score (%) |
|---|---|---|---|---|---|---|
| Specific Raw Material Cost | $C_{SRM}$ | 0.55 | 0.34 | US $/kg prod. | 95.37 | 97.45 |
| Total Material Cost | $C_{mat,\ tot.}$ | 177.29 | 111.55 | $10^6$ US $/ yr | 94.09 | 96.28 |
| Total Product Cost | TPC | 318.63 | 191.56 | $10^6$ US $/ yr | 72.69 | 94.11 |
| Total Energy Cost | $C_{E,\ tot}$ | 26.35 | 10.14 | $10^6$ US $/yr | 47.3 | 79.71 |
| Manufacturing Cost | REV | 254.91 | 153.25 | $10^6$ US $/yr | 77.64 | 93.58 |
| Production Cost | Epc | 178.33 | 112.33 | $10^6$ US $/yr | 71.48 | 86.15 |
| Capital Cost | $C_{TM}$ | 3.43 | 2.73 | $10^6$ US $/yr | 82.92 | 86.41 |

Intelligence Services (ICIS), Alibaba, and Thermo Fisher Scientific. Equipment costs can be estimated from tools such as CAPCOST or obtained from process simulation programs including Aspen, HYSYS, and CHEMCAD (Turton et al. 2018). Economic efficiency results for this process are available in Figure 14.12.

Next, the indicator names, actual values, and scores used in Figure 14.12 are shown in Table 14.6.

In this process analysis, the advantages of the optimized case become apparent when looking at the economic efficiency indicators. In this section, expenses are calculated considering several variables, in which in general, a lower value calculated in either case (i.e., base, or optimized case) is considered more sustainable. The specific raw material cost provides a score that evaluates the cost of raw materials normalized by the product stream flow rate. Scores of 95.4% and 97.4% were obtained for this indicator. These two indicator scores relating to raw material cost show the optimized case is more cost-effective in terms of raw material usage. The total material cost scores of 94.1% and 96.3%, respectively, result entirely from the reduction in methyl iodide in the optimized case. This was performed to improve the separation downstream while maintaining the same production rate of acetic acid. The total energy cost indicator considers expenses required to provide energy in the form of utilities to meet production targets. The main contributions to the larger cost of energy in the base case include the high requirement for low-pressure steam, refrigerated water, and the electricity demand in the additional compressor unit. Scores of 47.3% and 79.7% were obtained for the base and optimized cases, respectively. The manufacturing cost indicator provides an additional level of insight into operating costs, as this indicator combines the raw material, utility, and waste treatment costs. An approach for waste treatment simulation, and equipment sizing before costing is not currently available in GREENSCOPE but available in the literature associated with our group (Li et al. 2018). Scores of 77.6% and 93.6% for base and optimized cases respectively were obtained. Moreover, the total product cost scores of 71.5% and 86.1% for base and optimized cases respectively consider the cost of manufacturing and additional estimates for administrative and research costs, which were obtained from correlations. Finally, the next economic indicator provides additional insight into the sustainable performance of these two process configurations. The capital cost indicator considers direct and indirect costs, notably including equipment, installation materials, and labor costs. Scores of 82.9% and 86.4% were obtained in the base and optimized cases, respectively. One of the reasons for this difference is the cost of additional equipment in the separation section; this includes an additional vessel, heat exchangers, and compressor in the base case process.

#### 14.3.1.5 Conclusions

In this section, two process design cases for acetic acid production were evaluated. This section showed some of the possibilities of GREENSCOPE in sustainability assessment for evaluating multiple process alternatives. Multiple different designs or iterations of the same design may be assessed. The results for the acetic acid cases showed better overall performance of the optimized case when compared to the base case. The acetic acid production processes with two different separation techniques were analyzed. Among the four sustainability categories, the economic sustainability indicators showed the most difference between the two process topologies. Similar input boundary conditions and production targets resulted in several similar indicator values mostly in material and environmental categories. In the energy category, the optimized case is more sustainable in terms of energy utilization resulting from an improved separation section with reduced steam and electricity demand. In addition, this improved separation section translates to increased performance in economic indicators due to a reduction of capital required to support the operating cost and additional required equipment. Overall, implementing the optimized case is more sustainable in terms of energy utilization and material footprint, as this allows for the resources saved from providing additional energy as steam, refrigerated water, etc., and capital for additional equipment from the base case to be available for other investment applications.

## 14.3.2 Biofuel Production via Novel Biorefinery Process

### 14.3.2.1 Introduction and Process Description

A biorefinery is a process that converts biomass into value-added industrial chemical products and biofuels. Proposed configurations for new biorefineries are expected to emit less when compared to current industrial processes (Katakojwala and Mohan 2021). Additionally, new raw materials that some researchers are proposing to use in biorefineries correspond to waste generated in industrial processes. This allows biorefineries to provide another opportunity to transform raw/waste materials to value-added products and biofuels (Ubando et al. 2021). Good candidate feedstocks for biorefineries are biomass compounds with high content of hemicellulose, cellulose, and lignin. Such feedstocks could be obtained from agricultural and forestry activities. Fermentation, dehydration, catalytic, and hydrogenation processes are generally employed to process this feedstock in the biorefinery into valuable products (Clauser et al. 2021). The crop *Opuntia* spp. is considered as a novel biorefinery feedstock. This crop has been studied by researchers as a potential feedstock to produce biogas and biofertilizer by anaerobic digestion (Homer, Varnero, and Bedregal 2020). Additionally, another approach to obtaining biofertilizer from *Opuntia* as a value-added product has been studied in the presence of microbial groups (Quintanar-Orozco et al., 2018). Other value-added products such as pectin and mucilage can be obtained in addition to biofuels and biofertilizers. A biorefinery process scheme associated with these products is evaluated here to provide the sustainable and economic performance of this new process. This evaluation enables identifying critical indicators to optimize this biorefinery process to improve overall performance in the future. The process flow diagram of this proposed novel biorefinery is depicted in Figure 14.13. In this process, the feedstock, *Opuntia*, is crushed in unit B1, and then separated into liquid and solid fractions. *Opuntia* is then combined with ethanol in B5, and the recovered products include mucilage, water, unreacted ethanol, and lignocellulosic biomass. Air is then used to dry mucilage to remove moisture. The mixture containing ethanol and the residue is separated in a column.

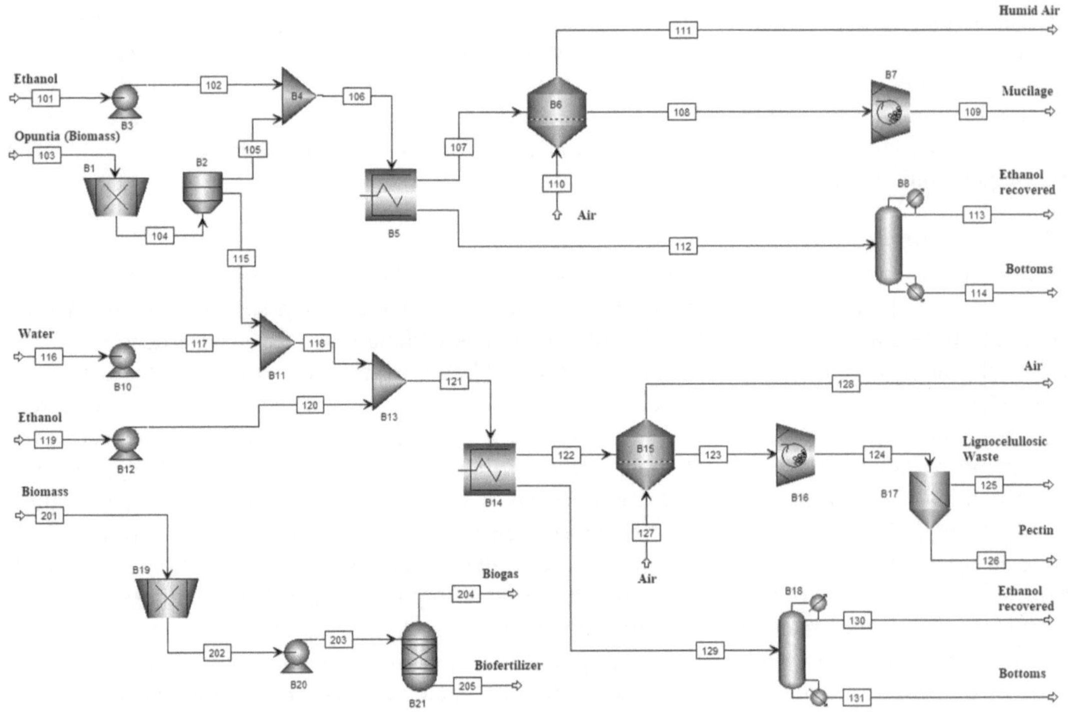


**FIGURE 14.13** Aspen plus flowsheet for Opuntia biorefinery.

**TABLE 14.7**
**Inventory of Biorefinery Process**

| Compound Name | Input (kg/h) | Output (kg/h) |
|---|---|---|
| Water | 106.00 | 2071.45 |
| Ethanol | 4294.00 | 4255.99 |
| *Opuntia* | 4230.00 | 107.72 |
| Pectin | 0.00 | 71.58 |
| Air | 208446.00 | 208501.00 |
| Mucilage | 0.00 | 40.70 |
| Biofertilizer | 0.00 | 1857.45 |
| Methane | 0.00 | 28.50 |
| Carbon Dioxide | 0.00 | 15.20 |
| Sulfhydryl Acid | 0.00 | 0.40 |
| Nitrogen | 0.00 | 0.40 |
| Oxygen | 0.00 | 0.40 |
| Hydrogen | 0.00 | 0.20 |

Water, ethanol, and biomass are combined to extract pectin, and lignocellulosic waste in unit B14. The pectin-lignocellulosic mixture is then recovered by first drying with air. Next, unit B17 is then used for the final separation of the pectin-lignocellulosic mixture. The ethanol and bottoms from unit B14 are separated with a column. Finally, biogas and biofertilizer are recovered by anaerobic digestion in unit B21 with crushed *Opuntia* in unit B19.

This designed biorefinery process was simulated in Aspen Plus (v.8), and the simulation data were imported into GREENSCOPE for the sustainability assessment. Table 14.7 shows the inventory necessary for this biorefinery scale, as well as the total amount of each product obtained per hour, such as biogas, biofertilizer, pectin, and mucilage. Note that in this process, ethanol is used as a solvent to extract pectin, and mucilage, and eliminate chlorophyll from the medium.

### 14.3.2.2 Sustainability Evaluation

#### *14.3.2.2.1 Material Efficiency*

The material efficiency indicators are the first set of indicators analyzed for this process. In this category, seven indicators are chosen. To evaluate all the indicators chosen, properties in addition to that available in the Aspen Plus flowsheet are obtained. Next, the best- and worst-case values of these selected indicators are adjusted for this type and scale of biorefinery. The results of this analysis are shown in Figure 14.14.

The indicator names and actual values used in Figure 14.14, including the best- and worst-case values, are shown in Table 14.8.

The mass intensity considers the ratio between the total mass of all inputs into the biorefinery process over the sales revenue of valuable products. A score of 98% represents a value of 1.04 kg/$, which is close to the best-case value selected. The 99% score for the physical return on investment represents good performance for this indicator, which implies the ratio of the value of the product to the additional quantity of resources needed to convert the raw materials *Opuntia*, air, water, and ethanol into products for this indicator is low. A good score for this indicator is obtained when either the revenue of the product is large or a small number of resources are used. Either will reduce the ratio value resulting in higher performance. The renewability-material index represents the ratio of the consumption of the renewable input (i.e., *Opuntia*, water, and air) relative to the total input materials. A 98% score was obtained for this indicator. Breeding-material factor score of 100% indicates excellent performance for this indicator, which results from the large percentage of renewable materials present

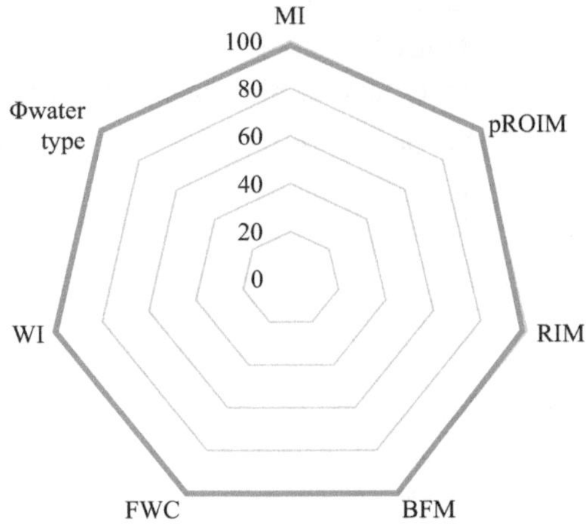


**FIGURE 14.14** Radar plot of selected material efficiency indicators for biorefinery process.

**TABLE 14.8**
**Summary of GREENSCOPE Material Efficiency Evaluation**

| Indicator | Symbol | Value | Unit | Best Case | Worst Case | Score (%) |
|---|---|---|---|---|---|---|
| Mass intensity | MI | 1.04 | kg/$ | 0 | 52 | 98.00 |
| Physical return on investment | $pROI_M$ | 0.01 | kg/kg | 0 | 40 | 99.90 |
| Renewability-material index | $RI_M$ | 0.98 | kg/kg | 1 | 0 | 98.00 |
| Breeding-material factor | $BF_M$ | 50.55 | kg/kg | 10 | 0 | 100.00 |
| Fractional water consumption | FWC | 4.54E-05 | $m^3$/kg | 0 | 2.95 | 100.00 |
| Water intensity | WI | 5.06E-07 | $m^3$/$ | 0 | 1.55 | 100.00 |
| Volume fraction of water type | $\Phi_{water\ type}$ | 0 | kg/ $m^3$ | 0 | 1 | 100.00 |

in the input streams of the process. The fractional water consumption score of 100% is due to the low volume of freshwater consumed relative to the biorefinery product generation and revenue. In addition, the water intensity score of 100% shows the ratio between the volume of fresh water and the product revenue. Next, the volume fraction of water type score of 100% measures the percentage of water used in the process with drinking water quality with respect to all other types of water qualities used in the biorefinery. The 100% score indicates that in this process 0 kg of water with drinking water quality is needed.

#### *14.3.2.2.2 Energy Efficiency*

The energy indicators for this biorefinery process consider the energy from streams, utility steam, cooling water, and electricity. Next, the energy generated and consumed by each piece of equipment is required. These required data can be obtained from the Aspen Plus simulation. In this process, electricity is used to operate the crushing and pump units while steam and cooling water are used to operate the columns. The results of the energy efficiency indicators are shown in Figure 14.15.

The indicator names and actual values used in Figure 14.15 including the best- and worst-case values are shown in Table 14.9.

The total energy consumption in the process had a value of 14743.13 MJ/h, this results in 94% for this indicator. Next, the specific energy intensity score of 99.7% indicates that a small ratio of energy was used in the process from sources such as steam, cooling water, and electricity required to provide heat and power relative to the unit mass of the main product (i.e., biogas). A value of 6.33 MJ/kg was obtained and used in this evaluation. Similarly, the energy intensity indicator evaluates the energy

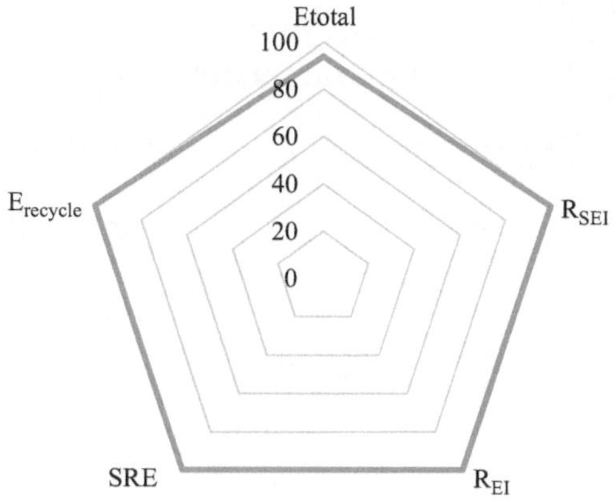


**FIGURE 14.15** Radar plot of selected energy efficiency indicators for biorefinery process.

**TABLE 14.9**
**Summary of GREENSCOPE Energy-efficiency Evaluation**

| Indicator | Symbol | Value | Unit | Best Case | Worst Case | Score (%) |
|---|---|---|---|---|---|---|
| Total energy consumption | Etotal | 14743.13 | MJ/h | 2867.00 | 28461.00 | 94.00 |
| Specific energy intensity | $R_{SEI}$ | 6.33 | MJ/kg | 0.00 | 1949.00 | 99.68 |
| Energy Intensity | $R_{EI}$ | 0.07 | MJ/$ | 0.00 | 37.30 | 99.81 |
| Solvent recovery energy | SRE | 0.00 | MJ/kg | 0.00 | 0.63 | 100.00 |
| Energy for recycle | $E_{recycl.}$ | 0.00 | MJ/kg | 0.00 | 0.63 | 100.00 |

used with respect to the product generation rate and sales revenue of the main product. The low value of 0.07MJ/$ implies low total energy usage relative to the product revenue. This results in an excellent indicator score of 99.8%. Equipment B16 removes moisture from the pectin and lignocellulosic mixture for further separation immediately downstream. The solvent recovery energy score with respect to units B6, B15, B18, and B8 was 100% as no additional energy was required to recover air and ethanol solvents used in this piece of equipment. The 100% score for the energy for recycling was because no additional energy was used for recycling operations.

#### *14.3.2.2.3 Environmental Efficiency*

The environmental efficiency indicators for this biorefinery evaluate the potential impact of biogas production from *Opuntia* on humans, animals, and the environment. This analysis considers the potential impact of the specific inputs and outputs to cause harm to humans, animals, and the environment. In this process, ethanol was classified as hazardous due to its flammability effects. Like in the acetic acid case study, additional properties needed are obtained from sources such as handbooks and/or online databases. The results of the analysis are summarized in Figure 14.16.

The indicator names and actual values used in Figure 14.16 including the best and worst-case values are shown in Table 14.10.

The score for-number of hazardous materials inputs was 80%. This is due to ethanol's classification as a hazardous material. In this biorefinery, however, ethanol makes up 20% of the total inputs, which include water, *Opuntia*, and air. This results in a 98.02% score for the mass of hazardous material input. The specific hazardous raw material input represents the total mass of hazardous substances fed to the process relative to the mass of the main product, with a score of 98% obtained for this indicator. The 98% score corresponds to 1.84 kg of hazardous material (i.e., ethanol) with respect to the product. The environmental quotient indicator characterizes the environmental

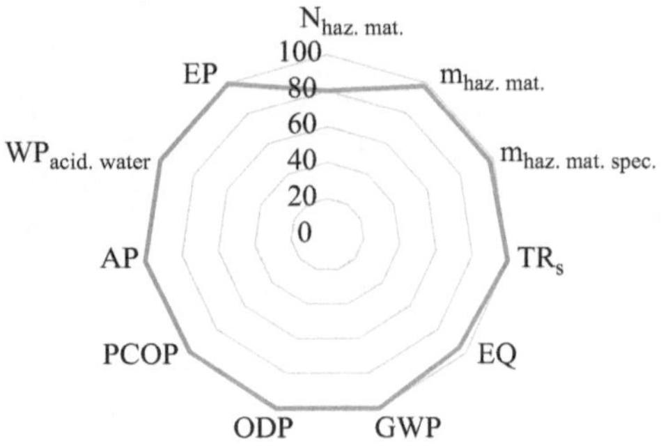


**FIGURE 14.16** Radar plot of selected environmental efficiency indicators for biorefinery process.

**TABLE 14.10**
**Summary of GREENSCOPE Environmental Efficiency Evaluation for Biorefinery Process**

| Indicator | Symbol | Value | Unit | Best Case | Worst Case | Score (%) |
|---|---|---|---|---|---|---|
| Number of hazardous materials input | $N_{haz.\ mat.}$ | 1.00 | – | 0 | 5.00 | 80.00 |
| Mass of hazardous materials input | $m_{haz.\ mat.}$ | 4294.00 | kg/h | 0 | 217076.00 | 98.02 |
| Specific hazardous raw materials input | $m_{haz.\ mat.\ spec.}$ | 1.84 | $kg_{haz.\ input}/kg_{prod}$ | 0 | 93.23 | 98.02 |
| Environmental quotient | EQ | 939.37 | $m^3/kg$ | 0 | 19500.00 | 95.18 |
| Global warming potential | GWP | 0.00 | $kg\ CO_{2\ equivalent}/kg_{prod}$ | 0 | 92.17 | 100.00 |
| Stratospheric ozone-depletion potential | ODP | 0.00 | $kg\ CFC\text{-}11_{equivalent}/kg_{prod}$ | 0 | 92.17 | 100.00 |
| Photochemical oxidation potential | PCOP | 0.75 | $kg\ C_2H_{2\ equivalent}/kg_{prod}$ | 0 | 91.23 | 99.18 |
| Atmospheric acidification potential | AP | 0.00 | $kg\ SO_{2\ equivalent}/time$ | 0 | 92.17 | 100.00 |
| Aquatic acidification potential | $WP_{acid.\ water}$ | 0.00 | $kg\ H^+_{equivalent}/time$ | 0 | 91.29 | 100.00 |
| Eutrophication potential | EP | 0.00 | $kg\ PO4^{3+}_{equivalent}/time$ | 0 | 91.29 | 100.00 |

unfriendliness of the produced waste streams. The 95.2% score implies good performance for this indicator, due to the value of the ratio of the waste relative to the product is close to being 0. A value of 0 is the best-case scenario as this implies no waste was generated and/or released into the environment. Next, the global warming potential score of 100% indicates none of the components in the output stream directly contributes to global warming. This is evident from the value of 0 total mass of carbon dioxide equivalents emitted relative to the product generated. The stratospheric ozone-depletion potential is based on the potential of each substance to deplete ozone relative to chlorofluorocarbon-11 (CFC-11). The 100% score obtained for this indicator results from no compound contributing to ozone depletion in this process. Additionally, photochemical oxidation potential evaluates the potential of each compound to contribute to ozone creation when released into the environment. This indicator is intended to assess this effect at low altitudes. The score of 100% obtained for this indicator implies the compounds that contribute to photochemical oxidation are not present. Atmospheric acidification potential is based on the potential of compounds to cause acid rain that eventually seeps into water bodies. For this indicator, a score of 100% was calculated as no compound in this process contributes to this effect.

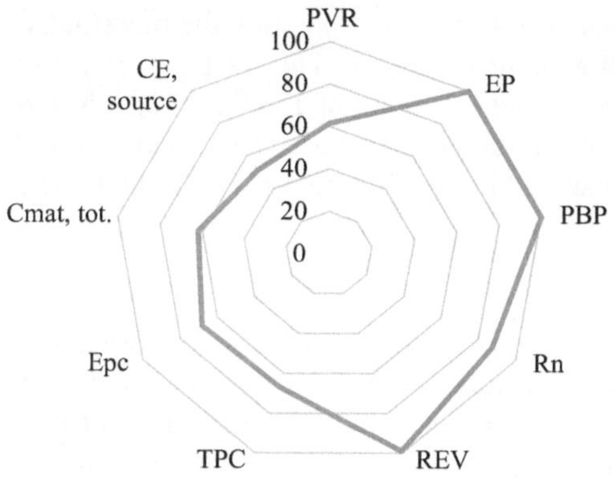


**FIGURE 14.17** Radar plot of selected economic efficiency indicators for biorefinery process.

### *14.3.2.2.4 Economic Efficiency*

Lastly, the economic efficiency indicators consider a variety of costs, a few of which include raw materials, utility generation, equipment, and other annuitized costs. The revenue associated with the sale of products is included in several indicator evaluations. A plot of the results of this analysis is shown in Figure 14.17, where a total of nine indicators are analyzed, with performance ranging from 50% to 100%.

The indicator names and actual values used in calculating the scores in Figure 14.17 including the best and worst-case values are shown in Table 14.11.

The present value ratio (PVR) indicator evaluates the profitability of the process. This indicator assumes the worst-case value as the break-even point. A PVR score of 61.43% implies that this process performs better than the break-even condition. The economic potential value of 66.69 $/kg product represents a 100% score for this indicator, which is mainly due to the high sales revenue from the product. Next, the net return indicator is a non-discounted profitability criterion of the biorefinery process to recover the total capital investment. A score of 87.5% was obtained for this process, with the indicator calculated over the total ten-year lifetime of the biorefinery. Considering the products from the biorefinery (i.e., biogas, biofertilizer, pectin, and mucilage), the revenue from eco-products indicator score of 99.2% with a value of 1816.72 x $10^6$ $ per year was obtained. This only considers the sales revenue of products. The sum of operating cost, distribution, and revenue from products comes to a value of 486.42 x $10^6$ $/yr, which represents the total production cost indicator with a percent score of 66.7%. This percent score is low due to the extra steps needed to extract pectin and mucilage. Similarly, the production cost of the biorefinery

**TABLE 14.11**
**Summary of GREENSCOPE Economic-efficiency Evaluation For Biorefinery Process**

| Indicator | Symbol | Value | Unit | Best Case | Worst Case | Score (%) |
|---|---|---|---|---|---|---|
| Present Value Ratio | PVR | 1244.79 | $/$ | 2025.89 | 1.00 | 61.43 |
| Economic Potential | EP | 66.69 | $/kg product | 1.00 | 0.00 | 100.00 |
| Net Return | Rn | 10086.22 | 1 x $10^6$ US $/yr | 2302.04 | 0.00 | 87.54 |
| Net Revenue | REV | 1816.72 | 1 x $10^6$ US $/yr | 491.18 | 0.00 | 99.19 |
| Total Product Cost | TPC | 486.42 | 1 x $10^6$ US $/yr | 225.22 | 944.90 | 66.73 |
| Production Cost | Epc | 312.00 | 1 x $10^6$ US $/yr | 87.53 | 767.10 | 68.24 |
| Total Material Cost | $C_{mat,\ tot}$ | 178.04 | 1 x $10^6$ US $/yr | 42.48 | 389.14 | 62.00 |
| Specific Energy Cost | $C_{E,\ source}$ | 0.01 | US $/ MJ | 0.00 | 0.02 | 51.47 |

describes raw materials, waste treatment, and labor costs associated with operating the biorefinery. This excludes cost of the equipment, which results in a score of 68.2%. The total material cost relates to the absolute cost of materials used in the process, with a value of 178.04 x $10^6$ US $/yr calculated and represents a score of 62%. The cost of energy source indicator evaluates the ratio of utility cost relative to the magnitude of total energy consumed in the biorefinery. This resulted in a score of 51% with a value of 0.01 US $/ MJ per day.

#### 14.3.2.3 Conclusions

Overall, the results for this biorefinery process indicate good performance in the material efficiency category. This is evident in the scores of the selected indicators being greater than 80%, which indicates efficient raw material transformation into valuable products. This implies the choice of raw materials, reaction process, catalysts, and separation process resulted in an efficient and sustainable process in general. Similarly, the indicators evaluated in the energy efficiency category resulted in scores greater than 80%, which shows good performance in this area. This implies an efficient use of utilities including steam, cooling water, and electricity to meet the energy demands of the process. Furthermore, the solvent dependent subprocesses do not require regeneration of the solvent and additional energy, which contributes to the good performance in this category. Next, good performance was obtained in the environmental category as well. Ethanol was the only material labeled as hazardous due to its flammable nature. In this category, the values calculated for many of the indicators were close to or at 100% values. This implies that from the selected indicators this process has limited harmful effects on the environment, humans, and animals. Finally, average performance was obtained for the economic efficiency evaluation; this is mainly due to the use of selected indicators that factor in the high cost of materials, utilities, and equipment. However, indicators such as return on investment, pay-back period, and revenue from eco-products reveal that over time, the revenue generated makes this process economically sustainable. The results presented for the biorefinery process showed the capability of GREENSCOPE as a sustainability assessment tool for novel processes. This enables the sustainability assessment in the four main categories i.e., materials, energy, environmental, and economic for sustainable decision-making and future improvement of new conceptual designs.

## 14.4 DYNAMIC PROCESS CASE STUDY

### 14.4.1 Gasification Process Optimization and Control

#### 14.4.1.1 Introduction

In this section, another application of sustainability evaluation is demonstrated via a dynamic process that requires control and optimization targets. Here, the sustainable process control framework is integrated with a multi-objective optimization method for dynamic process improvement (Li et al. 2018; Li, Ruiz-Mercado, and Lima 2020). Similar ideas have been explored for other processes, and the results of these works show the sustainability decision-making capability of this approach for evaluating different processes (Li et al. 2016; Lima et al. 2016). The objective is to enable a process that meets the desired optimal control setpoints while maintaining specified sustainability indicator constraints. Such sustainability constraints are chosen from economic and/or environmental sustainability indicators available in GREENSCOPE. In the performed analysis, there are three main steps: process and sustainability assessment model construction, multi-objective optimization formulation, and implementation of the advanced control strategy. Typically, a rigorous dynamic/steady-state process model can be developed in commercial platforms such as Aspen HYSYS, and then a derived reduced model is built from it for control and optimization purposes. Appropriate GREENSCOPE indicators will be employed to represent the sustainability model in this work. To maximize the economic performance while minimizing the environmental impact, a non-dominated sorting genetic algorithm-III (NSGA-III) is used to solve such multi-objective

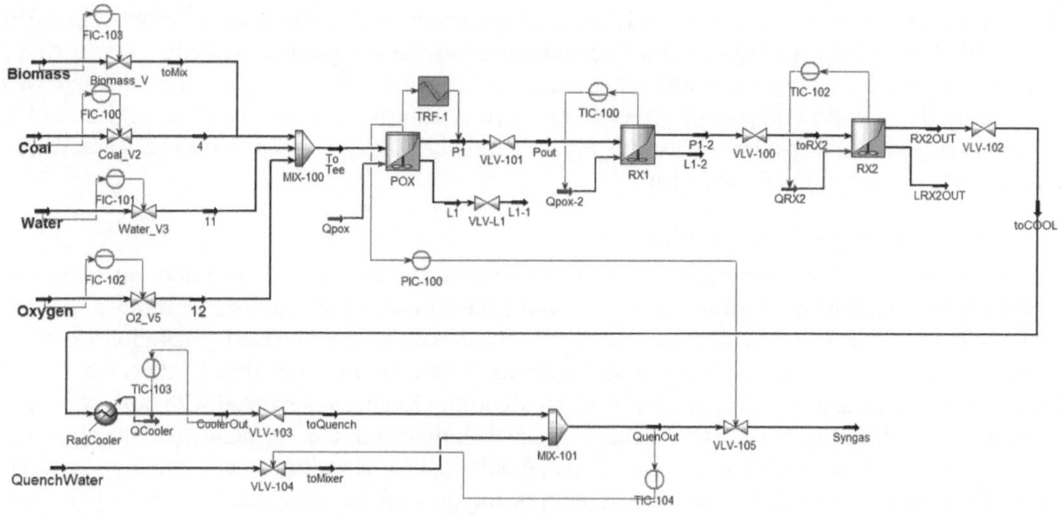


**FIGURE 14.18** Aspen HYSYS flowsheet of gasification process.

optimization (MOO) problem. Lastly, a Model Predictive Control (MPC) approach is implemented to take the system to the selected setpoints determined by the MOO algorithm, while meeting the sustainability constraints (Li 2019). Details about each of the three steps are provided next.

### 14.4.1.2 Gasification Process Model in Aspen HYSYS

Gasification technology is used to convert different forms of hydrocarbons into synthesis gas (syngas) in the presence of water and oxygen. The operating conditions and source of hydrocarbon impact the produced syngas composition and process efficiencies. In this work, for a coal and biomass-based co-gasification process, entrained flow gasifier is modeled based on the General Electric (GE) gasification technology, including partial oxidizer (POX), radiant synthesis gas cooler (RSC), and water quench sections. As shown in Figure 14.18, the Aspen HYSYS flowsheet shows three continuous stirred tank reactors (CSTRs) to approximate a POX process. Equations 14.7–14.12 represent the reaction kinetics of this process.

$$\mathrm{CO} + 0.5\mathrm{O_2} \rightarrow \mathrm{CO_2} \quad r_1 = 1 \times 10^{17.6}\exp(-20130/\mathrm{T})[\mathrm{CO}][\mathrm{H_2O}]^{0.5}[\mathrm{O_2}]^{0.5} \tag{14.7}$$

$$CH_4 + 2O_2 \rightarrow CO_2 + 2H_2O \quad r_2 = 5.44 \times 10^{12}\exp(-24358/T)[\mathrm{CH4}]^{0.3}[\mathrm{O2}]^{1.3} \tag{14.8}$$

$$\mathrm{H_2} + 0.5\mathrm{O_2} \rightarrow \mathrm{H_2O} \quad r3 = 2.85 \times 10^{16}\exp(-20130/T)[\mathrm{H2}]^{0.25}[\mathrm{O2}]^{1.5} \tag{14.9}$$

$$\mathrm{C_{18}H_{20}} + 9\mathrm{O_2} \rightarrow 18\mathrm{CO} + 10\mathrm{H_2} \quad \mathrm{r_4} = 1.0 \times 10^6 \tag{14.10}$$

$$\begin{aligned} \mathrm{CO} + \mathrm{H_2O} \leftrightarrow \mathrm{CO_2} + \mathrm{H_2} \quad \mathrm{r}_{5f} &= 1.612 \times 10^{-6}\exp(-47400/T)P_{CO}\cdot P_{H_2O} \\ \mathrm{r}_{5r} &= 1.224 \times 10^{-3}\exp(-85460/T)P_{CO_2}\cdot P_{H_2} \end{aligned} \tag{14.11}$$

$$\begin{aligned} \mathrm{CO} + 3\mathrm{H_2} \leftrightarrow \mathrm{CH_4} + \mathrm{H_2O} \quad r_{6f} &= 312\exp(-30000/T)C_{CO}(C_{H2})^3 \\ r_{6r} &= 6.09 \times 10^{14}\exp(-257000/T)C_{CH4}\cdot C_{H2O} \end{aligned} \tag{14.12}$$

The RSC heat transfer system is modeled with a heat exchanger. A direct water quench is included and modeled as a direct mixing process. This further cools the gas product while increasing water content. To adequately represent and simulate a typical gasification process, the coal fed was modeled as the pseudo component $C_{18}H_{20}$. The pyrolysis process consists of several complex chemical and physical transformations. The biomass fed goes through this pyrolysis process, which releases volatile matter (VM) and char.

#### 14.4.1.3 Multi-objective Optimization

Genetic algorithms (GAs) can handle multiple objectives simultaneously. The concept of GA is inspired by the mechanism of natural selection, where the current set of solutions (chromosomes) are influenced/updated using crossover and mutation. In the crossover step, the next-generation solutions are updated based on combining two parents' solutions. However, mutation directly alters the current solution to explore the new search spaces. The GA algorithm has been integrated with other selection algorithms to efficiently search for Pareto-optimal solutions for the formulated multi-objective optimization problem. NSGA-II is one of the modified GA algorithms that employ the non-dominated ranking and elite-preservation concepts for passing the diversified solution from one generation to the next (Tavana et al. 2016). In the NSGA-II algorithm, parent and offspring populations are combined to avoid inter-direction competition. Non-dominated sorting is applied to the solution. From the Pareto frontier, solutions pass to the next parent generation. Often, the algorithm will hit a boundary that has more individuals than the remaining spaces in the next population of parents. In this situation, the points of the Pareto frontier are normalized using extreme values of the current population, associated with the reference direction. Finally, to fill the next parent generation, a niching method is performed to obtain solutions attached to underrepresented reference directions. The obtained operating points in the Pareto frontier can be ranked by assigning preference weights for each objective value and then selecting the optimization solutions as the desired conditions.

The implemented NSGA-III calculates the optimal feeding ratio for biomass/coal. This optimal input ratio is obtained with respect to constraints on environmental impact and operating cost objectives, which are defined in equations 14.13–14.16.

$$f_1 = w_1 SI_{GWP} + w_2 SI_{ms,\ spec} + w_3 SI_{VL,\ spec} \tag{14.13}$$

$$f_2 = \text{(Operating Cost)/(Syngas Production Rate)} \tag{14.14}$$

$$\text{s.t. Aspen HYSYS Steady–State Model} \tag{14.15}$$

$$x(t) \in [x(t)^{lb},\ x(t)^{ub}] \tag{14.16}$$

Where $f_1$ and $f_2$ are the two sustainability-related objective functions. In $f_1$, the variables $SI_{GWP}$, $SI_{ms,\ spec}$, and $SI_{VL,\ spec}$ are the sustainability indices associated with global warming potential (GWP), specific solid waste mass ($m_{s,\ spec}$), and specific liquid waste volume ($V_{L,\ spec}$), respectively. The next set of indicators of interest corresponds to reaction yield (RY), economic potential (EP), and specific energy intensity ($R_{SEI}$). These are not included in the optimization algorithm but are calculated and changed because of the optimization. The results of the Pareto frontier are shown in Figure 14.19. This was obtained by solving the MOO problem with a population size of 120 and a generation number of 150. This Pareto trend shows a direct relationship between economic performance and environmental emissions. This implies that improved economic performance results in higher environmental emissions.

The trend can be explained by the fact that coal gasification has higher environmental waste index and lower operating cost index than biomass gasification due to the pretreating of biomass is

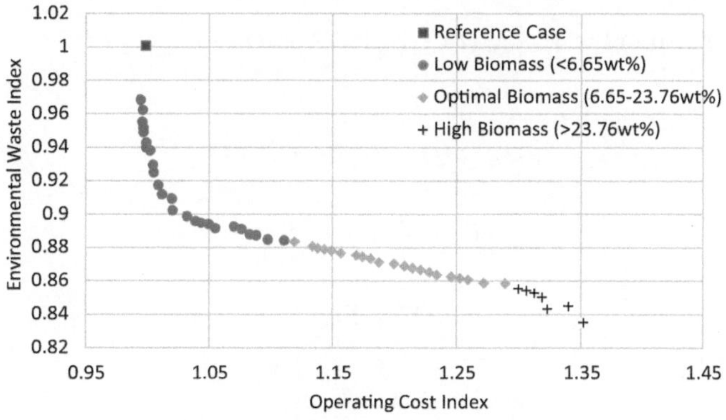


**FIGURE 14.19** Pareto frontier for biomass/coal co-gasification process.

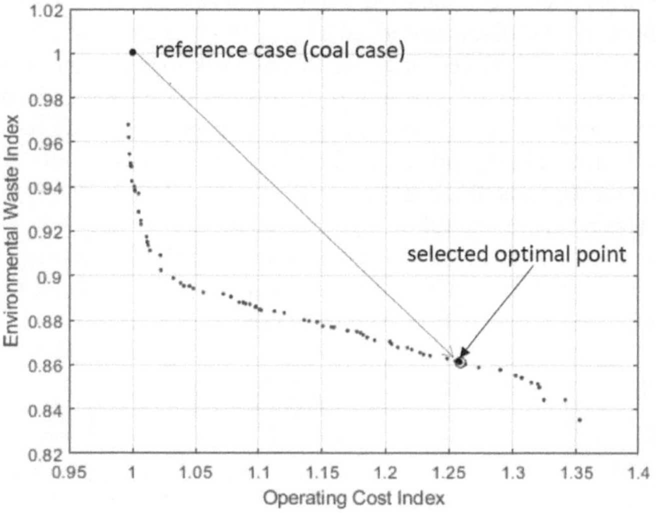


**FIGURE 14.20** Selected operating condition for the MPC controller.

very costly. Such results can be clearly shown in Figure 14.19. For example, adding < 6.65 wt% biomass causes a decrease in the environmental waste index up to 11.5% while operating cost does not increase significantly. A linear relationship was obtained for the environmental waste and operating cost indices from the biomass content in the feed between 6.65~23.76%. In addition, the effect of high biomass content (>23.76%) results in operating cost increases greater than 30%.

For a case study of minimizing the environmental impact, the environmental and operating cost weights are selected as 0.7 and 0.3, respectively. The selected point will be used as the setpoint for the control problem, which will be shown next. Note that different weights for environmental and economic performance would result in different desired operating points for the implemented controller (Fig. 14.20).

### 14.4.1.4 Model Predictive Control Implementation Results

With the chosen optimal operating points, an MPC is formulated with the sustainability indicator index as shown in the following optimization problem:

$$\min_{u^*(t)} J = \int_{\tau_i}^{\tau_s} \left( \|y(\tau) - y_{sp}\|_{w1}^2 + \|u(\tau) - u^-\|_{w2}^2 \right) d\tau \tag{14.17}$$

$$\text{s.t.} \quad \dot{x}(t) = Ax(t) + Bu(t) \tag{14.18}$$

$$y(t) = Cx(t) + Du(t) \tag{14.19}$$

$$SI_i \geq SI_{th} \tag{14.20}$$

$$x(t) \in [x(t)^{lb}, x(t)^{ub}] \tag{14.21}$$

$$u(t) \in [u(t)^{lb}, u(t)^{ub}] \tag{14.22}$$

In which, $u(t)$, $x(t)$, $y(t)$, and $\tau$ are the input, state, output variables, and time, respectively. The optimal input trajectory for this control problem is defined as $u(t)$, which is calculated over the sampling time $t \in [\tau_i, \tau_s]$. The control model is a reduced-order model obtained from the high-fidelity Aspen HYSYS model. This control model is a state-space model that relates the inputs and outputs using several states. The calculated control action depends on the state-space process model, sustainability constraints specified by a sustainability index, and boundary constraints on $u(t)$ and $x(t)$. Based on previous experience in the analysis of this gasification process, a 3 by 3 control strategy was selected for use in the MPC. This results in the following: coal, oxygen, and water flow rates as manipulated variables and syngas production rate, gasifier temperature, and $H_2$/CO ratio as controlled variables. The setpoints for the controlled variables are obtained from the decision-making optimization algorithm. Figure 14.21 shows the controller input and output trajectories results from this study.

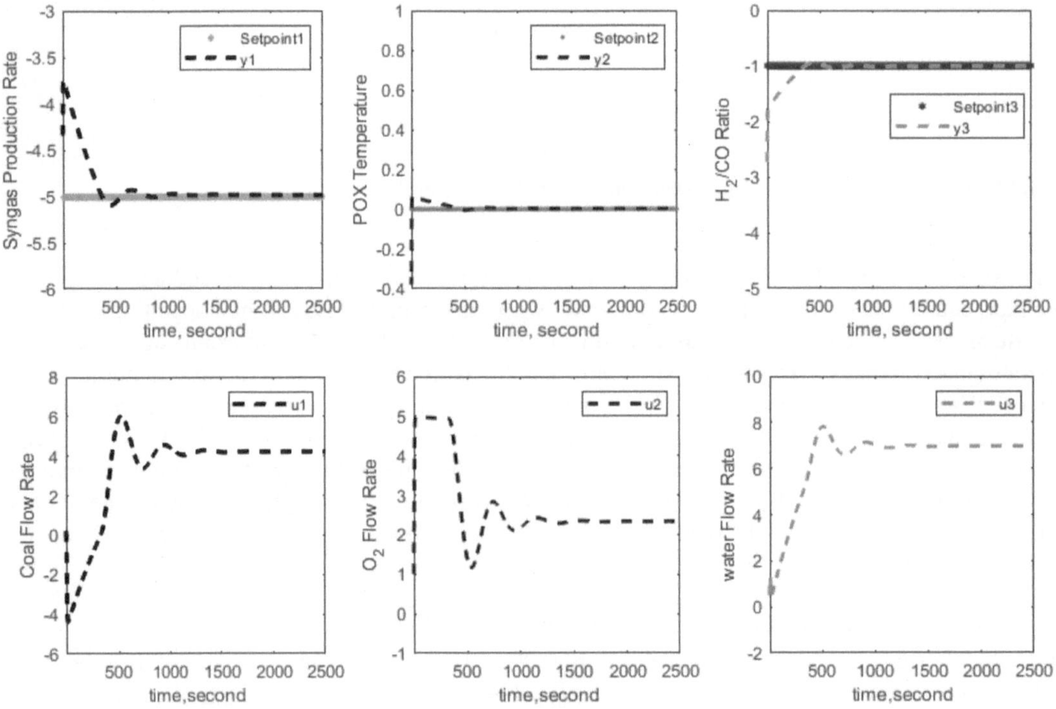


**FIGURE 14.21** MPC results: output (top) and input (bottom) profiles.

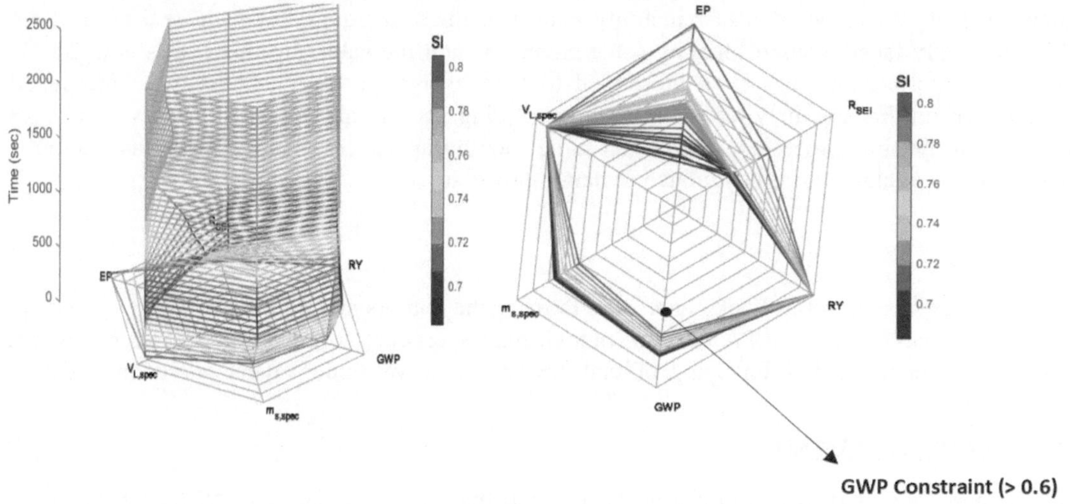


**FIGURE 14.22** Sustainability performance of closed-loop simulation for co-gasification process. Side view of time evolution of sustainability indicators (left), top view of time evolution of sustainability indicators (right).

It can be observed that the controller is able to achieve the selected setpoint while ensuring the transient trajectory meets a selected sustainability constraint as in Figure 14.22. The sustainability constraint selected here was global warming potential sustainability score > 60% (0.6).

From the results in Figure 14.21 and Figure 14.22, this combined algorithm enables achieving desired setpoints relative to sustainability objectives. The evolution of the sustainability indicators in Figure 14.22 shows the improvement in sustainability indicators over time. From this result, although the initial sustainability performance for economic potential (EP) performance was very good, the time evolution shows a dynamic trade-off especially where economic potential is penalized to initially decrease sustainability performance between 500s–1,000s. However, after a steady state (>1,500s), GWP and $V_{L,\ spec}$ show the most improvement.

### 14.4.1.5 Conclusions

This section showed the use of a sustainability-oriented control strategy with a multi-objective optimization. Specifically, A NSGA-II algorithm was employed to solve the multi-objective optimization solution that considers conflicting objective functions. Then an optimal operating condition can be selected by assigning weights to different objective values. The selected point is then sent to an MPC controller to take the system to sustainable operating conditions while satisfying predefined sustainable criteria. The effectiveness of the proposed framework was demonstrated with the co-gasification process, a complex process with known trade-offs between economic and sustainability objectives.

## 14.5 CHALLENGES, CONCLUSIONS, AND FUTURE WORK

The results showed in this chapter, for the acetic acid, biorefinery, and coal/biomass co-gasification production processes, demonstrate the capability of GREENSCOPE in evaluating the sustainability performance of steady-state, dynamic, conventional, and novel processes. Such analysis provides insight into an approach for comprehensive plant-wide performance evaluation in terms of raw material and energy usage, the impact of the process on human health and the environment, in addition to its economic performance. This evaluation shows specific aspects of a process that can be improved and provides metrics that enhance decision-making in terms of sustainability. The main challenge encountered in sustainability analysis is the data required to evaluate the indicators. This is usually faced when evaluating some environmental indicators; to this end, a new GREENSCOPE interface that has been developed to help overcome this challenge was introduced. Future work in this area may focus further on exploring appropriate sets of indicators for already operational dynamic processes, in addition to developing approaches that improve integration of sustainability evaluation into simulated dynamic processes.

## Disclaimer

The views expressed in this book chapter are those of the authors and do not necessarily reflect the views or policies of the U.S. EPA. Mention of trade names, products, or services does not convey, and should not be interpreted as conveying, official U.S. EPA approval, endorsement, or recommendation.

## ACKNOWLEDGEMENT

The authors thank Krishna Murthy Busam for contributions to the narrative of this work.

## REFERENCES

Chemstations. 2017. CHEMCAD 7.

Clauser, N.M., F.E. Felissia, M.C. Area, and M.E. Vallejos. 2021. A Framework for the Design and Analysis of Integrated Multi-Product Biorefineries from Agricultural and Forestry Wastes. *Renewable and Sustainable Energy Reviews* 139, no. July 2020.

Homer, I., M.T. Varnero, and C. Bedregal. 2020. Nopal (Opuntia Ficus-Indica) Energetic Potential Cultivated in Arid and Semi-Arid Zones of Chile: An Assessment. *Idesia* 38, no. 2: 119–127.

Katakojwala, R., and V.S. Mohan. 2021. A Critical View on the Environmental Sustainability of Biorefinery Systems. *Current Opinion in Green and Sustainable Chemistry* 27: 100392.

Li, S. 2019. Development of a Sustainability Evaluation and Control Framework for Chemical Processes. *Graduate Theses, Dissertations, and Problem Reports* (January 1). https://researchrepository.wvu.edu/etd/7382

Li, S., Y. Feliachi, S. Agbleze, G.J. Ruiz-Mercado, R.L. Smith, D.E. Meyer, M.A. Gonzalez, and F. V. Lima. 2018. A Process Systems Framework for Rapid Generation of Life Cycle Inventories for Pollution Control and Sustainability Evaluation. *Clean Technologies and Environmental Policy* 20, no. 7 (September 1): 1543–1561. https://pubmed.ncbi.nlm.nih.gov/30245612/

Li, S., G. Mirlekar, G. J. Ruiz-Mercado, and F. V. Lima. 2016. Development of Chemical Process Design and Control for Sustainability. *Processes* 4, no. 3: 23. https://doaj.org/article/6437f256b93b4bdb80455a7964bfa42a

Li, S., G.J. Ruiz-Mercado, and F. V. Lima. 2020. A Visualization and Control Strategy for Dynamic Sustainability of Chemical Processes. *Processes 2020, Vol. 8, Page 310*, no. 3 (March 7): 310. https://www.mdpi.com/2227-9717/8/3/310/htm

Lima, F. V., S. Li, G. V. Mirlekar, L.N. Sridhar, and G. Ruiz-Mercado. 2016. Modeling and Advanced Control for Sustainable Process Systems. In *Sustainability in the Design, Synthesis and Analysis of Chemical Engineering Processes*, G. Ruiz-Mercado and H. Cabezas (eds.), Elsevier: 115–139.

Pal, P., and J. Nayak. 2017. Acetic Acid Production and Purification: Critical Review Towards Process Intensification. *Separation & Purification Reviews* 46, no. 1: 44–61. http://www.tandfonline.com/doi/abs/10.1080/15422119.2016.1185017

Ruiz-Mercado, G.J., A. Carvalho, and H. Cabezas. 2016. Using Green Chemistry and Engineering Principles to Design, Assess, and Retrofit Chemical Processes for Sustainability. *ACS Sustainable Chemistry and Engineering* 4, no. 11 (November 7): 6208–6221. https://pubs.acs.org/doi/full/10.1021/acssuschemeng.6b02200

Ruiz-Mercado, G.J., M.A. Gonzalez, and R.L. Smith. 2013. Sustainability Indicators for Chemical Processes: III. Biodiesel Case Study. *Industrial and Engineering Chemistry Research* 52, no. 20 (May 22): 6747–6760. https://pubs.acs.org/doi/full/10.1021/ie302804x

Ruiz-Mercado, G.J., M.A. Gonzalez, and R.L. Smith. 2014. Expanding GREENSCOPE beyond the Gate: A Green Chemistry and Life Cycle Perspective. *Clean Technologies and Environmental Policy* 16, no. 4: 703–717.

Ruiz-Mercado, G.J., R.L. Smith, and M.A. Gonzalez. 2012a. Sustainability Indicators for Chemical Processes: I. Taxonomy. *Ind. Eng. Chem. Res* 51: 2309–2328. https://pubs.acs.org/doi/10.1021/ie102116e

Ruiz-Mercado, G.J., R.L. Smith, and M.A. Gonzalez. 2012b. Sustainability Indicators for Chemical Processes: II. Data Needs. *Industrial & Engineering Chemistry Research* 51, no. 5: 2329–2353. https://pubs.acs.org/doi/full/10.1021/ie200755k

Smith, R.L., G. Ruiz-Mercado, D.E. Meyer, M.A. Gonzalez, J.P. Abraham, W.M. Barrett, and P.M. Randall. 2017. Coupling Computer-Aided Process Simulation and Estimations of Emissions and Land Use for Rapid Life Cycle Inventory Modeling. *ACS Sustainable Chemistry & Engineering* 5, no. 5: 3786–3794. 10.1021/acssuschemeng.6b02724

Smith, R.L., G.J. Ruiz-Mercado, and M.A. Gonzalez. 2015. Using GREENSCOPE Indicators for Sustainable Computer-Aided Process Evaluation and Design. *Computers and Chemical Engineering* 81 (October 5): 272–277.

Tavana, M., Z. Li, M. Mobin, M. Komaki, and E. Teymourian. 2016. Multi-Objective Control Chart Design Optimization Using NSGA-III and MOPSO Enhanced with DEA and TOPSIS. *Expert Systems with Applications* 50 (May 15): 17–39.

Turton, R., R.C. Baille, J.A. Shaeiwitz, D. Bhattacharyya, and W. Whiting. 2018. CAPCOST.

Ubando, A.T., A.J.R. Del Rosario, W.H. Chen, and A.B. Culaba. 2021. A State-of-the-Art Review of Biowaste Biorefinery. *Environmental Pollution* 269: 116149.

Yoneda, N., S. Kusano, M. Yasui, P. Pujado, and S. Wilcher. 2001. Recent Advances in Processes and Catalysts for the Production of Acetic Acid. *Applied Catalysis A: General* 221, no. 1–2 (November): 253–265. http://linkinghub.elsevier.com/retrieve/pii/S0926860X01008006

# 15 Germany's Industrial Climate Transformation Strategy and the Role of Carbon Capture and Utilization as a Building Block
## *Targets, Pathways, Policies, and Societal Acceptance*

*Sonja Thielges and Kristina Fürst*[1]
Research Institute for Sustainability – Helmholtz Centre Potsdam, Germany

**CONTENTS**



## 15.1 CLIMATE PROTECTION IN GERMANY: GOALS AND PLANNED PATHWAYS FOR 2030 AND 2050

Germany has traditionally been regarded as an international climate frontrunner with its "Energiewende" (energy transition). In 2021, Germany achieved a reduction in greenhouse gas

DOI: 10.1201/9781003167693-15

emissions (GHG) of 39 percent below 1990 levels – from 1,249 million tons of $CO_2$ equivalent (million t $CO_{2e}$) in 1990 to 762 million of $CO_{2e}$ in 2021. The country has assumed a leadership role in global climate protection efforts by setting ambitious climate targets: The German Climate Law of 2021, updated from its previous, less ambitious version of 2019 after a court ruling, sets the target of reducing GHG emissions by 65% below 1990 levels by the year 2030. For 2045, the target is climate neutrality (Die Bundesregierung, 2022).

The German energy transition and the country's climate targets enjoy broad support from the public and across party lines. Large majorities of the population support the government-set climate targets and individual measures, such as binding requirements for solar PV panels on buildings or increased investments in railway infrastructure (Wolf et al., 2022). This broad acceptance of ambitious climate targets and policies, however, has its limitations when it comes to different technologies. Germany has historically had a strong anti-nuclear movement, and after the Fukushima nuclear disaster in 2011, the German government confirmed its plans to phase out all nuclear generation in Germany by the year 2022 (Appunn, 2021). Following the energy crisis set on by Russia's invasion of Ukraine in February 2022, the German government postponed the final phase-out date for the remaining three operating nuclear power plants to April 2023. Biofuels for the transport sector is another example of an approach that is met with considerable skepticism from different stakeholders in Germany. Civil society groups, as well as think tanks, underline their questionable sustainability and climate mitigation potential due to the associated land-use changes in Germany and abroad (World Wildlife Fund, n.d). Another example from the late 2000s is carbon capture and storage (CCS). Through CCS, $CO_2$ is captured from industrial flue gas (point source $CO_2$ capture). This approach drew heavy protests from local actors who organized to form protest groups and launch online campaigns. The protesters framed CCS as a means to prolong coal-based electricity generation in Germany. The government responded by abandoning its search for appropriate storage sites and passing a national law on CCS (KSpG) in 2012. CCS on an industrial scale was practically banned for use in all sectors (Otto et al., 2022).

Despite these limitations with regard to technology-openness in Germany, the country has been pursuing its climate targets with renewed ambition since the inauguration of the current coalition government between the Social Democratic Party (SPD), the Greens, and the Liberal Party (FDP) in 2021. This ambition is framed by the targets and strategies of the European Union (EU). Since 2005, an EU-wide greenhouse gas emissions trading system (EU ETS) has been in operation. The ETS covers the energy sector and energy-intensive industries, as well as aviation. Its GHG emissions cap has declined over the years, and carbon prices increased to above 80 euros per ton of $CO_2$ in 2022. In 2019, the EU Commission introduced the European Green Deal, a strategy to achieve climate neutrality. Following that, in 2021, the EU passed the European Climate Law, which raises the EU's climate target for the year 2030 to a 55% emission reduction compared to 1990. By 2050, the EU aims to achieve climate neutrality. With its "Fit-for-55-Package", the EU Commission introduced various legislative proposals to reach these goals. This includes, for instance, the inclusion of transport and heating into the ETS, an increase in renewable energy targets, and measures to prevent companies from moving to countries with lower environmental standards (carbon leakage) through a so-called Carbon Border Adjustment Mechanism (CBAM), which would put a tax on goods imported into the EU from countries with less stringent climate policies (Council of the European Union, 2022). As a response to the Russian invasion of Ukraine, the EU Commission further introduced its REPowerEUPlan. It aims to accelerate the EU's energy transition and rapidly reduce dependence on Russian fossil fuel imports. Among the proposed measures are increases of the current EU renewable energy target from 40 to 45% by 2030 and the energy efficiency target from 9 to 13% below 2020 levels, as well as increased uptake of green hydrogen, biogas, and biomethane in the industry (European Commission, 2022).

These EU targets set minimum requirements for emission reductions in all sectors. With its goal of achieving climate neutrality by 2045, however, Germany exceeds the ambitions of the overall targets of the EU. In order to achieve these national climate targets, the German climate law assigns responsibilities to different economic sectors by setting binding emission thresholds for the year 2030. The two biggest emitters in Germany are the energy sector (30% of total GHG emissions) and the industrial sector (24%), followed by transport (19%) and buildings (16%) (Bundesministerium fur Umwelt, Naturschutz und nukleare Sicherheit, 2021). The climate law's goal for the energy sector is a reduction of 77% below 1990 levels by 2030. The industrial sector is required to reduce its $CO_2$ emissions by 58%, buildings by 68%, and transport by 48% (Umweltbundesamt, 2022a).

The German energy sector has seen the greatest energy transition success so far. Emissions dropped to 47% below 1990 levels in 2021. This is mainly due to a fuel switch from coal to renewable energy and natural gas in electricity production (Agora Energiewende, 2021). Germany has set the binding target of phasing out coal by 2038 at the latest. Current plans by the government aim for a complete phase-out by 2030 (Schiffer et al., 2022).

While decarbonization keeps unfolding in the energy sector, emission reductions in the industry sector have been considerably less dynamic. A large share of its roughly 37% emission reductions which were achieved by 2020 was realized as a consequence of the German reunification in the 1990s when many emission-intensive factories were shut down in the East German states (Umweltbundesamt, 2020). A 2021 foresight report released by the German government shows that the industrial sector is currently not on track towards reaching its 2030 emission threshold (Bundesministerium fur Umwelt, Naturschutz und nukleare Sicherheit, 2021). As the industry is the second-biggest source of emissions after the energy sector, this lacking dynamic has important implications for the achievement of Germany's climate targets.

To enhance the mitigation efforts of the industrial sector, the German government pursues an industrial decarbonization strategy that consists of several different building blocks. This chapter introduces these building blocks. In a case study, it discusses the role of one building block, in particular: Carbon Capture and Utilization (CCU). CCU is a technological approach that has received only little attention in public debates on climate protection in Germany thus far. Yet, CCU is highlighted in government strategy as a means to address hard-to-abate emissions in industrial processes such as cement production.

## 15.2 THE BUILDING BLOCKS OF GERMANY'S INDUSTRIAL DECARBONIZATION STRATEGY

Germany's industrial decarbonization strategy aims at putting Germany on track towards achieving climate neutrality as well as at positioning the country internationally as a green industrial leader. It also seeks to make a contribution to reducing the dependence on imports of natural gas and oil, particularly from Russia, given the current political dynamics as a consequence of Russia's invasion of Ukraine in the spring of 2022. The building blocks include: 1) The further expansion of renewable energy; 2) electrification and energy efficiency improvements; 3) green hydrogen; 4) moving towards a circular economy; and 5) CCS and CCU.

The building blocks, on the one hand, address energy-related GHG emissions caused by the combustion of fuels, for instance, for electricity or heat generation. On the other hand, they tackle process-related emissions that are not related to energy use.

### 15.2.1 Building Block 1: Expansion of Renewable Energy for Electricity Production

The use of renewable energy for electricity generation as well as heating and cooling is a means to tackle energy-related industrial emissions and provides an important foundation for the other building blocks of the industrial decarbonization strategy. It is therefore, arguably, the central element of industrial transformation and benefits from a range of different support schemes.

Germany has, in fact, already made progress in reducing its dependence on conventional fuels in the power sector. In 2021, renewable energy, mainly wind and solar, had a share of 39.7 percent of Germany's electricity mix (AG Energiebilanzen, 2022). Germany is one of only a few countries lacking abundant hydropower resources that have such a high share of renewables in their generation mix (Schiffer et al., 2022). Capacity additions, mostly from solar and wind, for power generation increased more than sixfold between 2000 and 2021 (Federal Ministry for Economic Affairs and Climate Action, 2022).

Renewable energy capacity development has been strongly driven by government programs in the past. One major governance scheme was a surcharge under the Renewable Energy Law (Erneuerbare Energien Gesetz, EEG), which expired in 2022. Through this EEG surcharge, electricity customers supported renewable energy expansion. Energy-intensive industries were subject to a reduced surcharge. In addition, there has been a priority feed-in for electricity from renewable energies.

The German government enacted its so-called Easter Package for the Expansion of Renewable Energy ("Osterpaket") in July of 2022 as a response to the current energy crisis in Germany. For instance, its solar acceleration package includes solar panel mandates for all new commercial premises. The goal is to expand photovoltaics to some 200 GW by 2030. In order to facilitate additional wind energy capacity, the federal government has designated 2% of German land for the construction of wind turbines (Deutscher Bundestag, 2022). Onshore wind power capacity increases, in particular, have been slowed down in recent years due to local protests to the construction of new windmills, long permitting processes and construction times, and the lack of sites designated for wind parks (Lintz and Liebenath, 2020). One of the responses of the federal government in the Easter Package is to improve the financial participation of municipalities in onshore wind and photovoltaics. Offshore, the goal is to build up at least 30 GW of wind energy capacity by 2030. In addition, the government is currently seeking to simplify and streamline renewable energy permitting to speed up capacity additions (Deutscher Bundestag, 2022; Harmsen, 2022).

The government places a heavy focus on wind and solar energy in its efforts. In the Easter Package, biomass is given a much more limited role with a goal of achieving an installed capacity of 8400 MW by 2030. It is mostly to be used for peak load power plants. The sustainability of biomass has long been questioned in German public debates, for instance, in the context of the above-mentioned debates around how the limited amounts of arable lands should be used in Germany ("food vs. fuel" debate) (Umweltbundesamt, 2022b). More recently, the EU's biomass imports have become subject to criticism, as well, with biomass sourced from trees, eliminating its emission mitigation benefits (Hurtes & Cai, 2022).

### 15.2.2 Building Block 2: Electrification and Energy Efficiency Improvements

Electrification poses one approach to the reduction of process-related industrial emissions—if it is based on the above-discussed green transformation of the electricity mix (Bundesministerium für Umwelt, Naturschutz und nukleare Sicherheit, 2021). Electrification can unfold its potential in electricity-based heat and industrial steam generation, the application of high-temperature heat pumps, as well as the electrification of industrial production processes. In addition to electrification in industry, the electrification of heat generation and transport in Germany is expected to cause a significant increase in electricity demand in Germany. Out of the 750 terra watt hours (TWh) of electricity demand calculated for 2030, the government aims for 600 TWh of electricity from renewable energy. For electrification processes, simultaneous energy efficiency improvements are therefore of particular importance in tackling energy-related industrial emissions and ensuring the efficient use of scarce renewable energy. Such improvements include the use of more energy-efficient technologies, the improvement of production processes, and the use of heat from industrial processes (Bundesministerium fur Umwelt, Naturschutz und nukleare Sicherheit, 2021).

Such demand increases will need to be met with an expansion of electricity transmission infrastructure. The electricity grid will need to transmit larger amounts of electricity from the production sites of renewable energy, which are mostly located in northeastern Germany, a scarcely populated region with favorable weather conditions for wind energy production, to industrial clusters in the densely populated south and west. This necessitates the expansion of transmission lines as well as interconnectors or smart grid applications to avoid line congestion in times of high energy demand or volatile renewable energy supply (Federal Ministry for Economic Affairs and Climate Action, n.d.). Grid expansion has already begun and has, once again, been met with local protests (e.g., Bundesverband Bürgerinitiativen gegen Suedlink, 2022) criticizing, for instance, changes in landscape caused by underground cables (Mieritz, 2019).

### 15.2.3 Building Block 3: Establishment of a Green Hydrogen Economy

In 2020, the German government issued its first hydrogen strategy. It places a focus on the production and import of green hydrogen, produced through electrolysis from renewable energy. Blue hydrogen, produced from natural gas in conjunction with CCS through steam methane reforming, is given only the role of a bridge solution. Green hydrogen has immediate potential for the reduction of process-related industrial emissions. In the steel industry, in particular, a switch to direct reduction of iron ore through green hydrogen can avoid $CO_2$ emissions that occur through the usual production in a blast furnace. As the steel industry is one of the biggest emitters in Germany, this is a highly relevant pathway for Germany's industrial transformation (Federal Ministry for Economic Affairs and Energy, 2020).

Green hydrogen, moreover, offers a partial solution to some of the challenges of renewables expansion in Germany and Europe as it can store renewable energy at times of over-supply. At the same time, the production of green hydrogen is an energy-intensive process which requires significant amounts of renewable energy.

Germany's plans for a green hydrogen economy are still recent. Globally, green hydrogen is not available at scale yet, and due to limited potentials for renewable energy, Germany will not become a major green hydrogen producer itself. Efforts have therefore focused on setting up partnerships with potential exporters of green hydrogen such as, the Canada-Germany Hydrogen Alliance (German Missions in Canada, 2022).

### 15.2.4 Building Block 4: A Circular Economy

Moving from the existing linear to a circular economy is a further building block of the German industrial transformation strategy. While the linear economy disposes of products and materials after use, thus exploiting raw materials and fossil fuels, a circular economy takes into account the limitations of raw material availability. The circular economy approach seeks to limit new natural resource extraction by keeping products and materials in use as long as possible and by recycling and reusing materials (Tan & Lamers, 2021). The German government underlines the potentials of the circular economy approach to reduce process-related industrial emissions. The goal is to reduce the need for new primary materials by improving recycling. Reusing materials, accordingly, also uses far less energy than using new materials (Bundesministerium fur Umwelt, Naturschutz und nukleare Sicherheit, 2021). Assuming that only renewable energy is used, a circular economy approach, therefore, also has the potential to reduce energy-related industrial emissions. So far, recycling for high-emission materials, including cement, metals, and plastics, is limited. Increasing the rate of recycling here would, for instance, require new product design to facilitate recycling. If successful, circularity could significantly reduce demand for new steel, aluminum, and polymers and achieve significant emission reductions (Agora Energiewende, 2018).

### 15.2.5 Building Block 5: Carbon Capture and Storage (CCS) and Carbon Capture and Utilization (CCU)

For industries in which other technological approaches, such as electrification or the use of hydrogen, cannot eliminate GHG emissions, the German government envisions CCS and CCU as a solution. It particularly sees both approaches as a solution for the cement industry with its hard-to-abate process-related $CO_2$ emissions (Bundesministerium fur Umwelt, Naturschutz und nukleare Sicherheit, 2021). In contrast to CCU, CCS, as noted above, is a contested topic in Germany – in broader society but also among political institutions. The federal Environmental Agency (Umweltbundesamt, UBA) acknowledges CCU as a measure of last resort that might be applied if changes in production processes or the substitution of emission-intensive raw materials are not possible (Umweltbundesamt, 2020). UBA has declared CCS, however, a potential disincentive for innovations in climate-friendly technologies and questioned its necessity on the road toward industrial climate neutrality. This negative assessment of CCS, in turn, contrasts with studies from industrial actors that regard CCS as a means to reduce some 50% of current industrial emissions in various industries (Agora Energiewende, 2018).

Although the German government has declared both CCU and CCS as an element of the German industrial decarbonization strategy, their role and interaction in the German industrial transformation towards climate neutrality remain unclear, as this brief discussion suggests. CCU so far remains a niche approach with less negative connotations than CCS. Therefore, CCU provides an interesting case study as one of the building blocks of the German strategy with an unclear potential.

## 15.3 DEEP DIVE CCU: A NICHE TOPIC IN GERMANY'S INDUSTRIAL DECARBONIZATION STRATEGY?

As the German industrial decarbonization strategy implies, a multifaceted approach will be necessary to achieve the 2030 targets and the 2045 net-zero targets for the industrial sector. A major challenge is that for many emission-intensive industries, $CO_2$ emissions will not be entirely avoided by electrifying, switching to renewable energy sources, and increasing energy efficiency. For instance, the chemical industry relies on carbon as a building block in produced chemicals which is transformed into $CO_2$ emissions at the end of a chemical product's life cycle. In the cement industry, $CO_2$ is released in the production of limestone, which provides a challenge for emission reductions (Hodgson & Hugues, 2022). Ultimately the aviation and maritime transportation sectors might have to rely on using carbon-based fuels due to their high energy density which are transformed into $CO_2$ emissions during their combustion. For these so-called hard-to-abate sectors, three technological approaches exist that allow the reduction of the remaining $CO_2$ emissions. (1) CCS; (2) direct air carbon capture and storage (DACCS): $CO_2$ is captured from the ambient air and stored underground; and (3) CCU: the captured $CO_2$ is being used as an (industrial) resource (Fajardy, 2022).

Different approaches and challenges exist for both capturing and utilizing $CO_2$. Among the capture technologies, DAC facilities struggle with efficiency challenges: $CO_2$ concentration in the ambient air is low (around 0.04%), and their energy consumption during operation is high. Capturing $CO_2$ from industrial flue gas (i.e., point source $CO_2$) is often regarded as more efficient because the $CO_2$ concentration is higher. However, here a challenge exists in the varying degree of purity of the $CO_2$, which is dependent on the industry itself and the capture process (e.g., pre- versus post-combustion capture) (Goto et al., 2013). With regard to utilization, $CO_2$ can be used physically (directly), for example, as a fire extinguishing medium or as an additive in the food and soft drink industry. Direct use of $CO_2$ is not new, but in these traditional applications, mostly natural $CO_2$ sources are being used. The most common understanding of CCU is, however, the indirect use of $CO_2$. This entails $CO_2$ capture from industrial fumes or the ambient air with the motivation to reduce $CO_2$ emissions. The recycled carbon is then used, for example, to produce methanol as a fuel or to produce polymers as a building block for secondary industries (Olfe-Kräutlein, 2022).

### 15.3.1 CCS and CCU as $CO_2$ Mitigation Strategies

Industrial processes with CCS or CCU mitigate $CO_2$ emissions. Point source $CO_2$ capture usually does not achieve carbon neutrality as the capture rate is limited to 90 to 95% of emissions, and the capture process itself has energy-related emissions. Carbon neutrality could only be achieved if all the emissions from the point sources and associated emissions of the CCU/CCS process were stored at 100%. This can be illustrated by the example of synthetic fuel produced with $CO_2$. During fuel combustion, temporarily stored $CO_2$ is re-emitted into the atmosphere. Only if $CO_2$ is captured from the ambient air and all processes ($CO_2$ capture, transportation, and transformation) are powered by renewable energies and do not emit $CO_2$ can the resulting synthetic fuel be $CO_2$-neutral. If the $CO_2$ source is of fossil origin (e.g., point source capture from a cement plant), this process will recycle fossil $CO_2$ emissions for one more use before releasing $CO_2$ through combustion, and hence mitigate $CO_2$ emissions, but not neutralize them (Zimmermann et al., 2020).

Consequently, for most CCU/CCS processes, additional carbon dioxide removal (CDR) measures will be necessary to bring emissions down to zero. CDR technologies not only neutralize $CO_2$ emissions but aim at producing net negative $CO_2$ emissions. With the CDR processes, $CO_2$ is captured from the ambient air (or the ocean) and stored permanently. Methods of CDR include, for instance, DACCS and enhanced weathering, which distributes $CO_2$ -consuming minerals onto land or sea at a large scale (Bach et al., 2019).

### 15.3.2 CCU in Germany and the EU: A Brief History

CCU is still a niche topic in Germany's industrial decarbonization approach, with a comprehensive CCU strategy lacking on the national level. CCU-specific policies exist mainly on the Research and Development (R&D) level, whereas questions about its large-scale implementation remain unspecified. Instead, regulations on the implementation of CCU technologies can be found in legal documents of related sectors that are applicable in the context of CCU. Such legal sources are, for instance, documents governing CCS (e.g., the Federal Immission Control Act, BlmSchV; or the federal law on $CO_2$-storage, KPsG), circular economy (e.g., the federal circular economy law, KrWG), or the transboundary transportation of $CO_2$ (e.g., the London Protocol).

In the 2010s, the Federal Ministry of Education and Research (BMBF) launched a large research framework program called FONA (Forschung für Nachhaltigkeit)—Research for sustainable development—as an executional measure for the New German High Tech Strategy. Most financing measures for CCU originate in the FONA research framework program. Moreover, Germany initiated a key research program for CCU: "Technologies for Sustainability and Climate Protection: Chemical Processes and $CO_2$ Use". Between 2010 and 2016, €100 million were allocated for 33 collaborative research and development initiatives, totaling 150 individual projects. Industrial partners contributed an additional €50 million to the research initiatives. This constituted one of the world's largest, specifically focused federal CCU support programs (Mennicken et al., 2016).

In the EU, as in Germany, there is no explicit strategy for CCU. Instead, CCU is supported through different policy instruments and funding schemes, many of which are undergoing change through the current reform proposals under "Fit for 55". Among the reform proposals are tax credits for the production of synthetic fuels for air and maritime transport as part of the Energy Taxation Directive. Moreover, while CCU was largely excluded from the EU ETS, the current reform proposals include CCU if it binds $CO_2$ in a product permanently. This could, for instance, incentivize plastics or mineralization processes in the cement industry. The reform proposals for the Renewable Energy Directive (RED II) also, in principle, incentivize $CO_2$-based fuels but at the same time raise concerns with regards to the efficiency of using limited green electricity to produce these "Renewable Fuels of Non-Biological Origin" (RFNBOs) (Thielges et al., 2022; Fajardy, 2022).

Aside from these reform proposals, the EU already provides infrastructure support for CCU through the Connecting Europe Facility, which funds $CO_2$ transportation infrastructure. CCU has, moreover, received significant R&D funding in the past through the Horizon 2020 funding scheme and its successor Horizon Europe. Moreover, funding for CCU(S) is one of the focal themes under the EU Innovation Fund, which currently funds several CCU(S) projects (Thielges et al., 2022).

### 15.3.3 $CO_2$ Capture: The Regulatory Framework in Germany

In general, German law allows the capture, inland transportation, and use of $CO_2$ despite the lack of specifically targeted CCU laws (Weber, 2022). The construction and operation of $CO_2$-capture facilities are governed by the Federal Immission Control Act (BlmSchV). The law requires public consultation for any $CO_2$-capture facility to be approved: Companies need to declare their ambitions with the planned CCU facility, display related documents, and give the public a chance to raise objections (Weber, 2022; § 10 para. 3 BlmSchG, in BGBl, 2022; Ministerim für Umwelt, Naturschutz und Verkehr des Landes Nordrhein-Westphalen, n.d.).

The federal law on CCS (KSpG) covers $CO_2$ capture and inland transportation for CCS purposes. However, it limits the amount of $CO_2$ that can be stored within Germany and allows storage for research purposes only (§2 para. 1 KSpG, in BGBl, 2012). Critics have also called the KSpG a "Non-CCS Act": The stated application deadline in the KSpG, to receive an allowance to build a CCS facility on German territory expired on December 31, 2016. The KSpG further allows the German federal states to prohibit CCS in their state territory altogether (Weber, 2022). Consequently, as of now, CCS is not possible on German soil. A revision of the KSpG is expected by the end of 2022 and will serve as a basis from which the German government aims to develop a CCU/CCS strategy (Tagesspiegel Background, 2022).

To govern CCU and CCS activities, many regions around the world have been developing so called carbon management strategies. In Germany, the state government of North Rhine-Westphalia (NRW), a region with a long history of industrial activity, was the first to adopt a comprehensive carbon management strategy at the subnational level. The NRW Carbon Management Strategy provides guidelines on how to achieve $CO_2$-neutrality until 2045 and implement full carbon circularity in the industrial sector: "*The goal is to cycle the carbon already present in the economic system to avoid the input of additional fossil carbon and the generation of $CO_2$*". The strategy also acknowledges that CCU will not always be the most energy- or $CO_2$-effective solution. For the estimated 31 million metric tons of surplus $CO_2$ emitted annually by the German industry sector, subtracting all decarbonization and CCU potential, the NRW Carbon Management Strategy envisages exterritorial CCS. For sub-seabed CCS, the strategy plans the implementation of regional and international $CO_2$ infrastructure. As a first step, the NRW Carbon Management Strategy asks the federal government to set the legal prerequisites for such operation (Ministerium für Wirtschaft, Innovation, Digitalisierung und Energie des Landes Nordrhein-Westfalen, 2022; Prognos, 2021).

### 15.3.4 $CO_2$ Infrastructure

Transporting $CO_2$ from its point of capture to where it will be utilized requires the construction of $CO_2$ infrastructure. At the German national level, the implementation and operation of $CO_2$ networks are governed by the national Carbon Capture and Storage law (Kohlenstoffdioxid-Speicherungsgesetz, KPsG), in conjunction with the Energy Industry Act (Energiewirtschaftsgesetz, EnWG) and the Environmental Impacts Assessment Act (Gesetz über die Umweltverträglichkeitsprüfung, UVPG). Depending on the mode of transport, the general pipeline law (Allgemeines Leitungsrecht) or the

Hazardous Goods Transport Act (Gefahrengutbeförderungsgesetze), for the transportation of $CO_2$ via ship or rail, become relevant, as well. Further regulatory requirements arise depending on what the $CO_2$ is finally (re)used for (Weber, 2022).

One existing regulation gap concerns the export of $CO_2$ from Germany's industry sites to its envisaged extraterritorial storage sites in other European countries like Great Britain, the Netherlands, or Norway. The transboundary transportation of $CO_2$, for the purpose of sub-seabed storage, is governed under the 1996 *Protocol to the London Convention on the Prevention of Marine Pollution by Dumping of Wastes and Other Matter*, which was itself adopted in 1972. While the original text of the London Protocol (also known as the London Convention) does not cover the transboundary export of $CO_2$, the text was amended in 2009 to include international transportation of $CO_2$. However, the 2009 amendments have yet to be ratified by two-thirds of the contracting parties to enter into force (International Energy Agency, 2011; Weber, 2022).

### 15.3.5 The Legal Status of $CO_2$: Waste Product or Industrial Resource?

Specific to any individual industrial production process, $CO_2$ is treated either as an industrial by-product, waste product (emission), or an industrial resource that is channeled back into the production cycle. The scale-up of CCU technologies increases the significance of $CO_2$ as an industrial resource and introduces novel $CO_2$-based value chains. Yet, redefining $CO_2$ emissions as a $CO_2$ resource comes with legal obstacles: Once $CO_2$ molecules are declared as waste, a legal discussion is required to revitalize and utilize $CO_2$ in industry. With the national waste management act (Kreislaufwirtschaftsgesetz, KrWG) from June 2012, the German government sought to improve the framework for the industrial recycling of waste (BMUV, Kreislaufwirtschaftsgesetz). Article 4 KrWG defines industrial by-products in distinction to waste products. However, there is no general legal regulation for CCU. Instead, several conditions need to be met for $CO_2$ to be regarded as an industrial resource: A case-to-case examination and approval is needed.

*A by-product is a substance that is produced in connection with the manufacture of another substance or product and is thus not the main focus of the manufacturing process. In order for an element to qualify as a by-product, it must also meet the following criteria:*

- *It must be possible to reuse the substance.*
- *Preprocessing exceeding the normal industry standard scope is unnecessary.*
- *Production of the substance is inherent to a manufacturing process.*
- *Reuse of the substance complies with all applicable laws concerning the following: product, environmental and health protection requirements; the substance not being an environmental or health hazard.*

*Article 4 (1), KrWG* (translation cited from Umweltbundesamt, 2014).[2]

Furthermore, the federal government has the responsibility to "*determine criteria by which certain substances are to be considered by-products*" (Article 4 para. 2, KrWG, in Bundesgesetzblatt, 2012).

Similarly, the Immission Control Act remains unclear on whether $CO_2$ can be categorized as a carbon resource. A legal report commissioned by the Bellona Foundation, an environmental think tank, concluded that the approval of $CO_2$ capture facilities only pertains to the building and operation of $CO_2$ capture plants for CCS purposes (Annex 1 of the 4th BImSchV, No. 10.4, in conjunction with Article 5 BlmSchV, in Bundesgesetzblatt, 2022). According to this legal study, $CO_2$ capture plants for the purpose of reusing $CO_2$ were not legally covered by the amendment (Altrock et al., 2022).

In practice, CCU technologies in Germany can be found in niche applications, where it is economically viable to apply CCU (Weber, 2022). As of today, in most cases, it remains more cost-effective to use carbon from fossil resources and either emit $CO_2$ or apply sub-seabed CCS. One of the few exemptions is Covestro, its headquarters based in North-Rhein Westphalia, a company that produces and sells polymers with carbon captured from a nearby chemical plant. In this rare case, the production with recycled carbon is more cost-effective than the production with crude oil. It is profitable to reuse the by-product $CO_2$ from the chemical plant as a resource in Covestro's manufacturing processes (Bezirksregierung Köln, 2021). As another example, the German drug store dm advertises its "climate-friendly" product line with cleaning products that come in a new synthetic packaging based on "$CO_2$-recycling" (Packaging 360°, n.d.). The synthetic packages are produced by LanzaTech, a U.S.-based company that has the ambition to reach a closed industrial carbon cycle. However, these are rare cases (niches) where CCU is commercially applied in Germany.

In sum, the regulatory framework for $CO_2$ capture and utilization allows the implementation of CCU technologies; however, it is currently not entirely suitable for the large-scale uptake of CCU in Germany. A comprehensive CCU regulatory framework is needed for investment security. Besides continued and coordinated funding of R&D projects, the industry would need legal foundations for the implementation of a national and EU-wide $CO_2$-transportation infrastructure (Agora Energiewende, 2021). Furthermore, not only the $CO_2$ infrastructure but also the building of CCU facilities is capital-intensive. Tax benefits or possible blending quotas for synthetic fuel produced with recycled carbon could further function as economic incentives for successful market entry. Another major challenge to CCU's business case is that in most industrial production today, the use of recycled carbon is more expensive than producing with fossil carbon and emitting $CO_2$. The current ETS framework largely does not reward CCU as a means to reduce emissions. $CO_2$ captured through CCS, in contrast, is considered not-emitted (European Commission, 2021). Increasing the rewards for CCU under the reformed ETS framework, which is expected at the end of 2022, could be an important incentive for the scale-up of CCU technologies.

### 15.3.6 CCU Policy Instruments in the EU and the United States: A Brief Comparison

In Germany and the EU, the discussion shows CCU is incentivized to a limited extent through the ETS and the Renewable Energy Directive and, to a larger extent, through major funding initiatives such as FONA in Germany and the Innovation Fund or Horizon Europe in the EU. The United States, one of the other major global players in CCU, has chosen a different approach. It mainly incentivizes the upscaling and market implementation of CCU through a market-based approach, namely tax credits under paragraph 45Q of the U.S. tax code. This is a technology-open approach, not limited to industrial applications. The tax credits are awarded based on the amount of captured $CO_2$. Funding instruments, however, also play an important role. Major R&D funding comes from the Office of Fossil Energy and Carbon Management of the U.S. Department of Energy. This funding is concentrated more on promoting CCS technologies than on CCU and includes funding for the latter. More recent legislation, such as the USE IT and the Infrastructure Investment and Jobs Act, increases funding specifically for CCU technologies for pathways such as chemicals, plastics, building materials and fuels. It also includes coal-utilization products, which draw heavy criticism in Germany. Another focus is on developing guidelines for the public procurement of CCU products, which is not part of the instrument mix in Germany yet (Thielges et al., 2022).

### 15.3.7 Societal Acceptance of CCU in Germany

Since CCU technologies remain by and large at the scale of laboratory studies or pilot projects, analyses on CCU remain mostly focused on technical and economic feasibility. Studies analyzing the societal and political feasibility of CCU are rare.

Generally speaking, studies on the societal acceptance of CCU/CCS technologies in Germany suggest low awareness levels of CCS and even more so of CCU (Simons et al., 2021) within the general public. CCU technologies tend to enjoy a more positive attitude from the general public than CCS. An online survey in Germany from 2017, however, concludes that publicly perceived risks associated with $CO_2$ storage and transportation hampered CCS acceptance. Likewise, the perceived risks associated with CCU products and disposal reduced CCU acceptance (Witte, 2021). A study by the Fraunhofer Institute Systems and Innovation Research investigated the societal acceptance factors for insulation boards made with $CO_2$-derived foam. The results showed that CCU insulation boards were mostly accepted rather than rejected and that a benefit perception was the common predictor for the acceptance measures (Simons et al., 2021). This suggests that both CCU and CCS acceptance could be hampered by public risk perceptions, but that CCU benefits from a "benefit framing" in the context of acceptance.

With increased awareness of CCU, the general perception of CCU is often muddled with CCS, with all the reservations concerning CCS in Germany ($CO_2$ leakage, environmental degradation, prolonging coal-based electricity generation) meeting CCU technologies, as well (Olfe-Kräutlein & Krämer, 2022). In addition, media analysis shows that energy experts and environmental activists criticize CCU technology mainly for its energy- and cost-intensity. Critics also point out that CCU can store $CO_2$ only temporarily, and once the product's life cycle ends, $CO_2$ is re-emitted (Olfe-Kräutlein, 2021). This alludes to the importance of a critical and comprehensive analysis of individual use cases, for instance through life-cycle assessments (LCAs), in order to address societal concerns and assess the viability of CCU as an industrial decarbonization strategy.

## 15.4 OUTLOOK AND CONCLUSIONS

The chapter has provided an overview of the five different building blocks of the German government's industrial decarbonization strategy: renewable energy expansion, electrification in conjunction with energy efficiency increases, green hydrogen, a circular economy, and CCS or CCU. Taken together, these building blocks are intended to put German industries on a path towards climate neutrality by 2045—a long way ahead, as current progress and the manifold challenges suggest.

Societal acceptance emerges as a cross-cutting challenge for industrial transformation in Germany. In the past, local protests have slowed down wind turbine construction significantly and challenged grid expansion, posing an obstacle to the expansion of renewable energy capacity. CCS has faced similar challenges to the extent that it was not pursued anymore as a climate strategy in the past decade.

Renewable energy expansion, electrification, and energy efficiency improvements can be seen as the main approaches for industrial decarbonization. Only for the hard-to-abate sectors, especially cement, the German government has, nevertheless, declared both CCS and CCU as options to achieve emission reductions. In these sectors, electrification and energy efficiency measures alone will not bring down emissions to net-zero.

In a case study on CCU, the chapter zooms in on a technological approach that has neither received much political attention in the past in Germany nor has faced large societal opposition so far. It remains a niche approach and its role in industrial transformation – for instance, whether it is envisioned as a solution for the cement industry only or other hard-to-abate industries as well – remains unclear. This is a challenge for these expensive, research-intensive technologies as there are no clear investment signals from the government. Political factors, as well as a multitude of economic, technical, societal, and ecological factors, will eventually determine the success of CCU. For consumers, moreover, there is so far hardly any information available on the sustainability of products made from "recycled" $CO_2$. In this context, the potential of CCU to contribute to emission reductions or even a net-zero economy is unclear. It depends on the specific technology in focus – synthetic fuels, for instance, have different properties than $CO_2$ mineralization. More research specifying the emission reduction potentials of other different technologies is needed. A brief comparison with the United States reveals a more technology-open approach to CCU in the United States, which draws major financial support from tax credits under the U.S. tax code ("45Q"). Germany has no similar incentive in place so far.

If it *is* desired politically and from a societal perspective in Germany, CCU will continue to need government support through R&D funding, a clearer CCU strategy, a more conducive legal and regulatory framework, as well as demand-side measures, for instance, through government procurement of CCU products. It will need this type of conducive environment not just in Germany but also in the EU as a whole (Thielges et al., 2022).

## NOTES

1 The contribution of Kristina Fürst to this chapter was generously funded by the German Federal Ministry of Education and Research (BMBF) within the context of the project $CO_2$WiN Connect.
2 Translation cited from: Umweltbundesamt, 2014.

## REFERENCES

AG Energiebilanzen e. V. (2022). Strommix in Deutschland. Retrieved November 16, 2022, from https://ag-energiebilanzen.de/wp-content/uploads/2022/04/STRERZ21_Abgabe-09-2022A11.pdf

Agora Energiewende. (2021). *Energiewende Deutschland Stand.* Retrieved November 16, 2022, from https://static.agora-energiewende.de/fileadmin/Projekte/2021/2021_11_DE-JAW2021/A-EW_247_Energiewende-Deutschland-Stand-2021_WEB.pdf

Agora Energiewende. (2018). *Klimaneutrale Industrie: Schlüsseltechnologien und Politikoptionen für Stahl. Chemie und Zement.* Retrieved November 16, 2022, from https://static.agoraenergiewende.de/fileadmin/Projekte/2018/Dekarbonisierung_Industrie/164_A-EW_Klimaneutrale-Industrie_Studie_WEB.pdf

Altrock, M., Däuper, O., Kliem, C., Braun, F. & Hausmann, N. (2022, April, 06) Rechtliche Rahmenbedingungen für Carbon Capture and Storage (CCS) in Deutschland. *Bellona & BBH.* Retrieved November 18, 2022, from Rechtliche Rahmenbedingungen für Carbon Capture and Storage (CCS) in Deutschland - Bellona.de

Appunn, K. (2021). The history behind Germany's nuclear phase-out. *Clean Energy Wire.* Retrieved November 16, 2022, from https://www.cleanenergywire.org/factsheets/history-behind-germanys-nuclear-phase-out

Bach, L. T., Gill, S. J., Rickaby, R. E., Gore, S., & Renforth, P. (2019). CO2 removal with enhanced weathering and ocean alkalinity enhancement: potential risks and co-benefits for marine pelagic ecosystems. *Frontiers in Climate, 1*, 7. 10.3389/fclim.2019.00007

Bezirksregierung Köln. (2021). Genehmigungsbescheid 53.0023/19/G16-JS. Retrieved November 22, 2022, from https://www.bezreg-koeln.nrw.de/brk_internet/brk_media/_industrieanlagen_genehmigungen/pub_bekanntmachungen_dormagenchempark/covestro_deutschland_20211221/bescheid.pdf

Bundesgesetzblatt (2022) Bundes-Immissionsschutzgesetz. Gesetz zum Schutz vor schädlichen Umwelteinwirkungen durch Luftverunreinigungen, Geräusche, Erschütterungen und ähnliche Vorgänge. *BGBl. I S. 1792 m.W.v. 26.10.2022.* Retreived November 24, 2022, from https://dejure.org/gesetze/BImSchG

Bundesgesetzblatt. (2012). Gesetz zur Demonstration und Anwendung von Technologien zur Abscheidung, zum Transport und zur dauerhaften Speicherung von Kohlendioxid. BGBI.1, 1726. Retrieved November 24, 2022, from https://dejure.org/BGBl/2012/BGBl._I_S._1726

Bundesverband Bürgerinitiativen gegen SuedLink (2022). JA zur Energiewende, NEIN zur Stromautobahn SuedLink. Retrieved November 22, 2022, from https://bundesverband-gegen-suedlink.de/

Bundesministerium für Umwelt. Naturschutz und nukleare Sicherheit. (2021). *Klimaschutz in Zahlen Fakten, Trends und Impulse deutscher Klimapolitik.* Retrieved November 16, 2022, from https://www.bmuv.de/fileadmin/Daten_BMU/Pools/Broschueren/klimaschutz_zahlen_2021_bf.pdf

Council of the European Union. (2022). *Fit for 55.* Retrieved November 16, 2022, from https://www.consilium.europa.eu/en/policies/green-deal/fit-for-55-the-eu-plan-for-a-green-transition/

Deutscher Bundestag. (2022). Osterpaket zum Ausbau erneuerbarer Energien beschlossen. Retrieved November 22, 2022, from https://www.bundestag.de/dokumente/textarchiv/2022/kw27-de-energie-902620

Die Bundesregierung. (2022). *Klimaschutzgesetz: Klimaneutralität Bis 2045: Bundesregierung*. Retrieved November 16, 2022, from https://www.bundesregierung.de/breg-de/themen/klimaschutz/klimaschutzgesetz-2021-1913672

European Commission. (2022). REPowerEU: A plan to rapidly reduce dependence on Russian fossil fuels and fast forward the green transition. Retrieved November 16, 2022, from https://ec.europa.eu/commission/presscorner/detail/en/IP_22_3131

European Commission. (2021). Carbon capture, use, and storage. Retrieved November 23, 2022, from https://climate.ec.europa.eu/eu-action/carbon-capture-use-and-storage_en

Fajardy, M. (2022). CO2 Capture and utilisation – analysis. *IEA*. Retrieved November 16, 2022, from https://www.iea.org/reports/co2-capture-and-utilisation

Federal Ministry for Economic Affairs and Climate Action. (2022). Time series for the development of renewable energy sources in Germany. Retrieved November 22, 2022, from https://www.erneuerbare-energien.de/EE/Redaktion/DE/Downloads/zeitreihen-zur-entwicklung-der-erneuerbaren-energien-in-deutschland-1990-2021-en.pdf?__blob=publicationFile&v=17

Federal Ministry for Economic Affairs and Climate Action. (n.d.). An electricity grid for the energy transition. Retrieved November 22, 2022, from https://www.bmwk.de/Redaktion/DE/Dossier/netze-und-netzausbau.html

Federal Ministry for Economic Affairs and Energy. (2020). The National Hydrogen Strategy. Retrieved November 16, 2022, from https://www.bmwk.de/Redaktion/EN/Publikationen/Energie/the-national-hydrogen-strategy.pdf?__blob=publicationFile&v=4

German Missions in Canada. (2022). Canada-Germany Hydrogen Alliance. Retrieved November 22, 2022, from https://canada.diplo.de/ca-en/vertretungen/german-consulate-general-vancouver/-/2549574

Goto, K., Kazama, S., Furukawa, A., Serizawa, M., Aramaki, S., & Shoji, K. (2013). Effect of CO2 purity on energy requirement of CO2 capture processes. *Energy Procedia*, *37*, 806–812. doi: 10.1016/j.egypro.2013.05.171

Harmsen, S. (2022). Paket für Windkraft und Planungsbeschleunigung verabschiedet. *Energy & Management*. Retrieved November 16, 2022, from https://www.energie-und-management.de/nachrichten/energiepolitik/detail/paket-fuer-windkraft-und-planungsbeschleunigung-verabschiedet-156418

Hodgson, D., & Hugues, P. (2022). Cement – analysis. *IEA*. Retrieved November 16, 2022, from https://www.iea.org/reports/cement

Hurtes, S., & Cai, W. (2022). Europe Is Sacrificing Its Ancient Forests for Energy. *The New York Times*. Retrieved November 20, 2022, from https://www.nytimes.com/interactive/2022/09/07/world/europe/eu-logging-wood-pellets.html

International Energy Agency. (2011). Carbon Capture and Storage and the London Protocol: Options for Enabling Transboundary CO2 Transfer. *OECD Publishing*, Retrieved November 22, 2022, from 10.1787/5kg3n27pfv30-en

Ministerium für Wirtschaft, Innovation, Digitalisierung und Energie des Landes Nordrhein-Westfalen. (n.d.). Carbon Management Strategie des Landes Nordrhein-Westfalen. Retrieved November 23, 2022, from https://www.wirtschaft.nrw/sites/default/files/documents/mwide_carbon_management_strategie_barrierefrei.pdf

Lintz, G., & Leibenath, M. (2020). The Politics of Energy Landscapes: The influence of local anti-wind initiatives on state policies in Saxony, Germany. *Energy, Sustainability and Society 10*, 5. 10.1186/s13705-019-0230-3

Ministerium für Umwelt, Naturschutz und Verkehr des Landes Nordrhein-Westfalen (n.d.) Zulassung und Genehmigung. Retrieved November 16, 2022, from https://www.umwelt.nrw.de/umwelt/umwelt-und-ressourcenschutz/immissionsschutz-und-anlagen/immissionsschutzrecht/zulassung-und-genehmigung

Mennicken, L., Janz, A., & Roth, S. (2016). The German R&D program for CO2 utilization—innovations for a green economy. *Environmental Science and Pollution Research*, *23*(11), 11386–11392. 10.1007/s11356-016-6641-1

Mieritz, T. (2019). Der Stromnetzausbau ist nicht immer gewollt. *NABU*. Retrieved November 22, 2022, from https://blogs.nabu.de/stromnetzausbau/

Olfe-Kräutlein, B. (2022). *CO2-Nutzung für Dummies*. Weinheim: Wiley.

Olfe-Kräutlein, B., & Krämer, D. (2022). Kommunikationsleitfaden: Hinweise und Ideen für die Öffentlichkeitsarbeit in CO2WIN. Retrieved November 22, 2022, from CO2WIN_Connect_Kommunikationsleitfaden.pdf (co2-utilization.net)

Olfe-Kräutlein, B. (2021). CO2-Nutzungstechnologien in den Medien: Zwischenbericht zur ersten Statuskonferenz am 8. und 9. Juni 2021 in Berlin. Retrieved November 22, 2022, from CO2WIN_Medienanalyse_2021.pdf (co2-utilization.net)

Otto, D, Pfeiffer, M., de Brito, M. & Groß, M. (2022). Fixed Amidst Change: 20 Years of Media Coverage on Carbon Capture and Storage in Germany. *Sustainability*. 14. 10.3390/su14127342

Packaging 360° (n.d.) dm spart weiteres Verpackungsmaterial ein. Retrieved November 24, 2022, from https://www.packaging-360.com/kosmetik/dm-spart-weiteres-verpackungsmaterial-ein/

Prognos (2021). How Germany can achieve its climate targets even before 2050. Retrieved November 16, 2022, from https://www.prognos.com/en/project/climate-neutral-germany-2045

Schiffer, HW., Thielges, S. & Unger, C. (2022) Taking Stock of the Energy and Climate Profile of Germany and the USA: New Potential for Cooperation. *Zeitschrift für Energiewirtschaft 46*, 159–174. 10.1007/s12398-022-00330-7

Simons, L., Ziefle, M., & Arning, K. (2021). The social acceptance factors for insulation boards produced with $CO_2$-derived foam. *Frontiers in Energy Research*, *9*. 10.3389/fenrg.2021.717975

Tagesspiegel Background. (2022). CCUS-Strategie wird Teil des Sofortprogramms. *Tagesspiegel Background Energie & Klima*, June 22, 2022.

Tan, E.C.D.& Lamers, P. (2021). Circular Bioeconomy Concepts – A Perspective. *Frontiers in Sustainability*, 2, 10.3389/frsus.2021.701509

Thielges, S., Olfe-Kräutlein, B., Rees A., Jahn J., Sick, V., & Quitzow, R. (2022) Committed to implementing CCU? A comparison of the policy mix in the US and the EU. *Frontiers in Climate 4*, 943387. doi: 10.3389/fclim.2022.943387

Umweltbundesamt (2022a). Treibhausgasminderungsziele Deutschlands. Retrieved November 20, 2022, from https://www.umweltbundesamt.de/daten/klima/treibhausgasminderungsziele-deutschlands#nationale-treibhausgasminderungsziele

Umweltbundesamt (2022b). Bioenergie. Retrieved November 22, 2022, from https://www.umweltbundesamt.de/themen/klima-energie/erneuerbare-energien/bioenergie#bioenergie-ein-weites-und-komplexes-feld

Umweltbundesamt (2020). Klimaschutz und Dekarbonisierung im Industriesektor. Retrieved November 16, 2022, from https://www.umweltbundesamt.de/themen/wirtschaft-konsum/klimaschutz-dekarbonisierung-im-industriesektor

Umweltbundesamt (2014). Waste regulations. Retrieved November 24, 2022, from https://www.umweltbundesamt.de/en/topics/waste-resources/waste-management/waste-regulations

Weber, R. (2022). Regulatorische Weiterentwicklung eines klimapolitischen Dilemmas: Der Einsatz von CCS und CCU als Negativemissionstechnologien. *Die Verwaltung*, *55*(2), 219–248. 10.3790/verw.55.2.219

Witte, K. (2021). Social acceptance of carbon capture and storage (CCS) from Industrial Applications. *Sustainability*, *13*(21), 12278. 10.3390/su132112278

Wolf, I., Huttarsch, J.-H., Fischer, A.-K. & Ebersbach, B. (2022). Soziales Nachhaltigkeitsbarometer der Energie-und Verkehrswende. Retrieved November 16, 2022, from https://snb.ariadneprojekt.de/start

World Wildlife Fund (n.d.). Biokraftstoffe. Retrieved November 16, 2022, from https://www.wwf.de/themen-projekte/landwirtschaft/bioenergie/biokraftstoffe

Zimmermann, A. W., Wunderlich, J., Müller, L., Buchner, G. A., Marxen, A., Michailos, S., Armstrong, K., Naims, H., McCord, S., Styring, P., Sick, V., Schomäcker, R. (2020): Techno-Economic Assessment Guidelines for CO2 Utilization. *Frontiers in Energy Research*, *8*, 5. 10.3389/fenrg.2020.00005

# Index